T0093862

Introduction to Nonextensive Statistical Mechanics

Preface

In 1902, after three decades that Ludwig Eduard Boltzmann formulated the first version of standard statistical mechanics, Josiah Willard Gibbs shares, in the Preface of his superb *Elementary Principles in Statistical Mechanics* [1, 2, 3]: *"Certainly, one is building on an insecure foundation ..."*. After such words by Gibbs, it is, still today, uneasy to feel really comfortable regarding the foundations of statistical mechanics from first principles. Since the time when I took the decision to write the present book, I would certainly second his words. Several interrelated facts contribute to this inclination.

First, the verification of the notorious fact that all branches of physics deeply related to the theory of probabilities, such as statistical mechanics and quantum mechanics, have exhibited, along history and up to now, endless interpretations, reinterpretations, and controversies. All this is fully complemented by philosophical and sociological considerations. As one among many evidences, let us mention the eloquent words by Gregoire Nicolis and David Daems [4]: *"It is the strange privilege of statistical mechanics to stimulate and nourish passionate discussions related to its foundations, particularly in connection with irreversibility. Ever since the time of Boltzmann, it has been customary to see the scientific community vacillating between extreme, mutually contradicting positions."*

Second, I am inclined to think that, together with the central geometrical concept of *symmetry*, virtually nothing more basically than *energy* and *entropy* deserves the qualification of *pillars* of modern physics. Both concepts are amazingly subtle. However, energy has to do with *possibilities*, whereas entropy with the *probabilities* of those possibilities. Consequently, the concept of entropy is, epistemologically speaking, one step further. One might remember, for instance, the illustrative dialog that Claude Elwood Shannon had with John von Neumann [5, 6]: *"My greatest concern was what to call it. I thought of calling it 'information', but the word was overly used, so I decided to call it 'uncertainty'. When I discussed it with John von Neumann, he had a better idea. Von Neumann told me, 'You should call it entropy, for two reasons. In the first place, your uncertainty function has been used in statistical mechanics under that name, so it already has a name. In the second place, and more important, nobody knows what entropy really is, so in a debate, you will always have the advantage.'"*. Frequently we hear and read diversified opinions about what *should* and what *should not* be considered as *"the physical entropy"*, its connections with heat, information, and so on.

Third, the dynamical foundations of the standard, Boltzmann-Gibbs (BG) statistical mechanics are, mathematically speaking, *not* yet fully established. It is known that, for classical systems, exponentially diverging sensitivity to the initial conditions (i.e., positive Lyapunov exponents almost everywhere, which typically imply mixing and ergodicity, properties that are consistent with Boltzmann's celebrated *Stoszzahl Ansatz*, "molecular chaos hypothesis") is a *sufficient* property for having a meaningful statistical theory. More precisely, one expects that this property implies, for many-body Hamiltonian systems attaining thermal equilibrium, central features such as the celebrated exponential weight, introduced and discussed in the 1870s by Ludwig Boltzmann (very especially in his 1872 [7] and 1877 [8] papers)[1] in the so-called μ-space, thus recovering, as particular instance, the velocity distribution published in 1860 by James Clerk Maxwell [11]. More generally, the exponential divergence typically leads to the exponential weight in the *full* phase space, the so-called Γ-space first proposed by Gibbs. However, hypothesis such as this exponentially diverging sensitivity, are they *necessary*? In the first place, are they, in some appropriate logical chain, *necessary* for having BG statistical mechanics? I would say *yes*. But are they also *necessary* for having a valid statistical mechanical description at all for *any* type of thermodynamic-like systems?[2] I would say *no*. In any case, it is within this belief that I write the present book. All in all, if such is today the situation for the successful, undoubtedly correct for a very wide class of systems, universally used, and centennial BG statistical mechanics and its associated thermodynamics ("a science with secure foundations, clear definitions, and distinct boundaries" as so well characterized by James Clerk Maxwell), what can we then expect for its possible generalization only a few decades after its first proposal, in 1988?

Fourth,—last but not least—no logical-deductive mathematical procedure exists, nor will presumably ever exist, for proposing a new physical theory or for generalizing a pre-existing one. It is enough to think about Newtonian mechanics, which has already been generalized along at least two completely different (and compatible) paths, which eventually led to the theory of relativity and to quantum mechanics. This fact is consistent with the evidence that there is no unique way of generalizing a coherent set of axioms. Indeed, the most obvious manner of generalizing it is to replace one or more of its axioms with weaker ones. And this can be done in more than one manner, sometimes in infinite manners. So, if the prescriptions of logics and mathematics are helpful only for analyzing the *admissibility* of a given generalization, how do generalizations of physical theories, or even scientific discoveries in

---

[1] English translation in [9]; see also [10].

[2] For example, we can read in a recent paper by Giulio Casati and Tomaz Prosen [12] the following sentence: *"While exponential instability is sufficient for a meaningful statistical description, it is not known whether or not it is also necessary."*

general, occur? Through all types of heuristic procedures, but mainly—I would say—through *methaphors* [13]. Indeed, theoretical and experimental scientific progress occurs all the time through all types of logical and heuristic procedures, but the particular progress involved in the generalization of a physical theory immensely, if not essentially, relies on some kind of metaphor.[3] Well-known examples are the idea of Erwin Schroedinger of generalizing Newtonian mechanics through a *wave*-like equation inspired by the phenomenon of optical interference, and the discovery by Friedrich August Kekule of the *cyclic* structure of benzene inspired by the shape of the mythological Ouroboros. In other words, generalizations not only use the classical logical procedures of *deduction* and *induction*, but also, and overall, the specific type of inference referred to as *abduction* (or *abductive reasoning*), which plays the most central role in Charles Sanders Peirce's *semiotics*. The procedures for theoretically proposing a generalization of a physical theory somehow crucially rely on the construction of what one may call *a plausible scenario*. The scientific value and universal acceptability of any such a proposal are of course ultimately dictated by its successful verifiability in natural and/or artificial and/or social systems. Having made all these considerations, I hope that it must by now be very transparent for the reader why, in the beginning of this Preface, I evoked Gibbs' words about the fragility of the basis on which we are founding.

Newton's decomposition of white light into rainbow colors, not only provided a deeper insight into the nature of what we know today to classically be electromagnetic waves, but also opened the door to the discovery of infrared and ultraviolet. While trying to follow the methods of this great master, it is my cherished hope that the present, nonextensive generalization of Boltzmann–Gibbs statistical mechanics, may provide a deeper understanding of the standard theory, in addition to proposing some extension of the domain of applicability of the powerful methods of statistical mechanics. The book is written at a graduate course level, and some basic knowledge of quantum and standard statistical mechanics, as well as thermodynamics, is assumed. The style is, however, slightly different from a conventional textbook, in the sense that not *all* the results that are presented are proved. The quick ongoing development of the field does not yet allow for such an ambitious task. Various points of the theory are presently only partially known and understood. So, here and there we are obliged to proceed with heuristic arguments. The book is unconventional also in the sense that here and there historical and other side remarks are included as well. Some sections of the book, the most basic ones, are presented with all details and intermediate steps; some others, more advanced or quite lengthy, are presented only through their main results, and the reader is referred to the original publications to know more. We hope, however, that a unified perception of statistical mechanics, its background, and its basic concepts does emerge.

The book is organized into four parts, namely Part I—*Basics or How the Theory Works*, Part II—*Foundations or Why the Theory Works*, Part III—*Applications or What for the Theory Works*, and Part IV—*Last (But Not Least)*. The first part constitutes a pedagogical introduction to the theory and its background (Chaps. 1–3). The

---

[3] I was first led to think about this by Roald Hoffmann in 1995.

second part contains the state of the art in its dynamical foundations, in particular how the index (indices) q can be obtained, in some paradigmatic cases, from microscopic first principles or, alternatively, from mesoscopic principles (Chaps. 4–6). The third part is dedicated to listing brief presentations of typical applications of the theory and its concepts, or at least of its functional forms, as well as possible extensions existing in the literature (Chap. 7). Finally, the fourth part constitutes an attempt to place the present—intensively evolving, open to further contributions and insights[4]—theory into contemporary science, by addressing some frequently asked or still unsolved current issues (Chap. 8). An Appendix with useful formulae has been added at the end, as well as another one discussing escort distributions and q-expectation values.

It may be useful to point out at this stage that this book can be quite conveniently read along two possible tracks. The first track concerns readers at a graduate student level. It basically consists of reading Chaps. 1, 2, 3, 5, and 8. The second track concerns readers at the research level with some practice in standard statistical mechanical methods. It basically consists of reading Chaps. 3–8. For both tracks, let us emphasize that Chap. 7 contains a large amount of diversified applications. The reader may focus on those of his/her main preference. In the present new edition, we have refined some concepts, included some recent analytical discussions, and added a considerable number of applications as well as experimental and numerical verifications. Let us mention also that it has been unfortunately impossible to unify the notations along all the chapters of the book, due to the fact that very many figures have been reproduced from a large number of papers in the literature.

Toward this end, it is a genuine pleasure to warmly acknowledge the contributions of M. Gell-Mann, *maître à penser*, with whom I have had frequent and delightfully deep conversations on the subject of nonextensive statistical mechanics ... as well as on many others. Many other friends and colleagues have substantially contributed to the ideas, results, and figures presented in this book. Those contributions range from insightful questions or remarks—sometimes fairly critical—to entire mathematical developments and seminal ideas. Their natures are so diverse that it becomes an impossible task to duly recognize them all. So, *faute de mieux*, I decided to name them in alphabetical order, being certain that I am by no means doing justice to their enormous and diversified intellectual importance. In all cases, my gratitude could not be deeper. They are S. Abe, F. C. Alcaraz, R. F. Alvarez-Estrada, S. Amari, G. F. J. Ananos, C. Anteneodo, V. Aquilanti, N. Ay, G. Baker Jr., F. Baldovin, M. Baranger, G. G. Barnafoldi, C. Beck, I. Bediaga, G. Bemski, T. S. Biro, A. R. Bishop, H. Blom, B. M. Boghosian, E. Bonderup, J. P. Boon, E. P. Borges, L. Borland, T. Bountis, E. Brezin, A. Bunde, L. F. Burlaga, B. J. C. Cabral, M. O. Caceres, S. A. Cannas, A. Carati, F. Caruso, M. Casas, G. Casati, N. Caticha, A. Chame, P.-H. Chavanis, C. E. Cedeño, L. J. L. Cirto, J. Cleymans, E. G. D. Cohen, A. Coniglio, M. Coutinho Filho, E. M. F. Curado, S. Curilef, J. S. Dehesa, S. A. Dias, A. Deppman, A. Erzan, L. R. Evangelista, J. D. Farmer, R. Ferreira, M. A. Fuentes, L. Galgani, J. P. Gazeau, P.-G. de Gennes, A. Giansanti, A. Greco, P. Grigolini, D. H. E. Gross, G. R. Guerberoff, E. Guyon, M. Hameeda, R. Hanel, H. J. Haubold, R. Hersh, H. J. Herrmann, H. J.

---

[4] For a regularly updated bibliography of the subject see http://tsallis.cat.cbpf.br/biblio.htm.

Hilhorst, R. Hoffmann, G. 't Hooft, K. Huang, M. Jauregui, H. J. Jensen, P. Jizba, L.
P. Kadanoff, G. Kaniadakis, T. A. Kaplan, S. Kawasaki, J. Korbel, D. Krakauer, P. T.
Landsberg, V. Latora, C. M. Lattes, E. K. Lenzi, S. V. F. Levy, H. S. Lima, J. A. S.
Lima, M. L. Lyra, S. D. Mahanti, A. M. Mariz, J. Marsh, A. M. Mathai, R. Maynard,
G. F. Mazenko, E. Megias, R. S. Mendes, L. C. Mihalcea, L. G. Moyano, J. Naudts,
K. P. Nelson, G. Nicolis, F. D. Nobre, J. Nogales, F. A. Oliveira, P. M. C. Oliveira, I.
Oppenheim, A. W. Overhauser, G. Parisi, R. Pasechnik, G. P. Pavlos, R. Piasecki, A.
Plastino, A. R. Plastino, A. Pluchino, D. Prato, P. Quarati, S. M. D. Queiros, A. K.
Rajagopal, A. Rapisarda, M. A. Rego-Monteiro, M. S. Ribeiro, A. Robledo, M. C.
Rocca, A. Rodriguez, S. Ruffo, G. Ruiz, M. Rybczynski, S. R. A. Salinas, Y. Sato, V.
Schwammle, L. Silva, L. R. da Silva, R. N. Silver, I. D. Soares, A. M. C. Souza, H.
E. Stanley, D. A. Stariolo, D. Stauffer, S. Steinberg, R. Stinchcombe, H. Suyari, F.
Vallianatos, H. L. Swinney, F. A. Tamarit, P. Tempesta, W. J. Thistleton, S. Thurner,
U. Tirnakli, L. Tisza, F. Topsoe, R. Toral, A. C. Tsallis, A. F. Tsallis, E. L. P. Tsallis, F.
Tsallis Viegas, S. Umarov, P. Van, M. E. Vares, M. C. S. Vieira, C. Vignat, J. Villain,
A. F. Vinas, S. Vinciguerra, G. Viswanathan, R. S. Wedemann, B. Widom, G. Wilk,
H. O. Wio, Z. Wlodarczyk, D. J. Zamora, and I. I. Zovko. Unavoidably, I must have
forgotten to mention some—this idea started developing over three decades ago!—to
them my most genuine apologies and gratitude. Finally, as in virtually all the fields of
science and very especially during the first stages of any new development, there are
also a few colleagues whose intentions have not been—I confess—very transparent
to me. But they have nevertheless—perhaps even unwillingly—contributed to the
progress of the ideas that are presented in this book. They surely know who they are.
My gratitude goes to them as well: it belongs to human nature to generate fruitful
ideas through all types of paths.

Along the years, I have relevantly benefited from the partial financial support
of various Agencies, especially the Brazilian CNPq, FAPERJ, PRONEX/MCT, and
CAPES, the USA NSF, SFI, SI International, AFRL, and John Templeton Foundation,
the Italian INFN and INFM, among others. I am indebted to all of them.

Finally, some of the figures that are presented in the present book have been reproduced from various publications indicated case by case. I gratefully acknowledge the
gracious authorization to do so from their authors.

Rio de Janeiro, Brazil through the periods 2004–2009 and 2018–2023

Rio de Janeiro, Brazil   Constantino Tsallis

Contents

Part I
Basics or How the Theory Works

Chapter 1
Historical Background and Physical Motivations

Beauty is the first test:
there is no permanent place in the world for ugly mathematics.
G.H. Hardy
(A Mathematician's Apology, 1941)

1.1 An Overall Perspective

Let us start by placing the content of the present book in the scenario of contemporary physics. Its present theoretical pillars are usually considered to be mechanics—classical, special relativity, general relativity, and quantum—Maxwell electromagnetism, and Boltzmann–Gibbs (BG) statistical mechanics, which essentially incorporates theory of probabilities within the realm of mechanics and electromagnetism. Moreover, the BG theory connects, for large systems, to classical thermodynamics.

In what concerns mechanics, the corresponding regions of physical applicability of those theories are as follows.

– Systems with masses m neither too small nor too large and velocities v small compared to the speed of light c (e.g., a falling apple) obey Newton's classical mechanics.
– Systems with masses neither too small nor too large and velocities v comparable to c (e.g., the Earth's orbit around the Sun, the Solar system's orbit around the galaxy, the surface of the Sun) obey Einstein's special relativity, which recovers classical mechanics as the limiting case $v/c \to 0$.
– Systems with large and compact masses and velocities comparable to c (e.g., neutron star, black hole) obey Einstein's general relativity, which recovers special relativity as the limiting case $Gm/(c^2 R) \to 0$, where G and R are, respectively, Newton's gravitation constant and its (linear) size [14].
– Systems with masses very small and velocities small compared to c (e.g., an atomic electron in orbits distant from the nucleus) obey nonrelativistic quantum

© Springer Nature Switzerland AG 2023
C. Tsallis, *Introduction to Nonextensive Statistical Mechanics*,
https://doi.org/10.1007/978-3-030-79569-6_1

3

mechanics, namely Schroedinger equation, which recovers classical mechanics as the limiting case $h/mv \to 0$, where h is Planck's constant.
- Systems with masses very small and velocities comparable to c (e.g., an atomic electron in orbits close to the nucleus) obey relativistic quantum mechanics, namely Klein–Gordon and Dirac equations, for spinless particles and particles with spin, respectively.

Let us focus now on statistical mechanics.

- Thermostatistical systems with *local* space-time correlations obey Boltzmann–Gibbs statistical mechanics, constructed upon the Boltzmann–Gibbs-von Neumann-Shannon *additive* entropy S_{BG}. Important examples include distinguishable particles with short-range interactions, indistinguishable particles with half-integer spin (Fermi–Dirac statistics), indistinguishable particles with integer spin (Bose–Einstein statistics), strongly chaotic nonlinear dynamical systems, second-order critical phenomena *close* to T_c, random geometrical systems such as bond percolation (Kasteleyn and Fortuin theorem [15]) and self-avoiding random walk (de Gennes isomorphism [16]).
- Thermostatistical systems with *nonlocal* (sometimes referred to as *global*) space-time correlations obey non-Boltzmann–Gibbs statistical mechanics, e.g., *nonextensive statistical mechanics* (*q-statistics* for short), constructed upon *nonadditive* entropies such as S_q with $S_1 = S_{BG}$. Typical examples include distinguishable particles with long-range interactions (e.g., gravitation, plasmas), weakly chaotic nonlinear dynamical systems, second-order critical phenomena *at* T_c, random geometrical systems such as (asymptotically) scale-invariant networks. The present book is dedicated to this generalization, which recovers BG statistical mechanics as the limiting case $(1 - q)/k_B \to 0$, where k_B is Boltzmann's constant.

The word *"nonextensive"* that—after some hesitation—I eventually adopted, in the title of the book and elsewhere, to refer to the present specific generalization of BG statistical mechanics may—and occasionally *does*—cause some confusion, and surely deserves clarification.

The whole theory is based on a single concept, namely the *entropy* noted S_q which, for the *entropic index q* equal to unity, reproduces the standard BG entropy.

The traditional functional S_{BG} is said to be *additive* [17]. Indeed, for a system composed of any two (probabilistically) *independent* subsystems, the entropy S_{BG} of the sum coincides with the sum of the entropies. The entropy S_q ($q \neq 1$) violates this property, and is therefore *nonadditive*. As we see, entropic additivity depends, from its very definition, only on the functional form of the entropy in terms of probabilities. The situation is generically quite different for the thermodynamical concept of *extensivity*. An entropy of a system or of a subsystem is said *extensive* if, for a large number N of its elements (probabilistically *independent or not*), the entropy is (asymptotically) proportional to N. Otherwise, it is *nonextensive*. This is to say, extensivity depends on *both* the mathematical form of the entropic functional *and* the correlations possibly existing within the elements of the system. Consequently, for a (sub)system whose elements are either independent or locally correlated, the additive entropy S_{BG} is extensive, whereas the nonadditive entropy S_q ($q \neq 1$) is nonextensive. In contrast, however, for a (sub)system whose elements are generically nonlocally correlated, the additive entropy S_{BG} is typically nonextensive, whereas the nonadditive entropy S_q ($q \neq 1$) can be extensive for a special value of q. Probabilistic systems exist such that S_q is *not* extensive for *any* value of q, either $q = 1$ or $q \neq 1$. All these statements are further detailed in the body of the book[1].

We shall also see that, consistently, the index q appears to characterize *universality classes of nonadditivity*, by phrasing this concept similarly to what is done in the standard theory of critical phenomena. Within each class, one expects to find infinitely many dynamical systems.

Coming back to the name *nonextensive statistical mechanics*, would it not be more appropriate to call it *nonadditive statistical mechanics*? Certainly yes if one focuses on the entropy that is being used. However, there is on one hand the fact that the expression *nonextensive statistical mechanics* is by now spread in thousands of papers. There is, on the other hand, the fact that important systems whose approach benefits from the present generalization of the BG theory are long-range-interacting many-body Hamiltonian systems. For such systems, the total energy is well known to be nonextensive, even if the extensivity of the entropy can be preserved by conveniently choosing the value of the index q. Summarizing, since Gibbs coined the expression, this branch of contemporary theoretical physics is referred to through the composition of two words, namely *statistical* and *mechanics*. The word *statistical* definitively refers to probabilities, and consistently to entropy; the word *mechanics* refers instead to mechanical concepts, hence, inevitably, to energy. For naming the present generalization, we could focus on entropy and call it *nonadditive statistical mechanics*, or alternatively focus on energy and call it *nonextensive statistical mechanics*. For historical reasons the latter prevailed.

Still at the linguistic and semantic levels, should we refer to S_q as an *entropy* or rather as an *entropic functional* or *entropic form*? And, even before that, why should such a minor-looking point have any relevance in the first place? The point is that, in physics, since more than one century, only one entropic functional has been unanimously considered "physical" in the thermodynamical sense, namely the BG one. In other areas, such as cybernetics, control theory, nonlinear dynamical systems, information theory, many other (by now over fifty!) entropic functionals differing from the Shannon one (which precisely coincides with the BG one) have been proposed, studied, and occasionally used. In the physical community only the BG form is unquestionably admitted as physically meaningful because of its deep connections

---

[1] During more than one century, physicists have primarily addressed locally interacting systems, and therefore the entropic form which satisfies the thermodynamical requirement of extensivity is S_{BG}. A regretful consequence of this fact is that entropic *additivity* and *extensivity* have been practically—and wrongly—considered as synonyms in many communities, thus generating all kinds of confusions and inadvertences. For example, by mere inadvertence, our own book entitled *Nonextensive Entropy—Interdisciplinary Applications* [18] should definitively have been more appropriately entitled *Nonadditive Entropy—Interdisciplinary Applications*! Indeed, already in its first chapter, an example is shown where the nonadditive entropy S_q ($q \neq 1$) is extensive.

with classical thermodynamics. So, what about S_q in this specific context? A variety of thermodynamical arguments—extensivity, Clausius inequality, zeroth, first and second principles of thermodynamics, Carnot cycle, and others—that are presented later on, definitively point S_q as a physical entropy in a strongly analogous sense that S_{BG} surely is. This issue is deeply related to the Einstein principle of likelihood factorization, to the Legendre transformation structure of thermodynamics, and to the large deviation theory, as we shall further elaborate in the body of the book.

Complexity is nowadays a frequently used yet poorly defined—at least quantitatively speaking—concept. It tries to embrace a great variety of scientific and technological approaches of all types of natural, artificial, and social systems. A name, *plectics*, has been coined by Murray Gell-Mann to refer to this emerging science [19]. One of the main—necessary but by no means sufficient—features of complexity has to do with the fact that both very ordered and very disordered systems are, in the sense of plectics, considered to be *simple*, not *complex*. Ubiquitous phenomena, such as the origin of life and languages, the growth of cities and computer networks, citations of scientific papers, co-authorships and co-actorships, displacements of living beings, financial fluctuations, turbulence, are frequently considered to be complex phenomena. They all seem to occur close, in some sense, to the frontier between order and disorder. Most of their basic quantities exhibit *nonexponential* behaviors, very frequently *power-laws*. It happens that the distributions and other relevant quantities that emerge naturally within the frame of nonextensive statistical mechanics are precisely of this type, becoming of the exponential type in the $q = 1$ limit. One of the most typical dynamical situations has to do with the *edge of chaos*, occurring in the frontier between regular motion and standard chaos. Since these two typical regimes would clearly be considered "simple" in the sense of plectics, one is strongly tempted to consider as "complex" the regime in between, which has some aspects of the disorder associated with strong chaos but also some of the order lurking nearby.[2] Nonextensive statistical mechanics turns out to be appropriate precisely for that intermediate region, thus suggesting that the entropic index q could be a convenient manner for quantifying some relevant aspects of complexity, surely not in all cases but probably so for vast classes of systems. Regular motion and chaos are time analogs for the space configurations occurring, respectively, in crystals and fluids. In this sense, the edge of chaos would be the analog of quasi-crystals, glasses, spin-glasses, and other amorphous, typically metastable, structures. One does not expect statistical concepts to be intrinsically useful for regular motions and regular structures. On the contrary, one naturally tends to use probabilistic concepts for chaos and fluids. These probabilistic concepts and their associated entropy, S_{BG}, would typically be the realm of BG statistical mechanics and standard thermodynamics. It appears that, in the marginal cases, or at least in very many of them, between

---

[2] It is frequently encountered nowadays the belief that complexity emerges typically at the edge of chaos. For instance, the final words of the Abstract of a lecture delivered in September 2005 by Leon O. Chua at the Politecnico di Milano were *"Explicit mathematical criteria will be given to identify a relatively small subset of the locally active parameter region called the edge of chaos where most of the complex phenomena emerge"* [20].

great order and great disorder, the statistical procedures can *still* be used. However, the associated natural entropy would not anymore be the BG one, but typically S_q with $q \neq 1$. It then appears quite naturally the scenario within which BG statistical mechanics is the microscopic thermodynamical description properly associated with smooth geometry (typically, although not necessarily, the plain Euclidean one), whereas nonextensive statistical mechanics would be the proper counterpart which has privileged connections with (multi)fractal and similar, hierarchical, statistically scale-invariant structures (at least asymptotically speaking). As already mentioned, a paradigmatic case would be those nonlinear dynamical systems whose largest Lyapunov exponent is neither negative (easily predictable systems) nor positive (strong chaos) but vanishing instead, i.e., the edge of chaos (weak chaos[3]). Standard, equilibrium critical phenomena also deserve a special comment. Indeed, I have always liked to think and say that "criticality is a little window through which one can see the nonextensive world". Many people have certainly had similar insights. Alberto Robledo, Filippo Caruso, and myself have exhibited some rigorous evidences—to be discussed later on—along this line. *Not* that there is anything wrong with the usual and successful use of BG concepts to discuss the neighborhood of criticality in many cooperative systems at thermal equilibrium! But, if one wants to make a delicate quantification and classification of some of the physical concepts *precisely at the critical point*, the nonextensive language appears to be a privileged one for this task. It may be so for many anomalous systems. Paraphrasing Angel Plastino's (A. Plastino Sr.) last statement in his lecture at the 2003 Villasimius meeting, "for different sizes of screws one must use different screwdrivers"! We cannot avoid remembering, at this point, Sadi Carnot's famous words "When a hypothesis no longer suffices to explain phenomena, it should be abandoned".

A proposal of a generalization of the BG entropy as the physical basis for dealing, in statistical-mechanical terms, with some wide classes of complex systems might— in the view of many—in some sense imply in a new paradigm, whose validity may be further consolidated by future progress and verifications. Indeed, we shall argue along the entire book that q is, for time evolving systems, determined *a priori* by the microscopic (or associated mesoscopic) dynamics of the system. This is in some sense less innocuous than it looks at first sight. Indeed, this means that the entropy to be used for thermostatistical purposes would be *not universal* but would *depend on the system* or, more precisely, *on the nonadditive universality class to which the system belongs*[4]. Whenever a new scientific viewpoint is proposed, either correct or wrong, it usually attracts quite extreme opinions. One of the questions that is regularly asked is the following: "Do I really need this? Is it not possible to work

---

[3] In the present book, the expression "weak chaos" is used in the sense of a sensitivity to the initial conditions diverging with time slower than exponentially, and *not* in other senses occasionally used in the theory of nonlinear dynamical systems.

[4] Leo Tolstoy's 1877 novel Anna Karenina begins: *All happy families are alike; each unhappy family is unhappy in its own way.* This dramatic sentence can be used as a transparent metaphor for the present generalization of the BG entropy. Indeed, there is only *one* manner of elements being essentially independent ($q = 1$), whereas there are *infinitely many manners* of being nontrivially correlated ($q \neq 1$).

all this out just with the concepts that we already have, and that have been lengthily tested?". This type of question is rarely easy to answer, because it involves the proof without ambiguity that some given result can *by no means* be obtained within the traditional theory.

However, let me present an analogy, basically due to Michel Baranger [21, 22], in order to clarify at least one of the aspects that are relevant for this nontrivial problem. Suppose one only knows how to handle straight lines and segments and wants to calculate areas delimited by curves. Does one really need the Newton-Leibnitz differential and integral calculus? Well, one might approach the result by approximating the curve with polygonals, and that works reasonably well in most cases. However, if one wants to better approach reality, one would consider more and more, shorter and shorter, straight segments. But one would ultimately want to take an *infinity* of such infinitely small segments. If one does so, then one has precisely jumped into the standard differential and integral calculus! How big, epistemologically speaking, was that step is a matter of debate, but its practicality is out of question. The curve that is handled might, in particular, be a straight line itself (or a finite number of straight pieces). In this case, there is of course no need to do the limiting process.

Let me present a second analogy, this one primarily due to Angel Ricardo Plastino (A. Plastino Jr.). It was known by ancient astronomers that the apparent orbits of stars are *circles*, a form that was considered geometrically "perfect". The problematic orbits were those of the planets, for instance that of Mars. Ptolemy proposed a very ingenious way out, the *epicycles*, i.e., circles turning around circles. The predictions became of great precision, and astronomers along centuries developed, with sensible success, the use of dozens of epicycles, each one on top of the previous one. It remained so until the proposal of Johannes Kepler: the orbits are well described by *ellipses*, a form which generalizes the circle by having an extra parameter, the *eccentricity*. The eccentricities of the various planets were determined through fitting with the observational data. Today we know, through Newtonian mechanics, that it is possible to compute numerically the planetary orbits, including the changes in the eccentricities, and in the other orbital elements, arising from the gravitational interaction between the planets. To do that, one needs to know the masses of the planets, and their positions and velocities at a given time (in practice, all these quantities can be regarded as fitting parameters to be determined from observations). Kepler and his immediate followers, lacking the necessary observational, theoretical, and computational resources, just fitted the orbital elements, using, however, the appropriate functional forms, i.e., the Keplerian ellipses. In few years, virtually all European astronomers abandoned the use of the complex Ptolemaic epicycles and adopted the simple Keplerian orbits. We know today, through Fourier transform, that the periodic motion on *one* ellipse is totally equivalent to an *infinite* number of specific circular epicycles. So we can proceed either way. It is clear, however, that an ellipse is by far more practical and concise, even if in principle it can be thought as infinitely many circles. We must concomitantly "pay the price" of an extra parameter, the elliptical eccentricity.

1.2 Introduction

Let us consider the free *surface* of a glass covering a table. And let us idealize it as being *planar*. What is its *volume*? Clearly *zero* since it has no height. An uninteresting answer to an uninteresting question. What is its *length*? Clearly *infinity*. One more uninteresting answer to another uninteresting question. Now, if we ask what is its *area*, we will have a meaningful answer, say $2\,\mathrm{m}^2$. A *finite* answer. Not zero, not infinity—correct but poorly informative features. A *finite* answer for a measurable quantity, as expected from good theoretical physics, good experimental physics, and good mathematics. Who "told" us that the interesting question for this problem was the *area*? *The system did!* Its planar geometrical nature did. If we were focusing on a fractal, the interesting question would of course be its measure in d_f dimensions, d_f being the corresponding *fractal* or *Hausdorff dimension*. Its measure in any dimension d larger than d_f is zero, and in any dimension smaller than d_f is infinity. Only the measure at precisely d_f dimensions yields a *finite* number. For instance, if we consider an ideal 10 cm long straight segment, and we proceed through the celebrated Cantor-set construction (i.e., eliminate the central third of the segment, and then also eliminate the central third of each of the two remaining thirds, and hypothetically continue doing this for ever) we will ultimately arrive to a remarkable geometrical set—the triadic Cantor-set—which is embedded in a one-dimensional space but whose Lebesgue measure is zero. The fractal dimension of this set is $d_f = \ln 2/\ln 3 = 0.6309...$ Therefore the interesting information about our present hypothetical system is that its measure is $(10\,\mathrm{cm})^{0.6309...} \simeq 4.275\,\mathrm{cm}^{0.6309}$. And, interestingly enough, the specific *nature* of this valuable geometric information was mandated by the system itself!

This entire book is written within precisely this philosophy: *it is the natural (or artificial or social) system itself which, through its geometrical-dynamical properties, determines the specific informational tool—entropy—to be meaningfully used for the study of its (thermo) statistical properties.* The reader surely realizes that this epistemological standpoint somehow involves what some consider as a kind of new paradigm for statistical mechanics and related areas. Indeed, the physically important entropy—a crucial concept—is *not* thought as being an universal functional that is given once for ever, but it rather is a delicate and powerful concept to be carefully constructed for classes of systems. In other words, we adopt here the viewpoint that the—simultaneously aesthetic and fruitful—way of thinking about this issue is the existence of *universality classes of systems*. These systems share the *same* functional connection between the entropy and the set of probabilities of their microscopic states. The most known such universality class is that which we shall refer to as the Boltzmann–Gibbs (BG) one. Its associated entropy is given (for a set of W discrete states) by

$$S_{BG} = -k \sum_{i=1}^{W} p_i \ln p_i, \tag{1.1}$$

with

$$\sum_{i=1}^{W} p_i = 1, \tag{1.2}$$

and where k is some conventional positive constant. This constant is taken to be Boltzmann constant k_B in thermostatistics, and is usually taken equal to unity for informational or computational purposes. In this book we shall use, without further clarification, one or the other of these two conventions, depending on the particular convenience. The reader will unambiguously detect which convention we are using within any specific context. For the particular case of *equal probabilities* (i.e., $p_i = 1/W$, $\forall i$), Eq. (1.1) becomes

$$S_{BG} = k \ln W, \tag{1.3}$$

which is carved on Boltzmann's grave in Vienna by suggestion of Planck. This celebrated expression, as well as Eq. (1.1), have been used in a variety of creative manners by Planck, Einstein, von Neumann, Shannon, Szilard, Jaynes, Tisza, and others. Eq. (1.1) has the following remarkable property. If we compose two *probabilistically independent* subsystems A and B (with numbers of states, respectively, denoted by W_A and W_B), i.e., such that the joint probabilities factorize, $p_{ij}^{A+B} = p_i^A p_j^B$ ($\forall (i, j)$), the entropy S_{BG} is *additive*[5] [17]. By this we mean that

$$S_{BG}(A + B) = S_{BG}(A) + S_{BG}(B), \tag{1.4}$$

where

$$S_{BG}(A + B) \equiv -k \sum_{i=1}^{W_A} \sum_{j=1}^{W_B} p_{ij}^{A+B} \ln p_{ij}^{A+B} \quad \text{(with } W_{A+B} = W_A W_B\text{)}, \tag{1.5}$$

$$S_{BG}(A) \equiv -k \sum_{i=1}^{W_A} p_i^A \ln p_i^A, \tag{1.6}$$

and

$$S_{BG}(B) \equiv -k \sum_{j=1}^{W_B} p_j^B \ln p_j^B. \tag{1.7}$$

Expression (1.1) was first proposed (in its form for simple continuous systems) by Boltzmann [7, 8] in the 1870s, and was then refined by Gibbs [1–3] for more general systems. It is the basis of the usual, *BG* statistical mechanics. In particular, its optimization under appropriate constraints (that we shall describe later on) yields, for a system in *thermal equilibrium* with a thermostat at *temperature T*, the celebrated BG *factor* or *weight*, namely

---

[5] The important mathematical distinction between *additive* and *extensive* is addressed later on.

$$p_i = \frac{e^{-\beta E_i}}{Z_{BG}} \qquad (1.8)$$

with

$$\beta \equiv 1/kT, \qquad (1.9)$$

$$Z_{BG} \equiv \sum_{j=1}^{W} e^{-\beta E_j}, \qquad (1.10)$$

and where $\{E_i\}$ denotes the *energy spectrum* of the system, i.e., the *eigenvalues* of the Hamiltonian of the system with the adopted boundary conditions; Z_{BG} is referred to as the *partition function*.

Expressions (1.1) and (1.8) are the landmarks of *BG* statistical mechanics, and are vastly and successfully used in physics, chemistry, mathematics, computational sciences, engineering, and elsewhere. Since their establishment, over 140 years ago, they constitute fundamental pieces of contemporary physics. Though notoriously applicable in very many systems and situations, we believe that they need to be modified (generalized) in others, in particular in most of the so-called *complex systems* (see, for instance, [19, 23–26]). In other words, there are nowadays strong reasons to believe that they are *not universal*, as somehow implicitly (or explicitly) thought until not long ago by many physicists. They *must* have in fact a restricted domain of validity, *as any other human intellectual construct*; as Newtonian mechanics, nonrelativistic quantum mechanics, special relativity, Maxwell electromagnetism, and all others. The basic purpose of this book is precisely to explore—the best that our present knowledge allows—for what systems and conditions the BG concepts become either inefficiently applicable or nonapplicable at all, and what might be done in such cases, or at least in (definitively wide) classes of them. The possibility of some kind of generalization of BG statistical concepts, or at least some intuition about the restricted validity of such concepts, already emerged along the years, in one way or another, in the mind of various physicists or mathematicians. This is, at least, what one is led to think from the various statements that we reproduce in the next Section.

1.3 Background and Indications in the Literature

We recall here interesting points raised along the years by various thinkers on the theme of the foundations and domain of validity of the concepts that are currently used in standard statistical mechanics.

Boltzmann himself wrote, in his 1896 *Lectures on Gas Theory* [27], the following words: (*The bold faces in this and subsequent quotations are mine.*)

> When the distance at which two gas molecules interact with each other noticeably is vanishingly small relative to the average distance between a molecule and its nearest neighbor—or,

as one can also say, when **the space occupied by the molecules (or their spheres of action) is negligible compared to the space filled by the gas**—then the fraction of the path of each molecule during which it is affected by its interaction with other molecules is vanishingly small compared to the fraction that is rectilinear, or simply determined by external forces. [...] The gas is "ideal" in all these cases.

Boltzmann is here referring essentially to the hypothesis of *ideal* gas. It shows nevertheless how clear it was in his mind the relevance of the *range* of the interactions for the thermostatistical theory he was putting forward.

Gibbs, in the Preface of his celebrated 1902 *Elementary Principles in Statistical Mechanics—Developed with Especial Reference to the Rational Foundation of Thermodynamics* [1–3], wrote the following touching words:

Certainly, **one is building on an insecure foundation**, who rests his work on hypotheses concerning the constitution of matter.

Difficulties of this kind have deterred the author from attempting to explain the mysteries of nature, and have forced him to be contented with the more modest aim of deducing some of the more obvious propositions relating to **the statistical branch of mechanics**.

In these lines, Gibbs not only shares with us his epistemological distress about the foundations of the science he, Maxwell, and Boltzmann are founding. He also gives a precious indication that, in his mind, this unknown foundation would certainly come *from mechanics*. Everything indicates that this was also the ultimate understanding of Boltzmann, who—unsuccessfully—tried his entire life (the so-called *Boltzmann's program*) to derive statistical mechanics from Newtonian mechanics. In fact, Boltzmann's program remains unconcluded until today!

As we see next, the same understanding permeates in the words of Einstein that we cite next, from his 1910 paper [28]:

Usually W is set equal to the number of ways (complexions) in which a state, which is incompletely defined in the sense of a molecular theory (i.e., coarse grained), can be realized. To compute W one needs a complete theory (something like a complete molecular-mechanical theory) of the system. For that reason it appears to be **doubtful whether Boltzmann's principle alone**, i.e., without a complete molecular-mechanical theory (Elementary theory) **has any real meaning. The equation** $S = k \log W + const.$ **appears [therefore]**, without an Elementary theory—or whatsoever one wants to call it—**devoid of any meaning from a phenomenological point of view.**[6]

---

[6] Most of the translation is due to E.G.D. Cohen [29]. A slightly different translation also is available: ["Usually W is put equal to the number of complexions... In order to calculate W, one needs a *complete* (molecular-mechanical) theory of the system under consideration. Therefore it is dubious whether the Boltzmann principle has any meaning without a complete molecular-mechanical theory or some other theory which describes the elementary processes. $S = \frac{R}{N} \log W +$ const. seems without content, from a phenomenological point of view, without giving in addition such an *Elementartheorie*." (Translation: Abraham Pais, *Subtle is the Lord...*, Oxford University Press, 1982)].

By *Boltzmann's principle*—expression coined apparently by Einstein himself—the author refers precisely to the logarithmic form for the entropy that he explicitly writes down a few words later. It is quite striking the crucial role that Einstein attributes to microscopic dynamics for giving a clear sense to that particular form for the entropy.

Coming back to Gibbs's book [1 3], in page 35 he writes:

> In treating of the canonical distribution, we shall always suppose the multiple integral in equation (92) *[the partition function, as we call it nowadays]* to have a **finite** value, as otherwise the coefficient of probability vanishes, and **the law of distribution becomes illusory**. This will exclude certain cases, but not such apparently, as will affect the value of our results with respect to their bearing on thermodynamics. It will exclude, for instance, cases in which the system or parts of it can be distributed in unlimited space [...]. **It also excludes many cases in which the energy can decrease without limit, as when the system contains material points which attract one another inversely as the squares of their distances.** [...]. For the purposes of a general discussion, it is sufficient to call attention to the assumption implicitly involved in the formula (92).

Clearly, Gibbs is well aware that the theory he is developing has limitations. It does not apply to anomalous cases such as gravitation.

Enrico Fermi, in his 1936 *Thermodynamics* [30] wrote:

> The entropy of a system composed of several parts is **very often** equal to the sum of the entropies of all the parts. This is true if the energy of the system is the sum of the energies of all the parts and if the work performed by the system during a transformation is equal to the sum of the amounts of work performed by all the parts. Notice that **these conditions are not quite obvious** and that in some cases **they may not be fulfilled**. Thus, for example, in the case of a system composed of two homogeneous substances, it will be possible to express the energy as the sum of the energies of the two substances only if we can neglect the surface energy of the two substances where they are in contact. The surface energy can generally be neglected only if the two substances are not very finely subdivided; otherwise, **it can play a considerable role**

So, Fermi says "very often", which virtually implies "not necessarily so"!

Ettore Majorana, mysteriously missing since 25 March 1938, wrote [31]:

> This is mainly because entropy is an addditive quantity as the other ones. In other words, the entropy of a system composed of several **independent** parts is equal to the sum of entropy of each single part. [...] Therefore one considers all possible internal determinations as equally probable. This is indeed a **new hypothesis** because the universe, which is far from being in the same state indefinitively, is subjected to continuous transformations. We will therefore **admit as an extremely plausible working hypothesis, whose far consequences could sometime not be verified, that all the internal states of a system are a priori equally probable** in specific physical conditions. **Under this hypothesis**, the *statistical ensemble* associated to each macroscopic state A turns out to be completely defined.

Like Fermi, Majorana leaves the door open to other, nonstandard, possibilities, which could be not inconsistent with the methods of statistical mechanics.

Claude Elwood Shannon, in his 1948/1949 *The Mathematical Theory of Communication* [32], justified the logarithmic entropy $k \ln W$ in these plain terms:

It is practically more **useful**. [...] It is nearer to our **intuitive** feeling as to the proper measure. [...] It is **mathematically** more **suitable**. [...].

And, after stating the celebrated axioms that yield, as unique answer, the entropy (1.1), he wrote:

This theorem and the assumptions required for its proof, **are in no way necessary** for the present theory. It is **given chiefly to lend a certain plausibility** to some of our later definitions. **The real justification of these definitions, however, will reside in their implications**.

It is certainly remarkable how wide Shannon leaves the door open to other entropies than the one that he brilliantly discussed.

Laszlo Tisza wrote, in 1961, in his *Generalized Thermodynamics* [33]:

The situation is different for the additivity postulate $P\,a2$, **the validity of which cannot be inferred from general principles**. We have to require that **the interaction energy between thermodynamic systems be negligible**. This assumption is closely related to the homogeneity postulate $P\,d1$. From the molecular point of view, **additivity** and homogeneity can be expected to be **reasonable approximations** for systems containing many particles, **provided that the intermolecular forces have a short-range character**.

Peter Landsberg wrote, in 1978/1990, in his *Thermodynamics and Statistical Mechanics* [34]:

The presence of long-range forces causes important amendments to thermodynamics, some of which are not fully investigated as yet.

And in 1984 he added [35]

[...] in the case of systems with long-range forces and which are therefore nonextensive (in some sense) **some thermodynamic results do not hold**. [...] The failure of some thermodynamic results, normally taken to be standard for black hole and other nonextensive systems has recently been discussed. [...] If two identical black holes are merged, the presence of long-range forces in the form of gravity leads to a more complicated situation, and **the entropy is nonextensive**.

Nico van Kampen, in his 1981 *Stochastic Processes in Physics and Chemistry* [36], wrote:

Actually an additional stability criterion is needed, see M.E. Fisher, Archives Rat. Mech. Anal. **17**, 377 (1964); D. Ruelle, *Statistical Mechanics: Rigorous Results* (Benjamin, New York 1969). A collection of point particles with **mutual gravitation** is an example where this criterion **is not satisfied**, and for which therefore **no statistical mechanics exists**.

L.G. Taff writes in his 1985 *Celestial Mechanics* [37]:

> [...] This means that the **total energy of any finite collection of self-gravitating mass points does not have a finite, extensive** (e.g., proportional to the number of particles) **lower bound. Without such a property there can be no rigorous basis for the statistical mechanics of such a system** (Fisher and Ruelle 1966). Basically it is that simple. One can **ignore the fact** that one knows that there is no rigorous basis for one's computer manipulations; one can **try to improve the situation**, or one can **look for another job**.

Needless to say that the very existence of the present book constitutes but an attempt *to improve the situation*!

The same point is addressed by W.C. Saslaw in his 1985 *Gravitation Physics of Stellar and Galactic Systems* [38]:

> When interactions are important **the thermodynamic parameters may lose their simple intensive and extensive properties** for subregions of a given system. [...] **Gravitational systems**, as often mentioned earlier, do not saturate and so **do not have an ultimate equilibrium state**.

Radu Balescu, in his 1975 *Equlibrium and Nonequilibrium Statistical Mechanics* [39], wrote:

> It therefore appears from the present discussion that **the mixing property of a mechanical system is much more important for the understanding of statistical mechanics than the mere ergodicity.** [...] **A detailed rigorous study** of the way in which the concepts of **mixing** and the concept of **large numbers of degrees of freedom** influence the macroscopic laws of motion **is still lacking**.

David Ruelle writes in page 1 of his 1978 *Thermodynamical Formalism* [40] (and maintains in page 1 of his 2004 Edition [41]):

> The formalism of equilibrium statistical mechanics—which we shall call *thermodynamic formalism*—has been developed since J.W. Gibbs to describe the properties of certain physical systems. [...] While **the physical justification of the thermodynamic formalism remains quite insufficient**, this formalism has proved remarkably successful at explaining facts.

as well as

> The mathematical investigation of the thermodynamic formalism is in fact not completed: the theory is a young one, with emphasis still more on imagination than on technical difficulties. This situation is reminiscent of pre-classic art forms, where inspiration has not been castrated by the necessity to conform to standard technical patterns. We hope that some of this juvenile freshness of the subject will remain in the present monograph!

He writes also, in page 3:

> **The problem of why the Gibbs ensemble describes thermal equilibrium** (at least for "large systems") when the above physical identifications have been made **is deep and incompletely clarified**.

The basic identification he is referring to is between β and the inverse temperature. Consistently, the first equation in both editions (page 3) is dedicated to define the entropy to be associated with a probability measure. The BG form is introduced after the words "we define its *entropy*" without any kind of justification or physical motivation.

The same theme is retaken by Floris Takens in the 1991 *Structures in Dynamics* [42]. Takens writes:

> The values of p_i are determined by the following **dogma**: if the energy of the system in the ith state is E_i and if the temperature of the system is T then: $p_i = e^{-E_i/kT}/Z(T)$, where $Z(T) = \sum_i e^{-E_i/kT}$, (this last constant is taken so that $\sum_i p_i = 1$). This choice of p_i is called the *Gibbs distribution*. **We shall give no justification for this dogma**; even a physicist like Ruelle disposes of this question as "deep and incompletely clarified".

We know that mathematicians sometimes use the word "dogma" when they do not have the theorem. Indeed, this is not widely known, but still today no theorem exists, to the best of our knowledge, stating the necessary and sufficient microscopic conditions for being legitimate the use of the celebrated BG weight!

Roger Balian wrote in his 1982/1991 *From Microphysics to Macrophysics* [43]:

> These various quantities are connected with one another through thermodynamic relations which make their extensive or intensive nature obvious, as soon as **one postulates**, for instance, for a fluid, **that the entropy**, considered as a function of the volume Ω and of the constants of motion such as U and N, **is homogeneous of degree 1**: $S(x\Omega, xU, xN) = xS(\Omega, U, N)$ (Eq. 5.43). [...] Two **counter-examples** will help us to feel why **extensivity is less trivial than it looks**. [...] **A complete justification** of the Laws of thermodynamics, starting from statistical physics, **requires a proof of the extensivity** (5.43), a property which was **postulated** in macroscopic physics. This proof is difficult and appeals to **special conditions which must be satisfied by the interactions between the particles**.

John Maddox wrote in 1993 an article suggestively entitled *When entropy does not seem extensive* [44]. He focused on a paper by Mark Srednicki [45] where the entropy of a black hole is addressed. Maddox writes:

> Everybody who knows about entropy knows that it is an extensive property, like mass or enthalpy. [...] Of course, there is more than that to entropy, which is also a measure of disorder. Everybody also agrees on that. But **how is disorder measured**? [...] So why is the entropy of a black hole proportional to the square of its radius, and **not** to the cube of it? **To its surface area rather than to its volume?**
> These comments and questions are of course consistent with the so-called black-hole Hawking entropy, whose value *per unit area* equals $1/(4\hbar\, Gk_B^{-1}c^{-3})$, a remarkable combination of four universal constants. Equivalently $S_{BG} = \frac{k_B}{4}\frac{A_{BH}}{l_P^2}$, where A_{BH} is the area of the horizon of events of the black hole, and l_P is Planck length, as defined in Eq. (3.276).

A suggestive paper by A.C.D. van Enter, R. Fernandez and A.D. Sokal appeared in 1993. It is entitled *Regularity Properties and Pathologies of Position-Space Renormalization-Group Transformations: Scope and Limitations of Gibbsian Theory* [46]. We transcribe here a few fragments of its content. From its Abstract:

> We provide a careful, and, we hope, pedagogical, overview of the theory of Gibssian measures as well as **(the less familiar) non-Gibbsian measures**, emphasizing the distinction between these two objects and the **possible occurrence of the latter in different physical situations**.

And from its Sect. 6.1.4 of [46] *Toward a Non-Gibbsian Point of View*:

> Let us close with some general remarks on the **significance of (non-)Gibbsianness and (non)quasilocality in statistical physics.** Our first observation is that **Gibbsianness has heretofore been ubiquitous in equilibrium statistical mechanics because it has been put in** *by hand*: **nearly all measures that physicists encounter are Gibbsian because physicists have** *decided* **to study Gibbsian measures!** However, we now know that natural operations on Gibbs measures can sometimes lead out of this class. [...] **It is thus of great interest to study which types of operations preserve, or fail to preserve, the Gibbsianness (or quasilocality) of a measure. This study is currently in its infancy.**
> [...] More generally, **in areas of physics where Gibbsianness is not put in by hand, one should expect non-Gibbsianness to be ubiquitous.** This is probably the case in nonequilibrium statistical mechanics.
> **Since one cannot expect all measures of interest to be Gibbsian, the question then arises whether there are** *weaker* **conditions that capture some or most of the "good" physical properties characteristic of Gibbs measures.** For example, the stationary measure of the voter model appears to have the critical exponents predicted (under the hypothesis of Gibbsianness) by the Monte Carlo renormalization group, even though this measure is probably non-Gibbsian.
> **One may also inquire whether there is a classification of non-Gibbsian measures according to their "degree of non-Gibbsianness".**

The authors make in this paper no reference whatsoever to nonextensive statistical mechanics (proposed in fact 5 years earlier [47]). It will nevertheless become evident that, interestingly enough, several among their remarks neatly apply to the content of the present book. Particularly, it will become obvious that $(q - 1)$ represents a possible measure of "non-Gibbsianness", where q denotes the entropic index to be soon introduced.

From the viewpoint of the dynamical foundations of statistical mechanics, a recent remark (already quoted in the Preface of this book) by Giulio Casati and Tomaz Prosen [12] is worthy to be reproduced at this point:

> While exponential instability is **sufficient** for a meaningful statistical description, it is not known whether or not it is also **necessary**.

Let us anticipate that it belongs to the aim of the present book to convince the reader precisely that it is *not* necessary: power-law instability appears to do the job similarly well, if we consistently adopt the appropriate entropy.

In his 2004 Boltzmann Award lecture *Boltzmann and Einstein: Statistics and dynamics—An unsolved problem*, E.G.D. Cohen wrote [29] (see also [48]):

> He (*Boltzmann*) used both a dynamical and a statistical method. However, Einstein strongly disagreed with Boltzmann's statistical method, arguing that a statistical description of a system should be based on the dynamics of the system. **This opened the way, especially for complex systems, for other than Boltzmann statistics. [...] It seems that perhaps a combination of dynamics and statistics is necessary to describe systems with complicated dynamics.**

And he concludes by quoting Boltzmann's Lecture *On recent developments of the methods of theoretical physics* in 1899 in München, Germany:

"Will the old (classical) mechanics with the old forces ... in its essence remain, or live on, one day, only in history ...superseded by entirely different notions?" [...] "Indeed interesting questions! One almost regrets to have to die long before they are settled. O! immodest mortal! Your fate is the joy of watching the ever-shifting battle!" (not to see the outcome).

Many more statements exist in the literature along similar lines. But we believe that the ones that we have selected are enough (both in quality and quantity!) for depicting, at least in a "impressionistic" way, the epistemological scenario within which we are evolving. A few basic interrelated points that emerge include:

(i)   No strict physical or mathematical reason appears to exist for not exploring the possible generalization of the BG entropy and its consequences.

(ii) The BG entropy and any of its possible generalizations should conform to the microscopical dynamical features of the system, very specifically to properties such as sensitivity to the initial conditions and mixing. The relevant rigorous necessary and sufficient conditions are still unknown. The ultimate justification of any physical entropy on theoretical grounds is expected to come from microscopic dynamics and detailed geometrical conditions.

(iii) No physical or mathematical reason appears to exist for not exploring, in natural, artificial and even social systems, distributions differing from the BG one, very specifically for stationary or quasi-stationary states differing from thermal equilibrium, such as metastable states, and other nonequilibrium ones.

(iv) Long-range microscopic interactions (and long-range microscopic memory), as well as interactions exhibiting severe (e.g., nonintegrable) singularities at the origin, appear as a privileged field for the exploration and understanding of anomalous thermostatistical behavior.

1.4   Symmetry, Energy, and Entropy

At this point, let us focus on some connections between three key concepts of physics, namely *symmetry*, *energy* and *entropy*: see Fig. 1.1. According to Plato, symmetry sits in *Topos Ouranos* (heavens), where sit all branches of mathematics—science of structures. In contrast, energy, and entropy should be assumed to sit in *Physis* (nature). Energy deals with the *possibilities* of the system; entropy deals with the *probabilities* of those possibilities. It is fair to say that energy is a quite subtle physical concept. Entropy is based upon the ingredients of energy, and therefore it is, epistemologically speaking, one step further. It is most likely because of this feature that entropy emerged, in the history of physics, well after energy. A coin has two faces, and can therefore fall in two possible manners, head and tail (if we disconsider the very unlike possibility that it falls on its edge). This is the world of the possibilities for this simple system. The world of its probabilities is more delicate. *Before* throwing the coin (assumed *fair* for simplicity), the entropy equals $k \ln 2$. *After* throwing it, *it still equals $k \ln 2$ for whoever does not know the result*, whereas *it equals zero for*

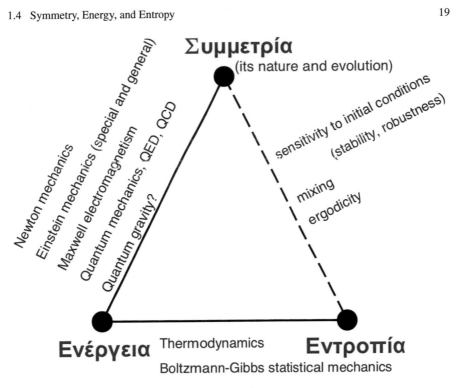

Fig. 1.1 Connections between symmetry, energy and entropy

whoever knows it. This example neatly illustrates the *informational nature* of the concept.

Let us now address the connections. Those between symmetry and energy are long and well known. Galilean invariance of the equations is central to Newtonian mechanics. Its simplest form of energy can be considered to be the kinetic one of a point particle, namely $p^2/2m$, p being the *linear momentum*, and m the *mass*. This energy, although having a unique form, is *not* universal; indeed it depends on the mass of the system. If we replace now the Galilean invariance by the Lorentzian one, this drastically changes the form itself of the kinetic energy, which now becomes $(p^2c^2 + m_0^2c^4)^{1/2}$, c being the speed of light in vacuum, and m_0 the mass at rest. In other words, this change of symmetry is far from innocuous; it does nothing less than changing Newtonian mechanics into special relativity! Maxwell electromagnetism is, as well known, deeply related to this same Lorentzian invariance, as well as to gauge invariance. The latter plays, in turn, a central role in quantum electrodynamics and quantum field theory. Quantum chromodynamics also is deeply related to symmetry properties. And so is expected to be quantum gravity, whenever it becomes reality. Summarizing, the deep connections between symmetry and energy are standard knowledge in contemporary physics. Changes in one of them are concomitantly followed by changes in the other one.

What about the connections between energy and entropy? Well, also these are quite known. They naturally emerge in thermodynamics (e.g., the possibility and manners for transforming work into heat, and, partially, the other way around). This obviously reflects on BG statistical mechanics itself.

But, what can we say about the possible connections between symmetry (its nature and evolution) and entropy? This topic has remained basically unchanged and quite unexplored during more than one century! *Why?* Hard to know. However, it is allowed to suspect that this intellectual lethargy comes, on one hand, from the "slopiness" of the concept of entropy, and, on the other one, from the remarkable fact that the unique functional form that has been used in physics is the BG one (Eq. (1.1) and its continuum or quantum analogs), *which depends on only one of the universal constants, namely Boltzmann constant k_B*. Within this intellectual landscape, generation after generation, the idea installed in the mind of very many physicists that the physical entropic functional itself *must* be universal, and that it is of course the BG one. In the present book, we try to convince the reader that it is *not* so, that many types of entropy can be physically and mathematically meaningful. Moreover, we shall argue that dynamical concepts such as the time dependence of the sensitivity to the initial conditions, mixing, and the associated occupancy and visitation networks they may cause in phase space, have so strong effects, that even the functional form of the entropy must be modified in order to conform to the empirical world. Naturally, the BG entropy will then still have a highly privileged position. It surely is the correct one when the microscopic nonlinear dynamics is controlled by *positive* Lyapunov exponents, hence *strong* chaos. If the system is such that strong chaos is absent (typically because the maximal Lyapunov exponent vanishes), then the physical entropy to be used appears to be a different one.

1.5   A Few Words on the Foundations of Statistical Mechanics

A mechanical foundation of statistical mechanics from *first principles* should essentially include, in one way or another, the following main steps [49].

(i) Adopt a *microscopic dynamics*. This dynamics typically is free from any phenomenological noise or stochastic ingredient, so that the foundation may be considered as *from first principles*. This dynamics could be Newtonian, or quantum, or relativistic mechanics (or some other mechanics to be found in future) of a many-body system composed by say N interacting elements or fields. It could also be conservative or dissipative coupled maps, or even cellular automata. Consistently, time t could be continuous or discrete. The same is valid for space. The quantity which is defined in space-time could itself be continuous or discrete. For example, in quantum mechanics, the quantity is a complex continuous variable (the wave function) defined in a continuous space-time. On the other extreme, we have cellular automata, for which all three relevant variables—time, space and the quantity therein defined—are discrete. In the case of a Newtonian mechanical system of particles, we may think of N Dirac delta functions localized in continuous spatial positions which depend on a continuous time. Langevin-like equations (and associated Fokker–Planck-like equations) are typically considered not microscopic, but *mesoscopic* instead. The reason of course is the fact that they include at their very formulation, i.e., in an essential manner, some sort of noise. Consequently, they should not be used as a starting point if we desire the foundation of statistical mechanics to be from first principles. A schematic ladder going from the microscopic, through the mesoscopic, to the macroscopic descriptions of the world is presented in Fig. 1.2.

(ii)   Then assume some set of *initial conditions* (either a single one or an ensemble of many of them) and let the system evolve in time. These initial conditions are, for say a classical system, defined in the so-called *phase space* of the microscopic configurations of the system, for example Gibbs' Γ space for a Newtonian N-

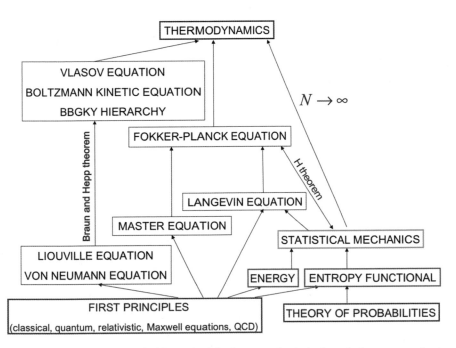

Fig. 1.2 From the microscopic (*first-principle* electro-mechanics), through the mesoscopic, to the macroscopic (thermodynamics) descriptions of nature in contemporary theoretical physics, including several, though obviously not all, relevant equations. The energy emerges from electro-mechanics as a constant of motion. The entropic functional emerges from first-principle electro-mechanics as an adequate function of probabilities directly related to the trajectories in say phase space or in Hilbert space or in Fock space, and the correlations they involve. The Braun and Hepp theorem that is mentioned in one of the links is the one proved in [50]. We only depict here a basic scheme. Further branches and connections do exist which are not detailed here (see for instance [51–57])

particle system (e.g., the Γ space for point masses has $2dN$ dimensions if the particles live in a d-dimensional space). These initial conditions typically (but not necessarily) involve one or more constants of motion. For example, if the system is a conservative Newtonian one of point masses, the initial total energy and the initial total linear momentum (d dimensional vector) are such constants of motion. The total angular momentum might also be a constant of motion. It is quite frequent to use coordinates such that both total linear momentum and total angular momentum vanish.

If the system consists of conservative coupled maps, the initial hypervolume of an ensemble of initial conditions near a given one is preserved through time evolution. By the way, in physics, such coupled maps are frequently obtained through Poincaré sections of Newtonian dynamical systems.

(iii) After some *sufficiently long evolution time* (which typically depends on both N and the spatial range of the interactions), the system might approach some *stationary* or *quasi-stationary* macroscopic state[7]. In such a state, the various regions of phase space are being visited with some probabilities. This set of probabilities either does not depend anymore on time or depends on it very slowly. More precisely, if it depends on time, it does so on a scale much longer (e.g., geological times) than the microscopic time scale. The visited regions of phase space that we are referring to typically correspond to a partition of phase space with a degree of (coarse or fine) graining that we adopt for specific purposes. These probabilities can be either insensitive or, on the contrary, very sensitive to the ordering in which the $t \rightarrow \infty$ (*asymptotic*) and $N \rightarrow \infty$ (*thermodynamic*) limits are taken. This can depend on various things such as the range of the interactions, or whether the system is on the ordered or on the disordered side of a continuous phase transition. Generically speaking, the influence of the ordering of the $t \rightarrow \infty$ and $N \rightarrow \infty$ limits is typically related to some kind of breakdown of symmetry, or of ergodicity, or the alike. In some cases, the result might even depend on the path along which the $t \rightarrow \infty$ and $N \rightarrow \infty$ limits are attained. For example, it might depend on whether the simultaneous limit $\lim_{(t,N)\rightarrow(\infty,\infty)} t/N^\gamma$ (with $\gamma > 0$) is above or below some model-dependent critical value.

The simplest nontrivial dynamical situation is expected to occur for an isolated many-body short-range-interacting classical Hamiltonian system (microcanonical ensemble); later on we shall qualify when an interaction is considered *short-ranged* in the present context. In such a case, the typical microscopic (nonlinear) dynamics is expected to be *strongly chaotic*, in the sense that the maximal Lyapunov exponent is positive almost everywhere in phase space. Such a system would necessarily be *mixing*, i.e., it would quickly visit virtually all the accessible phase space (more precisely, very close to almost all the accessible phase space) for almost *any* possible initial condition. Furthermore, it would necessarily be *ergodic* with respect to some measure in the full phase space, i.e., *time averages*

---

[7] When the system exhibits some sort of *aging*, the expression *quasi-stationary* is preferable to *stationary*.

and *ensemble averages* would coincide. In most of the cases this measure is expected to be uniform in phase space, i.e., the *hypothesis of equal probabilities* first thought of by Boltzmann would be satisfied.

A slightly more complex situation is encountered for those systems which exhibit a continuous phase transition. Let us consider the simple case of a ferromagnet which is invariant under inversion of the hard axis of magnetization, e.g., the $d = 3$ XY classical nearest-neighbor ferromagnetic model on simple cubic lattice. If the system is in its disordered (paramagnetic) phase, the limits $t \to \infty$ and $N \to \infty$ commute, and the entire phase space is expected to be equally well visited. If the system is in its ordered (ferromagnetic) phase, the situation is expected to be more subtle. The $\lim_{N \to \infty} \lim_{t \to \infty}$ set of probabilities is, as before, equally distributed all over the entire phase space *for almost any initial condition*. But this is not expected to be so for the $\lim_{t \to \infty} \lim_{N \to \infty}$ set of probabilities. The system probably lives, in this case, only in part of the entire phase space. Indeed, if the initial condition is such that the initial magnetization is *nonzero* along a given direction in the XY plane, even if infinitesimally so (for instance, under the presence of a vanishingly small external magnetic field), then the system is expected to be ergodic *but only in the part of phase space associated with a magnetization along that direction, even if the external magnetic field becomes vanishingly small*. This is usually referred to as a *spontaneous breakdown of symmetry*. This illustrates, already in this simple example, the importance that the ordering of the $t \to \infty$ and $N \to \infty$ limits can have.

A considerably more complex situation is expected to occur if we consider a *long-range-interacting* model, e.g., the same $d = 3$ XY classical ferromagnetic model on simple cubic lattice as before, but now with a coupling constant which decays with distance as $1/r^\alpha$, where r is the spin-to-spin distance measured in crystal units, and $\alpha \geq 0$ (the nearest-neighbor model that we just discussed corresponds to the $\alpha \to \infty$ limit, which is the extreme case of the (quasi) *short*-ranged domain $\alpha/d > 1$). The $0 \leq \alpha/d \leq 1$ *long*-ranged model also appears to have a continuous phase transition. In the disordered phase, the system possibly is ergodic over the entire phase space. But in the ordered phase the result can strongly depend on the ordering of the two limits, or on the path along which these two limits are accomplished. The $\lim_{N \to \infty} \lim_{t \to \infty}$ set of probabilities corresponds to the system living in the entire phase space. In contrast, the $\lim_{t \to \infty} \lim_{N \to \infty}$ set of probabilities for the same (conveniently scaled) total energy might be considerably more complex. It seems that, for this ordering, phase space could exhibit many macroscopic basins of evolution. One of them leads essentially to part of the phase space where the system lives in the $\lim_{N \to \infty} \lim_{t \to \infty}$ ordering, i.e., that part of phase space which is associated with a given alignment of the magnetization, which coincides with the direction of the initial magnetization along an external magnetic field. Other basins would correspond to living in a very complicated, hierarchical-like, geometrical structure. This structure could be a zero Lebesgue measure one (in the full multidimensional phase space), somewhat similar to that of an airlines company, say Air France, whose central hub is located in Paris, or British Airways, whose central hub is located in London. The specific *location* of the structure in phase space would depend on the particular initial condition within that special basin, but the *geometrical-dynamical nature* of the structure would be virtually the same for a generic initial condition within phase space. At this point, let us warn the reader that the scenario that we have depicted here is only conjectural, and remains to be proved. It is however based on various numerical evidences (see, e.g., [58–62] and references therein, as well as Sect. 5.3). It is expected to be caused by a possibly *vanishing* maximal Lyapunov exponent in the $N \to \infty$ limit. In other words, one would possibly have, instead of strong, only *weak* chaos.

(iv) Now let us focus further on the specific role played by the *initial conditions*. If the system is strongly chaotic, hence mixing, hence ergodic, this point is irrelevant. We can make or not averages over initial conditions, we can take almost any initial condition, the outcome for sufficiently long times will be the same, in the sense that the set of probabilities in phase space will be the same. But if the system is only weakly chaotic, the result can drastically change from initial condition to initial condition. If two initial conditions belong to the same "basin of evolution", the difference at the macroscopic level could be quite irrelevant. If they belong however to different basins, the results can be sensibly different. For some purposes we might wish to stick to a specific initial condition within a certain class of initial conditions. For other purposes, we might wish to average over all initial conditions belonging to a given basin, or even over all possible initial conditions of the entire phase space. The macroscopic result obtained after averaging might considerably differ from that corresponding to a single initial condition. Moreover, the traditional statistical-mechanical classification of *ensemble average* (which does not depend on time, as soon as a possible initial transient is overcome) versus *time average* (which does not depend on the chosen single initial initial condition, as long as it is a generic one) might become insufficient. For example, coming back to our air companies analogy, if we are within the Air France evolution basin we will observe an hierarchical-like geometric-dynamical structure centered on Paris, whereas if we are in the British Airways evolution basin we will observe a *similar* hierarchical-like structure, but this time centered on London. For some analysis, it might be interesting to make averages over the evolution basin of *one* air company; for other purposes, it might be useful to make averages over *all air companies within a given class* (say inter-continental flights); still for other purposes it might be useful to make averages over *all* flights of *all* air companies; and so on. A really interesting statistical mechanics should be able, through its various limits (e.g., infinite size, infinite number of initial conditions, infinite time, infinite precision) and their orderings, to correctly address at least some of these various relevant averages. As we shall see along this book, abandoning the traditional but unnecessary restriction to additive entropic functionals opens a big door toward this goal.

(v) Last but not least, the mathematical form of the *entropy functional* must be addressed. Strictly speaking, if we have deduced (from microscopic dynamics) the probabilities to be associated with every cell in phase space, we can in principle calculate useful averages of *any* physical quantity of interest which is defined in that phase space. In this sense, we do not need to introduce an entropic functional which is defined precisely in terms of those probabilities. Especially if we take into account that *any* set of physically relevant probabilities can be obtained through *extremization* (typically *maximization*) of an infinite number of entropic functionals (monotonically depending one onto the other, such as, for instance, $[-\sum_i p_i \ln p_i]$ and say its cube), given any set of physically and mathematically meaningful constraints. However, if we wish to make contact with classical thermodynamics, we certainly need to know the mathematical form of such entropic functional. This functional is expected to match, in the appropriate limits, the classical, macroscopic, entropy *à la Clausius*. In particular, one expects it to satisfy the Clausius property of extensivity, i.e., essentially to be proportional to the weight or mass of the system. In statistical-mechanical terms, we expect it to be proportional to N for large N. This simple requirement immediately admits, in the realm of BG statistical mechanics, the entropic functional $[-\sum_i p_i \ln p_i]$ and definitively excludes say its cube.[8]

The foundations of any valuable statistical mechanics are, as already said, expected to satisfactorily cover basically all of the above points. There is a wide-spread vague belief among some physicists that these steps have already been satisfactorily accomplished since long for the standard, BG statistical mechanics. *This is not so!* Not so surprising after all, given the enormity of the corresponding task! For example, as already mentioned, at this date there is no available deduction, from and only from microscopic dynamics, of the celebrated BG exponential weight (1.8). Neither exists the deduction from microscopic dynamics, without further assumptions, of the BG entropy (1.1).

For standard systems, there is not a single reasonable doubt about the correctness of the expressions (1.1) and (1.8), and of their relationships. But, from the logical-deductive viewpoint, there is still pretty much work to be done! This is, in fact, kind of easy to notice. Indeed, all the textbooks, without exception, introduce the BG factor and/or the entropy S_{BG} in some kind of phenomenological manner, or as self-evident, or within some axiomatic formulation. None of them introduces them as (and only as) a rational consequence of say Newtonian, or quantum mechanics, using theory of probabilities. This is in fact sometimes referred to as the *Boltzmann program*. Boltzmann himself died without succeeding its implementation. Although important progress has been accomplished in these last 140 years, Boltzmann program still remains in our days as a basic intellectual challenge. Were it not the genius of scientists like Boltzmann and Gibbs, were we to exclusively depend on

---

[8] Let us anticipate that it has been recently shown [63–66] that, if we impose a Poissonian distribution for visitation times in phase space, in addition to the first and second principles of thermodynamics, we obtain the BG functional form for the entropy. If a conveniently deformed Poissonian distribution is imposed instead, we obtain the S_q functional form. These results in themselves can *not* be considered as a justification from first principles of neither the BG nor the nonextensive, statistical mechanics. Indeed, the visitation distributions are phenomenologically introduced, and the first and second principles are just imposed. This connection is nevertheless extremely clarifying, and can help producing a full justification.

mathematically well-constructed arguments, one of the monuments of contemporary physics—BG statistical mechanics— could just not exist, or it would have taken much longer to emerge!

Many anomalous natural, artificial, and social systems exist for which BG statistical concepts appear to be inapplicable. Typically because they live in peculiar stationary or quasi-stationary states that are quite different from thermal equilibrium, where BG statistics reigns. Nevertheless, as we shall see, not few of them can still be handled within statistical-mechanical methods, but with a more general entropy, namely S_q, to be introduced later on [47, 67–69].

It should naturally be clear that, whatever is not yet mathematically justified in BG statistical mechanics, it is even less justified in the generalization to which the present book is dedicated. In addition to this, some of the points that are relatively well understood in the standard theory, can be still unclear in its generalization. In other words, the theory we are presenting here is still in intense evolution (sets of reviews can be found in [18, 70–92]).

Chapter 2
Learning with Boltzmann-Gibbs Statistical Mechanics

Πᾶν μέτρον ἄριστον
Kleoboulos of Lindos (6th century B.C.)

2.1 Boltzmann–Gibbs Entropy

2.1.1 Entropic Forms

The entropic forms (1.1) and (1.3) that we have introduced in Chap. 1 correspond to the case where the (microscopic) states of the system are *discrete*. There are, however, cases in which the appropriate variables are *continuous*. For these, the BG entropy takes the form

$$S_{BG} = -k \int dx \, p(x) \, \ln[\sigma p(x)],$$ (2.1)

with

$$\int dx \, p(x) = 1,$$ (2.2)

Ancient and popular Greek expression. A possible translation into English is "All things in their proper measure are excellent". The expression is currently attributed to Kleoboulos of Lindos (one of the seven philosophers of Greek antiquity), although in a more laconic—and logically equivalent—version, namely Μέτρον ἄριστον. Indeed, it is this abridged form that Clement of Alexandria (*Stromata*, 1.14.61) and Diogenes Laertius (*Lives of Philosophers*, Book 1.93, Loeb Series) attribute to Kleoboulos. This expression addresses what I consider the basis of all *variational principles*, in my opinion the most elegant form in which physical laws can be expressed. The *Principle of least action* in mechanics, Fermat's *Principle of least time* in optics, and the *Optimization of the entropy* in statistical mechanics, are but such realizations.

© Springer Nature Switzerland AG 2023
C. Tsallis, *Introduction to Nonextensive Statistical Mechanics*,
https://doi.org/10.1007/978-3-030-79569-6_2

27

where $x/\sigma \in \mathbb{R}^D$, $D \geq 1$ being the dimension of the full space of microscopic states (called *Gibbs* Γ *phase space* for classical Hamiltonian systems).[1] Typically x carries physical units. The constant σ carries the same physical units as x, so that x/σ is a dimensionless quantity (we adopt from now on the notation $[x] = [\sigma]$, hence $[x/\sigma] = 1$). For example, if we are dealing with an isolated classical N-body Hamiltonian system of point masses interacting among them in d dimensions, we may use $\sigma = \hbar^{Nd}$. This standard choice comes of course from the fact that, at a sufficiently small scale, Newtonian mechanics becomes incorrect and we must rely on quantum mechanics. In this case, $D = 2dN$, where each of the d pairs of components of momentum and position of each of the N particles has been taken into account (we recall that $[momentum][position] = [\hbar]$). For the case of *equal probabilities* (i.e., $p(x) = 1/\Omega$, where Ω is the hypervolume of the admissible D-dimensional space), we have

$$S_{BG} = k \ln(\Omega/\sigma). \qquad (2.3)$$

A particular case of $p(x)$ is the following one:

$$p(x) = \sum_{i=1}^{W} p_i\, \delta(x - x_i) \quad (W \equiv \Omega/\sigma), \qquad (2.4)$$

where δ denotes Dirac's distribution. In this case, Eqs. (2.1), (2.2), and (2.3) precisely recover Eqs. (1.1), (1.2), and (1.3).

In both discrete and continuous cases that we have addressed until now, we were considering classical systems in the sense that all physical observables are real quantities and *not operators*. However, for intrinsic quantum systems, we must generalize the concept. In that case, the BG entropic form is to be written (as first introduced by von Neumann [93–95]) in the following manner:

$$S_{BG} = -k\, Tr \rho \ln \rho, \qquad (2.5)$$

with

$$Tr \rho = 1, \qquad (2.6)$$

where ρ is the *density matrix* acting on a W-dimensional Hilbert vectorial space (typically associated with the solutions of the Schroedinger equation with the chosen boundary conditions; in fact, quite frequently we have $W \to \infty$).

A particular case of ρ is when it is *diagonal*, i.e., the following one:

$$\rho_{ij} = p_i\, \delta_{ij}, \qquad (2.7)$$

---

[1] Strictly speaking, we are using here an oversimplified notation to denote that the vector $x \equiv (x_1, x_2, \dots, x_W)$ is to be divided, component by component, by $(\sigma_1, \sigma_2, \dots, \sigma_W)$ with $\sigma = \prod_i^W \sigma_i$.

where δ_{ij} denotes Kroenecker's delta function. In this case, Eqs. (2.5) and (2.6) exactly recover Eqs. (1.1) and (1.2).

All three entropic forms (1.1), (2.1), and (2.5) will be generically referred in the present book as *BG-entropy* because they all constitute a logarithmic measure of *uncertainty* or *lack of information*. Although we shall use one or the other for specific purposes, we shall mainly address the simple form expressed in Eq. (1.1).

2.1.2 Properties

2.1.2.1 Non-negativity

If we know with *certainty* the state of the system, then $p_{i_0} = 1$, and $p_i = 0$, $\forall i \neq i_0$. Then it follows that $S_{BG} = 0$, where we have taken into account that $\lim_{x \to 0}(x \ln x) = 0$. In any other case we have $p_i < 1$ for at least two different values of i. We can therefore write Eq. (1.1) as follows:

$$S_{BG} = -k \langle \ln p_i \rangle = k \left\langle \ln \frac{1}{p_i} \right\rangle, \tag{2.8}$$

where $\langle \cdots \rangle \equiv \sum_{i=1}^{W} p_i (...)$ is the standard *mean value*. Since $\ln(1/p_i) > 0$ $(\forall i)$, it clearly follows that S_{BG} is *positive*.

2.1.2.2 Maximal at Equal Probabilities

Energy is a concept which definitively takes into account the physical nature of the system. *Not exactly so, in some sense, the BG entropy.*[2] This entropy depends of course on the total number of possible microscopic configurations of the system, but it is insensitive to its specific physical support; it only takes into account the (abstract) probabilistic information on the system. Let us make a trivial illustration: a spin that can be up or down (with regard to some external magnetic field), a coin that comes head or tail, a computer bit which can be 0 or 1, are all equivalent in what concerns the concept of entropy. Consequently, entropy is expected to be a functional which is *invariant with regard to any permutations of the states*. Indeed, expression (1.1) exhibits this invariance through the form of a sum. Consequently, if $W > 1$, the entropy must have an extremum (maximum or minimum), and this must occur for equal probabilities. Indeed, this is the unique possibility for which the entropy is invariant with regard to the permutation of *any two* states. It is easy to verify that, for

---

[2] This statement is to be revisited for the more general entropy S_q. Indeed, as we shall see, the index q does depend on some universal aspects of the physical system, e.g., the type of inflection of a dissipative unimodal map, or, possibly, the type of power-law decay with a distance of two-body long-range interactions for many-body Hamiltonian systems.

S_{BG}, it is a maximum, and not a minimum. In fact, the identification as a maximum (and not a minimum) will become obvious when we shall prove, later on, that S_{BG} is a *concave* functional. Of course, the expression that S_{BG} takes for equal probabilities has already been indicated in Eq. (1.3).

2.1.2.3   Expansibility

Adding to a system new possible states with *zero* probability should not modify the entropy. This is precisely what is satisfied by S_{BG} if we take into account the already mentioned property $\lim_{x \to 0}(x \ln x) = 0$. So, we have that

$$S_{BG}(p_1, p_2, ..., p_W, 0) = S_{BG}(p_1, p_2, ..., p_W). \tag{2.9}$$

2.1.2.4   Additivity

Let $\mathcal{O}$ be a physical quantity associated with a given system, and let A and B be two probabilistically independent subsystems. We shall use the term *additive* if and only if $\mathcal{O}(A + B) = \mathcal{O}(A) + \mathcal{O}(B)$. If so, it is clear that if we have N independent equal systems, then $\mathcal{O}(N) = N\mathcal{O}(1)$, where the notation is self-explanatory. A weaker condition is $\mathcal{O}(N) \sim N\Omega$ for $N \to \infty$, with $0 < |\Omega| < \infty$, i.e., $\lim_{N \to \infty} \mathcal{O}(N)/N$ is finite (generically $\Omega \neq \mathcal{O}(1)$). In this case, the expression *asymptotically additive* might be used. Clearly, any observable which is additive with regard to a given composition law, is asymptotically additive (with $\Omega = \mathcal{O}(1)$), but the opposite is not necessarily true.

It is straightforwardly verified that, if A and B are *independent*, i.e., if the *joint probability* satisfies $p_{ij}^{A+B} = p_i^A p_j^B$ ($\forall (ij)$), then

$$S_{BG}(A + B) = S_{BG}(A) + S_{BG}(B). \tag{2.10}$$

Therefore the entropy S_{BG} is *additive*.

2.1.2.5   Concavity

Let us assume two arbitrary and different probability sets, namely $\{p_i\}$ and $\{p_i'\}$, associated with a single system having W states. We define an *intermediate* probability set as follows:

$$p_i'' = \lambda p_i + (1 - \lambda)p_i' \quad (\forall i; \ 0 < \lambda < 1). \tag{2.11}$$

The functional $S_{BG}(\{p_i\})$ (or any other functional in fact) is said *concave if and only if*

$$S_{BG}(\{p_i''\}) > \lambda S_{BG}(\{p_i\}) + (1 - \lambda)S_{BG}(\{p_i'\}). \tag{2.12}$$

This is indeed satisfied by S_{BG}. The proof is straightforward. Because of its *negative second derivative*, the (continuous) function $-x \ln x$ $(x > 0)$ satisfies

$$- p_i'' \ln p_i'' > \lambda(-p_i \ln p_i) + (1 - \lambda)(-p_i' \ln p_i') \quad (\forall i; \, 0 < \lambda < 1). \quad (2.13)$$

Applying $\sum_{i=1}^{W}$ on both sides of this inequality, we immediately obtain Eq. (2.12), i.e., the *concavity* of S_{BG}.

2.1.2.6 Lesche-Stability or Experimental Robustness

An entropic form $S(\{p_i\})$ (or any other functional of the probabilities, in fact) is said *Lesche-stable* or *experimentally robust* [96][3] if and only if it satisfies the following continuity property. Two probability distributions $\{p_i\}$ and $\{p_i'\}$ are said *close* if they satisfy the metric property:

$$D \equiv \sum_{i=1}^{W} |p_i - p_i'| \leq d_\epsilon, \quad (2.14)$$

where d_ϵ is a small real number. Then, experimental robustness is verified if, for any $\epsilon > 0$, a d_ϵ exists such that $D \leq d_\epsilon$ implies

$$R \equiv \left| \frac{S(\{p_i\}) - S(\{p_i'\})}{S_{max}} \right| < \epsilon, \quad (2.15)$$

where S_{max} is the maximal value that the entropic form can achieve (assuming its extremum corresponds to a maximum, not a minimum). For S_{BG} the maximal value is of course $\ln W$.

Condition 2.15 should be satisfied under all possible situations, including for $W \to \infty$. This implies that the condition

$$\lim_{d_\epsilon \to 0} \lim_{W \to \infty} \left| \frac{S(\{p_i\}) - S(\{p_i'\})}{S_{max}} \right| = 0 \quad (2.16)$$

should *also* be satisfied, in addition to $\lim_{W \to \infty} \lim_{d_\epsilon \to 0} \left| \frac{S(\{p_i\}) - S(\{p_i'\})}{S_{max}} \right| = 0$, which is of course always satisfied.

What this property essentially guarantees is that *similar* experiments performed onto *similar* physical systems should provide *similar* results (i.e., small percentage discrepancy) for the measured physical functionals. Lesche showed [96] that

---

[3] Lesche himself called *stability* this property. Two decades later, in personal conversation, I argued with him that that name could be misleading in the sense that it seems to suggest some relation with *thermodynamical stability*, with which it has no mathematical connection (thermodynamical stability has, in fact, connection with concavity). I suggested the use of *experimental robustness* instead. Lesche fully agreed that it is a better name. Consistently, I use it preferentially since then.

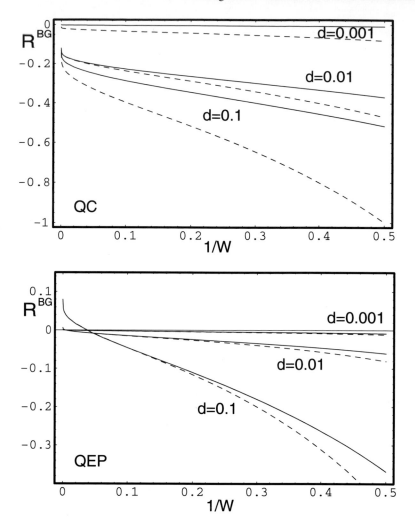

Fig. 2.1 Illustration of the Lesche-stability of S_{BG}. QC and QEP stand for *quasi-certainty* and *quasi-equal-probabilities*, respectively: see details in [97, 98]. From [98]

S_{BG} is experimentally robust, whereas the Renyi entropy $S_q^R \equiv \frac{\ln \sum_{i=1}^{W} p_i^q}{1-q}$ is not. See Fig. 2.1.

It is in principle possible to use, as a concept for *distance*, a quantity different from that used in Eq. (2.14). We could use for instance the following definition:

$$D_\mu \equiv \left[\sum_{i=1}^{W} |p_i - p_i'|^\mu \right]^{1/\mu} \quad (\mu > 0). \qquad (2.17)$$

Equation (2.14) corresponds to $\mu = 1$. The Pythagorean metric corresponds to $\mu = 2$. What about other values of μ? It happens that only for $\mu \geq 1$ the triangular inequality is satisfied in addition to the other traditional requirements for distance, and consequently it does constitute a metric. Still, why not using the values of $\mu > 1$? Because only for $\mu = 1$ the distance D does *not* depend on W, which makes it convenient for a generic property [99]. We come back in Sect. 3.2.2 onto this interesting property introduced by Lesche.

2.1.2.7 Shannon Uniqueness Theorem

Let us assume that an entropic form $S(\{p_i\})$ satisfies the following properties:

$$(i) \; S(\{p_i\}) \; is \; a \; continuous \; function \; of \; \{p_i\}; \tag{2.18}$$

$$(ii) \; S(p_i = 1/W, \forall i) \; monotonically \; increases \; with \; the \; total$$
$$number \; of \; possibilities \; W; \tag{2.19}$$

$$(iii) \; S(A+B) = S(A) + S(B) \quad if \quad p_{ij}^{A+B} = p_i^A p_j^B \; \forall (i,j), \tag{2.20}$$

$$where \; S(A+B) \equiv S(\{p_{ij}^{A+B}\}), \; S(A) \equiv S(\{p_i^A\}) \; (p_i^A \equiv \sum_{j=1}^{W_B} p_{ij}^{A+B}),$$

$$and \quad S(B) \equiv S(\{p_j^B\}) \; (p_j^B \equiv \sum_{i=1}^{W_A} p_{ij}^{A+B});$$

$$(iv) \; S(\{p_i\}) = S(p_L, p_M) + p_L S(\{p_i/p_L\}) + p_M S(\{p_i/p_M\}) \tag{2.21}$$

$$with \; p_L \equiv \sum_{L \, terms} p_i \; , \; p_M \equiv \sum_{M \, terms} p_i \; ,$$

$$L + M = W \; , \; and \; p_L + p_M = 1.$$

Then and only then [32]

$$S(\{p_i\}) = -k \sum_{i=1}^{W} p_i \ln p_i \quad (k > 0). \tag{2.22}$$

It is therefore very clear in what sense the functional (2.22) is unique. This neatly differs from the fallacious, and yet not rare, statement that form (2.22) is the unique physically admissible entropic functional.[4]

---

[4] Some authors prefer the notation $S(A \times B)$ instead of $S(A + B)$ in order to emphasize the fact that the phase space of the total system is the tensor product of the space phases of the subsystems A and B.

2.1.2.8 Khinchin Uniqueness Theorem

Let us assume that an entropic form $S(\{p_i\})$ satisfies the following properties:

$$(i)\ S(\{p_i\})\ is\ a\ continuous\ function\ of\ \{p_i\}; \tag{2.23}$$

$$(ii)\ S(p_i = 1/W, \forall i)\ monotonically\ increases\ with\ the\ total$$
$$number\ of\ possibilities\ W; \tag{2.24}$$

$$(iii)\ S(p_1, p_2, ..., p_W, 0) = S(p_1, p_2, ..., p_W); \tag{2.25}$$

$$(iv)\ S(A + B) = S(A) + S(B|A)\ , \tag{2.26}$$

$$where\ S(A + B) \equiv S(\{p_{ij}^{A+B}\}),\ S(A) \equiv S(\{p_i^A\})\ \ (p_i^A \equiv \sum_{j=1}^{W_B} p_{ij}^{A+B}),$$

$$and\ the\ conditional\ entropy\ S(B|A) \equiv \sum_{i=1}^{W_A} p_i^A S(\{p_{ij}^{A+B}/p_i^A\}).$$

Then and only then [100]

$$S(\{p_i\}) = -k \sum_{i=1}^{W} p_i \ln p_i \quad (k > 0). \tag{2.27}$$

It follows then that the Shannon and the Khinchin sets of axioms are equivalent.

2.1.2.9 Composability

A dimensionless entropic form $S(\{p_i\})$ (i.e., whenever expressed in appropriate conventional units, e.g., in units of k) is said *composable* [71, 72] (see also [101–104]) if the entropy $S(A + B)/k$ corresponding to a system composed of two *probabilistically independent* subsystems A and B can be expressed in the form

$$\frac{S(A + B)}{k} = F\left(\frac{S(A)}{k}, \frac{S(B)}{k}; \{\eta\}\right), \tag{2.28}$$

where $F(x, y; \{\eta\})$ is a smooth function of (x, y) which depends on a (typically small) set of universal indices $\{\eta\}$ defined in such a way that $F(x, y; \{0\}) = x + y$ (*additivity*), and which satisfies $F(x, 0; \{\eta\}) = x$ (*null-composability*), $F(x, y; \{\eta\}) = F(y, x; \{\eta\})$ (*symmetry*), $F(x, F(y, z; \{\eta\}); \{\eta\}) = F(F(x, y; \{\eta\}), z; \{\eta\})$ (*associativity*). For thermodynamical systems, this associativity appears to be consistent with the 0th Principle of Thermodynamics.

In other words, the whole concept of composability is constructed upon the requirement that the entropy of $(A + B)$ does *not* depend on the microscopic configurations of A and of B. Equivalently, we are able to macroscopically calculate the entropy of the composed system without any need of entering into the knowledge of the microscopic states of the subsystems. This property appears to be a natural one for an entropic form if we desire to use it as a basis for a statistical mechanics which would naturally connect to thermodynamics.

The entropy S_{BG} is composable since it satisfies Eq. (2.10). In other words, we have $F_{BG}(x, y) = x + y$. Being S_{BG} nonparametric, no index exists in F_{BG}.

2.1.2.10 Sensitivity to the Initial Conditions, Entropy Production Per Unit Time, and a Pesin-Like Identity

For a one-dimensional dynamical system (characterized by the variable x) the *sensitivity to the initial conditions* ξ is defined as follows:

$$\xi \equiv \lim_{\Delta x(0) \to 0} \frac{\Delta x(t)}{\Delta x(0)}, \tag{2.29}$$

$\Delta x(t)$ being the discrepancy, at time t, between two initially close trajectories.

It can be shown [105–110] that ξ paradigmatically satisfies the equation

$$\frac{d\xi}{dt} = \lambda_1 \xi, \tag{2.30}$$

whose solution is given by

$$\xi = e^{\lambda_1 t}. \tag{2.31}$$

(The meaning of the subscript 1 will become transparent later on). If the *Lyapunov exponent* $\lambda_1 > 0$ ($\lambda_1 < 0$), the system will be said to be *strongly chaotic (regular)*. The case where $\lambda_1 = 0$ is sometimes called *marginal* or *subexponential* behavior and will be extensively addressed later on.

At this point let us briefly review, without proof, some basic notions of nonlinear dynamical systems. If the system is d-dimensional (i.e., it evolves in a phase space whose d-dimensional Lebesgue measure is finite), it has d Lyapunov exponents: d_+ of them are positive, d_- are negative, and d_0 vanish, hence $d_+ + d_- + d_0 = d$. Let us order them all from the largest to the smallest: $\lambda_1^{(1)} \geq \lambda_1^{(2)} \geq \ldots \geq \lambda_1^{(d_+)} > \lambda_1^{(d_++1)} = \lambda_1^{(d_++2)} = \ldots = \lambda_1^{(d_++d_0)} = 0 > \lambda_1^{(d_++d_0+1)} \geq \lambda_1^{(d_++d_0+2)} \geq \ldots \geq \lambda_1^{(d)}$. An infinitely small *segment* (having then a vanishing one-dimensional Lebesgue measure) diverges like $e^{\lambda_1^{(1)} t}$; this precisely is the case focused in Eq. (2.31). An infinitely small *area* (having then a vanishing two-dimensional Lebesgue measure) diverges like $e^{(\lambda_1^{(1)} + \lambda_1^{(2)}) t}$. An infinitely small *volume* diverges like $e^{(\lambda_1^{(1)} + \lambda_1^{(2)} + \lambda_1^{(3)}) t}$, and so on. An infinitely small d-dimensional hypervolume evolves like $e^{[\sum_{r=1}^{d} \lambda_1^{(r)}] t}$. If the system is *conservative*, i.e., if the infinitely small d-dimensional hypervolume remains constant with time, consistently with $\sum_{r=1}^{d} \lambda_1^{(r)} = 0$. An important particular class of conservative systems is constituted by the so-called *symplectic* ones. For these, d is an even integer, and the Lyapunov exponents are coupled two by two as follows: $\lambda_1^{(1)} = -\lambda_1^{(d)} \geq \lambda_1^{(2)} = -\lambda_1^{(d-1)} \geq ... \geq \lambda_1^{(d_+)} = -\lambda_1^{(d_+ + d_0 + 1)} \geq \lambda_1^{(d_+ + 1)} = ... = \lambda_1^{(d_+ + d_0)} = 0$. Consistently, such systems have $d_+ = d_-$ and d_0 is a even integer. The most popular illustration of symplectic systems is the Hamiltonian system. They are conservative, which precisely is what the classical Liouville theorem states!

Do all these degrees of freedom contribute, as time evolves, to the erratic exploration of the phase space? *No, they do not.* Only those associated with the d_+ positive Lyapunov exponents, and some of the d_0 vanishing ones, do. Consistently, it is only these which contribute to our loss of information, as time evolves, about the location in phase space of a set of initial conditions. As we shall see, these remarks enable an intuitive understanding of the so-called Pesin identity, that we will soon state.

Let us now address the interesting question of the BG entropy production as time t increases. More than one entropy production can be defined as a function of time. Two basic choices are the so-called *Kolmogorov–Sinai entropy* (or *KS entropy rate* or *metric entropy*) [111], namely one which is based on a single trajectory in phase space, the other one being associated to the evolution of an *ensemble of initial conditions*. We shall preferentially use here the latter, because of its sensibly larger computational tractability. In fact, except for pathological cases, they both coincide.

Let us schematically describe the *KS entropy rate* concept [110–112]. We first partition the phase space into two regions, noted A and B. Then we choose a generic initial condition (the final result will not depend on this choice) and, applying the specific dynamics of the system at equal and finite time intervals τ, we generate a long string (infinitely long in principle), say $ABBBAABBABABAAA$.... Then we analyze words of length $l = 1$. In our case, there are only two such words, namely A and B. The analysis consists in running along the string a window whose width is l, and determining the probabilities p_A and p_B of the words A and B, respectively. Finally we calculate the entropy $S_{BG}(l = 1) = -p_A \ln p_A - p_B \ln p_B$. Then we repeat for words whose length equals $l = 2$. In our case, there are four such words, namely AA, AB, BA, BB. Running along the string a $l = 2$ window letter by letter, we determine the probabilities $p_{AA}, p_{AB}, p_{BA}, p_{BB}$, hence the entropy $S_{BG}(l = 2) = -p_{AA} \ln p_{AA} - p_{AB} \ln p_{AB} - p_{BA} \ln p_{BA} - p_{BB} \ln p_{BB}$. Then we repeat for $l = 3, 4, ...$ and calculate the corresponding values for $S_{BG}(l)$. We then choose another two-partition, say A' and B', and repeat the whole procedure. Then we do in principle for all possible two-partitions. Then we go to three-partitions, i.e., the alphabet will be now constituted by three letters, say A, B, and C. We repeat the previous procedure for $l = 1$ (corresponding to the words A, B, C), then for $l = 2$ (corresponding to the words $AA, AB, AC, BA, BB, BC, CA, CB, CC$), etc. Then we run windows with $l = 3, 4, ...$. We then consider a different three-partition, say A', B', and C'. Then we consider four-partitions, and so on. Of all these entropies we retain the *supremum*. In the appropriate limits of infinitely fine partitions and $\tau \to 0$ we obtain finally the largest rate of increase of the BG entropy. This is basically the Kolmogorov–Sinai entropy rate.

It is not necessary to insist on how deeply inconvenient this definition can be for any computational implementation! Fortunately, a different type of entropy production can be defined [113], whose computational implementation is usually very simple. It is defined as follows. First partition the phase space into W little cells (normally equal in size) and denote them with $i = 1, 2, ..., W$. Then randomly place M initial conditions in one of those W cells (if $d_+ \geq 1$, normally the result will not depend on this choice). And then follow, as time evolves, the number of points $M_i(t)$ in each cell ($\sum_{i=1}^{W} M_i(t) - M$). Define the probability set $p_i(t) \equiv M_i(t)/M$ ($\forall i$), and calculate finally $S_{BG}(t)$ through Eq. (1.1). The *entropy production per unit time* is defined as

$$K_1 \equiv \lim_{t \to \infty} \lim_{W \to \infty} \lim_{M \to \infty} \frac{S_{BG}(t)}{t}. \tag{2.32}$$

The Pesin identity [114], or more precisely the Pesin-like identity that we shall use here, states, for large classes of dynamical systems, [113]

$$K_1 = \sum_{r=1}^{d_+} \lambda_1^{(r)}. \tag{2.33}$$

As it will become gradually clear along the book, this relationship (and its q-generalization) will play an important role in the determination of the particular entropic form which is adequate for a given nonlinear dynamical system.

2.2 Kullback–Leibler Relative Entropy

In many problems the question arises on how different are two probability distributions p and $p^{(0)}$; for reasons that will become clear soon, $p^{(0)}$ will be referred to as the *reference*. It becomes therefore interesting to define some sort of "distance" between them. One possibility is of course the distance introduced in Eq. (2.17). In other words, we can use

$$D_\mu(p, p^{(0)}) \equiv \left[\int dx \, |p(x) - p^{(0)}(x)|^\mu \right]^{1/\mu} \quad (\mu > 0). \tag{2.34}$$

In general we have that $D_\mu(p, p^{(0)}) = D_\mu(p^{(0)}, p)$, and that $D_\mu(p, p^{(0)}) = 0$ if and only if $p(x) = p^{(0)}(x)$ almost everywhere. We remind, however, that the triangular inequality is satisfied only for $\mu \geq 1$. Therefore, only then the distance constitutes a metric. If $p(x) = \sum_{i=1}^{W} p_i \, \delta(x - x_i)$ and $p^{(0)}(x) = \sum_{i=1}^{W} p_i^{(0)} \, \delta(x - x_i)$, Eq. (2.34) leads to

$$D_\mu(p, p^{(0)}) \equiv \left[\sum_{i=1}^{W} |p_i - p_i^{(0)}|^\mu \right]^{1/\mu} \quad (\mu > 0), \tag{2.35}$$

which precisely recovers Eq. (2.17).

For some purposes, this definition of distance is quite convenient. For others, the *Kullback–Leibler entropy* [115] has been introduced (see, for instance, [116, 117] and references therein). It is also called *cross entropy*, or *relative entropy*, or *mutual information*, and it is defined as follows:

$$I_1(p, p^{(0)}) \equiv \int dx\, p(x) \ln\left[\frac{p(x)}{p^{(0)}(x)}\right] = -\int dx\, p(x) \ln\left[\frac{p^{(0)}(x)}{p(x)}\right]. \quad (2.36)$$

It can be proved, by using $\ln r \geq 1 - 1/r$ (with $r \equiv p(x)/p^{(0)}(x) > 0$), that $I_1(p, p^{(0)}) \geq 0$, the equality being valid if and only if $p(x) = p^{(0)}(x)$ almost everywhere. It is clear that in general $I_1(p, p^{(0)}) \neq I_1(p^{(0)}, p)$. This inconvenience is sometimes overcome by using the symmetrized quantity $[I_1(p, p^{(0)}) + I_1(p^{(0)}, p)]/2$.

$I_1(p, p^{(0)})$ (like the distance (2.34)) has the property of being invariant under variable transformation. Indeed, if we make $x = f(y)$, the measure preservation implies $p(x)dx = \tilde{p}(y)dy$. Since $p(x)/p^{(0)}(x) = \tilde{p}(x)/\tilde{p}^{(0)}(x)$, we have $I_1(p, p^{(0)}) = I_1(\tilde{p}, \tilde{p}^{(0)})$, which proves the above-mentioned invariance. The BG entropy in its continuous (*not* in its discrete) form $S_{BG} = -\int dx\, p(x) \ln p(x)$ lacks this important property. Because of this fact, the BG entropy is advantageously replaced, in some calculations, by the Kullback–Leibler one. Depending on the particular problem, the referential distribution $p^{(0)}(x)$ is frequently taken to be a standard distribution such as the uniform, or Gaussian, or Lorentzian, or Poisson, or BG ones. When $p^{(0)}(x)$ is chosen to be the uniform distribution on a compact support of Lebesgue measure W, we have the relation

$$I_1(p, 1/W) = \ln W - S_{BG}(p). \quad (2.37)$$

Because of relations of this kind, in some contexts, the minimization of the Kulback–Leibler entropy is sometimes used instead of the maximization of the Boltzmann–Gibbs–Shannon entropy.

Although convenient for a variety of purposes, $I_1(p, p^{(0)})$ has a disadvantage. It is needed that $p(x)$ and $p^{(0)}(x)$ *simultaneously* vanish, if p_0 vanishes for certain values of x (this property is usually referred to as p being *absolutely continuous* with respect to p_0). Indeed, it is evident that otherwise the quantity $I_1(p, p^{(0)})$ becomes ill-defined. To overcome this annoying difficulty, a different distance has been defined along the lines of the Kullback–Leibler entropy. We refer to the so-called *Jensen–Shannon divergence*. Although interesting in many respects, its study would take us too far from our present line. Details can be seen in [118, 119] and references therein.

Let us mention also that, for discrete probabilities, definition (2.36) leads to

$$I_1(p, p^{(0)}) \equiv \sum_{i=1}^{W} p_i \ln\left[\frac{p_i}{p_i^{(0)}}\right] = -\sum_{i=1}^{W} p_i \ln\left[\frac{p_i^{(0)}}{p_i}\right]. \quad (2.38)$$

Various other interesting related properties can be found in [120–123]. The quantum version of the Kulback–Leibler entropy is sometimes referred to as the Umegaki entropy [125].

A simple illustration consists in considering a binary variable (with probabilities p and $(1 - p)$), using as reference the equal probabilities case ($p = 1/2$). We then have

$$I_1(p, p^{(0)}) = p \ln \frac{p}{1/2} + (1 - p) \ln \frac{1 - p}{1/2}, \tag{2.39}$$

which, with $z \equiv p - 1/2$, can be equivalently rewritten as

$$I_1(p, p^{(0)})(z) = \ln 2 + \frac{1 + 2z}{2} \ln \frac{1 + 2z}{2} + \frac{1 - 2z}{2} \ln \frac{1 - 2z}{2} \tag{2.40}$$

$$\sim 2z^2 + \frac{4}{3} z^4 + \frac{32}{15} z^6 + \frac{32}{7} z^8 \quad (z \to 0). \tag{2.41}$$

hence $I_1(p, p^{(0)})(z) \in [0, \ln 2]$.

2.3 Constraints and Entropy Optimization

Let us now work out the most basic optimization cases.

2.3.1 Imposing the Mean Value of the Variable

In addition to

$$\int_0^\infty dx \, p(x) = 1, \tag{2.42}$$

we might know the mean value of the variable, i.e.,

$$\langle x \rangle \equiv \int_0^\infty dx \, x \, p(x) = X^{(1)} > 0. \tag{2.43}$$

By using the Lagrange method to find the optimizing distribution, we define

$$\Phi[p] \equiv -\int_0^\infty dx \, p(x) \ln p(x) - \alpha \int_0^\infty dx \, p(x) - \beta^{(1)} \int_0^\infty dx \, x \, p(x), \tag{2.44}$$

and then impose $\delta\Phi[p]/\delta p(x) = 0$. We straightforwardly obtain $1 + \ln p_{opt} + \alpha + \beta^{(1)} x = 0$ (*opt* stands for *optimal*), hence

$$p_{opt} = \frac{e^{-\beta^{(1)}x}}{\int_0^\infty dx\, e^{-\beta^{(1)}x}} = \beta^{(1)}\, e^{-\beta^{(1)}x}, \tag{2.45}$$

where we have used condition (2.42) to eliminate the Lagrange parameter α. By using condition (2.43) we obtain the following relation for the Lagrange parameter $\beta^{(1)}$:

$$\beta^{(1)} = \frac{1}{X^{(1)}}, \tag{2.46}$$

hence, replacing in (2.45),

$$p_{opt} = \frac{e^{-x/X^{(1)}}}{X^{(1)}}. \tag{2.47}$$

2.3.2   Imposing the Mean Value of the Squared Variable

Another simple and quite frequent case is when we know that $\langle x \rangle = 0$. In such case, in addition to

$$\int_{-\infty}^{\infty} dx\, p(x) = 1, \tag{2.48}$$

we typically know the mean value of the squared variable, i.e.,

$$\langle x^2 \rangle \equiv \int_{-\infty}^{\infty} dx\, x^2 p(x) = X^{(2)} > 0. \tag{2.49}$$

By using, as before, the Lagrange method to find the optimizing distribution, we define

$$\Phi[p] \equiv -\int_{-\infty}^{\infty} dx\, p(x) \ln p(x) - \alpha \int_{-\infty}^{\infty} dx\, p(x) - \beta^{(2)} \int_{-\infty}^{\infty} dx\, x^2 p(x), \tag{2.50}$$

and then impose $\delta\Phi[p]/\delta p(x) = 0$. We straightforwardly obtain $1 + \ln p_{opt} + \alpha + \beta^{(2)}x^2 = 0$, hence

$$p_{opt} = \frac{e^{-\beta^{(2)}x^2}}{\int_{-\infty}^{\infty} dx\, e^{-\beta^{(2)}x^2}} = \sqrt{\frac{\beta^{(2)}}{\pi}}\, e^{-\beta^{(2)}x^2}, \tag{2.51}$$

where we have used condition (2.48) to eliminate the Lagrange parameter α. By using condition (2.49) we obtain the following relation for the Lagrange parameter $\beta^{(2)}$:

$$\beta^{(2)} = \frac{1}{2X^{(2)}}, \tag{2.52}$$

hence, replacing in (2.51),

$$p_{opt} = \frac{e^{-x^2/(2X^{(2)})}}{\sqrt{2\pi X^{(2)}}}. \tag{2.53}$$

We thus see the very basic connection between Gaussian distributions and BG entropy.

2.3.3 Imposing the Mean Values of Both the Variable and Its Square

Let us unify here the two preceding subsections. We impose

$$\int_{-\infty}^{\infty} dx\, p(x) = 1, \tag{2.54}$$

and, in addition to this, we know that

$$\langle x \rangle \equiv \int_{-\infty}^{\infty} dx\, x\, p(x) = X^{(1)}, \tag{2.55}$$

and

$$\langle (x - \langle x \rangle)^2 \rangle \equiv \int_{-\infty}^{\infty} dx\, (x - \langle x \rangle)^2\, p(x) = X^{(2)} - (X^{(1)})^2 \equiv M^{(2)} > 0. \tag{2.56}$$

By using once again the Lagrange method, we define

$$\Phi[p] \equiv -\int_{-\infty}^{\infty} dx\, p(x) \ln p(x) - \alpha \int_{-\infty}^{\infty} dx\, p(x)$$
$$-\beta^{(1)} \int_{-\infty}^{\infty} dx\, x\, p(x) - \beta^{(2)} \int_{-\infty}^{\infty} dx\, (x - \langle x \rangle)^2\, p(x), \tag{2.57}$$

and then impose $\delta\Phi[p]/\delta p(x) = 0$. We straightforwardly obtain $1 + \ln p_{opt} + \alpha + \beta^{(1)}x + \beta^{(2)}(x - \langle x \rangle)^2 = 0$, hence

$$p_{opt} = \frac{e^{-\beta^{(1)}x - \beta^{(2)}(x - \langle x \rangle)^2}}{\int_{-\infty}^{\infty} dx\, e^{-\beta^{(1)}x - \beta^{(2)}(x - \langle x \rangle)^2}} = \sqrt{\frac{\beta^{(2)}}{\pi}}\, e^{-\beta^{(2)}\left[x - \left(\langle x \rangle - \frac{\beta^{(1)}}{2\beta^{(2)}}\right)\right]^2}, \tag{2.58}$$

where we have used condition (2.54) to eliminate the Lagrange parameter α. By using conditions (2.55) and (2.56) we obtain the following relations for the Lagrange parameters $\beta^{(1)}$ and $\beta^{(2)}$:

$$\beta^{(1)} = 0, \forall X^{(1)}, \tag{2.59}$$

and

$$\beta^{(2)} = \frac{1}{2[X^{(2)} - (X^{(1)})^2]}.$$ (2.60)

Replacing (2.59) and (2.60) in (2.58), we finally obtain

$$p_{opt} = \frac{e^{-\frac{\left(x - X^{(1)}\right)^2}{2\left[X^{(2)} - (X^{(1)})^2\right]}}}{\sqrt{2\pi[X^{(2)} - (X^{(1)})^2]}}.$$ (2.61)

We see that the only effect of a nonzero mean value $X^{(1)}$ is to re-center the Gaussian, the width around this mean value is exactly preserved. The fact that we have obtained $\beta^{(1)} = 0$ comes from the fact that we have imposed the second nontrivial constraint as indicated in (2.56), i.e., on $[X^{(2)} - (X^{(1)})^2]$, and not, alternatively, on $X^{(2)}$. Along both paths, the final result is of course one and the same, namely as given in Eq. (2.61).

2.3.4 Others

A quite general situation would be to impose, in addition to

$$\int dx\, p(x) = 1,$$ (2.62)

the constraint

$$\int dx\, f(x)\, p(x) = F,$$ (2.63)

where $f(x)$ is some known function and F a known *finite* number. We obtain

$$p_{opt} = \frac{e^{-\beta f(x)}}{\int dx\, e^{-\beta f(x)}}.$$ (2.64)

It is clear that, by appropriately choosing $f(x)$, we can force $p_{opt}(x)$ to be virtually *any* distribution we wish. For example, by choosing $f(x) = |x|^\gamma$ ($\gamma \in \mathbb{R}$), we obtain a generic stretched exponential $p_{opt}(x) \propto e^{-\beta |x|^\gamma}$; by choosing $f(x) = \ln x$, we formally obtain for $p_{opt}(x)$ a straight power law (which is in fact non-normalizable in the current interval $[0, \infty)$). But the use of such procedures hardly has any epistemological interest at all, since it provides no hint onto the underlying nature of the problem. Only choices such as $f(x) = x$ or $f(x) = x^2$ are generically sound since such constraints correspond to robust informational features, namely the *location of the center* and the *width* of the distribution. Other choices are, unless some exceptional fact enters into consideration (e.g., $f(x)$ being a constant of motion of the system), quite *ad hoc* and uninteresting. Of course, such mathematical features are by no means exclusive of S_{BG}: analogous ones hold for virtually *any* entropic form.

2.4 Boltzmann–Gibbs Statistical Mechanics and Thermodynamics

There are many formal manners for deriving the BG entropy and its associated probability distribution for thermal equilibrium. *None of them uses exclusively first principle arguments*, i.e., arguments that entirely remain at the level of mechanics (classical, quantum, relativistic, or any other). That surely was, as previously mentioned, one of the central scientific goals that Boltzmann pursued his entire life, but, although he probably had a strong intuition about this point, he died without succeeding. The difficulties are so heavy that even today we do not know how to do this. At first sight, this might seem surprising given the fact that S_{BG} and the BG weight enjoy the universal acceptance that we all know. So, let us illustrate our statement more precisely. Assume that we have a quite generic many-body short-range-interacting Hamiltonian. We currently know that its thermal equilibrium is described by the BG weight. What we still do not know is *how* to derive this important result from purely mechanical and statistical logical steps, i.e., without using a priori generic dynamical hypothesis such as *ergodicity*, or a priori postulating the validity of macroscopic relations such as some or all of the principles of thermodynamics. For example, Fisher et al [126–128] proved long ago, for a vast class of short-range-interacting Hamiltonians, that the thermal equilibrium physical quantities are computable within standard BG statistical mechanics. Such a proof, no matter how precious might it be, does *not* prove also that this statistics *indeed* provides the correct description at thermal equilibrium. Rephrasing, it proves that BG statistics *can* be the correct one, but it does *not* prove that it *is* the correct one. Clearly, there is no reasonable doubt today that, for such systems, BG is the correct one. It is nevertheless instructive to outline the logical implications of the available proofs.

In a similar vein, even for the case of long-range-interacting Hamiltonians (e.g., infinitely long-range interactions), the standard BG calculations can, for many of them, still be performed through convenient renormalizations of the coupling constants (e.g., *à la* Kac, or through the usual mean field approximation recipe of artificially dividing the coupling constant by the number N of particles raised to some appropriate power). The possibility of computability does *by no means* prove, strictly speaking, that BG statistics is the correct description. And certainly, it does not enlighten us on what the necessary and sufficient *first-principle* conditions could be for the BG description to be not only mathematically admissible but precisely the correct one.

In spite of all these mathematical difficulties, some nontrivial examples are available in the literature for both short- and long-range interactions (see Sect. 5.3). For short-ranged ones, for example, see [129], where it has been possible to *numeri-*

cally exhibit the BG weight by *exclusively* using Newton's $\mathbf{F} = m\mathbf{a}$ as microscopic dynamics, *with no thermostatistical or entropic assumption of any kind.*

Let us anticipate that these and even harder difficulties do exist for the considerably more subtle situations focused on in nonextensive statistical mechanics.

In what follows, we conform to the traditional, though epistemologically less ambitious, paths. We shall primarily follow the Gibbs' elegant lines of first *postulating* an entropic form, and then using it, *without proof*, as the basis for a variational principle including appropriate constraints. The philosophy of such path is quite clear. It is a form of *Occam's razor*, where we use all that we know and not more than we know. This is obviously extremely attractive from a conceptual standpoint. However, its mathematical implementation is to be done with a *given specific entropic functional* with *given specific constraints*, and the rigorous correctness of this procedure is of course far from trivial! After 150 years of impressive success, there can be no doubt that BG concepts and statistical mechanics provide the correct connection between microscopic and macroscopic laws for a vast class of physical systems. But—we insist—the mathematically *precise* microscopic qualification of this class remains, still today, an open question.

2.4.1   Isolated System—Microcanonical Ensemble

In this and subsequent subsections, we briefly review BG statistical mechanics (see, for instance, [43]). We consider a quantum Hamiltonian system constituted by N interacting particles under specific boundary conditions, and denote by $\{E_i\}$ its energy eigenvalues.

The microcanonical ensemble corresponds to an isolated N-particle system whose total energy U is known within some precision δU (to be in fact taken at its zero limit at the appropriate mathematical stage). The number of states i with $U \leq E_i \leq U + \delta U$ is denoted by W. *Assuming* that the system is such that its dynamics lead to ergodicity at its stationary state (thermal equilibrium), we assume that all such states are equally probable, i.e., $p_i = 1/W$, and the entropy is given by Eq. (1.3). The temperature T is introduced through

$$\frac{1}{T} \equiv \frac{\partial S_{BG}}{\partial U} = k \frac{\partial \ln W}{\partial U}. \tag{2.65}$$

2.4.2   In the Presence of a Thermostat—Canonical Ensemble

The canonical ensemble corresponds to a N-particle system defined in a Hilbert space whose dimension is denoted W, and which is in long-standing thermal contact with a (infinitely large) thermostat at temperature T. Its exact energy is unknown, but its mean energy U is known since it is determined by the thermostat. We must optimize the entropy given by Eq. (1.1) with the norm constraint (1.2), and with the energy constraint[5]

$$\sum_{i=1}^{W} p_i E_i = U. \tag{2.66}$$

Following along the lines of Sect. 2.3 we obtain the celebrated BG weight

$$p_i = \frac{e^{-\beta E_i}}{Z_{BG}}, \tag{2.67}$$

with the *partition function* given by

$$Z_{BG} \equiv \sum_{i=1}^{W} e^{-\beta E_i}, \tag{2.68}$$

the Lagrange parameter β being related with the temperature through $\beta \equiv 1/(kT)$.
    We can prove also that

$$\frac{1}{T} = \frac{\partial S_{BG}}{\partial U}, \tag{2.69}$$

that the *Helmholtz free energy* is given by

$$F_{BG} \equiv U - T S_{BG} = -\frac{1}{\beta} \ln Z_{BG}, \tag{2.70}$$

that the *internal energy* is given by

$$U = -\frac{\partial}{\partial \beta} \ln Z_{BG}, \tag{2.71}$$

and that the specific heat is given by

$$C \equiv T \frac{\partial S}{\partial T} = \frac{\partial U}{\partial T} = -\frac{\partial^2 F(\beta)}{\partial T^2} \geq 0. \tag{2.72}$$

---

[5] This type of constraint needs to be revisited if the system is in a relativistic regime, i.e., involving velocities approaching that of light. Indeed, the energy is not Lorentz-invariant in special relativity. Consequently, the arguments advanced by Jüttner in 1911 [130] might lead to a *wrong* (even if popular) relativistic generalization of the Maxwellian distribution of velocities. The correct $d = 1$ Lorentz-invariant relativistic generalization is given in [1251] $P(v) = \frac{\sqrt{\beta/\pi c^2}}{1-v^2/c^2} e^{-\beta \tanh^{-1}(v/c)^2}$ with $\beta \equiv m_0 c^2/(2k_B T)$. The standard Maxwellian distribution is recovered in the limit $\beta \to \infty$. Consistently, the $d = 1$ Lorentz-invariant q-generalized relativistic distribution is given by $P_q(v) \propto \frac{\sqrt{\beta/c^2}}{1-v^2/c^2} e_q^{-\beta \tanh^{-1}(v/c)^2}$.

In the limit $T \to \infty$ we recover the microcanonical ensemble.

2.4.3 Others

The system may be exchanged with the thermostat not only energy, so that the temperature is that of the thermostat, but also particles, so that also the chemical potential is fixed by the reservoir. This physical situation corresponds to the so-called *grand-canonical ensemble*. This and other similar physical situations can be treated along the same path, as shown by Gibbs. We shall not review here these types of systems, which are described in detail in [43], for instance.

Another important physical case, which we do not review here either, is when the particles cannot be considered as distinguishable. Such is the case of *bosons* (leading to Bose–Einstein statistics), *fermions* (leading to Fermi–Dirac statistics), and the so-called *gentilions* (leading to Gentile statistics, also called *parastatistics* [131–133], which unifies those of Bose–Einstein and Fermi–Dirac).

All these various physical systems, and even others, constitute what is currently referred to as BG statistical mechanics, essentially because at its basis we find, in one way or another, the entropic functional S_{BG}. It is this entire theoretical body that in principle we intend to generalize in the rest of the present book, through the generalization of S_{BG} itself.

Chapter 3
Generalizing What We Learnt: Nonextensive Statistical Mechanics

Don Quijote me ha revelado íntimos secretos suyos que no reveló a Cervantes

Víctor Goti

(Prólogo de *Niebla* de Miguel de Unamuno, 1935)

3.1 Playing with Differential Equations—A Metaphor

As we already emphasized, there is no logical-deductive procedure for generalizing *any* physical theory. This occurs through all types of paths that, in one way or another, are ultimately but metaphors. Let us present here a possible metaphor for generalizing the BG entropy.

The simplest ordinary differential equation can be considered to be

$$\frac{dy}{dx} = 0 \quad (y(0) = 1). \tag{3.1}$$

Its solution is

$$y = 1, \tag{3.2}$$

whose symmetric curve with regard to the bisector axis is

$$x = 1. \tag{3.3}$$

As the second simplest differential equation, we might consider

$$\frac{dy}{dx} = 1 \quad (y(0) = 1). \tag{3.4}$$

© Springer Nature Switzerland AG 2023
C. Tsallis, *Introduction to Nonextensive Statistical Mechanics*,
https://doi.org/10.1007/978-3-030-79569-6_3

Its solution is

$$y = 1 + x, \qquad (3.5)$$

whose inverse function is

$$y = x - 1. \qquad (3.6)$$

We may next wish to consider the following one:

$$\frac{dy}{dx} = y \quad (y(0) = 1), \qquad (3.7)$$

whose solution is

$$y = e^x. \qquad (3.8)$$

Its inverse function is

$$y = \ln x, \qquad (3.9)$$

and satisfies of course

$$\ln(x_A \, x_B) = \ln x_A + \ln x_B. \qquad (3.10)$$

Is it possible to unify the above three differential Eqs. ((3.1), (3.4), and (3.7))? Yes indeed. It is enough to consider

$$\frac{dy}{dx} = a + by \quad (y(0) = 1), \qquad (3.11)$$

and play with the *two* parameters a and b. Is it possible to unify the same three differential equations with only *one* parameter? Yes, it is, ... *out of linearity*! Just consider

$$\frac{dy}{dx} = y^q \quad (y(0) = 1; q \in \mathbb{R}). \qquad (3.12)$$

Its solution is

$$y = [1 + (1 - q)x]^{1/(1-q)} \equiv e_q^x \quad (e_1^x = e^x). \qquad (3.13)$$

referred to as the *q-exponential* function [134]. [1] Its inverse is

$$y = \frac{x^{1-q} - 1}{1 - q} \equiv \ln_q x \quad (x > 0; \ln_1 x = \ln x), \qquad (3.14)$$

referred to as the *q-logarithmic* function [134]. It satisfies the following property:

---

[1] By considering the analytical extension $[1 + (1 - q)z]^{1/(1-q)}$ with $z \in \mathbb{C}$, it is possible to naturally define q-trigonometric and q-hyperbolic functions [135], which have proved to be useful in various occasions.

$$\ln_q (x_A\, x_B) = \ln_q x_A + \ln_q x_B + (1 - q)(\ln_q x_A)(\ln_q x_B). \qquad (3.15)$$

These functions will play an important role throughout the entire theory. We may, in fact, anticipate that *virtually all* the generic expressions associated with BG statistics and its (nonlinear) dynamical foundations will, remarkably enough, turn out to be generalized essentially just by replacing the standard exponential and logarithm forms by the above q-generalized ones. Let us add that, whenever the $1 + (1 - q)x$ argument of the q-exponential is negative, the function is defined to vanish. In other words, the definition is $e_q^x \equiv [1 + (1 - q)x]_+^{1/(1-q)}$, where $[z]_+ = \max\{z, 0\}$. However, for simplicity, we shall, most of the time, avoid this notation. Typical representations of the q-exponential function are illustrated in Figs. 3.1 and 3.2. It is immediately verified that the $q \to -\infty$, $q = 0$ and $q = 1$ particular instances precisely recover the cases presented in Eqs. (3.1), (3.4) and (3.7), respectively.

3.2 Nonadditive Entropy S_q

3.2.1 Definition

Through the metaphor presented above, and because of various other reasons that will gradually emerge, we may *postulate* the following generalization of Eq. (1.3):

$$S_q = k \ln_q W \quad (S_1 = S_{BG}). \qquad (3.16)$$

See Fig. 3.3 for the illustration of this q-generalization of the celebrated Boltzmann formula for equal probabilities (referred to by Einstein as the *Boltzmann principle*). Let us address next the general case, i.e., for arbitrary $\{p_i\}$. We saw in Eq. (2.8) that S_{BG} can be written as the mean value of $\ln(1/p_i)$. This quantity is called *surprise* [136] or *unexpectedness* [137] by some authors. This is in fact quite appropriate. If we have *certainty* ($p_i = 1$ for some value of i) that something will happen, when it does happen we have *no surprise*. On the opposite extreme, if something is *very unexpected* ($p_i \simeq 0$), if it eventually happens, we are certainly *very surprised*! Along this line, it is certainly admissible to consider the quantity $\ln_q (1/p_i)$ and call it *q-surprise* or *q-unexpectedness*. It then appears as quite natural to *postulate* the simultaneous generalization of Eqs. (2.8) and (3.16) as follows:

$$S_q = k \langle \ln_q (1/p_i) \rangle. \qquad (3.17)$$

If we use the definition (3.14) in this expression, we straightforwardly obtain

$$S_q = k \frac{1 - \sum_{i=1}^{W} p_i^q}{q - 1}. \qquad (3.18)$$

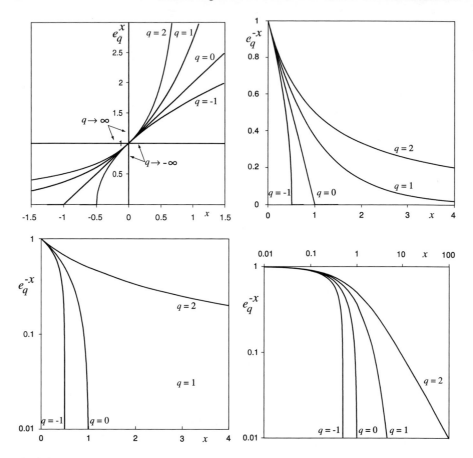

Fig. 3.1 *Top-left:* The q-exponential function e_q^x for typical values of q. For $q > 1$, it is defined in the interval $(-\infty, (q-1)^{-1})$; it diverges if $x \to (q-1)^{-1} - 0$. For $q < 1$, it is defined $\forall x$, and vanishes for all $x < -(1-q)^{-1}$. In the limit $x \to 0$, it is $e_q^x \sim 1 + x$ ($\forall q$). *Top-right:* The q-exponential function e_q^{-x} for typical values of q: linear-linear scales. For $q > 1$, it vanishes like $[(q-1)x]^{-1/(q-1)}$ for $x \to \infty$. For $q < 1$, it vanishes for $x > (1-q)^{-1}$ (*cutoff*). *Bottom-left:* The q-exponential function e_q^{-x} for typical values of q: log-linear scales. It is convex (concave) if $q > 1$ ($q < 1$). For $q < 1$ it has a vertical asymptote at $x = (1-q)^{-1}$. *Bottom-right:* The q-exponential function e_q^{-x} for typical values of q: log-log scales. For $q > 1$, it has an asymptotic slope equal to $-1/(q-1)$

This is precisely the form *postulated* in [47] as a possible basis for generalizing BG statistical mechanics. See Table 3.1. One possible manner for checking that $S_1 \equiv \lim_{q \to 1} S_q = S_{BG}$ is to directly replace into Eq. (3.18) the equivalence $p_i^q = p_i p_i^{q-1} = p_i \, e^{(q-1)\ln p_i} \sim p_i \, [1 + (q-1)\ln p_i]$.

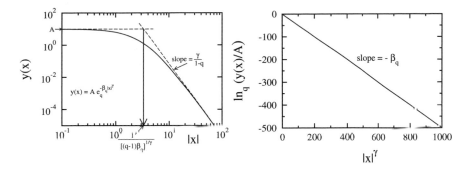

Fig. 3.2 *Left*: Construction based on the parameters (q, γ, β_q, A) of the function $y = A\, e_q^{-\beta_q |x|^\gamma}$ in log—log representation (Illustration: $A = 10$, $q = 4/3$, $\gamma = 3/2$; $\beta = 1/2$, hence the crossing point is located at $6^{2/3} = 3.3019\ldots$). *Right*: The same in linear $[\ln_q(y/A)]$—linear$[|x|^\gamma]$ representation. Courtesy by E.P. Borges

Fig. 3.3 The equiprobability entropy $S_q = \ln_q W$ as a function of the number of states W (with $k = 1$), for typical values of q. For $q > 1$, S_q saturates at the value $1/(q - 1)$ if $W \to \infty$; for $q \le 1$, it diverges. For $q \to \infty$ ($q \to -\infty$), it coincides with the abscissa (ordinate)

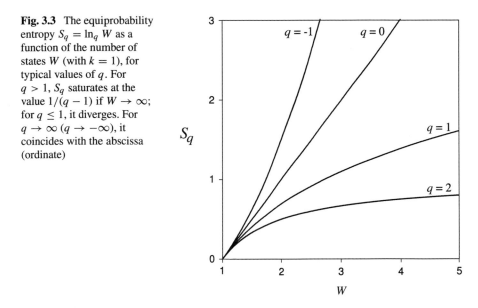

It turned out that this generalized entropic form, first with a *different* and then with the *same* multiplying pre-factor, had already appeared outside the literature of physics, namely in cybernetics and control theory [138].[2] It was rediscovered

---

[2] Havrda and Charvat [138] were apparently the first to ever introduce—for possible cybernetic purposes—the entropic form of Eq. (3.18), though with a different pre-factor, adapted to binary variables. Vajda [139] further studied this form, quoting [138]. Daroczy [140] rediscovered this form (he quotes neither [138] nor [139]). Lindhard and Nielsen [141] rediscovered this form (they quote none of the predecessors) through the property of entropic composability. Sharma, Taneja, and Mittal [147–149] introduced a two-parameter form which reproduces both S_q and Renyi entropy [150] as particular cases. Aczel and Daroczy [142] quote [138, 139], but not [141]. Wehrl [143]

Table 3.1 S_{BG} and S_q entropies ($S_1 = S_{BG}$)

Entropy	Equal probabilities $(p_i = 1/W, \forall i)$	Generic probabilities $(\forall \{p_i\})$
S_{BG}	$k \ln W$	$-k \sum_{i=1}^{W} p_i \ln p_i = k \sum_{i=1}^{W} p_i \ln(1/p_i)$
$S_q (q \in \mathbb{R})$	$k \ln_q W$	$k \dfrac{1 - \sum_{i=1}^{W} p_i^q}{q-1} = k \sum_{i=1}^{W} p_i \ln_q (1/p_i)$ $= -k \sum_{i=1}^{W} p_i^q \ln_q p_i$ $= -k \sum_{i=1}^{W} p_i \ln_{2-q} p_i$

independently in [47] when it was for the first time proposed as a starting point to generalize the standard statistical mechanics itself. This was done for the canonical ensemble, by optimizing S_q in the presence of an additional constraint, namely that related to the mean value of the energy. We shall focus on this calculation later on. The form (3.18) turns out to be in fact directly related to a generalized metric proposed in 1952 by Hardy, Littlewood and Polya [156], whose $q = 2$ particular case corresponds to the Pythagorean metric. Also, the same functional form S_q is used in ecology and referred to as *diversity* in [157–159].

A different path for arriving at the entropy (3.18) is the following one. This was in fact the original path, inspired by multifractals, that led to the postulate adopted in [47]. The entropic index q introduces a *bias* in the probabilities. Indeed, given the fact that generically $0 < p_i < 1$, we have that $p_i^q > p_i$ if $q < 1$, and $p_i^q < p_i$ if $q > 1$. Therefore, $q < 1$ (relatively) enhances the rare events, those which have

---

mentions the form of S_q in page 247, quotes [140], but ignores [138, 139, 141, 142, 147–149]. In 1984 an entropic divergence was defined by Cressie and Read [154] (details about the connection to the q-entropy can be found in [155]), who refer to Renyi but not to the other predecessors. I independently introduced this form in 1985 with the specific aim of generalizing Boltzmann–Gibbs statistical mechanics itself, and quote none of the predecessors in my 1988 paper [47]. Indeed, I started knowing the details of the whole story quite a few years later thanks to S.R.A. Salinas and R.N. Silver, who were the first to provide me with the corresponding information. Such rediscoveries can by no means be considered particularly surprising. Indeed, this happens in science more frequently than usually realized. This point is lengthily and colorfully developed by Stigler [144]; on page 284, a most interesting example is described, namely that of the celebrated *normal distribution*. It was first introduced by Abraham De Moivre in 1733, then by Pierre Simon Laplace in 1774, then by Robert Adrain in 1808, and finally by Carl Friedrich Gauss in 1809, nothing less than 76 years after its first publication! This distribution is universally called *Gaussian* because of the remarkable insights of Gauss concerning the theory of errors, applicable in all experimental sciences. A somewhat less glamorous illustration of the same phenomenon, but nevertheless interesting in the present context, is that of Renyi entropy ([150, 151] and references therein). According to Csiszar [145], the Renyi entropy had already been essentially introduced by Schutzenberger [146]. For Renyi's primordial proposal see [152, 153].

probabilities close to zero, whereas $q > 1$ (relatively) enhances the frequent events, those whose probability is close to unity. This property can be directly checked if we compare p_i with $p_i^q / \sum_{j=1}^{W} p_j^q$.

So, it appears appealing to introduce an entropic form based on p_i^q. Also, we want the form to be invariant under permutations. So the simplest assumption is to consider $S_q = f(\sum_{i=1}^{W} p_i^q)$, where f is some continuous function to be found. The simplest choice is the linear one, i.e., $S_q = a + b \sum_{i=1}^{W} p_i^q$. Since any entropy should be a measure of disorder or ignorance, we want that certainty corresponds to zero entropy. This immediately imposes $a + b = 0$, hence $S_q = a(1 - \sum_{i=1}^{W} p_i^q)$. But, since we are seeking for a generalization (and *not* an arbitrary alternative), for $q = 1$ we want to recover S_{BG}. Therefore, in the $q \rightarrow 1$ limit, a must be asymptotically proportional to $1/(q - 1)$ (we remind the equivalence indicated in the previous paragraph). The simplest way for this to occur is just to be $a = k/(q - 1)$, with $k > 0$, which immediately leads to Eq. (3.18).

We shall next address the properties of S_q. But before doing that, let us clarify a point which has a generic relevance. If $q > 0$, expression (3.18) is well defined whether or not one or more states have *zero* probability. Not so if $q < 0$. In this case, it must be understood that the sum indicated in Eq. (3.8) runs *only* over states with *positive* probability. For simplicity, we shall not explicitly indicate this fact along the book. But it is to be *always* taken into account.

3.2.2 Properties

Non-negativity

If we have certainty about the state of the system, then one of the probabilities equals unity and all the others vanish. Consequently, the entropy S_q vanishes for all q.

If we do not have certainty, at least two of the probabilities are smaller than unity. Therefore, for those, $1/p_i > 1$, hence $\ln_q(1/p_i) > 0$, $\forall i$ (see also Fig. 3.3). Consequently, using Eq. (3.17), it immediately follows that $S_q > 0$, for all q.

Extremal at equal probabilities

For the same reason indicated in the BG case (invariance of the entropy par rapport to *any* permutation of states), at equiprobability S_q must be extremal. It turns out to be a *maximum* for $q > 0$ and a *minimum* for $q < 0$. The proof will be completed as soon as we establish that $S_q(\{p_i\})$ is *concave* (*convex*) for $q > 0$ ($q < 0$), which will be done hereafter. The $q = 0$ case is marginal: the entropy is a constant. In that case, we have that

$$S_0 = k(W - 1) \quad (\forall\{p_i\}) \tag{3.19}$$

Expansibility

It is straightforwardly verified that S_q is *expansible*, $\forall q$, since

$$S_q(p_1, p_2, ..., p_W, 0) = S_q(p_1, p_2, ..., p_W). \tag{3.20}$$

This property trivially follows from the definition (3.18) if $q > 0$. For $q < 0$, it follows from the fact that the sum in (3.18) runs only for states whose probability is positive.

Nonadditivity

It is straightforwardly verified that, if A and B are *independent*, i.e., if the *joint probability* satisfies $p_{ij}^{A+B} = p_i^A p_j^B$ ($\forall(ij)$), then

$$\frac{S_q(A+B)}{k} = \frac{S_q(A)}{k} + \frac{S_q(B)}{k} + (1-q)\frac{S_q(A)}{k}\frac{S_q(B)}{k}. \tag{3.21}$$

It is due to this property that, for $q \neq 1$, S_q is said to be *nonadditive*.[3] However, drastic modifications occur when the subsystems A and B are correlated in a special manner. We shall see that in this case, a value of q might exist such that, either strictly or asymptotically ($N \to \infty$), $S_q(A+B) = S_q(A) + S_q(B)$. In other words, the nonadditive entropy S_q can be extensive for $q \neq 1$! This is a nontrivial issue that will be addressed in detail in Sect. 3.3.

Still, given the nonnegativity of S_q, it follows that, for *independent* subsystems, $S_q(A+B) \geq S_q(A) + S_q(B)$ if $q < 1$, and $S_q(A+B) \leq S_q(A) + S_q(B)$ if $q > 1$. Consistently, the $q < 1$ and $q > 1$ cases are occasionally referred in the literature as the *superadditive* and *subadditive* ones respectively.

Concavity and convexity

We refer to the concepts introduced in Eqs. (2.11)–(2.13), which naturally extend for arbitrary q. The second derivative of the (continuous) function $x(1 - x^{q-1})/(q - 1)$ is *negative* (*positive*) for $q > 0$ ($q < 0$). Consequently, for $q > 0$, we have, for the intermediate probability given in Eq. (2.11)

$$\frac{p_i''[1 - (p_i'')^{q-1}]}{q-1} > \lambda \frac{p_i[1 - p_i^{q-1}]}{q-1} + (1-\lambda)\frac{p_i'[1 - (p_i')^{q-1}]}{q-1} \quad (\forall i; 0 < \lambda < 1). \tag{3.22}$$

---

[3] During many years, this property has been referred in the literature as entropic *nonextensivity*. This is, in fact, kind of unfortunate. Indeed, it will become clear that, for a vast class of systems, a special value of q exists for which the nonadditive entropy S_q is extensive.

Fig. 3.4 The p-dependence of the $W = 2$ entropy $S_q/k = [1 - p^q - (1 - p)^q]/(q - 1)$ for typical values of q (with $S_1 = -p \ln p - (1 - p) \ln(1 - p)$). From Wolfram Mathematica

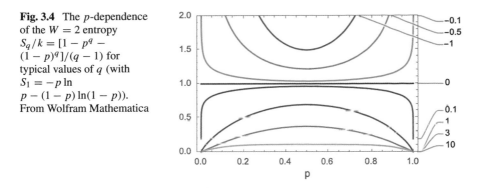

Applying $\sum_{i=1}^{W}$ on both sides of this inequality, we immediately obtain that

$$S_q(\{p_i''\}) > \lambda \, S_q(\{p_i\}) + (1 - \lambda) \, S_q(\{p_i'\}) \quad (q > 0). \quad (3.23)$$

These inequalities are obviously reversed for $q < 0$. It is, therefore, proved that S_q is *concave* (*convex*) for $q > 0$ ($q < 0$). An immediate corollary is, as announced previously, that the case of equal probabilities corresponds to a *maximum* for $q > 0$, whereas it corresponds to a *minimum* for $q < 0$. See in Fig. 3.4 an illustration of this property. See also Fig. 3.5.

Connection with Jackson derivative

One century ago, the mathematician Jackson generalized [160] the concept of *derivative* of a generic function $f(x)$. He introduced his differential operator D_q as follows:

$$D_q f(x) \equiv \frac{f(qx) - f(x)}{qx - x}. \quad (3.24)$$

We immediately verify that $D_1 f(x) = df(x)/dx$. For $q \neq 1$, this operator replaces the usual (infinitesimal) *translation* operation on the abscissa x of the function $f(x)$ by a *dilatation* operation.

Abe noticed a remarkable property [161]. In the same way, we can easily verify that

$$S_{BG} = -\frac{d}{dx} \sum_{i=1}^{W} p_i^x |_{x=1}, \quad (3.25)$$

we can verify that, $\forall q$,

$$S_q = -D_q \sum_{i=1}^{W} p_i^x |_{x=1}. \quad (3.26)$$

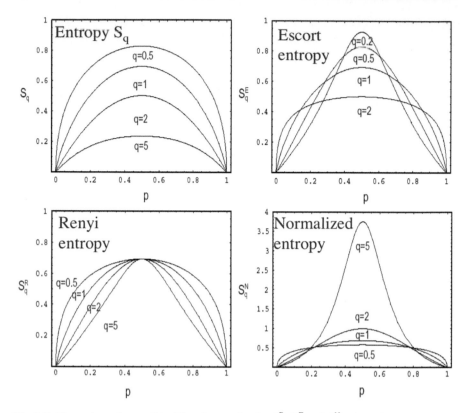

Fig. 3.5 The p-dependence of the $W = 2$ entropies S_q, S_q^R, S_q^E and S_q^N, where the *Renyi entropy* $S_q^R(\{p_i\}) \equiv \frac{\ln \sum_{i=1}^W p_i^q}{1-q} = \frac{\ln[1+(1-q)S_q(\{p_i\})]}{1-q}$, the *escort entropy* $S_q^E(\{p_i\}) \equiv S_q\left(\left\{\frac{p_i^q}{\sum_{j=1}^W p_j^q}\right\}\right) = \frac{1-[\sum_{i=1}^W p_i^{1/q}]^{-q}}{q-1}$, and the *Landsberg-Vedral-Rajagopal-Abe entropy*, or just *normalized entropy* $S_q^N(\{p_i\}) \equiv \frac{S_q(\{p_i\})}{\sum_{i=1}^W p_i^q} = \frac{S_q(\{p_i\})}{1+(1-q)S_q(\{p_i\})}$. We verify that, among these four entropic forms, only S_q is concave for all $q > 0$. From [98]

We consider this as an inspiring property, where the usual *infinitesimal* translational operation is replaced by a *finite* operation, namely, in this case, by the one which is basic for *scale-invariance*. Since the postulation of the entropy S_q was inspired by multifractal geometry, the least one can say is that this property is very welcome.

Lesche-stability or experimental robustness

Let us start by emphasizing that this property is totally independent of concavity. For example, Renyi entropy $S_q^R \equiv \frac{\ln \sum_{i=1}^W p_i^q}{1-q}$ is concave for $0 < q \leq 1$ and is neither concave nor convex for $q > 1$. However, it is Lesche-unstable for all $q > 0$ (excepting of course for $q = 1$) [96].

It has been proved [97, 98] that the definition of experimental robustness, i.e., Eq. (2.15), is satisfied for S_q for $q > 0$. See Fig. 3.6.

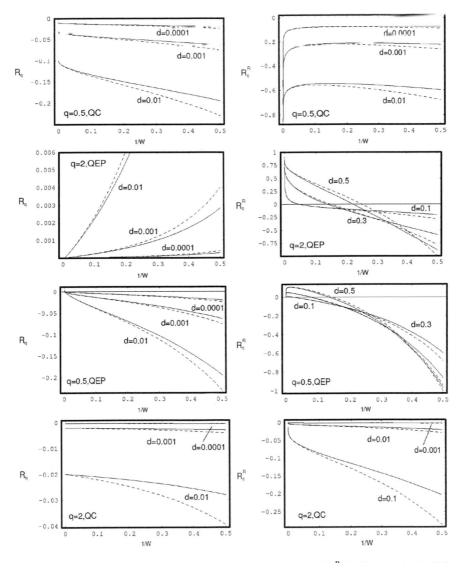

Fig. 3.6 Illustration of the dependence on (W, d) of the ratios R_q and R_q^R for the entropies S_q (*left*) and S_q^R (*right*) respectively. QC and QEP denote *quasi-certainty* and *quasi-equal-probabilities* (see the text). We see that $\lim_{d \to 0} \lim_{W^{-1} \to 0} R_q = 0$ in all four cases, whereas it is violated for R_q^R for the cases $(QC, q < 1)$ and $(QEP, q > 1)$. Not so for the two last cases, $(QC, q > 1)$ and $(QEP, q < 1)$, for which we do have $\lim_{d \to 0} \lim_{W^{-1} \to 0} R_q^R = 0$. The dashed (continuous) curves correspond to metric $\mu = 1$ ($\mu = 2$). From [98]

Conditional nonadditive entropy, q-expectations values, and escort distributions

Let us consider the entropy 3.18 and divide the set of W possibilities in K nonintersecting subsets respectively containing W_1, W_2, ..., W_K elements, with $\sum_{k=1}^{K} W_k = W$ ($1 \leq K \leq W$) [162]. We define the probabilities

$$\pi_1 \equiv \sum_{\{W_1 \ terms\}} p_i \,,$$

$$\pi_2 \equiv \sum_{\{W_2 \ terms\}} p_i \,, \dots \qquad (3.27)$$

$$\pi_K \equiv \sum_{\{W_K \ terms\}} p_i \,,$$

hence $\sum_{k=1}^{K} \pi_k = 1$. It is straightforward to verify the following property:

$$S_q(\{p_i\}) = S_q(\{\pi_k\}) + \sum_{k=1}^{K} \pi_k^q S_q(\{p_i/\pi_k\}) \,, \qquad (3.28)$$

where, consistently with Bayes' formula, $\{p_i/\pi_k\}$ are the *conditional* probabilities, and satisfy $\sum_{\{W_k \ terms\}}(p_i/\pi_k) = 1$ ($k = 1, 2, ..., K$). Property 3.28 recovers, for $q = 1$, Shannon's celebrated *grouping* relation

$$S_{BG}(\{p_i\}) = S_{BG}(\{\pi_k\}) + \sum_{k=1}^{K} \pi_k S_{BG}(\{p_i/\pi_k\}). \qquad (3.29)$$

This property constitutes in fact the fourth axiom of the Shannon theorem.

The nonnegative entropies $S_q(\{p_i\})$, $S_q(\{\pi_k\})$ and $S_q(\{p_i/\pi_k\})$ depend respectively on W, K and W_k probabilities. Equation 3.28 can be rewritten as

$$S_q(\{p_i\}) = S_q(\{\pi_k\}) + \langle S_q(\{p_i/\pi_k\})\rangle_q^{(u)}, \qquad (3.30)$$

where the *unnormalized q-expectation value* (u stands for *unnormalized*) of the conditional entropy is defined as

$$\langle S_q(\{p_i/\pi_k\})\rangle_q^{(u)} \equiv \sum_{k=1}^{K} \pi_k^q S_q(\{p_i/\pi_k\}), \qquad (3.31)$$

Also, since the definition of $S_q(\{\pi_k\})$ implies

$$\frac{1 + (1-q)S_q(\{\pi_k\})}{\sum_{k'=1}^{K} \pi_{k'}^q} = 1, \qquad (3.32)$$

Equation 3.28 can be rewritten as follows:

$$S_q(\{p_i\}) = S_q(\{\pi_k\}) + \sum_{k=1}^{K} \pi_k^q \, \frac{1 + (1-q)S_q(\{\pi_k\})}{\sum_{k'=1}^{K} \pi_{k'}^q} \, S_q(\{p_i/\pi_k\}). \tag{3.33}$$

Consequently

$$S_q(\{p_i\}) = S_q(\{\pi_k\}) + \langle S_q(\{\, p_i/\pi_k\})\rangle_q + (1-q)\, S_q(\{\pi_k\}) \, \langle S_q(\{p_i/\pi_k\})\rangle_q, \tag{3.34}$$

where the *normalized q-expectation value* of the conditional entropy is defined as

$$\langle S_q(\{p_i/\pi_k\})\rangle_q \equiv \sum_{k=1}^{K} \Pi_k \, S_q(\{p_i/\pi_k\}), \tag{3.35}$$

with the *escort probabilities* [163]

$$\Pi_k \equiv \frac{\pi_k^q}{\sum_{k'=1}^{K} \pi_{k'}^q} \quad (k = 1, 2, \dots, K). \tag{3.36}$$

Property 3.34 is, as we shall see later on, a very useful one, and it exhibits a most important fact, namely that the definition of the nonadditive entropic form 3.18 naturally leads to normalized q-expectation values and to escort distributions.

Let us further elaborate on Eq. (3.34). It can be also rewritten in a more symmetric form, namely as

$$1 + (1-q)S_q(\{p_i\}) = [1 + (1-q)S_q(\{\pi_k\})][1 + (1-q)\langle S_q(\{\, p_i/\pi_k\})\rangle_q]. \tag{3.37}$$

Since the *Renyi entropy* (associated with the probabilities $\{p_i\}$) is defined as $S_q^R(\{p_i\}) \equiv (\ln \sum_{i=1}^{W} p_i^q)/(1-q)$, we can conveniently define the (monotonically increasing) function $\mathcal{R}_q[x] \equiv \ln[1 + (1-q)x]/[1-q] = \ln\{[1 + (1-q)x]^{[1/(1-q)]}\}$ (with $\mathcal{R}_1[x] = x$), hence, for any distribution of probabilities, we have $S_q^R = \mathcal{R}_q[S_q]$. Equation 3.37 can now be rewritten as

$$\mathcal{R}_q[S_q(\{p_i\})] = \mathcal{R}_q[S_q(\{\pi_k\})] + \mathcal{R}_q[\langle S_q(\{\, p_i/\pi_k\})\rangle_q], \tag{3.38}$$

or equivalently,

$$S_q^R(\{p_i\}) = S_q^R(\{\pi_k\}) + \mathcal{R}_q[\langle \mathcal{R}_q^{-1}[S_q^R(\{\, p_i/\pi_k\})]\rangle_q]. \tag{3.39}$$

where the inverse function is defined as $\mathcal{R}_q^{-1}[y] \equiv [(e^y)^{(1-q)} - 1]/[1 - q]$ (with $\mathcal{R}_1^{-1}[y] = y$). Notice that, in general, $\mathcal{R}_q[\langle...\rangle_q] \neq \langle \mathcal{R}_q[...]\rangle_q$.

Everything we have said in this Section is valid for *arbitrary* partitions (in K nonintersecting subsets) of the ensemble of W possibilities. Let us from now on address the particular case where the W possibilities correspond to the joint possibilities of two subsystems A and B, having respectively W_A and W_B possibilities (hence $W = W_A W_B$). Let us denote by $\{p_{ij}\}$ the probabilities associated with the total system $A + B$, with $i = 1, 2, ..., W_A$, and $j = 1, 2, ..., W_B$. The *marginal probabilities* $\{p_i^A\}$ associated with subsystem A are given by $p_i^A = \sum_{j=1}^{W_B} p_{ij}$, and those associated with subsystem B are given by $p_j^B = \sum_{i=1}^{W_A} p_{ij}$. A and B are said to be *independent* if and only if $p_{ij} = p_i^A p_j^B$ ($\forall (i, j)$). We can now identify the K subsets we were previously analyzing with the W_A possibilities of subsystem A, hence the probabilities $\{\pi_k\}$ correspond, respectively, to $\{p_i^A\}$. Consistently, Eq. (3.34) implies now

$$S_q[A + B] = S_q[A] + S_q[B|A] + (1 - q)S_q[A]S_q[B|A], \qquad (3.40)$$

where $S_q[A + B] \equiv S_q(\{p_{ij}\})$, $S_q[A] \equiv S_q(\{p_i^A\})$ and the conditional entropy

$$S_q[B|A] \equiv \frac{\sum_{i=1}^{W_A}(p_i^A)^q S_q[B|A_i]}{\sum_{i=1}^{W_A}(p_i^A)^q} \equiv \langle S_q[B|A_i]\rangle_q, \qquad (3.41)$$

where

$$S_q[B|A_i] \equiv \frac{1 - \sum_{j=1}^{W_B}(p_{ij}/p_i^A)^q}{q - 1} \qquad (i = 1, 2, ..., W_A), \qquad (3.42)$$

with $\sum_{j=1}^{W_B}(p_{ij}/p_i^A) = 1$. Symmetrically, Eq. (3.40) can be also written as

$$S_q[A + B] = S_q[B] + S_q[A|B] + (1 - q)S_q[B]S_q[A|B]. \qquad (3.43)$$

If A and B are independent, then $p_{ij} = p_i^A p_j^B$ ($\forall (i, j)$), hence $S_q[A|B] = S_q[A]$ and $S_q[B|A] = S_q[B]$, therefore, both Eqs. (13) and (16) yield the well-known pseudo-additivity property of the nonadditive entropy S_q, namely

$$S_q[A + B] = S_q[A] + S_q[B] + (1 - q)S_q[A]S_q[B]. \qquad (3.44)$$

We thus see that Eqs. (3.40) and (3.43) nicely compress into *one* property two important properties of the entropic form S_q, namely Eqs. (3.28) and (3.44). The axiomatic implications of these relations have been discussed by Abe [164].

Santos uniqueness theorem

The Santos theorem [165] generalizes that of Shannon (addressed in Sect. 2.1.2).
Let us assume that an entropic form $S(\{p_i\})$ satisfies the following properties:

(i) $S(\{p_i\})$ *is a continuous function of* $\{p_i\}$; $\qquad$ (3.45)

(ii) $S(p_i - 1/W, \forall i)$ *monotonically increases with the total*
number of possibilities W; $\qquad$ (3.46)

(iii) $\dfrac{S(A+B)}{k} = \dfrac{S(A)}{k} + \dfrac{S(B)}{k} + (1-q)\dfrac{S(A)}{k}\dfrac{S(B)}{k}$ $\qquad$ (3.47)
if $p_{ij}^{A+B} = p_i^A p_j^B \forall(i,j),$ *with* $k > 0$;

(iv) $S(\{p_i\}) = S(p_L, p_M) + p_L^q S(\{p_i/p_L\}) + p_M^q S(\{p_i/p_M\})$ $\qquad$ (3.48)
with $p_L \equiv \displaystyle\sum_{L\,terms} p_i$, $p_L \equiv \displaystyle\sum_{M\,terms} p_i$,
$L + M = W,$ *and* $p_L + p_M = 1.$

Then and only then [165]

$$S(\{p_i\}) = k\frac{1 - \sum_{i=1}^{W} p_i^q}{q-1}.$$ $\qquad$ (3.49)

Abe uniqueness theorem

The Abe theorem [164] generalizes that of Khinchin (addressed in Sect. 2.1.2).
Let us assume that an entropic form $S(\{p_i\})$ satisfies the following properties:

(i) $S(\{p_i\})$ *is a continuous function of* $\{p_i\}$; $\qquad$ (3.50)

(ii) $S(p_i = 1/W, \forall i)$ *monotonically increases with the total*
number of possibilities W; $\qquad$ (3.51)

(iii) $S(p_1, p_2, ..., p_W, 0) = S(p_1, p_2, ..., p_W)$; $\qquad$ (3.52)

(iv) $\dfrac{S(A+B)}{k} = \dfrac{S(A)}{k} + \dfrac{S(B|A)}{k} + (1-q)\dfrac{S(A)}{k}\dfrac{S(B|A)}{k}$ $\qquad$ (3.53)

where $S(A+B) \equiv S(\{p_{ij}^{A+B}\})$, $S(A) \equiv S(\{\displaystyle\sum_{j=1}^{W_B} p_{ij}^{A+B}\})$, *and the*

conditional entropy $S(B|A) \equiv \dfrac{\sum_{i=1}^{W_A}(p_i^A)^q S(\{p_{ij}^{A+B}/p_i^A\})}{\sum_{i=1}^{W_A}(p_i^A)^q}$ $\quad (k > 0)$

Then and only then [164][4]

$$S(\{p_i\}) = k \frac{1 - \sum_{i=1}^{W} p_i^q}{q - 1}. \tag{3.54}$$

Notice that, interestingly enough, what enters in the definition of conditional entropy is the escort distribution, and *not* the original one. Notice also that Eq. (3.52) only holds for $q > 0$. Therefore, the expression (3.54) for $q < 0$ can only be defined for *strictly positive* values of $\{p_i\}$, and it is to be understood as an analytical extension of the $q > 0$ case. Let us finally emphasize that both Santos axioms and Abe axioms are necessary and sufficient conditions for the emergence of S_q. Consequently, those two sets of axioms are mathematically equivalent.

The axiomatic justification of diverse entropic functionals has in fact deserved great attention in both recent and not so recent literature. Let us summarize here, along lines close to Jizba and Korbel [168], the present status of this interesting path of research. Three main consistent lines of analysis exist, namely generalized Shannon–Khinchine axioms of the type of [164, 165], the Shore and Johnson axioms [169], and the Uffink class of entropies [170]. All three lead to the same set of admissible entropies, which include S_q and monotonic functions of it. The particular line related to the Shore–Johnson axioms deserves a special mention because it has been the object of a neat controversy. Indeed, Presse et al [171–173] lengthily insisted that the Shore–Johnson axioms exclude entropies such as S_q. Their arguments were addressed in [168, 174–176], either exhibiting what is argued to be their fallacies [168, 174, 176], or further supporting them [175].[5]

Composability

The entropy S_q is, like the BG one, composable (see also [177]). Indeed, it satisfies Eq. (3.21). In other words, we have $F(x, y; q) = x + y + (1 - q)x y$.

The Renyi entropy S_q^R is composable since it is additive. In other words, in that case, we have $F(x, y; q) = x + y, \forall q$.

As examples of the various *noncomposable* entropic forms that exist in the literature, we may cite the Curado b-entropy S_b^C [178, 179] and the Anteneodo–Plastino entropy S^{AP} [180]. Since these two forms have some quite interesting mathematical properties, it would in principle be thermodynamically valuable to construct entropies following along the lines of these ones, but which would be composable instead. Let us also mention at this point that an entropic functional which is a function of composable entropic functionals is not necessarily composable. It can be composable,

---

[4] The possibility of the existence of such a theorem through the appropriate generalization of Khinchin's fourth axiom had already been considered by Plastino and Plastino [166, 167]. Abe established [164] the precise form of this generalized fourth axiom and proved the theorem.

[5] The title *Conceptual inadequacy of the Shore and Johnson axioms for wide classes of complex systems* of [174] constitutes a sort of inadvertence. It should have been *Conceptual inadequacy of the Presse et al interpretation of the Shore and Johnson axioms for wide classes of complex systems*.

like the Renyi entropy (which is a function of the composable entropy S_q), or not, like the Kaniadakis entropy (which is a function of the composable entropies $S_{1+\kappa}$ and $S_{1-\kappa}$, as described in equality (3.289) later on). Let us remind at this point that further issues about composability were already presented in Sect. 2.1.2.

Sensitivity to the initial conditions, entropy production per unit time, and the q-generalized Pesin-like identity

Let us focus on a one-dimensional nonlinear dynamical system (characterized by the variable x) whose Lyapunov exponent λ_1 *vanishes* (e.g., the *edge of chaos* for typical unimodal maps such as the logistic one). The sensitivity to the initial conditions ξ defined in Eq. (2.29) is *conjectured* to satisfy the equation

$$\frac{d\xi}{dt} = \lambda_q \, \xi^q, \tag{3.55}$$

whose solution is given by

$$\xi = e_q^{\lambda_q t}. \tag{3.56}$$

The paradigmatic case corresponds to $\lambda_q > 0$ and $q < 1$. In this case, we have

$$\xi \propto t^{1/(1-q)} \quad (t \to \infty), \tag{3.57}$$

(see also [181–185]) and we refer to it as *weak chaos*, in contrast to *strong chaos*, associated with positive λ_1. To be more precise, Eq. 3.56 has been shown to be the *upper bound* of an entire family of such relations at the edge of chaos of unimodal maps. For each specific (strongly or weakly) chaotic one-dimensional dynamical system, we generically expect to have a couple $(q_{sen}, \lambda_{q_{sen}})$ (where *sen* stands for *sensitivity*) such that we have

$$\xi = e_{q_{sen}}^{\lambda_{q_{sen}} t}. \tag{3.58}$$

Clearly, strong chaos is recovered here as the particular instance $q_{sen} = 1$.

Let us now address the interesting question of the S_q entropy production as time t increases. By using S_q instead of S_{BG}, we could follow the same steps already indicated in Sect. 2.1.2, and attempt the definition of a q-generalized Kolmogorov-Sinai entropy rate. We will not follow along this line, but we shall rather q-generalize the entropy production K_1 introduced in Sect. 2.1.2. We define now

$$K_q \equiv \lim_{t \to \infty} \lim_{W \to \infty} \lim_{M \to \infty} \frac{S_q(t)}{t}. \tag{3.59}$$

We *conjecture* that generically a unique value of q exists, noted q_{ent} (where *ent* stands for *entropy*) such that (the upper bound of) $K_{q_{ent}}$ is *finite* (i.e., positive), whereas K_q vanishes (diverges) for $q > q_{ent}$ ($q < q_{ent}$).

We further conjecture for one-dimensional systems that

$$q_{ent} = q_{sen},\tag{3.60}$$

and that

$$K_{q_{ent}} = K_{q_{sen}} = \lambda_{q_{sen}}.\tag{3.61}$$

As already mentioned, strong chaos is recovered as a particular case, and we obtain the Pesin-like identity $K_1 = \lambda_1$. Conjectures (3.58), (3.60) and (3.61) were first introduced in [186], and have been analytically proved and/or numerically verified in a considerable number of examples [187, 188, 190–200].

If our weakly chaotic system has ν positive q-generalized Lyapunov coefficients $\lambda_{q_{sen}}^{(1)}, \lambda_{q_{sen}}^{(2)}, \ldots, \lambda_{q_{sen}}^{(\nu)}$, we expect [201]

$$\frac{1}{1 - q_{ent}} = \sum_{k=1}^{\nu} \frac{1}{1 - q_{sen}^{(k)}}.\tag{3.62}$$

This yields, if all the q_{sen}'s are equal (e.g., a probabilistic model with exchangeable random variables),

$$q_{ent} = 1 - \frac{1 - q_{sen}}{\nu},\tag{3.63}$$

hence $q_{ent} = 1$ in the $\nu \to \infty$ limit. If $\nu = 1$ we recover Eq. (3.60). If $q_{sen} = 0$, we obtain

$$q_{ent} = 1 - \frac{1}{\nu},\tag{3.64}$$

which implies $q_{ent} = 1$ in the limit $\nu \to \infty$. We illustrated here the connection between the sensitivity to the initial conditions and the entropy production per unit time when the BG entropy does not apply, but we shall lengthily come back onto this important question in Chap. 5.[6] Moreover, it is possible to generalize Eq. (3.61) in the presence of an *escape rate*, i.e., a mechanism which makes the number of available microscopic states to decrease along time [202].

---

[6] In order to quantify the expansion in space-phase for the weakly chaotic case where $\xi \sim a\,t^b$ ($a > 0$, $b > 0$), it was proposed [185] $\lambda_{MHHK} = \lim_{t \to \infty} \frac{\ln \xi(t)}{\ln t}$. We adopt here the alternative $\lambda_q = \lim_{t \to \infty} \frac{\ln_q \xi(t)}{t}$. Both yield $\lambda_{MHHK} = \lambda_q = b$, but the latter has two important advantages, namely, (i) it elegantly unifies the present case with the strongly chaotic one ($q = 1$), (ii) it provides a *finite linear rate per unit time*, which is highly convenient to q-generalize the important Pesin identity.

3.3 Correlations, Occupancy of Phase Space, and Extensivity of S_q

3.3.1 The Thermodynamical Limit

Let us assume a classical mechanical many-body system characterized by the following Hamiltonian:

$$\mathcal{H} = K + V = \sum_{i=1}^{N} \frac{p_i^2}{2m} + \sum_{i \neq j} V(r_{ij}), \tag{3.65}$$

where the two-body potential energy $V(r)$ presents no mathematical difficulties near the origin $r = 0$ (e.g., in the $r \rightarrow 0$ limit, either it is repulsive, or, if it is attractive, it is nonsingular or at least integrable), and which behaves at long distances ($r \rightarrow \infty$) like

$$V(r) \sim -\frac{A}{r^\alpha} \quad (A > 0; \ \alpha \geq 0). \tag{3.66}$$

A typical example would be the $d = 3$ Lennard-Jones gas model, for which $\alpha = 6$. Were it not the nonintegrable singularity at the origin, another example would have been Newtonian $d = 3$ gravitation, for which $\alpha = 1$; this difficulty is sometimes avoided by introducing a cutoff of some sort near the origin.

Let us analyze the characteristic average potential energy U_{pot} per particle

$$\frac{U_{pot}(N)}{N} \propto -A \int_1^\infty dr \, r^{d-1} \, r^{-\alpha}, \tag{3.67}$$

where we have integrated from a typical distance (taken equal to unity) on. This is the typical energy one would calculate within a BG approach at vanishing temperature. We see immediately that this integral *converges* for $\alpha/d > 1$ (hereafter referred to as *short-range interactions* for classical systems) but *diverges* for $0 \leq \alpha/d \leq 1$ (hereafter referred to as *long-range interactions*). This already indicates that some anomaly might be present.[7] By the way, it is historically fascinating the fact that

---

[7] This is essentially the very same reason which virtually all statistical mechanics textbooks discuss paradigmatic systems like a particle in a square well, the harmonic oscillator, the rigid rotator, and a spin 1/2 in the presence of an external magnetic field, *but not the non-ionized Hydrogen atom!* *All* these simple systems, as well as the Hydrogen atom, are discussed in the quantum mechanics textbooks. But, in what concerns statistical mechanics, the non-ionized Hydrogen atom constitutes an illustrious absence. Amazingly enough, and in spite of the existence of a centennial-related literature [203–216], this highly important system passes without any comments in almost all the textbooks on thermal statistics. The—understandable but not justifiable—reason for such an absence is of course that, since the system involves the long-range Coulombian attraction between electron and proton, the energy spectrum exhibits an accumulation point at the ionization energy (frequently taken to be zero), which makes the BG partition function to diverge.

Gibbs himself was aware of the possibility of such difficulty! (see, in Sect. 1.2, Gibbs' remarks concerning long-range interactions).

On a vein slightly differing from the standard BG recipe, which would demand integration up to infinity in Eq. (3.67), let us assume that the N-particle system is roughly homogeneously distributed within a limited sphere. Then Eq. (3.67) has to be replaced by the following one:

$$\frac{U_{pot}(N)}{N} \propto -A \int_1^{N^{1/d}} dr\, r^{d-1}\, r^{-\alpha} = -\frac{A}{d} N^*, \tag{3.68}$$

with

$$N^* \equiv \frac{N^{1-\alpha/d} - 1}{1 - \alpha/d} = \ln_{\alpha/d} N \sim \begin{cases} \dfrac{1}{\alpha/d - 1} & \text{if } \alpha/d > 1; \\[2mm] \ln N & \text{if } \alpha/d = 1; \\[2mm] \dfrac{N^{1-\alpha/d}}{1 - \alpha/d} & \text{if } 0 < \alpha/d < 1. \end{cases} \tag{3.69}$$

Therefore, in the $N \to \infty$ limit, $\frac{U_{pot}(N)}{N}$ approaches a constant ($\propto -A/(\alpha - d)$) if $\alpha/d > 1$, and diverges like $N^{1-\alpha/d}/(1 - \alpha/d)$ if $0 \leq \alpha/d < 1$ (it diverges logarithmically if $\alpha/d = 1$).[8] In other words, the energy is *extensive for short-range interactions* ($\alpha/d > 1$), and *nonextensive for long-range interactions* ($0 \leq \alpha/d \leq 1$). Satisfactorily enough, Eq. 3.69 recover the characterization with Eq. 3.67 in the limit $N \to \infty$, but they have the great advantage of providing, for *finite N*, a *finite* value. This fact will be now shown to enable to properly scale the macroscopic quantities in the thermodynamic limit ($N \to \infty$), *for all values of $\alpha/d \geq 0$* and not only for $\alpha/d > 1$. See Figs. 3.7 and 3.8.

A totally similar situation occurs if we have, playing the role of Hamiltonian 3.65, say N coupled planar or three-dimensional rotators, or oscillators, localized on a d-dimensional lattice. We further focus on these cases later on.

We are now prepared to address the thermodynamical consequences of the microscopic interactions being short- or long-ranged ([220], and references within [221]). To present a slightly more general illustration, we shall assume from now on that our

---

[8] These results turn out afterward to be consistent with those discussed in relation to Eq. (1.67) of [217], in the frame of how strongly can N random variables be correlated, and be still applicable the standard Central Limit Theorem, in the sense that the corresponding attractor be a Gaussian distribution.

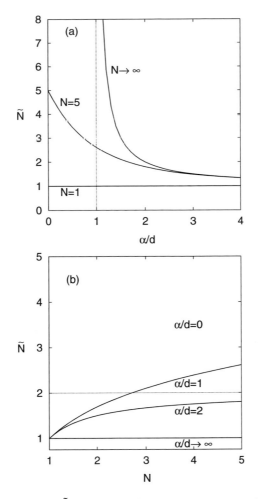

Fig. 3.7 The rescaling function $\tilde{N}(N, \alpha/d) \equiv N^*(N, \alpha/d) + 1$ versus α/d for typical values of N (a), and versus N for typical values of α/d (b). For fixed $\alpha/d \geq 0$, $\tilde{N}$ monotonically increases with N increasing from 1 to ∞; for fixed $N > 1$, $\tilde{N}$ monotonically decreases for α/d increasing from 0 to ∞. $\tilde{N}(N, 0) = N$, thus recovering the Mean Field Approximation usual rescaling; $\lim_{N \to \infty} \tilde{N}$ diverges for $0 \leq \alpha/d \leq 1$, thus separating the extensive from the nonextensive region; $\tilde{N}(\infty, \alpha/d) = (\alpha/d)/[(\alpha/d) - 1]$ if $\alpha/d > 1$; $\lim_{\alpha/d \to \infty} \tilde{N} = 1$, thus recovering *precisely* the traditional intensive and extensive thermodynamical quantities; $\tilde{N}(N, 1) = \ln N$. From [101]

say d-dimensional homogeneous and isotropic classical fluid constituted by magnetic particles is in thermodynamical equilibrium. Its Gibbs free energy is then given by

$$G(N, T, p, \mu, H) = U(N, T, p, \mu, H) - TS(N, T, p, \mu, H)$$
$$+ pV(N, T, p, \mu, H) - \mu N - HM(N, T, p, \mu, H), \quad (3.70)$$

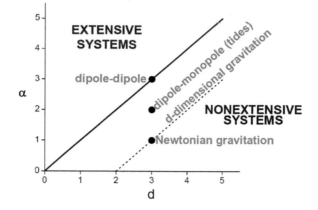

Fig. 3.8 The so-called *extensive systems* ($\alpha/d > 1$ for the classical ones) typically involve *absolutely convergent series*, whereas the so-called *nonextensive systems* ($0 \le \alpha/d < 1$ for the classical ones) typically involve *divergent series*. The marginal systems ($\alpha/d = 1$ here) typically involve *conditionally convergent series*, which, therefore, depend on the boundary conditions, i.e., typically on the external shape of the system. Capacitors constitute a notorious example of the $\alpha/d = 1$ case. The model usually referred to in the literature as the *Hamiltonian-Mean-Field* (HMF) one [218] lies on the $\alpha = 0$ axis ($\forall d > 0$). The models usually referred to as the d-dimensional α-XY [219], α-Heisenberg and α-Fermi-Pasta-Ulam models lie on the vertical axis at abscissa d ($\forall \alpha \ge 0$). The standard Lennard-Jones gas is located at $(d, \alpha) = (3, 6)$

where (T, p, μ, H) correspond, respectively, to the temperature, pressure, chemical potential, and external magnetic field, U is the internal energy, S is the entropy, V is the volume, N is the number of particles, and M the magnetization.

If the interactions are short-ranged (i.e., if $\alpha/d > 1$), we can divide this equation by N and then take the $N \to \infty$ limit. We obtain

$$g(T, p, \mu, H) = u(T, p, \mu, H) - Ts(T, p, \mu, H) + pv(T, p, \mu, H) - \mu - Hm(T, p, H), \quad (3.71)$$

where $g(T, p, \mu, H) \equiv \lim_{N\to\infty} G(N, T, p, \mu, H)/N$, and analogously for the other variables of the equation.

If the interactions are long-ranged instead (i.e., if $0 \le \alpha/d \le 1$), all the terms of expression (3.71) *diverge*, hence thermodynamically speaking they are nonsense. Consequently, the generically correct procedure, i.e. $\forall \alpha/d \ge 0$, must conform to the following lines:

$$\lim_{N\to\infty} \frac{G(N, T, p, \mu, H)}{NN^\star} = \lim_{N\to\infty} \frac{U(N, T, p, \mu, H)}{NN^\star} - \lim_{N\to\infty}\left[\frac{T}{N^\star}\frac{S(N, T, p, \mu, H)}{N}\right]$$
$$+ \lim_{N\to\infty}\left[\frac{p}{N^\star}\frac{V(N, T, p, \mu, H)}{N}\right] - \lim_{N\to\infty}\frac{\mu}{N^\star} - \lim_{N\to\infty}\left[\frac{H}{N^\star}\frac{M(N, T, p, \mu, H)}{N}\right] \quad (3.72)$$

hence

$$g(T^\star, p^\star, \mu^\star, H^\star) = u(T^\star, p^\star, \mu^\star, H^\star) - T^\star s(T^\star, p^\star, \mu^\star, H^\star)$$
$$+ p^\star v(T^\star, p^\star, \mu^\star, H^\star) - \mu^\star - H^\star m(T^\star, p^\star, \mu^\star, H^\star), , \tag{3.73}$$

where the definitions of $T^\star$ and of all the other variables are self-explanatory (e.g., $T^\star \equiv \lim_{N\to\infty}[T/N^\star]$ and $s(T^\star, p^\star, \mu^\star, H^\star) \equiv \lim_{N\to\infty}[S(N, T, p, \mu, H)/N]$). In other words, in order to have *finite* thermodynamic equations of states, we must in general express them in the $(T^\star, p^\star, \mu^\star, H^\star)$ variables. If $\alpha/d > 1$, this procedure recovers the usual equations of states, and the usual *extensive* (G, U, S, V, N, M) and *intensive* (T, p, μ, H) thermodynamic variables. But, if $0 \leq \alpha/d \leq 1$, the situation is more complex, and we realize that *three*, instead of the traditional *two*, classes of thermodynamic variables emerge. We may call them *extensive* (S, V, N, M), *pseudo-extensive* (G, U) (superextensive in the present case) and *pseudo-intensive* (T, p, μ, H) (super-intensive in the present case) variables. All the energy-type thermodynamical variables (G, F, U) give rise to pseudo-extensive ones, whereas those which appear in the usual Legendre thermodynamical pairs give rise to pseudo-intensive ones (T, p, μ, H) and extensive ones (S, V, N, M). Let us emphasize that (S, V, N, M) are *extensive in all cases!* Let us also emphasize that, consistently, (i) The ratio of any two pseudo-intensive variables $(T, p, \mu, H, \dots)$, e.g., p/T, is *intensive in all cases*; (ii) The ratio of any pseudo-extensive variable (G, F, U) with any pseudo-intensive variable, e.g., U/T, is *extensive in all cases*. A most important implication of these facts is that, in expressions such as $e_q^{-\beta_q \mathcal{H}}$, the argument $\beta_q \mathcal{H}$ is *extensive in all cases*. This plays a crucial role in the possible q-generalization of the Large Deviation Theory, as we shall see in Sect. 4.10. Indeed, the extensivity of $\beta_q \mathcal{H}$ mirrors the extensivity in all cases of the total entropy involved in $r_q N$, r_q being the ratio function. See Figs. 3.8 and 3.9.

The possibly long-range interactions within Hamiltonian 3.65 refer to the *dynamical variables themselves*. There is another important class of Hamiltonians, where the possibly long-range interactions refer to the *coupling constants between localized dynamical variables*. Such is, for instance, the case of the following classical Hamiltonian:

$$\mathcal{H} = K + V = \sum_{i=1}^{N} \frac{L_i^2}{2I} - \sum_{i\neq j} \frac{J_x s_i^x s_j^x + J_y s_i^y s_j^y + J_z s_i^z s_j^z}{r_{ij}^\alpha} \quad (\alpha \geq 0), \tag{3.74}$$

where $\{L_i\}$ are the angular momenta, I the moment of inertia, $\{(s_i^x, s_i^y, s_i^z)\}$ are the components of classical rotators, (J_x, J_y, J_z) are coupling constants, and r_{ij} runs over all distances between sites i and j of a d-dimensional lattice. For example, for a simple hypercubic lattice with unit crystalline parameter, we have $r_{ij} = 1, 2, 3, \dots$ if $d = 1$, $r_{ij} = 1, \sqrt{2}, 2, \dots$ if $d = 2$, $r_{ij} = 1, \sqrt{2}, \sqrt{3}, 2, \dots$ if $d = 3$, and so on. For such a case, we have that

$$N^\star \equiv \sum_{i=2}^{N} r_{1i}^{-\alpha}, \tag{3.75}$$

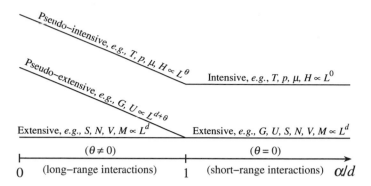

Fig. 3.9 Representation of the different scaling regimes for classical d-dimensional systems. For attractive long-range interactions (*i.e.*, $0 \leq \alpha/d \leq 1$, α characterizing the interaction range in a potential with the form $1/r^{\alpha}$; for example, Newtonian gravitation corresponds to $(d, \alpha) = (3, 1)$) we may distinguish *three* classes of thermodynamic variables, namely, those scaling with L^{θ}, named *pseudo-intensive* (L is a characteristic linear length, θ is a system-dependent parameter), those scaling with $L^{d+\theta}$ with $\theta = d - \alpha$, the *pseudo-extensive* ones (the energies), and those scaling with L^{d} (which are always extensive). For short-range interactions (*i.e.*, $\alpha > d$) we have $\theta = 0$ and the energies recover their standard L^{d} extensive scaling, falling within the same class of S, N, V, M, etc, whereas the previous pseudo-intensive variables become truly intensive ones (independent of L); this is the region, with only *two* classes of variables, that is covered by the traditional textbooks of thermodynamics. From [89, 103, 104, 222–224]

which has in fact the same asymptotic behaviors as indicated in Eq. 3.69. In other words, here again, $\alpha/d > 1$ corresponds to short-range interactions, and $0 \leq \alpha/d \leq 1$ corresponds to long-range ones.

For example, the $\alpha/d = 0$ particular case corresponds to the usual mean field approach. Indeed, in this case, we have $N^* = N - 1 \sim N$, which is equivalent to the usual rescaling of the microscopic coupling constant through division by N (see also [219]). In fact, to accommodate the common use of dividing by N (instead of $N - 1$) for the $\alpha/d = 0$ case, it it sometimes practical to use, as done in Fig. 3.7,

$$\tilde{N} \equiv N^* + 1 = \frac{N^{1-\alpha/d} - (\alpha/d)}{1 - \alpha/d} \, . \tag{3.76}$$

For short-range interactions, $N^* \to constant$, consequently we recover the usual *extensivity* of Gibbs, Helmholtz, and internal thermodynamical energies, entropy, volume, number of particles, and magnetization, as well as the *intensivity* of temperature, pressure, chemical potential, and magnetic field. But for long-range interactions, N^* diverges with N, therefore, the situation is quite more subtle. Indeed, in order to have *nontrivial equations of states* we must express the *nonextensive* Gibbs, Helmholtz, and internal thermodynamical energies, as well as the *extensive* entropy, volume, number of particles, and magnetization in terms of the rescaled variables (T^*, p^*, μ^*, H^*). In general, i.e. $\forall (\alpha/d)$, we see that the variables that are intensive when the interactions are short-ranged remain a *single class* (although scaling with N^*) in the presence of long-ranged interactions. But, in what concerns the variables that are extensive when the interactions are short-ranged, the situation is more complex. Indeed, they split into *two classes*. One of them contains all types of thermodynamical energies (G, F, U), which scale with NN^* (superextensive, in this case). The other one contains all those variables (S, V, N, M) that appear *in pairs* in the thermodynamical energies. These variables remain extensive, in the sense that they scale with N.[9]

By no means this implies that thermodynamical equilibrium between two systems occurs in general when they share the same values of say (T^*, p^*, μ^*, H^*). It only means that, in order to have *finite* mathematical functions for their *equations of states*, the variables (T^*, p^*, μ^*, H^*) must be used. Although this remains to be verified, it cannot be excluded that thermodynamical equilibrium might still be directly related to sharing the usual variables (T, p, μ, H).

The correctness of the present generalized thermodynamical scalings has already been specifically checked in many physical systems, such as a ferrofluid-like model [226], Lennard-Jones-like fluids [227], magnetic systems [219, 221, 228–230], anomalous diffusion [231], percolation [232, 233], and also on general grounds [234].

In addition to this, if a phase transition occurs in the system at a temperature T_c, it is expected to happen for a *finite* value of $\lim_{N\to\infty} T_c/\tilde{N}$. This implies that (i) in the limit $\alpha/d \to 1+0$, $T_c \propto 1/(\alpha/d - 1)$, thus recovering a result known since long (for instance for the n-vector ferromagnet, including the Ising one); (ii) for $0 \le \alpha/d < 1$, $T_c \propto N^{1-\alpha/d}$. In the latter context, let us mention that, for the $\alpha = 0$ models (i.e., mean field like models), it is largely spread in the literature to divide by N (in general, by $N^{1-\alpha/d}$ if $0 \le \alpha/d < 1$) the interaction term of the Hamiltonian in order to make it extensive *by force*. Although mathematically admissible (see [219] for an isomorphism involving a simple rescaling of time t), this obviously is physically quite bizarre. Indeed it implies a microscopic coupling constant which depends on N. What we have described here turns out to be the thermodynamically proper and unified way of eliminating the mathematical difficulties emerging in the models whenever long-range interactions are present.[10]

Notice also that it belongs to the essence of thermodynamics the following property. If we know, for a large system Σ, quantities such as $U(\Sigma)$, $S(\Sigma)$, $F(\Sigma)$, etc, we should be able to easily calculate the same quantities for say an even larger such system $(\lambda\Sigma)$, with $N(\lambda\Sigma) = \lambda N(\Sigma)$ $(\lambda > 1)$. It is indeed so in the present case. For example, for $N \gg 1$ we have

---

[9] Consequently, for $0 \le \alpha/d < 1$, we expect $U(N, T) \sim N^{2-\alpha/d} u(T/N^{1-\alpha/d})$, $S(N, T) \sim N s(T/N^{1-\alpha/d})$, the specific heat $C(N, T) \sim N c(T/N^{1-\alpha/d})$, etc.

[10] Let us illustrate this point on a d-dimensional n-vector ferromagnet whose microscopic coupling constant decays with distance as J/r_{ij}^α $(J > 0, 0 \le \alpha/d < 1)$. The critical temperature is given by $T_c = \mu J N^{1-\alpha/d}/[(1 - \alpha/d) k_B]$, where the pure number $\mu \simeq 1$. This is the thermodynamically correct result. What is instead customary to do in the literature is to (unphysically) replace J by $J/N^{1-\alpha/d}$, thus obtaining $T_c = \mu J/[(1 - \alpha/d) k_B]$, which remains *finite* for $N \to \infty$.

$$\frac{U(\lambda \Sigma)}{U(\Sigma)} = \lambda \frac{(\lambda N)^{1-\alpha/d} - 1}{N^{1-\alpha/d} - 1} \sim \begin{cases} \lambda & \text{if } \alpha/d \geq 1; \\ \lambda^{2-\alpha/d} & \text{if } 0 < \alpha/d < 1. \end{cases} \tag{3.77}$$

We see, therefore, that, for short-range interactions, the result depends on no microscopic detail at all, thus confirming the concept usually emphasized in textbooks of thermodynamics. This is, however, not true for long-range interactions, where we can see that, although, in a mathematically very simple manner, the result *does* depend on the microscopic ratio α/d.

It is clear that all these notions are quite subtle and yet a subject of active research. Nevertheless, they constitute a strong indication that, no matter the range of the interactions, S_{BG} should be generalized *preserving its extensivity*, i.e., as introduced on macroscopic grounds by Clausius. What we present in the next subsections is precisely consistent with this expectation.

3.3.2 Einstein Likelihood Principle

We address here, along the lines of [235], an important issue, namely the likelihood factorization required by Einstein [28].

Einstein presented in 1910 [28] an interesting argument on why he liked Boltzmann's connection between the classical thermodynamic entropy introduced by Clausius and the probabilities of microscopic configurations. This argument is based on the *factorization* of the likelihood function of *independent systems* (A and B), namely

$$\mathcal{W}(A + B) = \mathcal{W}(A)\mathcal{W}(B). \tag{3.78}$$

This is a very powerful epistemological reason since it reflects the basic procedure of all sciences, namely that, in order to study any given natural, artificial and social system, theoretical approaches typically start by focusing on a certain subset of relevant degrees of freedom of the universe, and, only at a more evolved stage of the theory, possible connections with degrees of freedom outside that subset are introduced as well, whenever necessary. In the present context, we shall refer to Eq. (3.78), as *Einstein likelihood principle* (see [29, 48] for related aspects).

From the celebrated Boltzmann principle in Eq. (1.3) we obtain, through Einstein's well known *reversal* of Boltzmann formula, the likelihood function

$$\mathcal{W}(\{p_i\}) \propto e^{S_{BG}(\{p_i\})/k}, \tag{3.79}$$

with $S_{BG}(\{p_i\})$ given in Eq. (1.1). For simplicity, we have used the case of discrete variables (instead of the continuous ones, that were of course used in the early times of statistical mechanics). Notice that W plays in Eqs. (1.3) and (1.1) the role of a total number of admissible microscopic configurations, whereas, through the Einstein reversal, $\mathcal{W}$ plays in Eqs. (3.78) and (3.79) the role of a likelihood function (for example, if we throw 100 coins, what is the probability $\mathcal{W}$ of obtaining 52 heads and 48 tails?). We remind that the entropy S_{BG} is *additive* according to Penrose's definition [17], i.e., if A and B are probabilistically *independent* systems (i.e., if $p_{ij}^{A+B} = p_i^A p_j^B$), we straightforwardly verify that $S_{BG}(A + B) = S_{BG}(A) + S_{BG}(B)$. By replacing this equality in Eq. (3.79), the Einstein requirement (3.78) is straightforwardly satisfied.

Before further proceeding, let us emphasize a point which not rare generates confusion. By definition [17], an entropic functional is said additive if, for any *independent systems* A and B, the total entropy equals the sum of the entropies of the parts. In other words, it is a property of the functional and by no means depends on the system (or subsystems) to which it may be applied. This is in neat contrast with entropic extensivity which depends on both the functional *and* the system to which it is being applied. This is why establishing whether a given entropic functional is additive or not is a mathematically trivial task. Not so for establishing whether a given entropic functional is thermodynamically extensive for a given system: this can be, and frequently is, mathematically extremely involved. This is no surprise since the answer depends on all the correlations existing between all the elements of the system.

Let us go on now and exhibit a crucial issue, namely that entropic additivity of S_{BG} *is sufficient but not necessary* for the Einstein principle (3.78) to be satisfied. Indeed, let us consider the generalized functional S_q [18, 47, 87, 103, 104]. If A and B are two probabilistically independent systems, it follows that $\frac{S_q(A+B)}{k} = \frac{S_q(A)}{k} + \frac{S_q(B)}{k} + (1-q)\frac{S_q(A)}{k}\frac{S_q(B)}{k}$. Consequently S_q is additive if $q = 1$, and *nonadditive* otherwise. As we know, if probabilities are all equal, we have $S_q = k \ln_q W$ hence Eq. (3.79) is generalized into

$$\mathcal{W}(\{p_i\}) \propto e_q^{S_q(\{p_i\})/k}. \tag{3.80}$$

Once again, now using the q-generalized property $e_q^{x+y+(1-q)xy} = e_q^x e_q^y$, we easily verify Einstein's principle (3.78) *for an arbitrary value of the index q*! This exhibits the most important fact that *entropic additivity is not necessary for satisfying Einstein's 1910 crucial requirement within the foundations of statistical mechanics.*

As a final comment, let us mention that this remarkable property can not be further generalized for arbitrary $\{p_i\}$ if we maintain the hypothesis that the entropic functional be *both composable and trace-form*. Indeed, it has been proved that, in this sense, S_q is unique [236]. If we release the requirement of validity for arbitrary $\{p_i\}$, then infinitely many entropic functionals can satisfy Einstein principle (3.78), as illustrated in [235]. Similarly, if we release the requirement of trace-form, it can be straightforwardly verified that the non-trace-form Rényi entropic functional also satisfies (3.78). The trace-form has particular implications at the thermodynamical level [237]. This is also deeply related to the fact that, even if two entropic functionals lead to the same optimizing distribution (as it is clearly the case when one of these functionals is a monotonic function of the other one, and the constraints that are used within the optimization are exactly the same), their ultimate thermodynamical consequences are different [237]. For example, the Rényi functional and S_q are always related through the monotonic function $S_q^R/k = \frac{\ln[1+(1-q)S_q/k]}{1-q}$, but if a value for $q \neq 1$ exists, for a specific system, such that $S_q(N) \sim N$ (i.e., extensive), then necessarily $S_q^R \sim \ln N$ (i.e., nonextensive). In other words, S_q conforms to the Legendre structure of classical thermodynamics, whereas S_q^R violates it.

3.3.3 The q-Product

In relation to the pseudo-additive property (3.44) of S_q, it has been recently introduced (independently and virtually simultaneously) [238, 239] a generalization of the product, which is called *q-product*. It is defined as follows:

$$x \otimes_q y \equiv \left[x^{1-q} + y^{1-q} - 1\right]_+^{\frac{1}{1-q}} \quad (x \geq 0, \ y \geq 0), \tag{3.81}$$

or,[11] equivalently,

$$x \otimes_q y \equiv e_q^{\ln_q x + \ln_q y} \tag{3.82}$$

Let us list some of its main properties:

(i) It recovers the *standard product* as a particular instance, i.e.

$$x \otimes_1 y = xy; \tag{3.83}$$

(ii) It is *commutative*, i.e.

$$x \otimes_q y = y \otimes_q x; \tag{3.84}$$

(iii) It is *additive under q-logarithm* (hereafter referred to as *extensive*)[12] i.e.

---

[11] It is in fact easy to get rid of the requirement of nonnegativity of x and y through the following extended definition: $x \otimes_q y \equiv sign(x)sign(y)\left[|x|^{1-q} + |y|^{1-q} - 1\right]_+^{\frac{1}{1-q}}$. The correct $q = 1$ limit is obtained by using $sign(x)|x| = x$ (and similarly for y).

[12] The word "extensive" is used as a kind of reminder of the equal-probability entropic property $\ln_q(W_A \otimes_q W_B) = \ln_q W_A + \ln_q W_B$, hence $\ln_q(W_{A_1} \otimes_q W_{A_2} \otimes_q \cdots \otimes_q W_{A_N}) = \sum_{k=1}^N \ln_q W_{A_k}$, hence $\ln_q(W_A \otimes_q W_A \otimes \cdots \otimes_q W_A) = N \ln_q W_A$. However, if we focus on generic correlated probabilistic systems A ($i = 1, 2, \ldots, W_A$) and B ($j = 1, 2, \ldots, W_B$)) such as $p_{ij} = \frac{p_i \otimes_q p_j}{\sum_{i'=1}^{W_A} \sum_{j'=1}^{W_B} p_{i'} \otimes_q p_{j'}}$ or say such as $(1/p_{ij}) = \frac{(1/p_i) \otimes_q (1/p_j)}{\sum_{i'=1}^{W_A} \sum_{j'=1}^{W_B} (1/p_{i'}) \otimes_q (1/p_{j'})}$, a simple relation between $S_q(\{p_i\})$, $S_q(\{p_j\})$ and $S_q(\{p_{ij}\})$ remains up to now elusive. Notice however that in constructions similar to these ones, the total number of states is $W_A W_B$, which is in significant contrast with the number of states $W_{A+B} \equiv W_A \otimes_q W_B =$. Indeed, in the former, $W_A W_B$ always is an integer number given integer W_A and W_B. In the latter, only special combinations of (W_A, W_B, q) can produce an integer value for W_{A+B}. For instance, for $A = B$ we have $W_{A+A} = [2(W_A)^{1-q} - 1]^{1/(1-q)}$. Consequently, for say $q = 1/2$ we have that only for $W_A = n^2$, ($n = 1, 2, 3, \ldots$), we have the integer $W_{A+A} = (2n-1)^2$ (whereas $W_A^2 = n^4$); for say $q = 2/3$ we have that only for $W_A = n^3$, ($n = 1, 2, 3, \ldots$), we have the integer $W_{A+A} = (2n-1)^3$ (whereas $W_A^2 = n^6$), and so on. In general, W_{A+B} is to be understood as an analytical extension in terms of (W_A, W_B, q).

$$\ln_q (x \otimes_q y) = \ln_q x + \ln_q y, \tag{3.85}$$

whereas we remind that

$$\ln_q (x \, y) = \ln_q x + \ln_q y + (1 - q)(\ln_q x)(\ln_q y). \tag{3.86}$$

Consistently

$$e_q^x \otimes_q e_q^y = e_q^{x+y}, \tag{3.87}$$

whereas

$$e_q^x \, e_q^y = e_q^{x+y+(1-q)xy}; \tag{3.88}$$

(iv) It has a $(2 - q)$-*duality* / *inverse* property, i.e.

$$1/(x \otimes_q y) = (1/x) \otimes_{2-q} (1/y); \tag{3.89}$$

(v) It is *associative*, i.e.

$$x \otimes_q (y \otimes_q z) = (x \otimes_q y) \otimes_q z = x \otimes_q y \otimes_q z = (x^{1-q} + y^{1-q} + z^{1-q} - 2)^{1/(1-q)}; \tag{3.90}$$

(vi) It admits *unity*, i.e.

$$x \otimes_q 1 = x; \tag{3.91}$$

(vii) It admits *zero* under certain conditions, more precisely,

$$x \otimes_q 0 = \begin{cases} 0 & \text{if } (q \geq 1 \text{ and } x \geq 0) \quad \text{or if } (q < 1 \text{ and } 0 \leq x \leq 1) , \\ \left(x^{1-q} - 1 \right)^{\frac{1}{1-q}} & \text{if } q < 1 \text{ and } x > 1 ; \end{cases} \tag{3.92}$$

For a special range of q, e.g., $q = 1/2$, the argument of the q-product can attain negative values, specifically at points for which $|x|^{1-q} + |y|^{1-q} - 1 < 0$. In these cases, and consistently with the cutoff for the q-exponential, we have set $x \otimes_q y = 0$. With regard to the q-product domain, and restricting our analysis of Eq. (3.81) to $x, y > 0$, we observe that for $q \to -\infty$ the region $\{0 \leq x \leq 1, 0 \leq y \leq 1\}$ leads to a vanishing q-product. As the value of q increases, the area of the vanishing region decreases, and when $q = 0$ we have the limiting line given by $x + y = 1$, for which $x \otimes_0 y = 0$. Only for $q = 1$, the *whole set* of real values of x and y has a defined value for the q-product. For $q > 1$, definition (3.81) yields a curve, $|x|^{1-q} + |y|^{1-q} = 1$, at which the q-product diverges. This undefined region increases as q goes to infinity. At the $q \to \infty$ limit, the q-product is only defined in $\{x > 1, y \leq 1\} \cup \{0 \leq x \leq 1, 0 \leq y \leq 1\} \cup \{x \leq 1, y > 1\}$. This entire scenario is depicted on the panels of Fig. 3.10. The profiles presented by $x \otimes_\infty y$ and $x \otimes_{-\infty} y$ illustrate the above features. To illustrate the q-product in another simple form, we show, in Fig. 3.11, a representation of $x \otimes_q x$ for typical values of q;

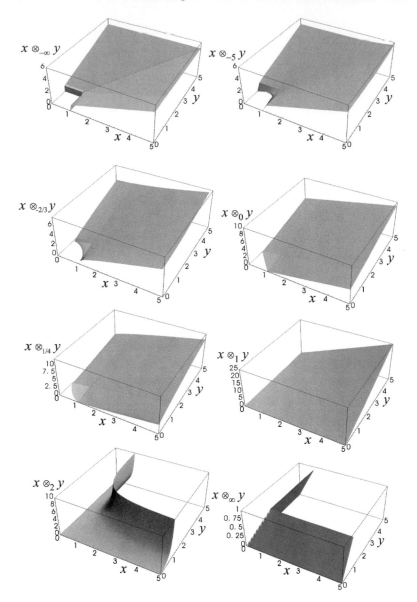

Fig. 3.10 Representation of the q-product, Eq. (3.81), for $q = -\infty, -5, -2/3, 0, 1/4, 1, 2, \infty$. As it is visible, the squared region $\{0 \leq x \leq 1, 0 \leq y \leq 1\}$ is gradually integrated into the nontrivial domain as q increases up to $q = 1$. From this value on, a new prohibited region appears, but this time coming from large values of $(|x|, |y|)$. This region reaches its maximum when $q = \infty$. In this case, the domain is composed of a horizontal and vertical strip of width 1. From [240]

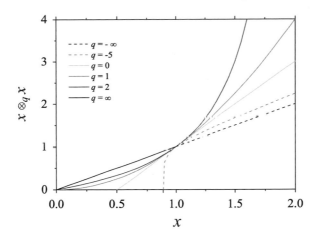

Fig. 3.11 Representation of the q-product, $x \otimes_q x$ for $q = -\infty, -5, 0, 1, 2, \infty$. Excluding $q = 1$, there is a special value $x^* = 2^{1/(q-1)}$, for which $q < 1$ represents the lower bound [in figure $x^* (q = -5) = 2^{-1/6} \simeq 0.89089$ and $x^* (q = 0) = 1/2$], and for $q > 1$ the upper bound [in figure $x^* (q = 2) = 2$]. For $q = \pm\infty$, $x \otimes_q x$ lies on the diagonal of bisection, but following the lower and upper limits mentioned above. From [240]

(viii) It satisfies

$$(x^q \otimes_{1/q} y^q)^{1/q} = x \otimes_{2-q} y, \tag{3.93}$$

or, equivalently,

$$x \otimes_{1/q} y = (x^{1/q} \otimes_{2-q} y^{1/q})^q ; \tag{3.94}$$

(ix) By q-multiplying n equal factors we can define the $n - th$ q-power as follows:

$$x^{\otimes_q^n} \equiv x \otimes_q x \otimes_q \dots \otimes_q x = [nx^{1-q} - (n-1)]^{1/(1-q)}, \tag{3.95}$$

which immediately suggests the following analytical extension

$$x^{\otimes_q^y} \equiv [yx^{1-q} - (y-1)]^{1/(1-q)}, \tag{3.96}$$

where both x can be y can be real numbers (with $y(x^{1-q} - 1) \geq -1$). From this, an interesting, *extensive*-like property follows, namely

$$\ln_q (x^{\otimes_q^y}) = y \ln_q x. \tag{3.97}$$

It will gradually become clear that the peculiar mathematical structure associated with the q-product appears to be at the "heart" of the *nonadditive* entropy S_q (which is nevertheless *extensive* for a special class of correlations) and its associated statistical mechanics (see also [241]).

3.3.4 The q-Sum and Related Issues

Analogously to the q-product we can define the q-sum

$$x \oplus_q y \equiv x + y + (1-q)xy. \tag{3.98}$$

It has the following properties:

(i) It recovers the *standard sum* as a particular instance, i.e.

$$x \oplus_1 y = x + y; \tag{3.99}$$

(ii) It is *commutative*, i.e.

$$x \oplus_q y = y \oplus_q x; \tag{3.100}$$

(iii) It is *multiplicative under q-exponential*, i.e.

$$e_q^{x \oplus_q y} = e_q^x \, e_q^y; \tag{3.101}$$

(iv) It is *associative*, i.e.

$$x \oplus_q (y \oplus_q z) = (x \oplus_q y) \oplus_q z = x \oplus_q y \oplus_q z$$
$$= x + y + z + (1-q)(xy + yz + zx) + (1-q)^2 xyz; \tag{3.102}$$

(v) It admits *zero*, i.e.

$$x \oplus_q 0 = x; \tag{3.103}$$

(vi) By q-summing n equal terms we obtain:

$$x^{\oplus_q^n} \equiv x \oplus_q x \oplus_q \ldots \otimes_q x = nx\left[\sum_{i=0}^{n-2}(1-q)^i x^i\right] + (1-q)^{n-1}x^n \ (n = 2, 3, \ldots); \tag{3.104}$$

(vii) It satisfies the following generalization of the *distributive* property of standard sum and product, i.e., of $a(x + y) = ax + ay$:

$$a(x \oplus_q y) = (ax) \oplus_{\frac{q+a-1}{a}} (ay); \tag{3.105}$$

(viii) Interesting *cross* properties emerge from the q-generalizations of the product and of the sum, for instance

$$\ln_q(x\ y) = \ln_q x \oplus_q \ln_q y, \tag{3.106}$$
$$\ln_q(x \otimes_q y) = \ln_q x + \ln_q y, \tag{3.107}$$

and, consistently,[13]

$$e_q^{x+y} = e_q^x \otimes_q e_q^y, \tag{3.108}$$

$$e_q^{x \oplus_q y} = e_q^x e_q^y, \tag{3.109}$$

hich, in a different context, repeats Eq. (3.101).

Let us make, at this point, a mathematical digression. If, to the relation $\ln(xy) = \ln x + \ln y$, we add relations 3.106 and 3.107, we feel tempted to find out whether a further generalized logarithmic function exists which would elegantly unify all of them in the form

$$\ln_{q,q'}(x \otimes_q y) = \ln_{q,q'} x \oplus_{q'} \ln_{q,q'} y. \tag{3.110}$$

It turns out that it *does* exist, and is given by [242]

$$\ln_{q,q'} x \equiv \ln_{q'} e^{\ln_q x} = \frac{1}{1-q'} \left[\exp\left(\frac{1-q'}{1-q} (x^{1-q} - 1) \right) - 1 \right]. \tag{3.111}$$

The relations $\ln_{q,1} x = \ln_{1,q} x = \ln_q x$ are easily recovered by evaluating Eq. (3.111) in the limits $q \to 1$ and $q' \to 1$. From Eq. (3.111), the inverse function $e_{q,q'}^x$ can be easily obtained as well [242].

The two-parameter generalized logarithm (3.111) has been further generalized into the following three-parameter one [243, 244]:

$$\ln_{q,q',r} x = \frac{1}{1-r} \left[\exp\left(\frac{1-r}{1-q'} (e^{(1-q')\ln_q x} - 1) - 1 \right) \right]. \tag{3.112}$$

We easily verify that $\ln_{q,q',1} x = \ln_{q,q'} x$. From this further generalized logarithm, it is straightforwardly defined the *Corcino-Corcino entropy*

$$S_{q,q',r}^{CC} = k \sum_i p_i \ln_{q,q',r} \frac{1}{p_i} ; \tag{3.113}$$

(ix) Let us mention here an interesting issue. Does an *associative* generalized sum $x \oplus^{(q)} y$ exist such as a q-generalized distributivity like the following holds?

$$x \otimes_q (y \oplus^{(q)} z) = (x \otimes_q y) \oplus^{(q)} (x \otimes_q z). \tag{3.114}$$

The answer is positive [238, 245] with the following definition:

---

[13] While both the q-sum and the q-product are mathematically interesting structures, they play a quite different role within the deep structure of the nonextensive theory. The q-product reflects an essential property, namely the extensivity of the entropy in the presence of special global correlations. The q-sum instead only reflects how the entropies would compose *if* the subsystems were independent, even if we know that in such a case we only actually need $q = 1$.

$$x \oplus^{(q)} y \equiv e_q^{\ln [e^{\ln_q x} + e^{\ln_q y}]} = \left\{ 1 + (1-q) \ln \left[e^{\frac{x^{1-q}-1}{1-q}} + e^{\frac{y^{1-q}-1}{1-q}} \right] \right\}^{1/(1-q)}.$$

$$(3.115)$$

The fact that this generalized sum is associative opens the door for having a q-generalized algebra. Analogously, does an *associative* generalized product $x \otimes^{(q)} y$ exist such as a q-generalized distributivity like the following holds?

$$x \otimes^{(q)} (y \oplus_q z) = (x \otimes^{(q)} y) \oplus_q (x \otimes^{(q)} z). \qquad (3.116)$$

As before, the answer is positive [238, 245] with the following definition:

$$x \otimes^{(q)} y \equiv \ln_q [e^{(\ln e_q^x)(\ln e_q^y)}] = \frac{e^{\frac{\left[\ln[1+(1-q)x]\right]\left[\ln[1+(1-q)y]\right]}{1-q}} - 1}{1-q}. \qquad (3.117)$$

Let us summarize the above by reminding that, for $q \neq 1$, the generalized product $x \otimes_q y$ is *not* distributive with regard to the generalized sum $x \oplus_q y$ but only with regard to the generalized sum $x \oplus^{(q)} y$, and the generalized product $x \otimes^{(q)} y$ is *not* distributive with regard to the generalized sum $x \oplus^{(q)} y$ but only with regard to the generalized sum $x \oplus_q y$;

(x) Let us finally focus on prime numbers [246]. As shown in [245], the q-number $\{n\}_q$ algebraically consistent with the q-product $m \otimes_q n = \left(m^{1-q} + n^{1-q} - 1 \right)^{\frac{1}{1-q}}$ is given by

$$\{n\}_q = e^{\ln_q n}, \qquad (3.118)$$

hence $\{n\}_1 = n$. It can be straightforwardly proved that

$$\{m\}_q \{n\}_q = \{m \otimes_q n\}_q, \qquad (3.119)$$

where we are now assuming that both m and n are natural numbers. All q-numbers satisfy the trivial factorization $\{n\}_q = \{n\}_q \{1\}_q$, but other factorizations are in many cases possible, e.g., $\{6\}_q = \{2\}_q \{3\}_q$. Those q-numbers $\{n\}_q$ with natural n that only admit the trivial factorization will be called q-*prime numbers*. With this definition, we immediately see a remarkable property: *the structure of prime numbers is q-invariant*! This leads to a natural q-generalization of the Riemann ζ function, namely

$$\zeta_q(s) \equiv \prod_{p \, prime} \frac{1}{1 - \{p\}_q^{-s}} = \frac{1}{1 - \{2\}_q^{-s}} \frac{1}{1 - \{3\}_q^{-s}} \frac{1}{1 - \{5\}_q^{-s}} \cdots, \qquad (3.120)$$

which recovers, for $q = 1$, Eq. (3.310).

3.3.5 Extensivity of S_q—Effective Number of States

Suppose we are composing the discrete states of two subsystems A and B, whose total numbers of states are respectively $W_A \geq 1$ and $W_B \geq 1$. To be more specific, $W_A(W_B)$ is the total number of states of A (B) whose associated probability is *not zero*. If A and B are (strictly or nearly) independent, the total number of *nonvanishing-probability* states of the system $A + B$ is then

$$W_{A+B} = W_A W_B. \tag{3.121}$$

For example, a set of a head-tail coin and a cubic dice has $2 \times 6 = 12$ possibilities. However, it might happen that strong correlations between A and B yield $W_{A+B} < W_A W_B$. Consequently, it will nearly always be[14]

$$W_{A+B} \leq W_{A+B}. \tag{3.122}$$

As said above, if A and B are independent, then the equality holds. The opposite is not true: correlation might exist between A and B and, nevertheless, the equality be satisfied. It is not, however, this kind of (weak) correlation that we are focusing on here. Our focus is on a *special* type of (strong) correlation, which *necessarily* decreases the number of joint states whose probability differs from zero. For example, let us focus on a correlation such that

$$W_{A+B} = W_A \otimes_q W_B = (W_A^{1-q} + W_B^{1-q} - 1)^{1/(1-q)} \quad (q \leq 1). \tag{3.123}$$

We can verify that, for $W_A > 1$ and $W_B > 1$, $W_{A+B}/W_A W_B$ monotonically decreases from unity to $1/\min\{W_A, W_B\}$ when q decreases from unity to $-\infty$.

Let us generalize the above to N subsystems $A_1, A_2, ..., A_N$ (they typically are the elements of the system) whose numbers of states (with nonzero probabilities) are respectively $W_{A_1}, W_{A_2}, ..., W_{A_N}$ with $W_{A_r} > 1$, $\forall r$. Then, it will nearly always be

$$W_{A_1+A_2+...+A_N} \leq \prod_{r=1}^{N} W_{A_r}. \tag{3.124}$$

Let us consider now the following strongly correlated case:

---

[14] Exceptional behavior, with $W_{A+B} > W_A W_B$, is nevertheless possible if the joint probability reflects the appearance of an enlarged (*exploding* or *superexponential*) phase space, e.g., $W(N) \sim N!$, or $W(N) \sim 2^{2^N}$. Such anomalous systems, focused on in [247–252], are sort of paradigmatic in what is occasionally referred to as *ontic openness* in biology, economics, culture, and other developmental processes [253, 254].

$$W_{A_1+A_2+\ldots+A_N} = W_{A_1} \otimes_q W_{A_2} \otimes_q \cdots \otimes_q W_{A_N}$$

$$= \left[\left(\sum_{r=1}^{N} W_{A_r}^{1-q} \right) - (N-1) \right]^{\frac{1}{1-q}}. \qquad (3.125)$$

We verify that $W_{A_1+A_2+\ldots+A_N} / \left[\prod_{r=1}^{N} W_{A_r} \right]$ monotonically decreases from unity to $1/\min\{W_{A_1}, W_{A_2}, \ldots, W_{A_N}\}$ when q decreases from unity to $-\infty$.

A frequent and important case is that in which the N subsystems are all equal (hence $W_{A_r} = W_{A_1} \equiv W_1$, $\forall r$). In such a case, we have

$$W(N) = [NW_1^{1-q} - (N-1)]^{1/(1-q)} \le W_1^N \quad (q \le 1), \qquad (3.126)$$

where the notation $W(N)$ is self-explanatory. Hence the following *suggestive* result is straightforwardly verified:

$$\ln_q[W(N)] = N \ln_q W_1, \ \forall q. \qquad (3.127)$$

If $q = 1$, $W(N) = W_1^N$, and this result is consistent with the well-known additivity of S_{BG} (and extensivity for such a system), i.e., $S_{BG}(N) = N S_{BG}(1)$ for the case of equal probabilities. Indeed, in the $q = 1$ case, the hypothesis of *simultaneously having equal probabilities* in each of the N equal subsystems as well as in the total system is *admissible*: the probability of each state of any single subsystem is $1/W_1$, and the probability of each state of the entire system is $1/W(N) = 1/W_1^N$.

The situation is more complex for $q \ne 1$, and here we focus on $q < 1$. Indeed, it appears (as we shall verify in the next Subsection) that, in such cases, a few or many of the states of the entire system become *forbidden* (in the sense that their corresponding probabilities vanish), either for finite N or in the limit $N \to \infty$. This is precisely why $W(N) < W_1^N$, thus being *subexponential*. So, if we assume that all states of each subsystem are *equally probable* (with probability $1/W_1$), then the states of the entire system are *not*. Reciprocally, if we assume that the *allowed* states of the entire system are *equally probable* (with probability $1/W(N) > 1/W_1^N$), then the states of each of the subsystems are *not*. We see here the seed of dynamical nonmixing and/or nonergodicity, hence the failure, for systems of this sort, of the BG statistical mechanical basic hypothesis.

Let us first consider the possibility in which the states of *each subsystem* are equally probable. Then $k \ln_q W_1$ is the entropy $S_q(1)$ associated with *one* subsystem. In other words Eq. (3.97) implies

$$k \ln_q[W(N)] = N S_q(1). \qquad (3.128)$$

Let us then consider the other possibility, namely that in which it is the allowed states of the *entire system* that are equally probable. Then $k \ln_q W(N)$ is the entropy $S_q(N)$ associated with the entire system. In other words Eq. (3.97) implies

$$S_q(N) = Nk \ln_q W_1. \tag{3.129}$$

We are now very close to answer a crucial question: *Can S_q for $q \neq 1$ generically be strictly or asymptotically proportional to N in the presence of these strong correlations, i.e., can it be extensive?* The examples that we present next exhibit that the answer is *yes*. By *generically*, we refer to the most common case, in which neither the states of each subsystem are equally probable, nor the allowed states of the entire system are equally probable. This is what we address in the next Section.

But before that, let us summarize the knowledge that we acquired in the present Subsection. We assume, for simplicity, that the $W(N)$ joint nonvanishing-probability states of a system are equally probable. Then [255]:

(i) *If* we are facing the family

$$W(N) \sim A\mu^N \quad (A > 0, \ \mu > 1, \ N \to \infty), \tag{3.130}$$

the basic entropy which is *extensive* is $S_{BG}(N) = k \ln W(N)$, i.e., $\lim_{N\to\infty} S_{BG}(N)/N = k \ln \mu \in (0, \infty)$. The nonadditive entropy S_q for $q \neq 1$ is, in contrast, *nonextensive*. It is primarily systems like this that are addressed within the BG scenario.

(ii) *If* we are facing the family

$$W(N) \sim BN^\rho \quad (B > 0, \ \rho > 0, \ N \to \infty), \tag{3.131}$$

the basic entropy which is *extensive* is $S_q(N) = k \ln_q W(N) \propto N^{\rho(1-q)}$ with

$$q = 1 - \frac{1}{\rho}, \tag{3.132}$$

i.e., $\lim_{N\to\infty} S_{1-(1/\rho)}(N)/N = kB^{1-q}/(1-q)$ is *finite*. For any other value of q (*including $q = 1$!*), S_q is *nonextensive* (e.g., $S_{BG} \sim k\rho \ln N$). It is primarily systems like this that are addressed within the nonextensive scenario.[15] This remark may be considered as some sort of *golden* reason for the present generalization of BG statistical mechanics!

(iii) *If* we are facing the family

$$W(N) \sim A\mu^{N^\gamma} \quad (A > 0, \ \mu > 1, \ 0 < \gamma < 1, \ N \to \infty), \tag{3.133}$$

---

[15] In our present examples, N typically is the *total* number of elements. But, as we shall see later in some applications related to quantum entanglement, the system whose entropy we are interested in might be *part* of a substantially larger system. In such a case, the expression *block entropy* is commonly used in the literature.

neither $S_{BG}(N)$ nor $S_q(N)$ $(q \neq 1)$ can be extensive.[16] This peculiar situation demands a special approach, which we briefly outline here. Let us consider the following entropic functional:

$$S_\delta(\{p_i\}) = k \sum_{i=1}^{W} p_i \left(\ln \frac{1}{p_i} \right)^\delta \quad (\delta \geq 1; \; S_1 = S_{BG}), \qquad (3.134)$$

whose equal-probability expression is given by

$$S_\delta = k(\ln W)^\delta. \qquad (3.135)$$

If we use for W the expression $W(N) \sim C N^\rho \mu^{N^\gamma}$ $(C > 0; \; \mu > 1; \; \rho \in \mathbb{R}$ if $0 < \gamma < 1$, and $\rho \leq 0$ if $\gamma = 1)$, we immediately verify that $S_{\delta=1/\gamma}(N) \sim k(\ln^{1/\gamma} \mu) N$, hence *extensive*, $\forall(C, \rho, \mu, \gamma)$. It can be straightforwardly verified that $S_\delta(\{p_i\})$ is nonnegative, expansible, concave for $0 < \delta \leq \delta_c(W)$ [256], where $\delta_c(W) > 1$ depends nontrivially on W; also, if $\delta \neq 1$, $S_\delta(\{p_i\})$ is nonadditive. Following along the path that we have used for S_q, it is clear that we can rewrite $S_\delta(\{p_i\}) = k \sum_{i=1}^{W} p_i \ln_\delta(1/p_i)$ with the δ-logarithmic function $\ln_\delta z \equiv (\ln z)^\delta$, hence, its inverse, the δ-exponential function $e_\delta^z \equiv e^{z^\delta}$. We may consistently define the δ-sum $x \oplus_\delta y \equiv (x^{1/\delta} + y^{1/\delta})^\delta$, and the δ-product $x \otimes_\delta y \equiv e^{[(\ln x)^\delta + (\ln y)^\delta]^{1/\delta}}$. We followed here along the lines of [103, 104], where we first introduced the entropic functional (3.134). Though with a different motivation, the same form was, simultaneously and independently as far as we know, introduced also in [257]. The functional form (3.134) was concretely applied, for the first time to a specific physical problem, in [222], namely for black holes. It has also been recently taken into consideration for various cosmological issues [258–269].

(iv) *If* we are facing the family

$$W(N) \sim D \ln N \; (D > 0, \; N \to \infty), \qquad (3.136)$$

neither $S_q(N)$ $(\forall q)$ nor $S_\delta(N)$ $(\forall \delta)$ can be extensive. Let us consider then [270]

$$S_\lambda^C(\{p_i\}) = k \left[e^{\lambda e^{\sum_{i=1}^{W} p_i \ln(1/p_i)}} - e^\lambda \right] \; (\lambda > 0), \qquad (3.137)$$

which is nonadditive and concave, and whose equal-probability expression is given by

$$S_\lambda^C = k \left[e^{\lambda W} - e^\lambda \right], \qquad (3.138)$$

---

[16] The families (i), (ii) and (iii) can be unified through the form $W(N) \sim A \left(1 + \frac{\ln \mu}{\rho} N^\gamma \right)^\rho = A e_{1-1/\rho}^{(\ln \mu) N^\gamma}$ $(A > 0, \mu > 1, 0 < \gamma \leq 1,$ and $1/\rho \geq 0)$. If $(1/\rho, \gamma) = (0, 1)$ we obtain $W(N) \sim A \mu^N$, i.e., family (i). If $1/\rho > 0$ and $\gamma = 1$ we obtain $W(N) \sim B N^\rho$ with $B \equiv A \left(\frac{\ln \mu}{\rho} \right)^\rho$, i.e., family (ii). If $1/\rho = 0$ and $0 < \gamma < 1$ we obtain $W(N) \sim A \mu^{N^\gamma}$, i.e., family (iii).

where C stands for *Curado*. If we use now Eq. (3.136), we obtain $S_\lambda(N) \sim kN^{\lambda D}$, hence $S_{\lambda=1/D}(N)$ is extensive, as thermodynamically required.[17] Notice that, in contrast with S_q and S_δ, S_λ^C is not trace-form and does not recover S_{BG} as a particular instance. Also, S_λ^C differs from another Curado entropy, namely S_b^C defined in Eq. (3.293).

Through the four illustrative families discussed here above, it becomes clear that S_{BG}, even being the most basic and historical entropic functional, can by no means cover all possible situations if thermodynamical extensivity is indeed required for the entropy. The situation is summarized in Fig. 3.12.

3.3.6 Extensivity of S_q—Binary Systems

We wish to address here, from a different perspective, that issue of central importance in statistical mechanics and thermodynamics, namely the extensivity of the entropy [271–275]. Let us start with the simple case of a system composed of N distinguishable subsystems, each of them characterized by a binary random variable.

N binary subsystems

If $N = 1$, we shall note p_1^A and p_2^A the probabilities of states 1 and 2 respectively. Of course, they satisfy $p_1^A + p_2^A = 1$.

If $N = 2$, we shall note p_{11}^{A+B}, p_{12}^{A+B}, p_{21}^{A+B} and p_{22}^{A+B} the corresponding joint probabilities. Of course, they satisfy $p_{11}^{A+B} + p_{12}^{A+B} + p_{21}^{A+B} + p_{22}^{A+B} = 1$. See Table 3.2.

If $N = 3$, we shall note p_{111}^{A+B+C}, p_{112}^{A+B+C}, p_{121}^{A+B+C}, ..., and p_{222}^{A+B+C} the corresponding joint probabilities. Of course, they satisfy $p_{111}^{A+B+C} + ... + p_{222}^{A+B+C} = 1$. See Table 3.3.

The joint probabilities corresponding to the general case are noted $p_{11...1}^{A_1+A_2+...+A_N}$, $p_{11...2}^{A_1+A_2+...+A_N}$, ..., and $p_{22...2}^{A_1+A_2+...+A_N}$. They satisfy

$$\sum_{i_1,i_2,...,i_N=1,2} p_{i_1 i_2...i_N}^{A_1+A_2+...+A_N} = 1 \quad (N = 1, 2, 3, ...), \tag{3.139}$$

and can be represented as a N-dimensional $2 \times 2 \times ... \times 2$ hypercube, which is associated with 2^N states. There are N sets of marginal probabilities where we have

---

[17] The entropy S_λ^C can be generalized as follows: $S_{\lambda,q}(\{p_i\}) = k\left[e^{\lambda e_q^{\sum_{i=1}^W p_i \ln_q (1/p_i)}} - e^\lambda\right]$ (with $S_{\lambda,1} = S_\lambda^C$), which is nonadditive, and whose equal-probability expression is Eq. (3.138), $\forall q$. Consequently, $S_{\lambda,q}(N)$ is extensive for $\lambda = 1/D$. Like Renyi entropy S_q^R, this entropy is a monotonic function of S_q, $\forall q$, and concave only within $0 < q \leq q_c \simeq 1$ (we remind that, for S_q^R, $q_c = 1$, $\forall W$). A trivial further generalization along this same line is possible, namely $S_{\lambda,q,q'}(\{p_i\}) = k\left[e_{q'}^{\lambda e_q^{\sum_{i=1}^W p_i \ln_q (1/p_i)}} - e_{q'}^\lambda\right]$, whose equal-probability expression is given by $S_{\lambda,q,q'}^C = k\left[e_{q'}^{\lambda W} - e_{q'}^\lambda\right]$, $\forall q$. However, at the present stage, further analysis of such mathematical possibility presents no particular interest.

SYSTEMS $W(N)$ [equiprobable]	ENTROPY S_{BG} (ADDITIVE)	ENTROPY S_q $(q \neq 1)$ (NONADDITIVE)	ENTROPY S_δ $(\delta \neq 1)$ (NONADDITIVE)	ENTROPY S_λ^C $(\lambda > 0)$ (NONADDITIVE)
$\sim A\,\mu^N$ $(A>0, \mu>1)$	EXTENSIVE	NONEXTENSIVE	NONEXTENSIVE	NONEXTENSIVE
$\sim B\,N^\rho$ $(B>0, \rho>0)$	NONEXTENSIVE	EXTENSIVE $(q=1-1/\rho)$	NONEXTENSIVE	NONEXTENSIVE
$\sim C\nu^{N^\gamma}$ $(C>0, \nu>1, 0<\gamma<1)$	NONEXTENSIVE	NONEXTENSIVE	EXTENSIVE $(\delta=1/\gamma)$	NONEXTENSIVE
$\sim D\,\ln N$ $(D>0)$	NONEXTENSIVE	NONEXTENSIVE	NONEXTENSIVE	EXTENSIVE $(\lambda=1/D)$

Fig. 3.12 Typical behaviors of $W(N)$ (number of *nonzero-probability states* of a system with N random variables) in the $N \to \infty$ limit and entropic functionals which, under the assumption of *equal probabilities* for all states with nonzero probability, yield extensive entropies for specific values of the corresponding (nonadditive) entropic indices. In what concerns the exponential class $W(N) \sim A\mu^N$, S_{BG} is not the unique entropy that yields entropic extensivity; the (additive) Renyi entropic functional S_q^R also is extensive for all values of q. Analogously, in what concerns the stretched exponential class $W(N) \sim C\nu^{N^\gamma}$, the (nonadditive) entropic functional S_δ is not unique; indeed, the Hanel-Thurner family $S_{c,d}^{HT}$ can also yield extensivity for a special couple of indices (c, d). All the entropic families illustrated in this Table contain S_{BG} as a particular case, excepting S_λ^C, which is appropriate for the logarithmic class $W(N) \sim D \ln N$. In the limit $N \to \infty$, the inequalities $\mu^N \gg \nu^{N^\gamma} \gg N^\rho \gg \ln N \gg 1$ are satisfied, hence $\lim_{N\to\infty} \nu^{N^\gamma}/\mu^N = \lim_{N\to\infty} N^\rho/\mu^N = \lim_{N\to\infty} \ln N/\mu^N = 0$. This exhibits that, in all these nonadditive cases, the occupancy of the full phase space corresponds to zero Lebesgue measure, similarly to (multi) fractals. If the *equal probabilities* hypothesis is not satisfied, the results might be different (see, for instance, the discussion Eq. (3.165))

summed over *one* subsystem. They are noted $p_{i_2...i_N}^{(A_1)+A_2+...+A_N}$, $p_{i_1i_3...i_N}^{A_1+(A_2)+...+A_N}$, ..., and $p_{i_1i_2...i_{N-1}}^{A_1+A_2+...+A_N}$. There are $N(N-1)/2$ sets of marginal probabilities where we have summed over *two* subsystems. And so on.

For future use, let us right away introduce the notation corresponding to the most general case of N distinguishable discrete subsystems. Subsystem A_r is assumed to have W_r states $(r = 1, 2, ..., N)$. The *joint* probabilities for the whole system are $\left\{ p_{i_1i_2...i_N}^{A_1+A_2+...+A_N} \right\}$, such that

$$\sum_{i_1=1}^{W_1} \sum_{i_2=1}^{W_2} \cdots \sum_{i_N=1}^{W_N} p_{i_1i_2...i_N}^{A_1+A_2+...+A_N} = 1 \quad (N = 1, 2, 3, ...) . \tag{3.140}$$

Table 3.2 Two binary subsystems A and B: *joint* probabilities p_{11}^{A+B}, p_{12}^{A+B}, p_{21}^{A+B} and p_{22}^{A+B}, and *marginal* probabilities $p_1^{A+(B)}$, $p_2^{A+(B)}$, $p_1^{(A)+B}$ and $p_2^{(A)+B}$

A	B		
	1	2	
1	p_{11}^{A+B}	p_{12}^{A+B}	$p_1^{A+(B)} \equiv$ $p_{11}^{A+B} + p_{12}^{A+B}$
2	p_{21}^{A+B}	p_{22}^{A+B}	$p_2^{A+(B)} \equiv$ $p_{21}^{A+B} + p_{22}^{A+B}$
	$p_1^{(A)+B} \equiv$ $p_{11}^{A+B} + p_{21}^{A+B}$	$p_2^{(A)+B} \equiv$ $p_{12}^{A+B} + p_{22}^{A+B}$	1

Table 3.3 Three binary subsystems: *joint* probabilities p_{ijk}^{A+B+C} ($i, j, k = 1, 2$). The quantities without (within) square brackets [ ] correspond to state 1 (state 2) of subsystem C. The *marginal* probabilities where we have summed over the states of B are defined as indicated in the Table. The *marginal* probabilities where we have summed over the states of A are defined as follows: $p_{11}^{(A)+B+C} \equiv p_{111}^{A+B+C} + p_{211}^{A+B+C}$, $p_{21}^{(A)+B+C} \equiv p_{121}^{A+B+C} + p_{221}^{A+B+C}$, $p_{12}^{(A)+B+C} \equiv p_{112}^{A+B+C} + p_{212}^{A+B+C}$ and $p_{22}^{(A)+B+C} \equiv p_{122}^{A+B+C} + p_{222}^{A+B+C}$. The *marginal* probabilities where we have summed over the states of *both* A and B are defined as follows: $p_1^{(A)+(B)+C} \equiv p_{111}^{A+B+C} + p_{121}^{A+B+C} + p_{211}^{A+B+C} + p_{221}^{A+B+C}$ and $p_2^{(A)+(B)+C} \equiv p_{112}^{A+B+C} + p_{122}^{A+B+C} + p_{212}^{A+B+C} + p_{222}^{A+B+C}$. Of course, $p_1^{(A)+(B)+C} + p_2^{(A)+(B)+C} = 1$

A	B		
	1	2	
1	p_{111}^{A+B+C} $\left[p_{112}^{A+B+C}\right]$	p_{121}^{A+B+C} $\left[p_{122}^{A+B+C}\right]$	$p_{11}^{A+(B)+C} \equiv$ $p_{111}^{A+B+C} + p_{121}^{A+B+C}$ $\left[p_{12}^{A+(B)+C} \equiv\right.$ $\left.p_{112}^{A+B+C} + p_{122}^{A+B+C}\right]$
2	p_{211}^{A+B+C} $\left[p_{212}^{A+B+C}\right]$	p_{221}^{A+B+C} $\left[p_{222}^{A+B+C}\right]$	$p_{21}^{A+(B)+C} \equiv$ $p_{211}^{A+B+C} + p_{221}^{A+B+C}$ $\left[p_{22}^{A+(B)+C} \equiv\right.$ $\left.p_{212}^{A+B+C} + p_{222}^{A+B+C}\right]$
	$p_{11}^{(A)+B+C}$ $\left[p_{12}^{(A)+B+C}\right]$	$p_{21}^{(A)+B+C}$ $\left[p_{22}^{(A)+B+C}\right]$	$p_1^{(A)+(B)+C}$ $\left[p_2^{(A)+(B)+C}\right]$

These probabilities can be represented in a $W_1 \times W_2 \times \ldots \times W_N$ hypercube. The *marginal* probabilities are obtained by summing over the states of at least one subsystem. For example, $p_{i_2 i_3 \ldots i_N}^{(A_1)+A_2+\ldots+A_N} \equiv \sum_{i_1=1}^{W_1} p_{i_1 i_2 \ldots i_N}^{A_1+A_2+\ldots+A_N}$, $p_{i_3 i_4 \ldots i_N}^{(A_1)+(A_2)+\ldots+A_N} \equiv \sum_{i_1=1}^{W_1} \sum_{i_2=1}^{W_2} p_{i_1 i_2 \ldots i_N}^{A_1+A_2+\ldots+A_N}$, and so on. The binary case that we introduced above corresponds of course to the particular case $W_r = 2$ ($\forall r$). Let us go now back to it.

Let us assume the simple case in which all N binary subsystems are equal. Tables 3.2 and 3.3 then become Tables 3.4 and 3.5, respectively.

Table 3.4 Two equal binary subsystems A and B. *Joint* probabilities r_{20}, r_{21} and r_{22}, with $r_{20} + 2r_{21} + r_{22} = 1$

A	B		
	1	2	
1	$r_{20} \equiv p_{11}^{A+B}$	$r_{21} \equiv p_{12}^{A+B} = p_{21}^{A+B}$	$r_{20} + r_{21}$
2	r_{21}	$r_{22} \equiv p_{22}^{A+B}$	$r_{21} + r_{22}$
	$r_{20} + r_{21}$	$r_{21} + r_{22}$	1

Table 3.5 Three equal binary subsystems. *Joint* probabilities r_{30}, r_{31}, r_{32} and r_{33}, with $r_{30} + 3r_{31} + 3r_{32} + r_{33} = 1$. The quantities without (within) square brackets [ ] correspond to state 1 (state 2) of subsystem C

A	B		
	1	2	
1	$r_{30} \equiv p_{111}^{A+B+C}$	$r_{31} \equiv p_{121}^{A+B+C} =$ $p_{211}^{A+B+C} = p_{112}^{A+B+C}$	$r_{30} + r_{31}$
	$[r_{31}]$	$[r_{32} \equiv p_{122}^{A+B+C} =$ $p_{212}^{A+B+C} = p_{221}^{A+B+C}]$	$[r_{31} + r_{32}]$
2	r_{31}	r_{32}	$r_{31} + r_{32}$
	$[r_{32}]$	$[r_{33} \equiv p_{222}^{A+B+C}]$	$[r_{32} + r_{33}]$
	$r_{30} + r_{31}$	$r_{31} + r_{32}$	$r_{30} + 2r_{31} + r_{32}$
	$[r_{31} + r_{32}]$	$[r_{32} + r_{33}]$	$[r_{31} + 2r_{32} + r_{33}]$

The general case of N equal subsystems has the joint probabilities $\{r_{Nn}\}$ ($n = 0, 1, 2, ..., N$), which satisfy

$$\sum_{n=0}^{N} \frac{N!}{(N-n)!\,n!} r_{Nn} = 1 \quad (N = 1, 2, 3, ...) . \qquad (3.141)$$

The probability r_{Nn} equals *all* the $\frac{N!}{(N-n)!\,n!}$ joint probabilities $\{p_{i_1 i_2 ... i_N}^{A_1 + A_2 + ... + A_N}\}$ that are associated with $(N - n)$ subsystems in state 1 and n subsystems in state 2, *in whatever order*.[18]

The instance of the N subsystems being equal admits a representation which is much simpler than the hypercubic one used up to now. They admit a "triangular" representation: see Table 3.6.

---

[18] If we consider outcomes 1 and 2 in a specific order, we can think of them as being a time series. In such a case, for say $N = 3$, the probabilities p_{112}, p_{121} and p_{211}, might not coincide due to memory effects. If they did, that would be a case in which we have no memory of their order of appearance. Within this interpretation, the case we are addressing above would correspond to having memory of how many $1's$ and $2's$ we have, but *not* having memory of their order. If we have no memory at all, that would correspond to *equal probabilities*, i.e., $r_{Nn} = 1/2^N$, $\forall n$. Normalization of these probabilities is in this case preserved through $\sum_{n=0}^{N} \frac{N!}{(N-n)!\,n!} = 2^N$.

Table 3.6 Merging of the Pascal triangle (*left* member of each pair) with the probabilities $\{r_{Nn}\}$ (*right* member of each pair) associated with N equal subsystems

$(N = 0)$	$(1, 1)$				
$(N = 1)$	$(1, r_{10})$	$(1, r_{11})$			
$(N = 2)$	$(1, r_{20})$	$(2, r_{21})$	$(1, r_{22})$		
$(N = 3)$	$(1, r_{30})$	$(3, r_{31})$	$(3, r_{32})$	$(1, r_{33})$	
$(N = 4)$	$(1, r_{40})$	$(4, r_{41})$	$(6, r_{42})$	$(4, r_{43})$	$(1, r_{44})$

Table 3.7 Merging of the Pascal triangle (the set of all *left* members) with the Leibnitz triangle [276] (the set of all *right* members) associated with N equal subsystems

$(N = 0)$	$(1, 1)$				
$(N = 1)$	$(1, \frac{1}{2})$	$(1, \frac{1}{2})$			
$(N = 2)$	$(1, \frac{1}{3})$	$(2, \frac{1}{6})$	$(1, \frac{1}{3})$		
$(N = 3)$	$(1, \frac{1}{4})$	$(3, \frac{1}{12})$	$(3, \frac{1}{12})$	$(1, \frac{1}{4})$	
$(N = 4)$	$(1, \frac{1}{5})$	$(4, \frac{1}{20})$	$(6, \frac{1}{30})$	$(4, \frac{1}{20})$	$(1, \frac{1}{5})$

A particular case of this probabilistic triangle is indicated in Table 3.7. The set of all the *left* members of the pairs constitute the so-called *Pascal triangle*, where each element equals the sum of its "North-West" and "North-East" neighbors. The set of all the *right* members of the pairs constitute the so-called *Leibnitz triangle* [276], where each element equals the sum of its "South-West" and "South-East" neighbors. In other words, Leibnitz triangle satisfies the rule

$$r_{N,n} + r_{N,n+1} = r_{N-1,n} \quad (\forall n, \ \forall N). \tag{3.142}$$

From now on we shall refer to this rule as "Leibnitz triangle rule", or simply "Leibnitz rule".[19] It should be clear that the Leibnitz triangle satisfies Leibnitz rule, but infinitely many different probabilistic triangles also satisfy it. As we shall see, this rule will turn out to play an important role in the discussion of the nature and applicability of the entropy S_q.

Let us answer the following question: *What is the probabilistic meaning of Leibnitz rule?* If we compare the triangle representation (Table 3.6) with the hypercubic

---

[19] This rule should not be confused with Kolmogorov's *consistency conditions* characterizing a *stochastic process* [277, 278]. Indeed, Kolmogorov conditions refer to the various marginal probabilities that are associated with a given set of N random variables (e.g., observing the probabilities associated with N' elements belonging to one and the same physical system with N elements, where $N' < N$), whereas the Leibnitz rule relates the marginal probabilities of a system with N variables with the joint probabilities of a *different* system with N' variables, where $N' < N$. Whereas Kolmogorov conditions are very generic, the Leibnitz rule is extremely restrictive.

By the way, another famous rule associated with Leibnitz is the so-called "Leibnitz chain rule" for derivation of a function. These two rules are in principle unrelated. However, they both have a recurrent structure. Is this just a coincidence, or does it provide a hint on the deep manner through which Leibnitz liked to think mathematics?

representation (e.g., Tables 3.4 and 3.5), we immediately verify that the Leibnitz rule means that *the marginal probabilities of the N-system coincide with the joint probabilities of the (N − 1)-system.* Generally speaking, if we calculate the marginal probabilities of the N-system where we have summed over the states of M subsystems, we precisely obtain the joint probabilities of the (N − M)-subsystem. This is a remarkable property which implies in a specific form of *scale-invariance.* This invariance is in fact quite close to that emerging within analytical procedures such as the renormalization group, successfully applied in critical phenomena and elsewhere [279–282]. Equation 3.142 will be referred to as *strict* scale-invariance. It can and does happen that this relation is only asymptotically true for large N, i.e.

$$\lim_{N \to \infty} \frac{r_{N,n} + r_{N,n+1}}{r_{N-1,n}} = 1 \quad (\forall n, \forall N). \tag{3.143}$$

In this case, we talk of *asymptotic* scale-invariance.

Leibnitz rule is in fact stronger than it might look at first sight. If we give, *for all* N, the value of the probability r_{Nn} for a *single* value of n, Leibnitz rule completely determines the entire set $\{r_{Nn}\}$ $\forall (N, n)$. A simple choice might be to give r_{N0}, $\forall N$.

For example, if we assume

$$r_{N0} = \frac{1}{N+1} \quad (N = 1, 2, 3, ...), \tag{3.144}$$

we straightforwardly obtain

$$r_{Nn} = \frac{1}{N+1} \frac{(N-n)! \, n!}{N!} \quad (n = 0, 1, 2, ..., N; \ N = 1, 2, 3, ...), \tag{3.145}$$

which precisely recovers the *Leibnitz triangle* itself, as exhibited in Table 3.7.

A second example is to assume the *exponential* family

$$r_{N0} = p^N \quad (0 \le p \le 1; \ N = 1, 2, 3, ...). \tag{3.146}$$

It then follows that

$$r_{Nn} = p^{N-n}(1 - p)^n \quad (n = 0, 1, 2, ..., N; \ N = 1, 2, 3, ...), \tag{3.147}$$

which recovers the basic case of N *independent* variables, with probabilities p and (1 − p) respectively for the two states of each individual variable. In particular, for p = 1/2, we obtain $r_{Nn} = 1/2^N$, $\forall n$, i.e., *equal probabilities.*[20]

A third example is to assume [274]

$$r_{N0} = p^{N^\alpha} \quad (0 \le p \le 1; \ \alpha \ge 0; \ N = 1, 2, 3, ...). \tag{3.148}$$

---

[20] Notice that, in the case of independent variables, r_{N0} decays *exponentially* with N, whereas, in the Leibnitz triangle, it decays much more slowly, as the $1/N$ power-law.

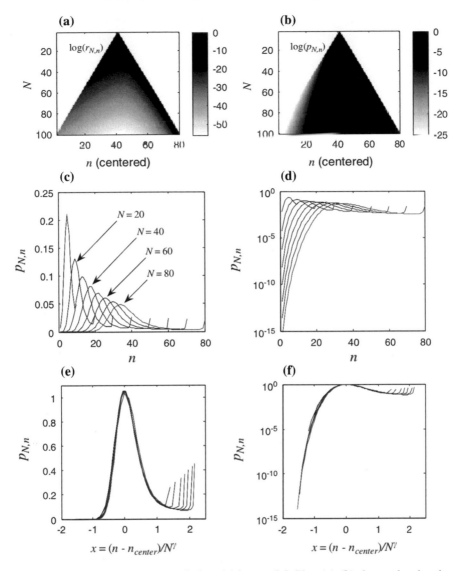

Fig. 3.13 Behavior of the TGS stretched model for $\alpha = 0.9$. Plots (**a**)–(**b**) show reduced probabilities $\log r_{N,n}$ and probabilities $\log p_{N,n}$ up to $N = 100$ as grayscale images. Plots (**c**)–(**d**) show uncentered and unscaled probabilities for $N = 10, 20, \ldots, 80$, in both linear and logarithmic scales. Plots (**e**)–(**f**) show centered and scaled probabilities, illustrating the attractor in probability space. We find $\mu \equiv \gamma = 0.70$. From [283]

We shall refer to this choice as the *stretched exponential* model. If $\alpha = 1$, it recovers the previous case, i.e., the *independent* model. If $\alpha = 0$ and $0 < p < 1$, we have that all probabilities vanish for fixed N, excepting $r_{N0} = p$ and $r_{NN} = 1 - p$. See Fig. 3.13 for an illustration.

All these models, with the unique exception of the independent model, involve correlations. These correlations might however be *not* strong enough in order to require an entropy different from S_{BG} if we seek for extensivity. Let us be more precise. The entropy of the N-system is given by

$$S_q(N) = \frac{1 - \sum_{n=0}^{N} \frac{N!}{(N-n)!\,n!}\,[r_{Nn}]^q}{q-1} \quad \left(S_1(N) = -\sum_{n=0}^{N} \frac{N!}{(N-n)!\,n!}\, r_{Nn} \ln r_{Nn} \right).$$

(3.149)

A question that we want to answer is the following one: *Is there a value of q such that $S_q(N)$ is extensive, i.e., such that $\lim_{N\to\infty} S_q(N)/N$ is finite?*

The answer is trivial for the independent model. The special value of q is simply unity. Indeed, in that case, we straightforwardly obtain

$$S_{BG}(N) = N S_{BG}(1) = -N[p \ln p + (1-p)\ln(1-p)].$$

(3.150)

In this simple case, the (additive) BG entropy is strictly extensive, i.e. $S_{BG} \propto N, \forall N$, and not only in the $N \to \infty$ asymptotic sense.

Numerical calculation has shown that the answer (for which value of q, S_q is extensive) still is $q = 1$ for the Leibnitz triangle, and for the stretched model with $p > 0$ and $\alpha > 0$. All these examples are illustrated in Fig. 3.14.

Is it possible to have particularly strong correlations that make S_q to be extensive only for $q \neq 1$? The answer is *yes*. Let us illustrate this on two examples [274]. The first of them is neither strictly nor asymptotically scale-invariant. The second one is asymptotically invariant. To construct both of them, we start from the Leibnitz triangle and then impose that most of the possible states have zero probability. Their initial probabilities are redistributed into a small number of all the other possible states, in such a way that norm is preserved. Notice in both Tables 3.8 and 3.9 that only a "left" strip of width $d + 1$ has nonvanishing probabilities. All the other probabilities are strictly zero. To complete the description of these models, we need to indicate the values of the nonvanishing probabilities.

The first model (Table 3.8), hereafter referred to as the *cutoff* one, has, for a fixed value of N, all nonvanishing r_{Nn} equal. This is to say

$$r_{N,n}^{(d)} = 1/2^N \qquad \text{(if } N \le d),$$

$$r_{N,n}^{(d)} = \frac{1}{W(N,d)} \quad \text{(if } N > d \text{ and } n \le d),$$

(3.151)

$$r_{N,n}^{(d)} = 0 \qquad \text{(if } N > d \text{ and } n > d),$$

with

$$W(N,d) = \sum_{n=0}^{d} \frac{N!}{(N-n)!\,n!},$$

(3.152)

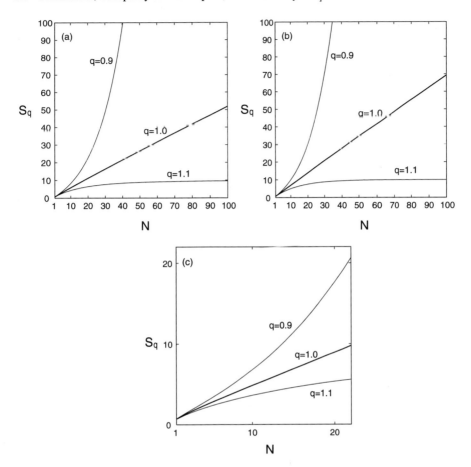

Fig. 3.14 $S_q(N)$ for **a** the Leibnitz triangle, **b** $p = 1/2$ independent subsystems, and **c** $r_{N,0} = (1/2)^{N^{1/2}}$. Only for $q = 1$ we have a *finite* value for $\lim_{N\to\infty} S_q(N)/N$; this ratio *vanishes* (*diverges*) for $q > 1$ ($q < 1$). From [274]

Table 3.8 Distribution of probabilities of the TGS cutoff model with $d = 1$ (*top*) and $d = 2$ (*bottom*). Notice that the number of triangle elements with nonzero probabilities grows like N, whereas that with zero probability grows like N^2

$(N = 0)$	$(1, 1)$
$(N = 1)$	$(1, 1/2)$ $(1, 1/2)$
$(N = 2)$	$(1, 1/3)$ $(2, 1/3)$ $(1, 0)$
$(N = 3)$	$(1, 1/4)$ $(3, 1/4)$ $(3, 0)$ $(1, 0)$
$(N = 4)$	$(1, 1/5)$ $(4, 1/5)$ $(6, 0)$ $(4, 0)$ $(1, 0)$
$(N = 0)$	$(1, 1)$
$(N = 1)$	$(1, 1/2)$ $(1, 1/2)$
$(N = 2)$	$(1, 1/4)$ $(2, 1/4)$ $(1, 1/4)$
$(N = 3)$	$(1, 1/7)$ $(3, 1/7)$ $(3, 1/7)$ $(1, 0)$
$(N = 4)$	$(1, 1/11)(4, 1/11)(6, 1/11)$ $(4, 0)$ $(1, 0)$

For example, $W(N, 0) = 1$, $W(N, 1) = N + 1$, $W(N, 2) = \frac{1}{2}N(N + 1) + 1$, $W(N, 3) = \frac{1}{6}N(N^2 + 5) + 1$, and so on. For fixed d and $N \to \infty$, we have that

$$W(N, d) \sim \frac{N^d}{d!} \propto N^d. \tag{3.153}$$

The entropy is given by

$$S_q(N) = \ln_q W(N, d). \tag{3.154}$$

Therefore, by using Eq. (3.132), we obtain that $S_q(N)$ is extensive if and only if

$$q = 1 - \frac{1}{d}. \tag{3.155}$$

If $q > 1 - \frac{1}{d}$ ($q < 1 - \frac{1}{d}$) we have that $\lim_{N \to \infty} S_q(N)/N$ vanishes (diverges). But this limit converges to a *finite* value for the special value of q. More precisely,

$$\lim_{N \to \infty} \frac{S_{1-1/d}(N)}{N} = \frac{d}{(d\,!)^{1/d}}. \tag{3.156}$$

Let us address now the second model (Table 3.9). The probabilities are given by

$$r_{N,n}^{(d,\epsilon)} = \begin{cases} \frac{1}{N+1} \frac{(N-n)!\,n!}{N!} + l_{N,n}^{(d,\epsilon)}\, s_N^{(d)} & (n \le d) \\ 0 & (n > d) \end{cases} \tag{3.157}$$

where the *excess probability* $s_N^{(d)}$ and the *distribution ratio* $l_{N,n}^{(d,\epsilon)}$ (with $0 < \epsilon < 1$) are defined through

Table 3.9 Anomalous probability sets: $d = 1$ (*top*), and $d = 2$ (*bottom*). The left number within parentheses indicates the multiplicity (i.e., Pascal triangle). The right number indicates the corresponding probability. The probabilities, noted $r_{N,n}$, asymptotically satisfy the Leibnitz rule, i.e., $\lim_{N \to \infty} \frac{r_{N,n} + r_{N,n+1}}{r_{N-1,n}} = 1$ ($\forall n$). In other words, the system is, in this sense, *asymptotically scale-invariant*. Notice that the number of triangle elements with nonzero probabilities grows like N, whereas that of zero probability grows like N^2

($N = 0$)	(1, 1)
($N = 1$)	(1, 1/2) (1, 1/2)
($N = 2$)	(1, 1/2) (2, 1/4) (1, 0)
($N = 3$)	(1, 1/2) (3, 1/6) (3, 0) (1, 0)
($N = 4$)	(1, 1/2) (4, 1/8) (6, 0) (4, 0) (1, 0)
($N = 0$)	(1, 1)
($N = 1$)	(1, 1/2) (1, 1/2)
($N = 2$)	(1, 1/3) (2, 1/6) (1, 1/3)
($N = 3$)	(1, 3/8) (3, 5/48) (3, 5/48) (1, 0)
($N = 4$)	(1, 2/5)(4, 3/40)(6, 3/60) (4, 0) (1, 0)

$$s_N^{(d)} \equiv \sum_{n=d+1}^{N} \frac{1}{N+1} \frac{(N-n)!\,n!}{N!} = \frac{N-d}{N+1},$$
(3.158)

$$l_{N,n}^{(d,\epsilon)} \equiv \begin{cases} (1-\epsilon)\,\epsilon^n\,\frac{(N-n)!\,n!}{N!} & (0 \le n < d) \\ \epsilon^d & (n = d). \end{cases}$$
(3.159)

The entropy is given by

$$S_q(N) = \frac{1 - \sum_{n=0}^{d} \frac{N!}{(N-n)!\,n!}\,[r_{Nn}^{(d,\epsilon)}]^q}{q-1}.$$
(3.160)

In Fig. 3.15, we have shown typical examples. Like for the previous example, the entropy is extensive if and only if q is given by Eq. (3.155): see Figs. 3.16 and 3.17.

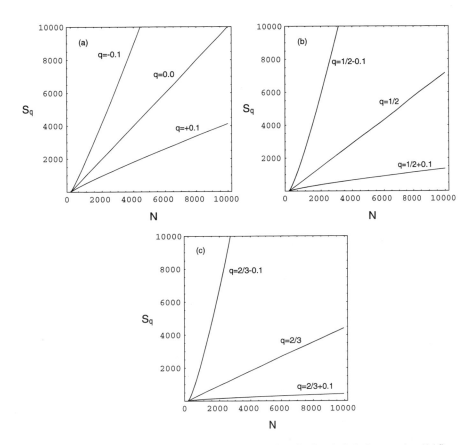

Fig. 3.15 $S_q(N)$ for anomalous systems: **a** $d = 1$, **b** $d = 2$, and **c** $d = 3$. Only for $q = 1 - (1/d)$ we have a *finite* value for $\lim_{N\to\infty} S_q(N)/N$; it *vanishes* (*diverges*) for $q > 1 + (1/d)$ ($q < 1 + (1/d)$). From [274]

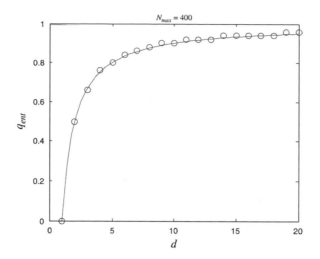

Fig. 3.16 The value of q_{ent} as a function of d for the TGS2 cutoff model. Numerical data were obtained by analyzing the linearity of $S_q(N)$ *versus* N plots, for N ranging up to $N_{max} = 400$. The open circles represent numerical data, and the solid curve is the theoretical result $q_{ent} = 1 - 1/d$. From [283]

Summarizing this Subsection, we have seen that, if the correlations are either strictly or asymptotically inexistent, the additive S_{BG} is extensive whereas the non-additive S_q ($q \neq 1$) is nonextensive. In contrast, when we have correlations so global that a large region of phase space is unoccupied, then S_q is extensive for a special value of q which differs from unity, whereas it is nonextensive for all other values of q, including $q = 1$. We have presented some models that basically satisfy Leibnitz rule. However, some of them yield $q = 1$ whereas others yield $q \neq 1$. The delicate balance responsible for all these features remains elusive at present time; we emphasize however that in those examples where $q < 1$, a large region of the full probability space has zero probability of occurrence. We shall come back to the subject when addressing the q-generalizations of the Central Limit Theorem (CLT) and of the Large Deviation Theory (LDT).

3.3.7 Extensivity of S_q—Physical Realizations

In the two previous subsections, we have discussed abstract, probabilistic, realizations of the extensivity of S_q for $q \neq 1$. Let us exhibit now physical realizations of the same property in Hamiltonian many-body systems. We shall focus on the block entropy of four quantum systems at temperature $T = 0$ [284–288]. The first three have a fermionic nature and consist of N elements on a linear chain ($d = 1$), with $N \to \infty$, and we focus on a block of L^d contiguous elements within the N, with $L >> 1$. The fourth system is a two-dimensional ($d = 2$) bosonic one.

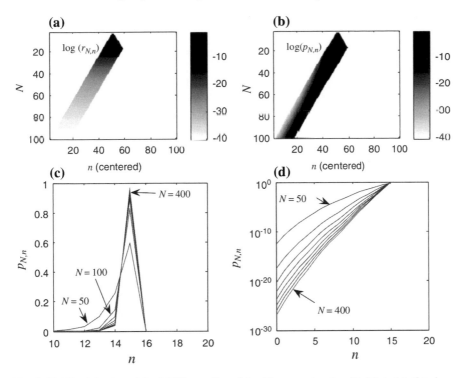

Fig. 3.17 The behavior of the TGS2 cutoff model with parameter $d = 15$. Plots (a)–(b) show reduced probabilities $\log r_{N,n}$ and probabilities $\log p_{N,n}$ up to $N = 100$ as grayscale images. Plots (c)–(d) show uncentered and unscaled probabilities for $N = 50, 100, \ldots, 400$, in both linear and logarithmic scales. We find $\mu/2 \equiv \gamma = 0$. From [283]

All four quantum systems are assumed at $T = 0$, hence they are in their ground state. Since this is a *pure state*, the density matrix ρ_N satisfies $Tr\rho_N^2 = 1$. Consequently, the entropy $S_q(N) = 0$, $\forall q > 0$. If we focus, however, on a block of L elements with $L^d < N$, and define $\rho_L \equiv Tr_{N-L}\rho_N$ (Tr_{N-L} denoting that we are tracing over all but L of the N elements), we will have (in the case of our quantum systems) $Tr\rho_L^2 < 1$, i.e., a *mixed state*. Therefore, the block entropy $S_q(L, N) > 0$. This fact is due to the nontrivial entanglement associated with quantum nonlocality. Our goal is to calculate for what value of the index q (noted q_{ent} if such value exists, where *ent* stands for *entropy*) the block entropy $S_{q_{ent}}(L) \equiv \lim_{N\to\infty} S_{q_{ent}}(L, N)$ is extensive. In other words, $S_{q_{ent}}(L) \sim s_{q_{ent}} L^d$ ($L \to \infty$, *after* we have taken $N \to \infty$), with the slope $s_{q_{ent}} \in (0, \infty)$.

Our first system [284] consists of the well-known linear chain of spin 1/2 XY ferromagnet with transverse magnetic field λ. The Hamiltonian is given by

$$\mathcal{H} = -\sum_{j=1}^{N-1}[(1+\gamma)\,\hat{\sigma}_j^x\hat{\sigma}_{j+1}^x + (1-\gamma)\,\hat{\sigma}_j^y\hat{\sigma}_{j+1}^y] - 2\lambda\sum_{j=1}^{N}\hat{\sigma}_j^z, \qquad (3.161)$$

where we assume periodic boundary conditions, i.e., we have a ring with N spins, and $(\sigma_j^x, \sigma_j^y, \sigma_j^z)$ are the Pauli matrices. For $|\gamma| = 1$ we have the *Ising ferromagnet*, for $0 < |\gamma| < 1$ we have the *anisotropic XY ferromagnet* (which, due to its symmetry, belongs to the Ising model universality class), and for $\gamma = 0$ we have the *isotropic XY ferromagnet* (or, simply, the *XY ferromagnet*). This model, being one-dimensional, has no phase transition at $T > 0$. But it does have a second order one at $T = 0$. More precisely, it is critical at $\lambda = 1$ if $\gamma \neq 0$, and at $0 \leq \lambda \leq 1$ if $\gamma = 0$. See [284] for the details of the numerical and analytical calculations. Central results are presented in Figs. 3.18 and 3.19. The numerics are consistent with the main present relation, namely $q_{entropy}$ as a function of the *central charge c*. This concept is long known in quantum field theory (see [291] and references therein). The central charge characterizes the critical universality class of vast sets of systems (more precisely, various critical exponents are shared between the systems that have the same value of c).

Reference [292] enables us to analytically confirm, at the critical point, the numerical results exhibited in the above figures. The continuum limit of a $(1 + 1)$-dimensional critical system is a conformal field theory with central charge c. In this quite different context, the authors re-derive the result

$$S_1(L) \sim (c/3) \ln L \qquad (3.162)$$

for a finite block of length L in an infinite critical system. To obtain this (clearly *nonextensive*) expression of the von Neumann entropy $S_1(L)$, they first find an ana-

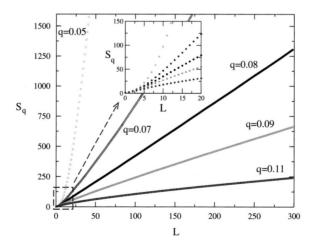

Fig. 3.18 Block q-entropy $S_q(\hat{\rho}_L)$ as a function of the block size L in a critical Ising chain ($\gamma = 1$, $\lambda = 1$), for typical values of q. Only for $q = q_{ent} \simeq 0.0828$, s_q is *finite* (i.e., S_q is *extensive*); for $q < q_{ent}$ ($q > q_{ent}$) s_q diverges (vanishes). The slope $s_{q_{ent}} = \lim_{L \to \infty}[S_{q_{ent}}(L)/L]$ equals $3.56\ldots$ for $\gamma = 1$, and is expected to monotonically decrease to zero when γ approaches zero. For $\gamma = 0$ the model makes a crossover from the Ising universality class ($0 < \gamma \leq 1$) to the isotropic XY one, for which we numerically obtain $q_{ent} \simeq 0.1623$. From [285]

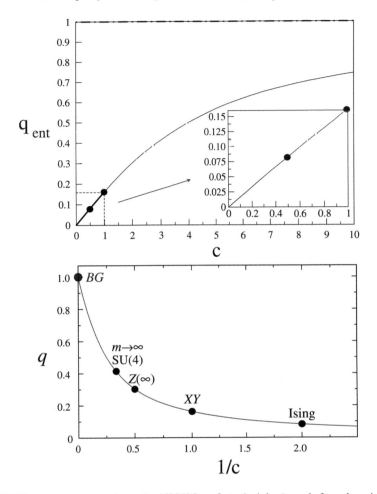

Fig. 3.19 *Top*: q_{ent} versus c, as determined [285] from first principles (namely from the universality class of the Hamiltonian) such that $S_{q_{ent}}(\hat{\rho}_L)$ is extensive, i.e., $\lim_{L\to\infty}\left[S_{\frac{\sqrt{9+c^2}-3}{c}}(\hat{\rho}_L)/L\right] < \infty$. When c increases from 0 to infinity, q_{ent} increases from 0 to unity (von Neumann entropy). The values $c = 1/2$ (hence $q_{ent} = \sqrt{37} - 6 \simeq 0.0828$) and $c = 1$ (hence $q_{ent} = \sqrt{10} - 3 \simeq 0.1623$), respectively, correspond to the Ising and XY ferromagnetic chains in the presence of the critical transverse field at $T = 0$; for $c = 4$, $q_{ent} = 1/2$, for $c = 6$, $q_{ent} = \frac{2}{\sqrt{5}+1} = \frac{1}{\Phi}$, where Φ is the golden mean, and, for $c \gg 1$, see Ref. [293]. For other models, see [289, 290]. Let us emphasize that this anomalous value of q occurs only and precisely at the zero-temperature second-order quantum critical point; anywhere else, the usual short-range-interaction BG behavior (i.e. $q = 1$) is valid. *Bottom*: The same versus $1/c$. From [235]

lytical expression, namely $\text{Tr}\hat{\rho}_L^q \sim L^{-c/6(q-1/q)}$. Here, this expression is used quite differently. We impose the *extensivity* of $S_q(L)$ finding the value of q for which $-c/6(q_{ent} - 1/q_{ent}) = 1$, *i.e.*

$$q_{ent} = \frac{\sqrt{9 + c^2} - 3}{c}. \tag{3.163}$$

Consequently, $\lim_{L \to \infty} \left[S_{\frac{\sqrt{9+c^2}-3}{c}}(L)/L \right] < \infty$. When c increases from 0 to infinity (see Fig. 3.19), q_{ent} increases from 0 to unity (von Neumann entropy). For $c = 4$ (dimension of physical space-time), $q = 1/2$; $c = 26$ corresponds to a 26-dimensional bosonic string theory, see [293]. It is well known that for critical quantum Ising and anisotropic XY models the central charge is equal to $c = 1/2$ (indeed they are in the same universality class and can be mapped to a free (nonlocal) fermionic field theory). For these models, at $\lambda = 1$, the value of q for which $S_q(L)$ is *extensive* is given by $q_{ent} = \sqrt{37} - 6 \simeq 0.0828$, in perfect agreement with our numerical results in Fig. 3.18. The critical isotropic XX model ($\gamma = 0$ and $|\lambda| \le 1$) is, instead, in another universality class, the central charge is $c = 1$ (free bosonic field theory) and $S_q(L)$ is *extensive* for $q_{ent} = \sqrt{10} - 3 \simeq 0.1623$, as found also numerically. We finally notice that, in the $c \to \infty$ limit, $q_{ent} \to 1$. The physical interpretation of this limit is not clearly understood. However, since c in some sense plays the role of a dimension (see [293]), this limit could correspond to some sort of mean field approximation. If so, it is along a line such as this one that a mathematical justification could emerge for the widely spread use of BG concepts in the discussion of mean field theories of spin-glasses (within the replica-trick and related approaches). Indeed, BG statistical mechanics is essentially based on the *ergodic* hypothesis. It is firmly known that real glassy systems (e.g., spin-glasses) precisely *violate ergodicity*, thus leading to an intriguing and fundamental question. Consequently, a mathematical justification for the use of BG entropy and associated energy distribution for such complex mean field systems would surely be welcome.

The Hamiltonian 3.161 can be generalized into the following quantum Heisenberg one:

$$\mathcal{H} = -\sum_{j=1}^{N-1} [(1 + \gamma)\, \hat{\sigma}_j^x \hat{\sigma}_{j+1}^x + (1 - \gamma)\, \hat{\sigma}_j^y \hat{\sigma}_{j+1}^y + \Delta \sigma_j^z \sigma_{j+1}^z] - 2\lambda \sum_{j=1}^{N} \hat{\sigma}_j^z. \tag{3.164}$$

For $\Delta = 1$ and $\lambda = 0$, there also occurs a critical phenomenon. Its associated value of c also is 1, hence $q_{ent} = \sqrt{10} - 3 \simeq 0.16$. If we include in this Hamiltonian say second-neighbor coupling (or, in fact, any short-range coupling which does not alter the ferromagnetic order parameter), the value of c, hence that of q_{ent}, remains the same. Not so with the slope $s_{q_{ent}}$, which depends on the details and not only on the symmetry which is being broken at criticality.

At this stage, let us briefly focus on a subtle issue that has been recently discussed in [288]. For the spin 1/2 Ising ferromagnet with transverse field ($c = 1/2$ criticality), it can be shown that the highly excited states loose relative weight in the $L \to \infty$ limit. Consistently, the number $\tilde{L}$ of thermodynamically relevant modes increases with the block size L not proportionally to L but roughly as $\tilde{L} \simeq 1.77 \ln L + 2.88$. It turns out that, due to this kind of bizarre (quantum-originated) behavior, the effective dimension of phase space (more precisely, Hilbert space) increases not as 2^L but rather as $2^{\tilde{L}}$. It is this unusual feature that makes S_{BG} to increase not linearly with L but rather proportionally to $\ln L$, thus loosing its expected thermodynamical extensivity (duly restored with S_q for a special value of $q < 1$). In this case, the use of BG statistical mechanics appears to remain legitimate but with an effective size which increases like $\tilde{L}$ instead of the real size L. The conjectural generic scenario which emerges is indicated in Fig. 3.20.

Before proceeding to our second quantum system, let us also mention another subtle issue, namely a possible inadvertence that can easily occur. It is known, for the above subsystems, that

$$\frac{S_{BG}(L)}{k} = \frac{c}{3} \ln L + \ln b + \dots = \ln[bL^{c/3}] + \dots \quad (b > 1). \tag{3.165}$$

But if we (wrongly) assume equal probabilities, we will (wrongly) use the *equal-probability* expression $S_{BG} = k \ln W$, thus identifying $W \sim bL^{c/3}$. This corresponds to the power-law class $W \propto L^\rho$ ($\rho > 0$; $L \to \infty$), whose extensive entropy is the nonadditive one S_q with $q = 1 - 1/\rho = 1 - 3/c$ [18, 274]. This expression is in variance with Eq. (3.163) and is, therefore, definitively wrong for all $c < \infty$! In this illustration, the error does not affect the class of nonadditive entropic functionals S_q, but it concerns the correct value of the index q. Physically speaking, this confusion arises from the fact that relation (3.165) is valid for the subsystem; let us recall that the kinetic temperature of the total system vanishes, whereas the effective temperature of a L-sized subsystem is *finite* [288]. This is in strong variance with *infinity*, which would correspond to the microcanonical ensemble and would, therefore, justify the use of the equal-probability hypothesis.

Our second quantum system consists in antiferromagnetic disordered spin-S Heisenberg chains [286]. The Hamiltonian is given by

$$\mathcal{H} = \sum_i J_i \vec{S}_i . \vec{S}_{i+1}, \tag{3.166}$$

where $\{J_i\}$ are random exchange couplings obeying a probability distribution $P(J)$ and $\{\vec{S}_i\}$ are spin-S operators, with periodic boundary conditions. The authors consider a L-sized block within an infinitely long chain, and once again they numerically seek for the value q_{ent} (which they note as q_{ext}) such that $S_{q_{ent}}$ is extensive, i.e., proportional to L. The result is indicated in Fig. 3.21, and corresponds to

$$q_{ent} \simeq 1 - \frac{5/3}{c_{eff}}, \tag{3.167}$$

where the *effective central charge* is given by $c_{eff} = \ln(2S + 1)$. Notice that, in the $S \to \infty$ limit, q_{ent} attains the BG value $q = 1$, as expected from the fact that in that limit the quantum nonlocal effects vanish. Notice also that, in variance with the previous case in [285], q_{ent} can here attain negative values.

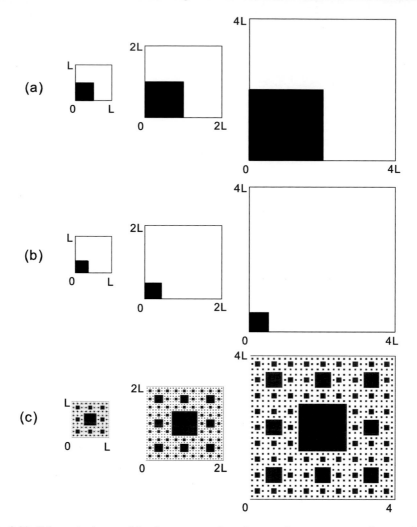

Fig. 3.20 Schematic classes of the phase-space trajectories covering a **a** *compact* subspace whose corresponding Lebesgue measure remains different from zero in the thermodynamic limit [$q_{ent} = 1$ and standard BG statistical mechanics surely is applicable, $\forall d$]; **b** *compact* subspace whose corresponding Lebesgue measure vanishes in the thermodynamic limit [$q_{ent} < 1$ and standard BG statistical mechanics appears to still be applicable, at least for $d = 1$; for $d > 1$, the thermodynamical entropic extensivity issue might still be solvable using S_q, or using some other nonadditive entropy such as $S_{(1-q)}$, and standard BG statistical mechanics might still be applicable], and; **c** *noncompact*, hierarchical-like, subspace whose corresponding Lebesgue measure vanishes in the thermodynamic limit [nonadditive entropies, for instance S_q with $q < 1$, as well as a statistical mechanics generalized beyond BG, for instance nonextensive statistical mechanics, become a must]. Three different size systems are presented, characterized by three linear sizes L, $2L$ and $4L$. Clearly, the central focus of the present book is the complex class (**c**). From [288]

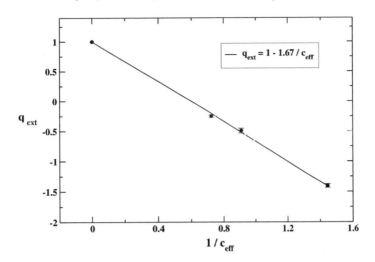

Fig. 3.21 Entropic index $q_{ent} \simeq 1 - \frac{5/3}{c_{eff}} = 1 - \frac{5/3}{\ln(2S+1)}$ such that that the entropy $S_{q_{ent}}$ of a L-sized block is extensive, i.e., $S_{q_{ent}}(L) \propto L$. From [286]

Our third quantum system consists in the generalized $d = 1$ isotropic Lipkin–Meshkov–Glick models with $su(m + 1)$ spin and long-range non-constant interactions, whose non-degenerate ground state is a Dicke state of $su(m + 1)$ type [287]. As in the two previous fermionic systems, the block entropy $S_{q_{ent}}(L)$ is extensive, i.e., $S_{q_{ent}}(L) \propto L$ for

$$q_{ent} = 1 - \frac{2}{m - k} \quad (m = 1, 2, \ldots; \ k = 0, 1, 2, \ldots; \ m - k \geq 3), \qquad (3.168)$$

where m is the number of internal degrees of freedom per particle (e.g., $m = 2S$, with spin size $S = 1/2, 1, 3/2\ldots$) and k is the number of fundamental-state vanishing magnon densities. As before, in the $S \to \infty$ limit (more precisely in the $m \to \infty$ limit for fixed k) we recover the BG value $q = 1$.

Let us address finally our fourth (and last in the present Section) quantum system, namely the bosonic one [285]. It is the bidimensional system of infinite coupled harmonic oscillators studied in Ref. [294], with Hamiltonian

$$\mathcal{H} = \frac{1}{2} \sum_{x,y} [\Pi_{x,y}^2 + \omega_0^2 \Phi_{x,y}^2 + (\Phi_{x,y} - \Phi_{x+1,y})^2 + (\Phi_{x,y} - \Phi_{x,y+1})^2], \qquad (3.169)$$

where $\Phi_{x,y}$, $\Pi_{x,y}$ and ω_0 are coordinate, momentum, and self-frequency of the oscillator at the site $\vec{r} = (x, y)$. The system has the dispersion relation

$$E(\vec{k}) = \sqrt{\omega_0^2 + 4 \sin^2 k_x/2 + 4 \sin^2 k_y/2}, \qquad (3.170)$$

hence a *gap* ω_0 at $\vec{k} = \vec{0}$. Applying the canonical transformation $b_i = \sqrt{\frac{\omega}{2}} \times (\Phi_i +$ $\frac{i}{\omega}\Pi_i)$ with $\omega = \sqrt{\omega_0^2 + 4}$, the Hamiltonian (3.169) is mapped into the quadratic canonical form

$$\mathcal{H} = \sum_{ij} \left[a_i^\dagger A_{ij} a_j + \frac{1}{2}(a_i^\dagger B_{ij} a_j^\dagger + h.c.) \right], \qquad (3.171)$$

where a_i are bosonic operators. It is found [294] that, for typical values of ω_0,

$$S_1(L) \propto L \quad (L >> 1) \qquad (3.172)$$

for square blocks of area L^2, i.e., the von Neumann entropy is nonextensive. This is so no matter how close to zero the gap energy is. In contrast, when we consider $q \neq 1$, it is found [285]

$$S_{q_{ent}}(L) \sim s_{q_{ent}}(\omega_0)L^2, \qquad (3.173)$$

i.e., an extensive entropy. As we can see Figs. 3.22 and 3.23, q_{ent} monotonically decreases from 0.87 down to 0.78 when the gap ω_0 increases from zero to 0.50.

Equation (3.172) can be seen as the $d = 2$ case of the so-called *area law* [295], namely

$$S_1(L) \propto L^{d-1} \quad (d > 1; \ L \to \infty). \qquad (3.174)$$

The $d = 3$ case recovers the celebrated Bekenstein-Hawking scaling for black holes, namely $S_1(L) \propto L^2$. Eqs. such as (3.162) and (3.174) can be unified as follows:

$$S_1(L) \propto \frac{L^{d-1} - 1}{d - 1} \equiv \ln_{2-d} L \quad (d \geq 1; \ L \to \infty), \qquad (3.175)$$

i.e., the Boltzmann–Gibbs–von Neumann entropy is nonextensive. Given the above results for fermionic and bosonic systems, a conjecture is quite plausible, namely that, for such systems, a value of $q < 1$ might exist such that

$$S_{q_{ent}}(L) \propto L^d \quad (d \geq 1; \ L \to \infty), \qquad (3.176)$$

i.e., the thermodynamic extensivity of the entropy is recovered.[21] The index q_{ent} is expected to depend on some generic parameters (symmetries, gaps, etc), but also on the dimension d. In particular, since the exponent $(d - 1)$ in Eq. (3.175) and the exponent d in Eq. (3.176) become closer and closer in the limit $d \to \infty$, we expect $\lim_{d\to\infty} q_{ent}(d) = 1$. As mentioned before, it is along lines such as this one that a justification could be found for the current use of BG statistical mechanics in systems like spin-glasses in the mean field approximation (replica trick), in spite of the well-known fact that real spin-glasses violate ergodicity.

---

[21] This issue is not yet sufficiently elucidated, and an entropic functional different from S_q, namely S_δ defined in Eq. (3.134), also constitutes, for $d > 1$, a candidate for solving this problem. Various aspects that are involved have been discussed in [222, 296].

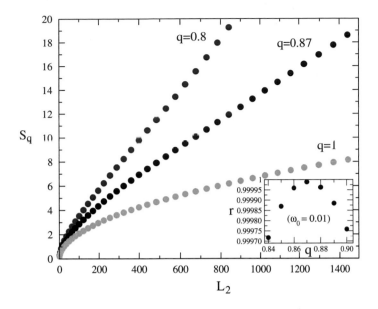

Fig. 3.22 Block q-entropy $S_q(\hat{\rho}_L)$ as a function of the square block area L^2 in a bosonic $d = 2$ array of infinite coupled harmonic oscillators at $T = 0$, for typical values of q. Only for $q = q_{ent} \simeq 0.87$, the slope s_q is *finite* (i.e., S_q is *extensive*); for $q < q_{ent}$ ($q > q_{ent}$) it diverges (vanishes). *Inset:* determination of q_{ent} through numerical maximization of the linear correlation coefficient r of $S_q(\hat{\rho}_L)$ when using the range $400 \leq L^2 \leq 1600$. From [285]

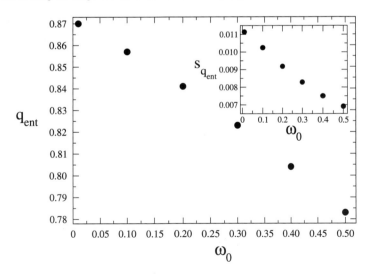

Fig. 3.23 The ω_0-dependence of the index q_{ent} in a bosonic $d = 2$ array of infinite coupled harmonic oscillators at $T = 0$. *Inset:* the ω_0-dependence of the q-entropic density $s_{q_{ent}} \equiv S_{q_{ent}}/L^2$. From [285]

3.3.8 An Epistemological Analogy

An interesting question may arrive: Why sacrifice the simplicity of the additivity of S_{BG} in Eq. (1.4) and extend it into Eq. (3.21)?

We may analogously ask: Why did Einstein accept to sacrifice the simplicity of the additivity of the Galilean composition of one-dimensional velocities

$$v_{13} = v_{12} + v_{23}, \qquad (3.177)$$

extending it into the (nonadditive) relativistic one?, namely

$$v_{13}/c = \frac{(v_{12}/c) + (v_{23}/c)}{1 + (v_{12}/c)\,(v_{23}/c)}, \qquad (3.178)$$

dramatically different from the Galilean one, since now, if one of the velocities is that of light, the composed velocity still is that of light, independently from the other velocity. The reason for this extension is well known. Indeed, through the invariance of c, Einstein wanted to unify nothing less than classical mechanics with Maxwell equations, which are invariant under the space-time Lorentz transformation, thus succeeding to formulate the physical laws of nature in an observer-independent form. We may argue that, within the importance of such a goal, the loss of the Galilean additivity represents a small price to pay.[22]

We may analogously argue that the preservation of thermodynamics and its Legendre structure (by succeeding in making the thermodynamical entropy to be extensive in all cases) is such an important issue that the loss of the BG entropic additivity can be seen as a small price to pay. Moreover, the BG additivity is recovered in the $k_B \to \infty$ limit, $\forall q$, in a similar manner that Eq. (3.178) recovers Eq. (3.177) in the $c \to \infty$ limit, for all velocities.

3.4 q-Generalization of the Kullback–Leibler Relative Entropy

The Kullback–Leiber entropy introduced in Sect. 2.2 can be straightforwardly q-generalized [116, 117]. The continuous version becomes

$$I_q(p, p^{(0)}) \equiv -\int dx\, p(x) \ln_q\left[\frac{p^{(0)}}{p(x)}\right] = \int dx\, p(x)\frac{[p(x)/p^{(0)}(x)]^{q-1} - 1}{q - 1}. \qquad (3.179)$$

With $r > 0$ we have that

---

[22] By the way, as a curiosity, we may observe that the special relativity celebrated relation $E^2 = m^2c^4 + p^2c^2$ can be rewritten as $E/mc^2 = e_{-1}^{(1/2)(p/mc)^2}$, i.e., a q-Gaussian with $q = -1$ and negative 'temperature' (see [297]).

$$\frac{r^{q-1} - 1}{q - 1} \geq 1 - \frac{1}{r} \quad \text{if} \quad q > 0,$$

$$= 1 - \frac{1}{r} \quad \text{if} \quad q = 0, \tag{3.180}$$

$$\leq 1 - \frac{1}{r} \quad \text{if} \quad q < 0.$$

Consequently, for say $q > 0$, we have that

$$\frac{[p(x)/p^{(0)}(x)]^{q-1} - 1}{q - 1} \geq 1 - \frac{p^{(0)}(x)}{p(x)}, \tag{3.181}$$

hence

$$\int dx\, p(x) \frac{[p(x)/p^{(0)}(x)]^{q-1} - 1}{q - 1} \geq \int dx\, p(x) \left[1 - \frac{p^{(0)}(x)}{p(x)} \right] = 1 - 1 = 0. \tag{3.182}$$

Therefore, we have

$$I_q(p, p^{(0)}) \geq 0 \quad \text{if} \quad q > 0,$$

$$= 0 \quad \text{if} \quad q = 0, \tag{3.183}$$

$$\leq 0 \quad \text{if} \quad q < 0.$$

It satisfies, therefore, the same basic property as the standard Kullback–Leibler entropy and can be used for the same purposes, while we have now the extra freedom of choosing q adequately for the specific system we are analyzing.

By performing the transformation $q - \frac{1}{2} \rightleftarrows \frac{1}{2} - q$ into the definition 3.179, we can easily prove the following property:

$$\frac{I_q(p, p^{(0)})}{q} = \frac{I_{1-q}(p^{(0)}, p)}{1 - q}. \tag{3.184}$$

Consequently, as a family of entropy-based testings, it is enough to consider $q \geq 1/2$, for which $I_q(p, p^{(0)}) \geq 0$ (the equality holding whenever $p(x) = p^{(0)}(x)$ almost everywhere). Also, as a corollary we have that only $I_{1/2}(p, p^{(0)})$ is generically symmetric with regard to permutation between p and $p^{(0)}$, i.e.

$$I_{1/2}(p, p^{(0)}) = I_{1/2}(p^{(0)}, p) = 1 - \int dx\, \sqrt{p(x)\, p^{(0)}(x)}. \tag{3.185}$$

This symmetry, that only exists for $q = 1/2$, opens the door for this quantity to constitute a metric. Moreover, the property $I_{1/2}(p, p^{(0)}) \geq 0$ implies

$$\int dx\, \sqrt{p(x)\, p^{(0)}(x)} \leq 1. \tag{3.186}$$

This expression can be interpreted as the continuous version of the scalar product between two unitary vectors, namely $\sqrt{p(x)}$ and $\sqrt{p^{(0)}(x)}$, and is directly related to the so-called *Fisher genetic distance* [298].

Let us also q-generalize Eq. (2.37). By choosing as $p^{(0)}(x)$ the uniform distribution on a compact support of length W, we easily establish the desired generalization,[23] i.e.

$$I_q(p, 1/W) = W^{q-1}[\ln_q W - S_q(p)]. \tag{3.187}$$

As in the $q = 1$ case, for $q > 0$, the *minimization* of the q- generalized Kulback-Leibler entropy I_q may be used instead of the *maximization* of the entropy S_q. More properties can be found in [117].

Let us finally mention an elegant property, referred to as the *triangle pseudo-equality* [120–123]. Through some algebra, it is possible to prove

$$I_q(p, p') = I_q(p, p'') + I_q(p'', p') + (q - 1)I_q(p, p'')I_q(p'', p'). \tag{3.188}$$

A simple corollary follows, namely

$$\begin{aligned} I_q(p, p') &\geq I_q(p, p'') + I_q(p'', p') \quad \text{if } q > 1, \\ &= I_q(p, p'') + I_q(p'', p') \quad \text{if } q = 1, \\ &\leq I_q(p, p'') + I_q(p'', p') \quad \text{if } q < 1. \end{aligned} \tag{3.189}$$

The name *triangle pseudo-equality* for Eq. (3.188) obviously comes from the $q = 1$ case, where we do have a strict equality. The inequality holding for $q = 1/2$ completes the standard requirements and $I_{1/2}(p, p')$ indeed constitutes a metric.

Let us now adapt our present main result, i.e., Eq. (3.183), to the problem of independence of random variables. Let us consider the two-dimensional random variable (x, y), and its corresponding *joint* distribution function $p(x, y)$, with $\int dx\, dy\, p(x, y) = 1$. The *marginal* distribution functions are then given by $h_1(x) \equiv \int dy\, p(x, y)$ and $h_2(y) \equiv \int dx\, p(x, y)$. The discrimination criterion for independence concerns the comparison of $p(x, y)$ with $p^{(0)}(x, y) \equiv h_1(x)\, h_2(y)$. The one-dimensional random variables x and y are independent if and only if $p(x, y) = p^{(0)}(x, y)$ (almost everywhere). The criterion (3.183) then becomes

$$\int dx\, dy\, p(x, y)\, \frac{\left[\frac{p(x,y)}{h_1(x)\, h_2(y)}\right]^{q-1} - 1}{q - 1} \geq 0 \quad (q \geq 1/2). \tag{3.190}$$

In the limit $q \to 1$, this criterion recovers the usual one, namely [299]

---

[23] This formula appears misprinted in Eq. (15) of the original paper [116]. This erratum was kindly communicated to me by R. Piasecki.

$$\int dx\, dy\, p(x, y) \ln p(x, y) - \int dx\, h_1(x) \ln h_1(x) - \int dy\, h_2(y) \ln h_2(y) \geq 0.$$
$$(3.191)$$

For $q = 1/2$, we obtain a particularly simple criterion, namely

$$\int dx\, dy\, p(x, y) \sqrt{p(x, y)\, h_1(x)\, h_2(y)} \leq 1. \tag{3.192}$$

For $q = 2$, we obtain

$$\int dx\, dy\, \frac{[p(x, y)]^2}{h_1(x)\, h_2(y)} \geq 1. \tag{3.193}$$

This can be considered as a satisfactory quadratic-like criterion, as opposed to the quantity introduced in [300]. We refer to the quantity frequently used in economics [300], namely, for $h_1 = h_2 \equiv h$,

$$\int dx\, dy\, [p(x, y)]^2 - \left\{ \int dx\, [h(x)]^2 \right\}^2. \tag{3.194}$$

This quantity has *not* a definite sign. In fact, if x and y are independent, this quantity vanishes. But, if it vanishes, x and y are *not* necessarily independent. In other words, its zero is *not* a necessary and sufficient condition for independence, and therefore, it does *not* constitute an optimal criterion. It could be advantageously replaced, in applications such as financial analysis and elsewhere, by the present criterion (3.193).

The generalization of criterion (3.190) for an arbitrary number d of variables (with $d \geq 2$) is straightforward, namely

$$I_q(p(x_1, x_2, ..., x_d), p^{(0)}(x_1, x_2, ..., x_d)) \geq 0 \quad (q \geq 1/2), \tag{3.195}$$

where

$$p^{(0)}(x_1, x_2, ..., x_d) \equiv \left[\int dx_2\, dx_3...dx_d\, p(x_1, x_2, ...x_d) \right]$$
$$\times \left[\int dx_1\, dx_3...dx_d\, p(x_1, x_2, ...x_d) \right]$$
$$\times ...$$
$$\times \left[\int dx_1\, dx_2...dx_{d-1}\, p(x_1, x_2, ...x_d) \right]. \tag{3.196}$$

Depending on the specific purpose, one might even prefer to use the symmetrized version of the criterion, i.e.

$$\frac{1}{2} \Big[I_q\big(p(x_1, x_2, ..., x_d), p^{(0)}(x_1, x_2, ..., x_d)\big)$$
$$+ I_q\big(p^{(0)}(x_1, x_2, ..., x_d), p(x_1, x_2, ..., x_d)\big) \Big] \geq 0 \quad (q \geq 1/2). \tag{3.197}$$

The equalities in (3.195) and (3.197) hold if and only if *all* the variables $x_1, x_2, ..., x_d$ are (almost everywhere) independent among them.

Before closing this Section, let us mention that the discrete version of definition (3.179) naturally is

$$I_q(p, p^{(0)}) \equiv \sum_{i=1}^{W} p_i \frac{[p_i/p_i^{(0)}]^{q-1} - 1}{q - 1}. \tag{3.198}$$

Notice that this quantity is generically well defined only if i runs over states for which there are nonvanishing values for both p_i and $p_i^{(0)}$. If this is not the case, we may use the q-generalized version [118, 119] of the so-called Jensen–Shannon divergence, defined to precisely avoid this difficulty.

By using the same illustration as in (2.40) and (2.41), we straightforwardly obtain

$$I_q(p, p^{(0)})(z) = \frac{1}{q-1}\left\{\frac{1}{2}[(1 + 2z)^q + (1 - 2z)^q] - 1\right\} \tag{3.199}$$

$$\sim 2qz^2 + \frac{2}{3}(3 - q)(2 - q)qz^4 + \frac{4}{45}(5 - q)(4 - q)(3 - q)(2 - q)qz^6$$

$$+ \frac{2}{315}(7 - q)(6 - q)(5 - q)(4 - q)(3 - q)(2 - q)qz^8$$

$$(z \to 0), \tag{3.200}$$

hence $I_q(p, p^{(0)})(z) \in [0, \ln_{2-q} 2]$. We verify that Eqs. (3.199) and (3.200) respectively recover, for $q \to 1$, Eqs. (2.40) and (2.41).

3.5 Constraints and Entropy Optimization

As we did with the BG entropy in Sect. 2.3, let us work out here the most simple entropic optimization cases.

3.5.1 Imposing the Mean Value of the Variable

In addition to

$$\int_0^\infty dx \, p(x) = 1, \tag{3.201}$$

we might know the following mean value of the variable (referred to as the *q-mean value*):

$$\langle x \rangle_q \equiv \int_0^\infty dx \, x \, P_q(x) = X_q^{(1)}, \tag{3.202}$$

where the *escort distribution* $P(x)$ is defined through [69, 163]

$$P_q(x) \equiv \frac{[p(x)]^q}{\int_0^\infty dx'\, [p(x')]^q}.$$ (3.203)

We immediately verify that also $P_q(x)$ is normalized, i.e.

$$\int_0^\infty dx\, P_q(x) = 1 \quad (\forall q).$$ (3.204)

The reasons for which we use $P_q(x)$ instead of $p(x)$ to express the constraint (3.202) are somewhat subtle and will be discussed later on. At the present stage, we just assume that, for whatever reason, what we know is the mean value of x with the escort distribution. We wish now to optimize S_q with the constraints (3.204) and (3.202), or, equivalently, with the constraints (3.201) and (3.202).

In order to use the Lagrange method to find the optimizing distribution, we define

$$\Phi[p] \equiv \frac{1 - \int_0^\infty dx\, [p(x)]^q}{q-1} - \alpha \int_0^\infty dx\, p(x) - \beta_q^{(1)} \frac{\int_0^\infty dx\, x\, [p(x)]^q}{\int_0^\infty dx\, [p(x)]^q},$$ (3.205)

where α and $\beta_q^{(1)}$ are the Lagrange parameters. We then impose $\partial \Phi[p]/\partial p = 0$, and straightforwardly obtain

$$p_{opt}(x) = \frac{e_q^{-\beta_q^{(1)}(x - X_q^{(1)})}}{\int_0^\infty dx'\, e_q^{-\beta_q^{(1)}(x' - X_q^{(1)})}}$$ (3.206)

where *opt* stands for *optimal*, and where we have used condition (3.201) to eliminate the Lagrange parameter α. Notice that the fact that the Lagrange parameter α *can* be factorized, and be, therefore, substituted by the "partition function" in the denominator, constitutes a quite remarkable and convenient mathematical property.

3.5.2 Imposing the Mean Value of the Squared Variable

Another simple and quite frequent case is when we know that $\langle x \rangle_q = 0$. In such case, in addition to the constraint

$$\int_{-\infty}^\infty dx\, p(x) = 1,$$ (3.207)

we might know the q-mean value of the squared variable, i.e.

$$\langle x^2 \rangle_q \equiv \int_{-\infty}^\infty dx\, x^2 P(x) = X_q^{(2)} > 0.$$ (3.208)

In order to use, as before, the Lagrange method to find the optimizing distribution, we define

$$\Phi[p] \equiv \frac{1 - \int_{-\infty}^{\infty} dx \, [p(x)]^q}{q-1} - \alpha \int_{-\infty}^{\infty} dx \, p(x) - \beta_q^{(2)} \frac{\int_{-\infty}^{\infty} dx \, x^2 \, [p(x)]^q}{\int_{-\infty}^{\infty} dx \, [p(x)]^q}.$$

(3.209)

We then impose $\partial \Phi[p]/\partial p = 0$, and straightforwardly obtain

$$p_{opt}(x) = \frac{e_q^{-\beta_q^{(2)}(x^2 - X_q^{(2)})}}{\int_{-\infty}^{\infty} dx' \, e_q^{-\beta_q^{(2)}(x'^2 - X_q^{(2)})}} \quad (q < 3),$$

(3.210)

where we have used condition (3.207) to eliminate the Lagrange parameter α. This distribution can be straightforwardly rewritten as

$$p_{opt}(x) = \frac{e_q^{-\beta_q^{(2)'} x^2}}{\int_{-\infty}^{\infty} dx' \, e_q^{-\beta_q^{(2)'} x'^2}} \quad (q < 3),$$

(3.211)

with

$$\beta_q^{(2)'} \equiv \frac{\beta_q^{(2)}}{1 + (1-q) \, \beta_q^{(2)} X_q^{(2)}}.$$

(3.212)

We thus see that, in the same way, Gaussians are deeply connected to S_{BG}, the present distributions (3.211), frequently referred to as *q-Gaussians*, are connected to the S_q entropy. Through different names, these distributions have already emerged phenomenologically in the past. In space research, they are sometimes referred to as κ distributions. They were first introduced informally by S. Olbert in 1966, as described in [301, 302]. Later on, they became known through [303]. Also, it is shown in [304] that, for special rational values of q, q-Gaussians recover Student's t- and r-distributions; equivalently, q-Gaussians may be seen as analytical extensions of the Student's t- and r-distributions.

3.5.3   Imposing the Mean Values of Both the Variable and Its Square

As normalization we use Eq. (3.207), and we add the following constraints

$$\langle x \rangle_q \equiv \frac{\int_{-\infty}^{\infty} dx \, x \, [p(x)]^q}{\int_{-\infty}^{\infty} dx' \, [p(x')]^q} = X_q^{(1)},$$

(3.213)

and

$$\langle x^2 \rangle_{2q-1} \equiv \frac{\int_{-\infty}^{\infty} dx \, x^2 \, [p(x)]^{2q-1}}{\int_{-\infty}^{\infty} dx' \, [p(x')]^{2q-1}} = X_{2q-1}^{(2)} > 0. \qquad (3.214)$$

The reason for using, as constraints, momenta with escort distributions whose q-indices are different is explained in detail in [305]. But it can be summarized here by saying that, in this specific manner, all three integrals (3.207), (3.213), and (3.214) are *simultaneously finite* for $q < 2$.[24]

It is convenient to introduce the variable $y \equiv x - \langle x \rangle_q$, hence

$$\int_{-\infty}^{\infty} dy \, p(y) = 1, \qquad (3.215)$$

$$\langle y \rangle_q = 0, \qquad (3.216)$$

and

$$\langle y^2 \rangle_{2q-1} = \langle x^2 \rangle_{2q-1} - 2 \langle x \rangle_{2q-1} \langle x \rangle_q + \langle x \rangle_q^2 > 0. \qquad (3.217)$$

We define now

$$\Phi[p] \equiv \frac{1 - \int_{-\infty}^{\infty} dy \, [p(y)]^q}{q - 1} - \alpha \int_{-\infty}^{\infty} dy \, p(y) - \beta_q^{(2)} \frac{\int_{-\infty}^{\infty} dy \, y^2 \, [p(y)]^q}{\int_{-\infty}^{\infty} dy \, [p(y)]^q}, \qquad (3.218)$$

and then impose $\partial \Phi[p]/\partial p = 0$. We obtain

$$p_{opt}(y) = \frac{e_q^{-\beta_q^{(1,2)} y^2}}{\int_{-\infty}^{\infty} dy' \, e_q^{-\beta_q^{(1,2)} y'^2}} \quad (q < 3). \qquad (3.219)$$

By replacing this expression into constraint (3.217), we obtain the relationship between $\langle y^2 \rangle_{2q-1}$ and $\beta_q^{(1,2)}$. Finally, we can rewrite the distribution (3.219) as follows:

$$p_{opt}(x) = \frac{e_q^{-\beta_q^{(1,2)} (x - \langle x \rangle_q)^2}}{\int_{-\infty}^{\infty} dy' \, e_q^{-\beta_q^{(1,2)} y'^2}} \quad (q < 3). \qquad (3.220)$$

3.5.4 Others

A quite general situation would be to impose, in addition to

$$\int dx \, p(x) = 1, \qquad (3.221)$$

---

[24] Notice that, for $q = 2$, the upper value for $(2q - 1)$ is 3, in accordance with the upper bound indicated in Eq. (3.211).

the constraint

$$\langle f(x) \rangle_q \equiv \int dx \, f(x) \, P_q(x) = F_q \,, \tag{3.222}$$

where $f(x)$ is some known function and F_q a finite known number. We obtain

$$p_{opt}(x) = \frac{e_q^{-\beta_q(f(x)-F_q)}}{\int dx' \, e_q^{-\beta_q(f(x')-F_q)}}. \tag{3.223}$$

As for the BG case, it is clear that, by appropriately choosing $f(x)$, we can (uninterestingly) force $p_{opt}(x)$ to be virtually *any* distribution we wish. For example, by choosing $f(x) = |x|^\gamma$ ($\gamma \in \mathbb{R}$), we obtain a generic *stretched q-exponential* $p_{opt}(x) \propto e_q^{-\beta|x|^\gamma}$. If we require this distribution to be normalizable, it must satisfy $q < 1 + \gamma$. If we also require its $m - th$ order momentum $\langle |x|^m \rangle$ to be finite, then it must be $q < 1 + \frac{\gamma}{1+m}$. In particular, the first momentum of the q-exponential $p(x) \propto e_q^{-\beta|x|}$ with $\beta > 0$ is finite for $q < 3/2$, and the second momentum of the q-Gaussian $p(x) \propto e_q^{-\beta x^2}$ with $\beta > 0$ is finite for $q < 5/3$.

3.6 Nonextensive Statistical Mechanics and Thermodynamics

We arrive now at the central goal of the present introduction to *nonextensive statistical mechanics*. This theory was first introduced in 1988 [47] as a possible generalization of Boltzmann-Gibbs statistical mechanics.[25] It is from the theory of multifractals that the notation q was adopted, although, as we shall soon see, these two 'q' s are *not* the same. In fact, to avoid confusion, we shall from now on denote by q_M the multifractal index, where M stands precisely for *multifractal*. Although different, the indices q and q_M ultimately turned out to have some relation. For example, in a class of dynamical systems [306, 307] that we discuss in Chap. 5, we can see that q (more precisely the index that will be noted q_{sen}) is a special value of q_M where some discontinuities (currently referred to as Mori's phase transitions) occur at the edge of chaos.[26]

---

[25] The idea first emerged in my mind in 1985 during a meeting in Mexico City. The inspiration was related to the geometrical theory of multifractals and its systematic use of powers of probabilities.

[26] q_M is a *running* index which takes values from $-\infty$ to ∞ and is useful to characterize the various scalings occurring in multifractal structures, whereas q is a *fixed* index which characterizes a particular physical system (or, more exactly, its universality class of nonextensivity). The so-called "Thermodynamics of chaotic systems" (see, for instance, [163]) addresses a convenient discussion of multifractal geometry and some of its aspects are isomorphic to BG statistical mechanics. Within this theory, one takes Legendre transforms on the index q_M. In contrast, within "Nonextensive thermodynamics", q is fixed once for ever for a given system and its Legendre transforms concern by no means q, but precisely the same variables that are normally used in classical thermodynamics.

The present theory—nowadays known as *nonextensive statistical mechanics* statistical mechanics—constitutes a generalization of, and *by no means an alternative to*, the standard BG statistical mechanics. Its goal is to enlarge the domain of applicability of the frame of the standard theory by extending the mathematical form of its entropy. More precisely, by generalizing the entropic functional which connects the microscopic world (i.e., the probabilities of the microscopic possibilities) with some of its macroscopic manifestations. The theory has substantially evolved during the last three decades, and naturally it is still evolving at the rhythm at which new insights emerge that enable a deeper understanding of its nature, its powers, and its limitations. Successive collections of mini-reviews are available in the literature: see [18, 70, 73–90, 224]. Some of its relevant mathematical foundations are explored in [225].

The theory starts by postulating the use of the *nonadditive entropy* S_q[27] as indicated (in its discrete form) in Eq. (3.18), with the norm constraint (1.2), i.e.

$$\sum_{i=1}^{W} p_i = 1. \tag{3.224}$$

If the system is *isolated*, no other constraint exists, and this physical situation is referred to as the *microcanonical ensemble*. All nonvanishing probabilities are equal and equal to $1/W$. Indeed, this uniform distribution is the one which extremizes S_q. The entropy is then given by expression (3.16).

If we want to formulate instead the statistical mechanics of the *canonical ensemble*, i.e., of a system *in longstanding contact with a large thermostat at fixed temperature*, we need to add one more constraint (or even more than one, in fact, for more complex systems), namely that associated with the energy. The expression of this constraint is less trivial than it seems at first sight! Indeed, it has been written in different forms since the first proposal of the theory. Let us describe here these successive forms since the underlying epistemological process is undoubtedly quite instructive.

The *first form* was that adopted in 1988 [47], namely the simplest possible one (Eq. (2.66)):

Its mathematics is, in variance with that of "Thermodynamics of chaotic systems", *not* isomorphic to the BG one, but rather contains it as a particular case.

[27] Let us emphasize that, although the index q is in principle chosen so that the total *nonadditive entropy* S_q is *extensive*, the theory is referred to as *nonextensive statistical mechanics*. This is due, on one hand, to historical reasons, and, on the other hand, to the fact that this thermostatistics primarily focuses on systems whose total energy is typically *nonextensive*. The systems to which this theory is, in one way or another, applicable are generically referred to as *nonextensive systems*. Let us however stress that referring to the *intrinsically* nonadditive entropy S_q as *nonextensive* constitutes a misnomer; it was so, by inadvertence, in the title of the book [18], and remains so today. Indeed, the extensivity of an entropy depends both on the specific entropic functional and on the system for which it is used.

$$\sum_{i=1}^{W} p_i \, E_i = U_q^{(1)}. \tag{3.225}$$

The extremization of S_q with constraints (3.224) and (3.225) yields

$$p_i^{(1)} \propto [1 - (q-1)\beta^{(1)} E_i]^{1/(q-1)} = e_{2-q}^{-\beta^{(1)} E_i}. \tag{3.226}$$

This expression already exhibits important facts of nonextensive statistics, namely the emergence of the q-exponential function which enables to recover the BG exponential weight at the $q \to 1$ limit, and the possibility for asymptotic power-law behavior (or *cutoff*) at high energies when $q \neq 1$. However, it can be seen that it does *not* allow for a simple connection with thermodynamics, in the sense that no partition function can be defined which would *not* depend on the Lagrange parameter α, but only on the parameter $\beta_q^{(1)}$. Moreover, $p_i^{(1)}$ is *not* invariant, for fixed $\beta_q^{(1)}$, with regard to a changement of zero of energies. Indeed, $e_q^{a+b} \neq e_q^a \, e_q^b$ (if $q \neq 1$), and therefore, as it stands, it is not possible to factorize the new zero of energy so that it becomes canceled between numerator and denominator.

The *second form* for the constraint was first indicated in [47] and developed in 1991 [67, 68]. It is written as follows:

$$\sum_{i=1}^{W} p_i^q \, E_i = U_q^{(2)}. \tag{3.227}$$

The extremization of S_q with constraints (3.224) and (3.227) yields

$$p_i^{(2)} \propto [1 - (1-q)\beta^{(2)} E_i]^{1/(1-q)} = e_q^{-\beta^{(2)} E_i}. \tag{3.228}$$

It can be seen that this result allows for a simple factorization of the Lagrange parameter α, hence a simple partition function emerges which, like in BG statistics, only depends on $\beta_q^{(2)}$. Consistently, a smooth connection with classical thermodynamics becomes possible. However, $p_i^{(2)}$ is still *not* invariant, for fixed $\beta_q^{(2)}$, with regard to a changement of zero of energies. Even more disturbing, the type of average used in Eq. (3.227) violates the (a priori reasonable) result that the average of a constant precisely coincides with that constant. For similar reasons, if we consider $E_{ij}^{A+B} = E_i^A + E_j^B$ with $p_{ij}^{A+B} = p_i^A \, p_j^B$,[28] we do *not* generically obtain $U_q^{(2)}(A+B) = U_q^{(2)}(A) + U_q^{(2)}(B)$. These features led finally to a new formulation of the energy constraint.

The *third form* for the constraint was introduced in 1998 [69]. It is written as follows:

---

[28] Notice however that $E_{ij}^{A+B} = E_i^A + E_j^B$ and $p_{ij}^{A+B} = p_i^A \, p_j^B$ are, of course, inconsistent with Eq. (3.228), unless $q = 1$.

$$\langle E_i \rangle_q \equiv \sum_{i=1}^{W} P_i \, E_i = U_q^{(3)}, \tag{3.229}$$

where we have used the escort distribution

$$P_i \equiv \frac{p_i^q}{\sum_{j=1}^{W} p_j^q}. \tag{3.230}$$

The extremization of S_q with constraints (3.224) and (3.229) yields

$$p_i^{(3)} = \frac{[1 - (1-q)\beta_q^{(3)}(E_i - U_q^{(3)})]^{1/(1-q)}}{\bar{Z}_q} = \frac{e_q^{-\beta_q^{(3)}(E_i - U_q^{(3)})}}{\bar{Z}_q}, \tag{3.231}$$

with

$$\beta_q^{(3)} \equiv \frac{\beta^{(3)}}{\sum_{j=1}^{W} [p_j^{(3)}]^q}, \tag{3.232}$$

and

$$\bar{Z}_q \equiv \sum_{i} e_q^{-\beta_q^{(3)}(E_i - U_q^{(3)})}, \tag{3.233}$$

$\beta^{(3)}$ being the Lagrange parameter associated with constraint (3.229). This formulation simultaneously solves *all* the difficulties mentioned above, namely (i) the α Lagrange parameter factorizes, hence we can define a partition function depending only on $\beta_q^{(3)}$, hence we can make a simple junction with thermodynamics; (ii) the average of a constant coincides with that constant; (iii) if we consider $E_{ij}^{A+B} = E_i^A + E_j^B$ with $p_{ij}^{A+B} = p_i^A \, p_j^B$, we generically obtain $U_q^{(3)}(A + B) = U_q^{(3)}(A) + U_q^{(3)}(B)$; and (iv) since the difference $(E_i - U_q^{(3)})$ does *not* depend on the choice of zero for energies, the probability $p_i^{(3)}$ is *invariant*, for fixed $\beta_q^{(3)}$, with regard to any arbitrary changement of that zero.

Because of all these remarkable properties, *the third form is frequently used nowadays*. Before enlarging its discussion and presenting its connection with thermodynamics, let us finish the brief review of this instructive evolution of ideas. A few years later, it was noticed [308] that the constraint (3.229) can be rewritten in the following compact manner:

$$\sum_{i=1}^{W} p_i^q \, (E_i - U_q) = 0. \tag{3.234}$$

This approach led to the so-called "optimal Lagrange multipliers", a twist which has some interesting properties. A question obviously arrives: *Which one is the correct one, if any of these?* The answer is quite simple: *basically all of them!* Indeed, as it was first outlined in [69], and discussed in detail recently [310], they can be transformed one into the other through simple operations redefining the q's and the β_q's. Further comments can be found in [309, 311–313].[29]

To avoid confusion, and also because of its convenient properties, we shall frequently (but not always) stick to the third form [69]. Consistently, we shall from now on use the simplified notation $(p_i^{(3)}, Z_q^{(3)\prime}, U_q^{(3)}, \beta^{(3)}, \beta_q^{(3)}) \equiv (p_i, Z_q', U_q, \beta, \beta_q)$. Let us rewrite Eqs. (3.231), (3.232) and (3.233) with this simplified notation:

$$p_i = \frac{[1 - (1 - q)\beta_q (E_i - U_q)]^{1/(1-q)}}{\bar{Z}_q} = \frac{e_q^{-\beta_q (E_i - U_q)}}{\bar{Z}_q}, \tag{3.235}$$

with

$$\beta_q \equiv \frac{\beta}{\sum_{j=1}^{W} p_j^q}, \tag{3.236}$$

and

$$\bar{Z}_q \equiv \sum_i^W e_q^{-\beta_q (E_i - U_q)}. \tag{3.237}$$

Notice that, from the definition of S_q,

$$\sum_{j=1}^{W} p_j^q = 1 + (1 - q) S_q / k, \tag{3.238}$$

and also that

$$\sum_{j=1}^{W} p_j^q = (\bar{Z}_q)^{1-q}. \tag{3.239}$$

Eq. (3.239) can be established from Eq. (3.238) by using

$$S_q = k \ln_q \bar{Z}_q, \tag{3.240}$$

which is proved a few lines further on.

The (meta)equilibrium or stationary-state distribution (3.235) can be rewritten as follows:

$$p_i = \frac{e_q^{-\beta_q' E_i}}{Z_q'}, \tag{3.241}$$

with

---

[29] One more form exists in fact and it has been used with operational advantages (see, for instance, [314]). It is based on the expression $S_{2-q} = -k \sum_i p_i \ln_{2-q} p_i$ (see Table 3.1) and it consists in optimizing S_{2-q} with the standard constraints (3.224) and (3.225).

$$Z'_q \equiv \sum_{j=1}^{W} e_q^{-\beta'_q E_j}, \tag{3.242}$$

and

$$\beta'_q \equiv \frac{\beta_q}{1 + (1-q)\beta_q U_q}. \tag{3.243}$$

The form (3.241) is particularly convenient for many applications where direct comparison with experimental, observational or computational data is involved.[30]

The connection to thermodynamics is established in what follows. It can be proved that

$$\frac{1}{T} = \frac{\partial S_q}{\partial U_q}, \tag{3.244}$$

with $T \equiv 1/(k\beta)$, where, for clarity, k has been restored into the expressions. Also, we prove, for the free energy

$$F_q \equiv U_q - T S_q = -\frac{1}{\beta} \ln_q Z_q, \tag{3.245}$$

where

$$\ln_q Z_q = \ln_q \bar{Z}_q - \beta U_q. \tag{3.246}$$

This relation takes into account the trivial fact that, in contrast with what is usually done in BG statistics, the energies $\{E_i\}$ are here referred to U_q in (3.229). From Eqs. (3.245) and (3.246), we immediately obtain the anticipated relation (3.240). It can also be proved

$$U_q = -\frac{\partial}{\partial \beta} \ln_q Z_q, \tag{3.247}$$

as well as relations such as

$$C_q \equiv T \frac{\partial S_q}{\partial T} = \frac{\partial U_q}{\partial T} = -T \frac{\partial^2 F_q}{\partial T^2}. \tag{3.248}$$

In fact, the entire Legendre transformation structure of thermodynamics is q-invariant, which no doubt is remarkable and welcome.

---

[30] We may rewrite in fact the distribution (3.235) with regard to *any* referential energy that we wish, say E_0. It just becomes $p_i = [1 - (1-q)\beta_q^{(0)}(E_i - E_0)]^{1/(1-q)}/Z_q^{(0)}$ with $\beta_q^{(0)} \equiv \beta_q/[1 + (1-q)\beta_q(U_q - E_0)]$ and $Z_q^{(0)} \equiv \sum_{j=1}^{W}[1 - (1-q)\beta_q^{(0)}(E_j - E_0)]^{1/(1-q)}$. If we choose $E_0 = U_q$ we get back (3.235); if we choose $E_0 = 0$ we recover (3.241); in some occasions, it is useful to identify $E_0 \equiv \mu$ with a chemical potential μ. The preference of a particular referential energy E_0 is dictated by convenience for specific applications. Notice also that these expressions, e.g., (3.235) are *self-referential* in the sense that Eq. (3.236) is itself expressed in terms of the set $\{p_i\}$. This implies of course in a slight operational complication. There are however in the literature several procedures that conveniently overcome this difficulty. One of those procedures is indicated in [69].

It is worth mentioning at this point more two general relations concerning U_q, namely[315]

$$U_q = -\frac{\frac{\partial \ln Z'_q}{\partial \beta'_q}}{1 - (1-q)\beta'_q \frac{\partial \ln Z'_q}{\partial \beta'_q}} \tag{3.249}$$

and

$$U_q = \frac{1}{(1-q)\beta'_q}\left[1 - \frac{Z'_q}{\sum_{j=1}^{W}\left(e_q^{-\beta'_q E_j}\right)^q}\right], \tag{3.250}$$

where β'_q is given by Eq. (3.243). The latter equation remains as it stands in the continuum limit by simply replacing the sum by the corresponding integral in the Gibbs $2dN$-dimensional Γ phase space of the d-dimensional system,

Let us stress an important fact. The temperatures $T \equiv 1/(k\beta)$ and $T_q \equiv 1/(k\beta_q)$ do *not* depend on the choice of the zero of energies, and are, therefore, susceptible of physical interpretation (even if they do not necessarily coincide). Not so the temperature $T'_q \equiv 1/(k\beta'_q)$. Obviously, all these temperatures converge onto the usual one in the limit $q \to 1$.

In addition to the Legendre structure, various other important theorems and properties exhibit a q-*invariant* structure. Let us briefly mention some of them:

(i) *H-theorem* (*macroscopic time irreversibility*). Under a variety of irreversible equations such as the master equation, Fokker-Planck equation, and others, it has been proved (see, for instance, [316–319]) that

$$q\frac{dS_q}{dt} \geq 0 \quad (\forall q), \tag{3.251}$$

the equality corresponding to (meta)equilibrium. In other words, the arrow time involved in the second principle of thermodynamics basically holds in the usual way. It is appropriate to remind at this point that, for $q > 0$ ($q < 0$), the entropy tends to attain its maximum (minimum) since it is a concave (convex) functional, as already shown. Analogous H-theorems are established in [320–324] focusing on the decreasing time evolution of the free energy, instead of the increasing time evolution of the $q > 0$ entropy as in Eq. (3.251). Within this realm, Boltzman's *Molecular Chaos Hypothesis*, based on the factorization $f_a f_b$, is generalized [318, 319] into the q-*Molecular Chaos Hypothesis*, based on the q-factorization $f_a \otimes_q f_b \equiv \left[f_a^{1-q} + f_b^{1-q} - 1\right]_+^{\frac{1}{1-q}}$. This hypothesis within the q-generalized Boltzmann equation precisely leads to the stationary-state q-exponential distribution of energies.

(ii) The *Clausius relation* is verified $\forall q$, and the second principle of thermodynamics remains the same [325].

(iii) *Ehrenfest theorem* (*correspondence principle between quantum and classical mechanics*). It can be shown [326] that

$$\frac{d\langle\hat{O}\rangle_q}{dt} = \frac{i}{\hbar}\langle[\hat{\mathcal{H}}, \hat{O}]\rangle_q \quad (\forall q), \tag{3.252}$$

where $\hat{O}$ is any observable of the system.

(iv) *Onsager reciprocity theorem* (*microscopic time reversibility*). It has been shown [327–329] that the reciprocal linear coefficients satisfy

$$L_{jk} - L_{kj} \quad (\forall q). \tag{3.253}$$

(v) *Kramers and Kronig relation* (*causality*). Its validity has been proved [328] for all values of q.

(vi) *Factorization of the likelihood function* (*thermodynamically independent systems*). This property generalizes [330–332] the celebrated one introduced by Einstein in 1910 [28] (*reversal of Boltzmann formula*). The q-invariance of this likelihood factorization principle has been established in Sect. 3.3.2.

(vii) *Pesin-like identity* (*relation between sensitivity to the initial conditions and the entropy production per unit time*). It has been conjectured [186] that the q-generalized entropy production per unit time (Kolmogorov-Sinai-like entropy rate) K_q and the q-generalized Lyapunov coefficient λ_q are related through

$$K_q = \begin{cases} \lambda_q & \text{if } \lambda_q > 0, \\ 0 & \text{otherwise.} \end{cases} \tag{3.254}$$

The actual validity of this relation has been analytically proved and/or numerically verified for various classes de models [187, 188, 190–200, 333]. We come back onto this identity later on. Indeed, as we shall see, Eq. (3.254) is in fact the simplest case among an infinite family of such relations.

Properties (iii), (iv), and (v) essentially reflect something quite basic. In the theory that we are presenting here, we have generalized *nothing* concerning mechanics, either classical, quantum or whatsoever. What we have generalized is the *concept of information* upon mechanics. Consistently, the properties whose essential origin lies in mechanics should be expected to be q-invariant, and we verify that indeed they are.

Some interesting physical interpretations of nonextensive statistics are available in the literature [334–336]. We come back onto this question later on, in particular in connection with the Beck-Cohen superstatistics [337].

Let us mention also that various procedures that are currently used in BG statistical mechanics have been q-generalized. These include the variational method [338–340], the Green-function methods [328, 339, 341, 344–347], the Darwin-Fowler steepest descent method [348, 351], the Khinchin large-numbers-law method [349, 352], the counting in the microcanonical ensemble [350, 353, 354], and the cumulant expansion [355].

In the continuous (classic) limit, Eqs. (3.235–3.237) take the form

$$p(\mathbf{p}, \mathbf{x}) = \frac{e_q^{-\beta_q[\mathcal{H}(\mathbf{p},\mathbf{x})-U_q]}}{\bar{Z}_q}, \tag{3.255}$$

with

$$\beta_q \equiv \frac{\beta}{\int d\mathbf{p}\, d\mathbf{x}\, [p(\mathbf{p}, \mathbf{x})]^q}, \tag{3.256}$$

and

$$\bar{Z}_q \equiv \int d\mathbf{p}\, d\mathbf{x}\, e_q^{-\beta_q[\mathcal{H}(\mathbf{p},\mathbf{x})-U_q]}, \tag{3.257}$$

$\mathcal{H}(\mathbf{p}, \mathbf{x})$ being the Hamiltonian of the system; $(\mathbf{p}, \mathbf{x})$ denotes a generic state in the Gibbs Γ-space of the system.

In the generic quantum case, Eqs. (3.235–3.237) take the form

$$\hat{\rho} = \frac{e_q^{-\beta_q(\hat{\mathcal{H}}-U_q)}}{\bar{Z}_q}, \tag{3.258}$$

with

$$\beta_q \equiv \frac{\beta}{Tr\,\hat{\rho}^q}, \tag{3.259}$$

and

$$\bar{Z}_q \equiv Tr\, e_q^{-\beta_q(\hat{\mathcal{H}}-U_q)}, \tag{3.260}$$

$\hat{\rho}$ and $\hat{\mathcal{H}}$ being respectively the density operator and the Hamiltonian of the system.

(viii) *Fluctuations.* Interesting relations focusing on nonequilibrium fluctuations, including those discussed by Jarzynski [51], have been q-generalized. We may mention [52–57] among others.

3.7 About the Escort Distribution and the q-Expectation Values

We have seen that the escort distributions play a central role in nonextensive statistical mechanics. Let us start by analyzing their generic properties. We shall focus on the discrete version, i.e.

$$P_i \equiv \frac{p_i^q}{\sum_{j=1}^{W} p_j^q} \quad (\sum_{i=1}^{W} p_i = 1; \; q \in \mathbb{R}). \tag{3.261}$$

We will note this transformation as follows:

$$\mathbf{P} \equiv T_q[\mathbf{p}], \qquad (3.262)$$

with the notation $\mathbf{p} \equiv (p_1, p_2, ..., p_W)$ and $\mathbf{P} \equiv (P_1, P_2, ..., P_W)$. With the notation

$$(T_q * T_{q'})[\mathbf{p}] \equiv T_q[T_{q'}[\mathbf{p}]] \qquad (3.263)$$

we can easily verify the following properties:

(i) *Unit.* The unit is given by T_1. Indeed

$$\mathbf{p} \equiv T_1[\mathbf{p}], \qquad (3.264)$$

(ii) *Inverse.* The inverse of T_q is given by $T_{1/q}$. Indeed,

$$T_{1/q} * T_q = T_q * T_{1/q} = T_1. \qquad (3.265)$$

(iii) *Commutativity.* This transformation is commutative. Indeed

$$T_q * T_{q'} = T_{q'} * T_q. \qquad (3.266)$$

(iv) *Associativity.* This transformation is associative. Indeed

$$T_q * (T_{q'} * T_{q''}) = (T_q * T_{q'}) * T_{q''} \equiv T_q * T_{q'} * T_{q''}. \qquad (3.267)$$

(v) *Cloture.* Indeed

$$T_q * T_{q'} = T_{qq'}. \qquad (3.268)$$

In other words, the set of transformations $\{T_q\}$ constitutes an Abelian continuous group.

Two more properties deserve to be stated.

(vi) *Certainty is a fixed point* of the transformation. Indeed, if one of the possible states has probability p equal to unity, hence all the others have probability zero, and the same happens with P.

(vii) *Equal probabilities is a fixed point* of the transformation. Indeed, if $p_i = 1/W$ ($\forall i$), then (and only then) $P_i = 1/W$ ($\forall i$).

We have seen in the previous Subsection that one of the most convenient manners[31] for performing the optimization of the entropy is to express the

---

[31] We have written "one of the most convenient manners", and not "the manner", because, as we have already seen in the previous Subsection, the calculation can be done through various equivalent paths. For example, optimizing S_q with fixed $\langle O \rangle_q$ is equivalent to optimizing S_{2-q} with fixed $\langle O \rangle_1 \equiv \langle O \rangle$ [312]. Both optimizations yield one and the same result, namely $p_i \propto e_q^{-\bar{\beta} O_i}$, where $\bar{\beta}$ is univocally determined by using say the constraint (3.269). This freedom is kind of reminiscent of the freedom one has in quantum mechanics, where we can equivalently include the time-dependence either in the eigenvectors (Schroedinger representation) or in the operators (Dirac representation), or even partially in both (Heisenberg representation).

constraints as q-expectation values, i.e., through the escort distributions. So, if we have an observable O whose possible values are $\{O_i\}$, the associated constraint is to be written as

$$\langle O \rangle_q \equiv \sum_{i=1}^{W} P_i \, O_i = O_q, \tag{3.269}$$

where O_q is a *known finite quantity*.

Regretfully, it is not yet totally transparent what is the geometrical/probabilistic reason which makes it convenient to express the constraints as q-expectation values. We do know, however, a set of properties that surely are directly related to this elusive reason. Let us, next, list some of them that are particularly suggestive:

(i) The derivative of e_q^x is *not* the same function (unless $q = 1$), but $(e_q^x)^q$. This simple property makes naturally appear P_i instead of p_i in the steepest descent method developed in [351].

(ii) The conditional entropy (3.41) naturally appears as a q-expectation value, without involving *any* optimizing operation.

(iii) The norm constraint involves the quantity $\sum_{i=1}^{W} P_i$ with $p_i \propto 1/[1 + (q - 1)\bar{\beta} E_i]^{1/(q-1)}$. A case which frequently appears concerns $W \to \infty$, with E_i increasingly large with increasing i. In such a case, we have that $p_i \propto 1/E_i^{1/(q-1)}$ for *high* values of E_i. Therefore, q must be such that $\sum_{i=i_0}^{\infty} E_i^{-1/(q-1)}$ is *finite*, where i_0 is some arbitrary value of the index i. Equivalently, in the continuous limit, q must be such that

$$\int_{constant}^{\infty} dE \, g(E) \, E^{-1/(q-1)} < \infty, \tag{3.270}$$

where $g(E)$ is the density of states. A typical case is $g(E) \propto E^\delta$ in the $E \to \infty$ limit. In such a case, the theory is mathematically well posed if $1/(q - 1) - \delta > 1$, i.e., if

$$q < \frac{2 + \delta}{1 + \delta}. \tag{3.271}$$

For the simplest case, namely for $\delta = 0$, this implies $q < 2$. Let us make the same analysis for the constraint $U_q = [\sum_{i=1}^{W} p_i^q E_i]/[\sum_{j=1}^{W} p_j^q]$. Under the same circumstances analyzed just above, we must have the finiteness of $\int_{constant}^{\infty} dE \, g(E) \, E \, E^{-q/(q-1)}$. But this equals $\int_{constant}^{\infty} dE \, g(E) \, E^{-1/(q-1)}$. Consequently, remarkably enough, we arrive at the *same* condition (3.270)! In other words, the *entire theory* is valid up to an *unique* value of q, namely that which guarantees condition (3.270). This nice property disappears if we impose the constraint using the standard expectation value. If we do that, the energy mean value diverges for a value of q different (*smaller* in fact) than that at which the norm diverges.

(iv) An interesting analysis is available [356] which exhibits that the relative entropy $I_q(p, p^{(0)})$ that was introduced in Sect. 3.4 is directly associated with differences of free energies calculated with the q-expectation values (i.e., ordinary expectation values but using $\{P_i\}$ instead of $\{p_i\}$), whereas some different specific relative entropy is directly associated with differences of free energies calculated with the ordinary expectation values (i.e., just using $\{p_i\}$). Then they show that $I_q(p, p^{(0)})$ satisfies three important properties that the other relative entropy violates. The first of these properties is to be *jointly convex* with regard to either p or $p^{(0)}$. The second of these properties is to be *composable*. And the third of these properties is to satisfy the Shore–Johnson axioms [169] for the principle of minimal relative entropy to be *consistent as a rule of statistical inference*. It is then concluded in [356] that these arguments select the q-expectation values, and exclude the ordinary expectation values whenever we wish to use the entropy S_q. These arguments clearly are quite strong. Some further clarification would however be welcome. Indeed, stated in this strong sense, there would be a contradiction with the arguments presented in [69, 310], which lead to the conclusion that the various existing formulations of the optimization problem using S_q are mathematically equivalent, in the sense that they can be transformed one into the other (as long as *all* the involved quantities are finite, of course).

(v) It has been discussed in various systems that the theory based on q-expectation values exhibits thermodynamic stability (see, for instance, [342, 343, 357, 358]).

(vi) The Beck–Cohen superstatistics [337] (see Chap. 6) is a theory which generalizes nonextensive statistics, in the sense that its stationary-state distribution contains the q-exponential one as a particular case. In order to go one step further along the same line, i.e., for this approach to become a statistical mechanics with a possible connection to thermodynamics, it also needs to have a corresponding entropy. This step was accomplished in [359–361] by generalizing the entropy S_q. But it became clear in this extension that generalizing the entropy was not enough: the mathematical form of the energy constraint had to be generalized as well. To be more precise, in order to make some contact with the macroscopic level, the only solution that was found was to simultaneously generalize the entropy *and* the form of the constraint. This fact suggests of course that, from an information-theoretical standpoint, it is kind of natural to generalize not only the entropic functional but *also* the expression of the constraints. Further results along this line have been established in [362, 363].

(vii) Let us anticipate that, in the context of the q-generalization of the central limit theorem that we present later on, a natural generalization emerges for the Fourier transform. This is given, for $q \geq 1$, by

$$F_q[p](\xi) \equiv \int_{-\infty}^{\infty} dx \, e_q^{i \, x \, \xi \, [p(x)]^{q-1}} \, p(x), \qquad (3.272)$$

where $p(x)$ can be a distribution of probabilities. We immediately verify that

$$F_q[p](0) = 1, \tag{3.273}$$

and

$$\left[\frac{d F_q[p](\xi)}{d\xi}\right]_{\xi=0} = i \int_{-\infty}^{\infty} dx \, x \, [p(x)]^q. \tag{3.274}$$

As we see, it is the numerator of the q-mean value, and *not* that of the standard mean value, which emerges naturally. As we shall see in due time (through the q-Fourier transform and [305]), Eqs. (3.273) and (3.274) are the two first elements of an infinite set of finite values which uniquely determine the distribution $p(x)$ itself.

(viii) Last but not least, let us rephrase property (iii) in very elementary terms. We assume that we have the simple case of a stationary-state q-exponential distribution $p(x) \propto e_q^{-\beta x}$ $(x \geq 0)$. The characterization of a distribution such as this one involves *two* important numbers, namely the *decay exponent* $1/(q - 1)$ of the tail, and the overall *width* $1/\beta$ of the distribution. It must be so for any value of $q < 2$ (upper bound for the existence of a norm). We easily verify that the standard mean value of x diverges in the region $3/2 \leq q < 2$, and it is, therefore, useless for characterizing the width of the distribution for that interval of q. The q-mean value instead is finite and uniquely determined by the width $1/\beta$ up to $q < 2$. In other words, the *robust* information about the width of the q-exponential distribution is conveniently provided precisely through the escort distribution. Totally analogous considerations are applicable to the q-Gaussian distributions, which is normalizable for $q < 3$, but whose standard variance remains finite only for $q < 5/3$: See Appendix B for a detailed discussion of this issue and its practical consequences.

3.8   About Universal Constants in Physics

I would mention at this place a point which epistemologically remains somewhat elusive. We shall exhibit and further comment that, *for any value of the entropic index $q \neq 1$ and all systems*, the stationary-state energy distribution within nonextensive statistical mechanics becomes that of BG statistical mechanics in the limit of vanishing inverse Boltzmann constant $1/k_B$. The physical interpretation of this property is, in my opinion, kind of intriguing and unavoidably reminds facts like quantum mechanics becoming Newtonian mechanics in the limit of vanishing $\hbar$, special relativity becoming once again Newtonian mechanics in the limit of vanishing $1/c$, and general relativity recovering the Newtonian flat space-time in the limit of vanishing G. While we may say that, for these three mechanical examples, the corresponding physical interpretations are reasonably well understood (see Fig. 3.24), it escapes to an equally clear perception of what kind of subtle informational meaning

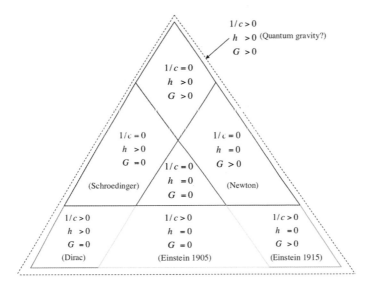

Fig. 3.24 Physical structure at the $1/k_B = 0$ plane. The full diagram involves 4 universal constants, and would be a tetrahedron. At the center of the tetrahedron, we have the case $c^{-1} = h = G = k_B^{-1} = 0$, and the overall tetrahedron corresponds to $1/c > 0, h > 0, G > 0, 1/k_B > 0$ (statistical mechanics of quantum gravity)

could be attributed to $1/k_B$ going to zero while q is kept fixed at an arbitrary value. The meaning of the four universal constants $\hbar$, c, G, k_B has been addressed by G. Cohen-Tannoudji in terms of *physical horizon* [364]. See also [365].

If we assume $k = k_B$ in Eq. (3.21), and cancel it on both sides, we obtain

$$S_q(A + B) = S_q(A) + S_q(B) + \frac{1-q}{k_B} S_q(A) S_q(B). \tag{3.275}$$

As we see, we go back to the BG situation if $(1 - q)/k_B = 0$. This can occur in two different manners, namely either $q = 1$ ($\forall k_B^{-1}$), or $k_B^{-1} = 0$ ($\forall q$). In this sense, any departure from the BG entropic composition law is equivalent to a departure from $k_B^{-1} \neq 0$.

In thermal equilibrium (as well as in other (quasi) stationary states) k_B always appears coupled together with the temperature T in the form $k_B T$. In other words, small values for k_B^{-1} are equivalent to the high-temperature limit. It seems reasonable to think that this connection is not unrelated to the fact that, for small $(k_B T)^{-1}$, the BG canonical and grand-canonical ensembles asymptotically recover the microcanonical ensemble. The same happens for Bose-Einstein and Fermi-Dirac quantum statistics, in fact for all Gentile parastatistics [131–133] which unifies the standard quantum ones. Even more, the same happens for all q-statistics (and similarly for Beck-Cohen superstatistics) if we take into account the property $e_q^x \sim 1 + x$, for $x \to 0$ and *all values of q*. In other words, for $(q - 1)/k_B T \to 0$, *all* the stationary-state statistics that we are focusing on asymptotically exhibit confluence onto a single behavior, namely that corresponding to the BG microcanonical ensemble, which corresponds to *even* occupancy of the admissible phase space. But even occupancy is associated to a Lebesgue measure which essentially factorizes into the Lebesgue measures corresponding to the various degrees of freedom. In other words, it corresponds to independence. The connection ends by recalling that the appropriate entropy for probabilistically independent subsystems precisely is S_{BG}, i.e., $q = 1$. So, from the entropic viewpoint, the k_B^{-1} plane represented in Fig. 3.24 equivalently corresponds to $q = 1$. Out of this plane, in some subtle sense, we start having information corresponding to nontrivially correlated subsystems (nonflat space in the language of Information Geometry [120, 122, 123]; see [124] for a comprehensive mathematical introduction of this theory).

In this context, it is interesting to focus again on Eq. (3.43). It is precisely the existence of the extra, bilinear term that enables [366], for special correlations, to recover the Clausius entropy thermodynamical extensivity $S_q(A + B) \sim S_q(A) + S_q(B)$ for large systems A and B.

Let us close this digression about the physical universal constants by focusing on the fact that all known constants used in contemporary physics (as well as in chemistry, engineering, biology, medicine, etc) can be expressed in terms of units of *length*, *time*, *mass*, and *temperature*. Equivalently, each of them can be expressed as a pure number multiplied by some combination of powers of c^{-1}, h, G, and k_B^{-1}. No further reduction below *four* universal constants is possible in contemporary physics. This point is however quite subtle, as can be seen in [367–370]. It is related to the fact that any fundamental discovery tends to reduce the number of units that are necessary to express the physical quantities. For example, in ancient times, there were independent units for *area* and *length*. The situation changed when it became clear that, in Euclidean geometry, an area can be expressed as the square of a length.

Consistently with the above, Planck introduced [371–373] four natural units for *length*, *mass*, *time*, and *temperature*, namely

$$l_P \equiv Planck\, length \qquad = \sqrt{\frac{hG}{c^3}} \qquad = 4.13 \times 10^{-33} cm \qquad (3.276)$$

$$m_P \equiv Planck\, mass \qquad = \sqrt{\frac{hc}{G}} \qquad = 5.56 \times 10^{-5} g \qquad (3.277)$$

$$t_P \equiv Planck\, time \qquad = \sqrt{\frac{hG}{c^5}} \qquad = 1.38 \times 10^{-43} s \qquad (3.278)$$

$$T_P \equiv Planck\, temperature = \frac{1}{k_B}\sqrt{\frac{hc^5}{G}} = 3.50 \times 10^{32} \,^{o}K \qquad (3.279)$$

There is no need to add to this list the elementary electric charge e. Indeed, it is related to the already mentioned constants through the *fine-structure constant* $\alpha \equiv 2\pi e^2 / hc = 1/137.035999206(11) \dots$ (pure numbers, like α and like say the proton-to-electron mass ratio $\beta \equiv m_p/m_e = 1836.152673\ldots$, appear to play crucial roles in the present context).

From the above considerations, it seems quite natural to think that the number of independent physical fundamental constants is, at the present stage of contemporary physics, four. However, this point is sometimes challenged. An excellent overview on this interesting issue has been published by Duff [374].

Let us close these remarks on the physical fundamental constants, by focusing on the concept of *stability of a theory* in the sense of Faddeev [375]. For example, Newtonian mechanics is *unstable* with regard to $1/c$ since, as soon as $1/c \neq 0$, the Lorentz invariance replaces the Galilean invariance. Special relativity is, in contrast, *stable* with regard to $1/c$ since there is no topological modification by changing the value of $1/c$ *if it is already different from zero*. Analogously, Newtonian mechanics is *unstable* with regard to h. Indeed, as soon as $h \neq 0$, the Poisson brackets are to be replaced by commutators, i.e., not all dynamical quantities can be precisely known simultaneously. Consistently, quantum mechanics is *stable* with regard to h. Similarly, special relativity is *unstable* with regard to G since $G \neq 0$ immediately implies in a curvature for space-time; consistently, general relativity is *stable* with regard to G. What about BG statistical mechanics? It is clear from Eq. (3.275) that the BG theory is *unstable* with regard to $1/k_B$, more precisely with regard to $(1-q)/k_B$ [365]. Indeed, as soon as we have $(1-q)/k_B \neq 0$, the additivity of the entropic functional with regard to the entropies of its parts (assumed to be independent) is lost. Consistently, nonextensive statistical mechanics is a theory which is *stable* with regard to $(1-q)/k_B$. Moreover, an interesting peculiarity emerges in the case of the q-generalized BG theory. Whereas all four fundamental constants $(1/c, h, G, 1/k_B)$ are either vanishing or positive, the quantity $(1-q)/k_B$ can also be negative! This reflects to the fact that the energy distribution at the stationary state has a cutoff for $q < 1$, instead of the asymptotic power law which emerges whenever $q > 1$. This is strongly reminiscent to say the isotropic Heisenberg ferromagnet at criticality. Indeed, the model can lose its isotropy in two very different manners: if it is oblate, the continuous symmetry of three-dimensional rotations jumps onto the discrete symmetry of the Ising ferromagnet, whereas if it is prolate it jumps onto the continuous symmetry of two-dimensional rotations (i.e., the isotropic XY ferromagnet).

3.9 Comparing Various Entropic Forms

For simplicity, we shall assume $k = 1$ in all the following interrelated definitions.

The *Renyi entropy* is defined as follows [150]:

$$S_q^R \equiv \frac{\ln \sum_{i=1}^W p_i^q}{1-q} = \frac{\ln[1+(1-q)S_q]}{1-q}, \quad (q \in \mathbb{R}). \tag{3.280}$$

This entropy can be q'-generalized as follows:

$$S_{q,q'}^{gR} \equiv \frac{\ln_{q'} \sum_{i=1}^{W} p_i^q}{1-q}, \quad ((q,q') \in \mathbb{R}^2), \tag{3.281}$$

with $\ln_q x$ given [134] by Eq. (3.14), or, even further, (q', q'')-generalized as follows:

$$S_{q,q',q''}^{gR} \equiv \frac{\ln_{q',q''} \sum_{i=1}^{W} p_i^q}{1-q}, \quad ((q,q',q'') \in \mathbb{R}^3), \tag{3.282}$$

with $\ln_{q,q'} x$ given [242] by Eq. (3.111).

The *Arimoto entropy* is defined as follows [376]:

$$S_q^{Ar} = \frac{q}{q-1} \Big[1 - \Big(\sum_{i=1}^{W} p_1^q \Big)^{1/q} \Big], \quad (q \in \mathbb{R}). \tag{3.283}$$

The *Sharma-Mittal* entropy (also called *Sharma-Taneja-Mittal* entropy) [147–149] (see also [377]) unifies S_q and S_q^R as follows:

$$S_{q,r}^{SM} = \frac{1}{1-r} \Big[\Big(\sum_{i=1}^{W} p_i^q \Big)^{\frac{1-q}{1-r}} - 1 \Big] = \frac{1}{1-r} \Big[\big(1 + (1-q)S_q \big)^{\frac{1-q}{1-r}} - 1 \Big], \quad ((q,r) \in \mathbb{R}^2), \tag{3.284}$$

with $\lim_{r \to 1} S_{q,r}^{SM} = S_q^R$ and $S_{q,q}^{SM} = S_q$. $S_{q,r}^{SM}$ is a monotonic function of $S_q, \forall(q,r) \in \mathbb{R}^2$.

Inspired by S_q, Borges and Roditi extended it in 1998 as follows [378]:

$$S_{q,q'}^{BR} = \frac{\sum_{i=1}^{W} p_i^q - \sum_{i=1}^{W} p_i^{q'}}{q'-q} = \frac{1-q}{q'-q} S_q + \frac{1-q'}{q-q'} S_{q'}, \quad ((q,q') \in \mathbb{R}^2), \tag{3.285}$$

with $S_{q,1}^{BR} = S_{1,q}^{BR} = S_q$, where BR stands for *Borges-Roditi*.

Abe introduced in 1997 the so-called *Abe entropy* as follows [161]:

$$S_q^{Ab} = \frac{\sum_{i=1}^{W} p_i^q - \sum_{i=1}^{W} p_i^{1/q}}{1/q - q}, \quad (q \in \mathbb{R}). \tag{3.286}$$

We can easily verify that $S_q^{Ab} = S_{q,1/q}^{BR}$.

The *Kaniadakis entropy*, also called the κ-*entropy*, is defined as follows [379]:

$$S_\kappa^K \equiv \sum_{i=1}^{W} p_i \ln_\kappa^K \frac{1}{p_i}, \quad (\kappa \in \mathbb{R}) \tag{3.287}$$

with

$$\ln_\kappa^K x \equiv \frac{x^\kappa - x^{-\kappa}}{2\kappa} \quad (\ln_0^K x = \ln x), \tag{3.288}$$

K standing for *Kaniadakis*. We trivially verify that $S_\kappa^K = S_{-\kappa}^K$, and also that $S_\kappa = S_{1+\kappa,1-\kappa}^{BR}$. Consistently, we also verify $\ln_\kappa^K x = \frac{1}{2}(\ln_q x - \ln_q \frac{1}{x}) = \frac{1}{2}(\ln_q x + \ln_{2-q} x)$ with $q = 1 + \kappa$, hence $\ln_\kappa^K x = \frac{1}{2}(\ln_{1+\kappa} x + \ln_{1-\kappa} x)$. Consequently, the definition (3.287) implies

$$S_\kappa^K = \frac{1}{2}(S_{1+\kappa} + S_{1-\kappa}). \qquad (3.289)$$

Let us mention that, as a mathematical possibility, the Abe and the Kaniadakis entropies can be unified by straightforwardly using transformation (4.106) (to be discussed later on) as follows:

$$S_q^{(z)} \equiv S_{q,\frac{z+2-(z+1)q}{z+1-zq}}^{BR} \qquad (z = 0, \pm 1, \pm 2, \pm 3, \dots). \qquad (3.290)$$

Indeed, for $z = -1$ and $z = 0$, we recover Abe and Kaniadakis entropies, respectively.

The Kaniadakis entropy (3.289) was extended in 2004 as follows [380]:

$$S_{\kappa,r}^{KLS} = -\sum_{i=1}^W p_i^{1+r} \frac{p_i^\kappa - p_i^{-\kappa}}{2\kappa}, \qquad ((\kappa, r) \in \mathbb{R}^2), \qquad (3.291)$$

where KLS stands for *Kaniadakis-Lissia-Scarfone*.

Let us also remark that, for $q = q'$ within $S_{q,q'}^{BR}$, we recover the *Shafee entropy* [381], namely

$$S_q^S \equiv \lim_{q/q' \to 1} S_{q,q'}^{BR} = -\sum_{i=1}^W p_i^q \ln p_i, \qquad (3.292)$$

where S stands for *Shafee*.[32, 33]

The *Curado b-entropy* is defined as follows [178, 179]:

$$S_b^C \equiv \sum_{i=1}^W (1 - e^{-bp_i}) + e^{-b} - 1 \quad (b \in \mathbb{R}; b > 0). \qquad (3.293)$$

The entropy introduced in [383], and which we shall from now on refer to as *exponential entropy*, is defined as follows:

$$S^E = \sum_{i=1}^W p_i \left(1 - e^{\frac{p_i - 1}{p_i}}\right). \qquad (3.294)$$

---

[32] This entropy can be expressed in terms of the escort distribution $P_i \equiv p_i^q / \sum_{j=1}^W p_j^q$. More precisely $S_q^S(\{p_i\}) = \frac{\sum_{j=1}^W p_j^q}{q}\left[S_{BG}(\{P_i\}) - \ln \sum_{j=1}^W p_j^q\right]$.

[33] For simplicity, we refer to it as Shafee entropy. However, in a rather cryptic form, it is already present in [382].

This entropic functional can be extended as follows:

$$S_c^E = \sum_{i=1}^{W} p_i \left(1 - e^{c \frac{p_i - 1}{p_i}}\right) \quad (c \in \mathbb{R}; c > 0).$$ (3.295)

This and the Curado entropy have the peculiarity of *not* having S_{BG} as a particular instance.

The *Anteneodo–Plastino entropy* is defined as follows [180]:

$$S_\eta^{AP} \equiv \sum_{i=1}^{W} \left[\Gamma\left(\frac{\eta+1}{\eta}, -\ln p_1\right) - p_i \Gamma\left(\frac{\eta+1}{\eta}\right)\right] \quad (\eta \in \mathbb{R}; \eta > 0),$$ (3.296)

where

$$\Gamma(\mu, t) \equiv \int_t^\infty dy \, y^{\mu-1} e^{-y} = \int_0^{e^{-t}} dx \, (-\ln x)^{\mu-1} \quad (\mu > 0)$$ (3.297)

is the *complementary incomplete Gamma function*, and $\Gamma(\mu) = \Gamma(\mu, 0)$ is the *Gamma function*. We verify that $S_1^{AP} = S_{BG}$.

The *Landsberg–Vedral–Rajagopal–Abe entropy*, or just *normalized S_q entropy*, is defined as follows [384, 385]:

$$S_q^{LVRA} \equiv S_q^N \equiv \frac{S_q}{\sum_{i=1}^{W} p_i^q} = \frac{1 - \left[\sum_{i=1}^{W} p_i^q\right]^{-1}}{1 - q} = \frac{S_q}{1 + (1-q)S_q}.$$ (3.298)

The so-called *escort entropy* is defined as follows [69]:

$$S_q^E \equiv \frac{1 - \left[\sum_{i=1}^{W} p_i^{1/q}\right]^{-q}}{q-1} = \frac{1 - [1 - \frac{1-q}{q} S_{1/q}]^{-q}}{q-1}, \quad (q \in \mathbb{R}).$$ (3.299)

Inspired by S_δ defined in (3.134), it has been introduced in [222], the following two-indices entropic form:

$$S_{q,\delta} = \sum_{i=1}^{W} p_i \left(\ln_q \frac{1}{p_i}\right)^\delta.$$ (3.300)

$S_{q,1}$ and $S_{1,\delta}$ respectively recover S_q and S_δ; $S_{1,1}$ recovers S_{BG}. Obviously, this entropy is nonadditive unless $(q, \delta) = (1, 1)$, and it is expansible (see, for instance, [103]), $\forall q > 0, \forall \delta > 0$. The region (q, δ) where $S_{q,\delta}$ is concave depends nontrivially on W [256].

For equal probabilities we have

$$S_{q,\delta} = (\ln_q W)^{\delta}. \tag{3.301}$$

Let us also mention at this point several other interrelated entropies with two, three, or more indices, which have been introduced with diverse mathematical motivations. One of them is as follows [242]:

$$S_{q,q'} = \frac{1 - \sum_{i=1}^{W} p_i e^{\left[\frac{(q'-1)(p_i^{q-1}-1)}{q-1}\right]}}{q' - 1}, \quad ((q,q') \in \mathbb{R}^2), \tag{3.302}$$

with $S_{q,1} = S_{1,q} = S_q$.

Another one is as follows [386, 387]:

$$S_{c,d}^{HT} = \frac{e \sum_{i=1}^{W} \Gamma(1+d, 1 - c \ln p_i)}{1 - c + cd} - \frac{c}{1 - c + cd}, \quad (c \in [0,1]; \ d \in \mathbb{R}), \tag{3.303}$$

Γ being the incomplete Gamma Function, and where HT stands for *Hanel–Thurner*. Still another one is as follows [388]:

$$S_{a,b,r}^{(\pm)} = \sum_{i=1}^{W} p_i \ln_{a,b,r}^{(\pm)} \frac{1}{p_i}, \quad ((a,b,r) \in \mathbb{R}^3), \tag{3.304}$$

where

$$\ln_{a,b,r}^{(\pm)}(x) \equiv \frac{2(x^r - 1)}{-a(x^r - 1) \pm \sqrt{a^2 + 4b}\,(x^r + 1)}, \quad (x > 0), . \tag{3.305}$$

For the intricate domains of validity of (a,b,r) refer to [388].

There is also the *Masi entropy* [389] defined as follows:

$$S_{q,r}^M = \frac{\left[1 + \frac{1-r}{1-q} \ln \sum_{i=1}^{W} p_i^q\right]^{\frac{1-q}{1-r}} - 1}{1 - q}, \quad ((q,r) \in \mathbb{R}^2), \tag{3.306}$$

where M stands for *Masi*. We verify that $\lim_{r \to 1} S_{q,r}^M = S_q$ and $\lim_{r \to q} S_{q,r}^M = S_q^R$. One more (with three independent indices) is as follows [390]:

$$S_{\alpha,\beta,q}^T = \frac{1}{1-q} \sum_{i=1}^{W} p_i \left(\alpha p_i^{-2(1-q)} + \frac{1}{2}(1 - 3\alpha + \beta) p_i^{-(1-q)}\right.$$
$$\left. + \frac{1}{2}(\alpha - 1 - 3\beta) p_i^{1-q} + \beta p_i^{2(1-q)}\right), \quad ((\alpha,\beta,q) \in \mathbb{R}^3) \tag{3.307}$$

where T stands for *Tempesta*.

It is worth mentioning at this point that, on the basis of group theory, a remarkable entropic functional has been introduced in 2011 by Tempesta [391], which is determined by an infinite number of indices, namely the set of coefficients of a Dirichlet series. The simplest Dirichlet series is the one whose coefficients are all equal to unity, and that is the Riemann celebrated zeta function $\zeta(s)$. Intriguingly, the corresponding entropy turns out to be S_q, the connection being

$$S_q \Leftrightarrow \frac{1}{(1-q)^{s-1}} \zeta(s) \quad (q < 1), \tag{3.308}$$

with

$$\zeta(s) \equiv \sum_{n=1}^{\infty} \frac{1}{n^s} \tag{3.309}$$

or, equivalently,

$$\zeta(s) = \prod_{p\ prime} \frac{1}{1-p^{-s}} = \frac{1}{1-2^{-s}} \frac{1}{1-3^{-s}} \frac{1}{1-5^{-s}} \cdots . \tag{3.310}$$

Another entropy, once again based on group theory, has been introduced in 2018 as follows [250]:

$$S_{\gamma,\alpha}^{JPPT} = exp\left[L\left(\frac{\ln \sum_{i=1}^{W} p_i^{\alpha}}{\gamma(1-\alpha)} \right) \right] - 1, \quad ((\gamma, \alpha) \in \mathbb{R}^2), \tag{3.311}$$

the *Lambert function* $L(x)$ being defined through $x = L(x)e^{L(x)}$, and where $JPPT$ stands for *Jensen-Pazuki-Pruessner-Tempesta*. Further analysis of the so-called exponential (typically $W(N) \sim \mu^N$ with $\mu > 1$), sub-exponential (typically $W(N) \sim N^{\rho}$ with $\rho > 0$ or $W(N) \sim \mu^{N^{\gamma}}$ with $\mu > 1$ and $0 < \gamma < 1$) and super-exponential class (e.g., $W(N) \sim N^{\bar{\gamma}N}$ with $\bar{\gamma} > 0$) is available at [251].

We straightforwardly verify that

$$\lim_{q \to 1} S_q = \lim_{q \to 1} S_q^R = \lim_{(q,r) \to (1,1)} S_{q,r}^{SM} = \lim_{(q,q') \to (1,1)} S_{q,q'}^{BR} = \lim_{q \to 1} S_q^L = \lim_{\kappa \to 0} S_\kappa^K$$

$$= \lim_{(\kappa,r) \to (0,0)} S_{\kappa,r}^{KLS} = \lim_{q \to 1} S_q^N = \lim_{q \to 1} S_q^E = \lim_{\eta \to 1} S_\eta^{AP} = \lim_{(1-q) \to 1} S_{(1-q)}$$

$$= \lim_{q \to 1} S_q^A = \lim_{(q,(1-q)) \to (1,1)} S_{q,(1-q)} = \lim_{(q,q') \to (1,1)} S_{q,q'} = \lim_{(c,d) \to (1,1)} S_{c,d}^{HT}$$

$$= \lim_{(a,b,r) \to (0,0,0)} S_{a,b,r} = \lim_{q \to 1} S_{\alpha,\beta,q}^T = S_{BG}. \tag{3.312}$$

We can also verify that each of S_q, S_q^R and $S_q^N \equiv S_q^{LVRA}$ is a monotonic function of each one of the others. Therefore, under the same constraints, they yield *one and the same extremizing probability distribution*. Indeed, optimization is preserved through monotonicity. Not so, by the way, for concavity, convexity, and other important properties. For various comparisons, see Figs. 2.1, 3.4, 3.5, 3.6.

As a reminder let us mention that many other extensions of the classical BG entropy are available in the literature that follow along related lines: see for instance [359, 381, 389, 392–395]. Some of them are founded on abstract mathematical grounds and can be characterized by countable infinitely many parameters. Such is the case of the so-called Z-entropies [390, 396] and their generalization referred to as the group entropies (in their univariate and multivariate forms [251, 397]). There are even examples that depend on quite generic functions, namely $f(\tau)$ [383] and $g(x)$ [397]. Let us finally mention an intriguing example based on the Riemann ζ function [398]. See Fig. 3.25 for a taxonomy of various entropic functionals in terms

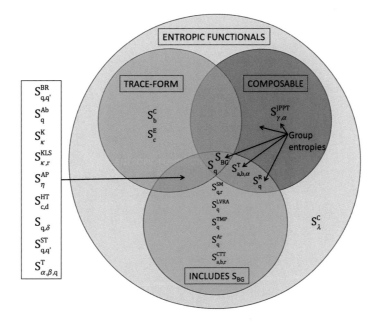

Fig. 3.25 It has been proved [236] that S_q is the unique entropic form which simultaneously is trace-form, composable, and recovers S_{BG} as a particular instance. S_q (hence S_{BG}), the Renyi entropy S_q^R [150], the Tempesta (a, b, α)-entropy $S_{a,b,\alpha}^T$ (Eq. (9.1) in [396]), the Jensen–Pazuki–Pruessner–Tempesta entropy $S_{\gamma,\alpha}^{JPPT}$ [250] and many others belong to the class of *group entropies* and are, therefore, composable. To facilitate the identification, we are here using the following notations: Sharma–Mittal entropy $S_{q,r}^{SM}$ [147], Landsberg–Vedral–Rajagopal–Abe entropy S_q^{LVRA} [384, 385], Tsallis–Mendes–Plastino entropy S_q^{TMP}, Arimoto entropy S_q^{Ar} [376], Curado–Tempesta–Tsallis entropy $S_{a,b,r}^{CTT}$ [388], Borges–Roditi entropy $S_{q,q'}^{BR}$ [378], Abe entropy S_q^{Ab} [161], Kaniadakis entropy S_κ^K [379], Kaniadakis–Lissia–Scarfone entropy $S_{\kappa,r}^{KLS}$ [380], Anteneodo–Plastino entropy S_η^{AP} [180], Hanel–Thurner entropy $S_{c,d}^{HT}$ [386, 387], $S_{q,\delta}$ [222], Schwammle-Tsallis entropy $S_{q,q'}^{ST}$ [242], the Tempesta (α, β, q)-entropy $S_{\alpha,\beta,q}^T$ [390], the Curado b-entropy S_b^C [178, 179], the Curado λ-entropy S_λ^C [270], and the exponential c-entropy S_c^E (see Eq. (3.295) and [383]). The entropic form S_λ^C is a rare case which does not include S_{BG} and is neither trace-form nor composable

of three important properties, namely whether it is of the trace-form, composable, and able to recover S_{BG} as a particular case. The unique functional simultaneously satisfying these three properties is S_q.

Before ending this Section, let us also mention that it is possible to considerably generalize and unify many of the above entropic forms, linking them to important informational quantities and properties such as Fisher information, Crámer-Rao inequality, de Bruijn identity, escort probabilities, and others [399, 400].

Part II
Foundations or Why the Theory Works

Chapter 4
Probabilistic and Stochastic Dynamical Foundations of Nonextensive Statistical Mechanics

Si l'action n'a quelque splendeur de liberté, elle n'a point de grâce ni d'honneur.

Montaigne, *Essais*

4.1 Introduction

In this chapter we primarily focus on *mesoscopic*-like nonlinear dynamical systems, in the sense that the time evolution includes, in addition to deterministic ingredients, *stochastic noise*.

A paradigmatic path in statistical physical systems is as follows. We assume the knowledge of the Hamiltonian of a classical or quantum many-body system. This is referred to as the *microscopic* level, or *microscopic* description. If the system is classical, the time evolution is given by Newton's law $\mathbf{F} = m\,\mathbf{a}$, and is therefore completely *deterministic*. The motion of the system is completely determined by the Hamiltonian and the initial conditions. However, it is in general tremendously difficult to solve the corresponding equations. So, as a simpler alternative, Langevin introduced the following *phenomenological* approach. We focus on *one* molecule or element of the system, and its motion is described in terms of the combination of two ingredients. The first ingredient is *deterministic*, coming typically from the existence of a possible external potential acting on the entire system, as well as from the average action of all the other molecules or elements. The second ingredient is *stochastic*, introduced in an *ad hoc* manner into the equations as a *noise*. This noise represents the rapidly fluctuating effects of the rest of the system onto the single molecule we are observing. This level and the associated description are referred to as the *mesoscopic* ones, and the basic equation is of course the *Langevin equation* (as well as the Kramers equation, of similar nature). The time evolution is determined by the initial conditions and the particular stochastic sequence. When we conveniently average over many initial conditions and many stochastic sequences [36],

C. Tsallis, *Introduction to Nonextensive Statistical Mechanics*,
https://doi.org/10.1007/978-3-030-79569-6_4

we obtain a probabilistic description of the system. More precisely, we obtain the time evolution of its probability distribution in the phase space of the system. The basic equation is the so-called *Fokker–Planck equation*, or, for quantum and discrete systems, the *master equation* (whose continuous limit recovers the Fokker–Planck equation). Finally, at a larger scale, we enter into the thermodynamical description, i.e., the level referred to as the *macroscopic* one. Statistical mechanics bridge from the microscopic level up to the macroscopic one.

For pedagogical reasons, we first focus on Fokker–Planck-like equations (Sects. 4.2, 4.3 and 4.5), and only later on the Langevin-like equations (Sect. 4.6). It should however be clear that, traditionally, the logical path is the other way around, i.e., from the Langevin equation level toward the Fokker–Planck level. It is only in Chap. 5 that we focus on the microscopic level, with its deterministic equations.

4.2 Diffusion

4.2.1 Normal Diffusion

The basic equation for normal diffusion is the so-called *heat equation*, first introduced in 1822 by Joseph Fourier. It is given, for $d = 1$, by

$$\frac{\partial p(x, t)}{\partial t} = D \frac{\partial^2 p(x, t)}{\partial x^2} \quad (D > 0), \tag{4.1}$$

where D is the *diffusion coefficient*. Let us assume the simplest initial condition, namely

$$p(x, 0) = \delta(x), \tag{4.2}$$

where $\delta(x)$ is Dirac' s delta distribution. The corresponding solution is given by

$$p(x, t) = \frac{1}{\sqrt{2\pi Dt}} e^{-x^2/2Dt} \quad (t \geq 0). \tag{4.3}$$

We can verify that

$$\int_{-\infty}^{\infty} dx \, p(x, t) = 1 \quad (t \geq 0), \tag{4.4}$$

and that

$$\langle x^2 \rangle \equiv \int_{-\infty}^{\infty} dx \, x^2 p(x, t) = Dt. \tag{4.5}$$

This corresponds to what is currently referred to as *normal diffusion*. Many types of functions $\langle x^2 \rangle (t)$ exist which increase nonlinearly with time (see, for instance, [401, 402]). But a very frequent one is

$$\langle x^2 \rangle \propto t^\mu \quad (\mu \geq 0; \ t \to \infty), \tag{4.6}$$

where x can be a d-dimensional quantity (and not necessarily the simple $d = 1$ case that we are focusing here). Diffusion is said *normal* or *anomalous* for $\mu = 1$ or $\mu \neq 1$, respectively. If $\mu < 1$ we have *subdiffusion* (*localization* implies $\mu = 0$); if $\mu > 1$ we have *superdiffusion* (the particular case $\mu = 2$ is also called *ballistic diffusion*).[1]

4.2.2 Anomalous Diffusion

Let us consider the following nonlinear generalization of (4.1), corresponding to correlated anomalous diffusion [403, 404],

$$\frac{\partial p(x,t)}{\partial t} = D \frac{\partial^2 [p(x,t)]^\nu}{\partial x^2} \quad (\nu \in \mathbb{R}). \tag{4.7}$$

This nonlinear equation is sometimes referred to as the *Porous medium equation* [405, 406], and has already been applied to various physical systems (see, for instance, references in [404]). In order to easily make the junction with nonextensive concepts, let us define

$$q \equiv 2 - \nu, \tag{4.8}$$

and rewrite Eq. (4.7) as follows:

$$\begin{aligned}
\frac{\partial p(x,t)}{\partial t} &= D_q \frac{\partial^2 [p(x,t)]^{2-q}}{\partial x^2} \\
&= (2-q) D_q \frac{\partial}{\partial x} \left\{ [p(x,t)]^{1-q} \frac{\partial p(x,t)}{\partial x} \right\} \\
&\quad (q \in \mathbb{R}; \ (2-q) D_q > 0).
\end{aligned} \tag{4.9}$$

Its solution for the initial condition (4.2) is given by

$$p_q(x,t) = p_q(x/[D_q t]^{\frac{1}{3-q}}), \tag{4.10}$$

where the normalized q-Gaussian is given by

---

[1] A weaker definition is sometimes used. It concerns the frequent cases where x^2 scales asymptotically like t^μ. Once again $\mu > 1$, $\mu = 1$, and $\mu < 1$ correspond, respectively, to superdiffusion, normal diffusion, and subdiffusion. The point is that x^2 scaling like t^μ is necessary but not sufficient for having a *finite* value of $\langle x^2 \rangle$ scaling like t^μ. A notorious example is the $q = 2$ case, i.e., the Cauchy-Lorentz distribution, whose width is *finite* while its variance *diverges*.

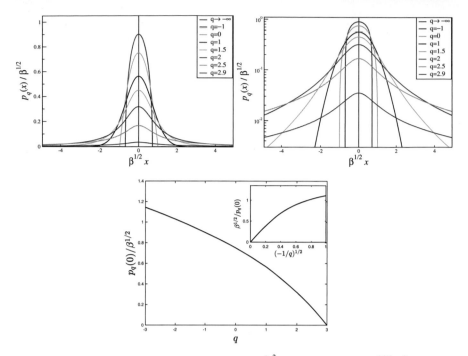

Fig. 4.1 *Top left*: Normalized distributions $p_q(x) \propto e_q^{-\beta x^2} = \left[1 - (1 - q)\,\beta\,x^2\right]^{1/(1-q)}$ for typical values of q, assuming $\beta > 0$; in the limit $q \to -\infty$, we recover the Dirac delta distribution. Some anomalous cases do exist for $\beta < 0$, that are described in [297]. *Top right*: The same in semi-log representation. *Bottom*: Values of $p_q(0)/\sqrt{\beta}$ as a function of q. The inset exhibits that, in the limit $q \to -\infty$, $p_q(0)/\beta \propto 1/(-q)^{1/2}$. This figure has been produced in co-authorship with U. Tirnakli

$$p_q(x) = \frac{1}{\sqrt{\pi}\,A_q}\,e_q^{-x^2} = \frac{1}{\sqrt{\pi}\,A_q}\,\frac{1}{\left[1 + (q - 1)x^2\right]^{\frac{1}{q-1}}}, \qquad (4.11)$$

with

$$A_q = \begin{cases} \dfrac{\sqrt{q-1}\,\Gamma\left(\frac{1}{q-1}\right)}{\Gamma\left(\frac{3-q}{2(q-1)}\right)} & \text{if } 1 < q < 3\,, \\[2ex] 1 & \text{if } q = 1\,, \\[2ex] \dfrac{\sqrt{1-q}\,(3-q)\,\Gamma\left(\frac{3-q}{2(1-q)}\right)}{2\,\Gamma\left(\frac{1}{1-q}\right)} & \text{if } q < 1\,. \end{cases} \qquad (4.12)$$

This family of distributions has been already focused on in various opportunities, e.g., [71, 297, 407], and is depicted in Figs. 4.1 and 4.2.

Let us make now several remarks.

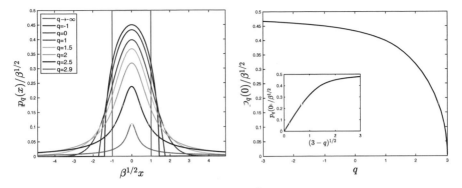

Fig. 4.2 *Left*: Normalized distributions $p_q(x) \propto e_q^{-\beta x^2/(3-q)} = \left[1 - \frac{1-q}{3-q}\beta x^2\right]^{1/(1-q)}$ for typical values of q, assuming $\beta > 0$. See [297] for details on the denominator $(3-q)$, which implies that the $q \to -\infty$ limit yields the uniform distribution. *Right*: The maximal value $p_q(0)/\sqrt{\beta}$ as a function of q. The inset exhibits that, in the limit $q \to 3$, $p_q(0)/\beta^{1/2} \propto (3-q)^{1/2}$. This figure has been produced in co-authorship with U. Tirnakli

(i) The upper bound $q = 3$ arrives from the imposition of normalization. In other words, $\int_{-\infty}^{\infty} dx\, p_q(x)$ diverges if $q \geq 3$, and converges otherwise.

(ii) If $1 \leq q < 3$, these distributions have an infinite support. If $q < 1$, they have a compact support; indeed, for $q < 1$, they vanish for $|x| > \sqrt{A_q/(1-q)}$.

(iii) The variance $\langle x^2 \rangle \equiv \int_{-\infty}^{\infty} dx\, x^2\, p_q(x)$ of these distributions is *finite* for $q < 5/3$ and *divergent* for $5/3 \leq q < 3$. This implies that, if we convolute them N times with $N \to \infty$, they approach a Gaussian distribution for $q < 5/3$ (more precisely the one that preserves the variance) and a Lévy distribution for $5/3 < q < 3$ (more precisely the one which preserves the asymptotic power-law with the same exponent). This corresponds to independence between the N random variables. The situation is completely different if a strong correlation exists between them. We focus again on this interesting and ubiquitous case later on, when q-generalizing the Central Limit Theorem.

(iv) In contrast with the standard variance, the q-variance $\langle x^2 \rangle_q \equiv \left\{\int_{-\infty}^{\infty} dx\, x^2\, [p_q(x)]^q\right\} / \left\{\int_{-\infty}^{\infty} dx\, [p_q(x)]^q\right\}$ of these distributions is *finite* for $q < 3$. Indeed, $\int_{constant\,>0}^{\infty} dx\, x^2\, x^{-2q/(q-1)}$ is finite for $q < 3$.

(v) These distributions extremize (maximize for $q > 0$, and minimize for $q < 0$) S_q under the appropriate constraints (see Sect. 3.5).

(vi) If $q = \frac{3+m}{1+m}$ where m is a *positive integer*, these distributions recover the *Student's t-distributions* with m degrees of freeedom.[2] If $q = \frac{n-6}{n-4}$ where $n > 4$ is a *positive integer*, these distributions recover the so-called *r-distributions* [304].

---

[2] In the financial literature, these q-Gaussian distributions with $q > 1$ emerge quite frequently ($p(x) \propto 1/(a^2 + x^2)^\eta$ with $\eta > 0$). They are referred to as Student's distributions for *any* real value of $\eta \equiv 1/(q-1)$. Strictly speaking, this is an abusive notation.

(vii) It is remarkable the relation $q \leftrightarrow (2 - q)$ existing between the nonlinearity of the Fokker–Planck-like equation and its q-exponential solution.

(viii) x scales with $t^{1/(3-q)}$ $(t \to \infty)$. Consequently, if $\langle x^2 \rangle$ is finite (i.e., if $q < 5/3$), it must be

$$\langle x^2 \rangle \propto t^{\frac{2}{3-q}} . \tag{4.13}$$

By using Eq. (4.6), we obtain

$$\mu = \frac{2}{3 - q} . \tag{4.14}$$

Similarly we obtain

$$\langle x^2 \rangle_q \propto t^{\frac{2}{3-q}} \quad (q < 3) . \tag{4.15}$$

Consequently, we have superdiffusion for $1 < q < 3$ (ballistic for $q = 2$), and subdiffusion for $q < 1$ (localization for $q \to -\infty$).

(ix) We see that $D_q > 0$ $(D_q < 0)$ if $q < 2$ $(2 < q < 3)$. The $q \to 2$ limit deserves a special comment. Equation (4.9) can be rewritten in the following form [408, 409, 411]:

$$\frac{\partial p(x, t)}{\partial t} = (2 - q)D_q \frac{\partial^2}{\partial x^2} \left[\frac{[p(x, t)]^{2-q} - 1}{2 - q} . \right] \tag{4.16}$$

In the limit $q \to 2$, this equation becomes

$$\frac{\partial p(x, t)}{\partial t} = \left\{ \lim_{q \to 2}[(2 - q)D_q] \right\} \frac{\partial^2}{\partial x^2} \ln p(x, t) , \tag{4.17}$$

with $\left\{ \lim_{q \to 2}[(2 - q)D_q] \right\} > 0$. This equation is known to have as solution the Cauchy-Lorentz distribution (more precisely, $p_2(x, t)$).

In the presence of an external drift, Eq. (4.9) is extended as follows [403, 404]:

$$\frac{\partial p(x, t)}{\partial t} = -\frac{\partial}{\partial x}[F(x)p(x, t)] + D_q \frac{\partial^2[p(x, t)]^{2-q}}{\partial x^2} , \tag{4.18}$$

where $F(x) = -dV/dx$ is an external force associated with the potential $V(x)$. This equation is currently referred to as the Plastino and Plastino one. In the context of this equation, Plastino and Plastino discovered the connection between the Sq entropies and non-linear diffusion phenomena [403]. We shall consider the particular (but important) case where

$$F(x) = k_1 - k_2 x \quad (k_2 \geq 0) ; \tag{4.19}$$

$k_2 = 0$ corresponds to the important case of external constant force, and $k_1 = 0$ corresponds to the Uhlenbeck–Ornstein process. For this simple force, it is possible to find the following analytical solution [404]:

$$p_q(x,t) = \frac{\{1 - (1-q)\,\beta(t)\,[x - x_M(t)]^2\}^{1/(1-q)}}{Z_q(t)} \,, \tag{4.20}$$

with

$$\frac{\beta(0)}{\beta(t)} = \left[\frac{Z_q(t)}{Z_q(0)}\right]^2 = \left[\left(1 - \frac{1}{K_2}\right)e^{-t/\tau} + \frac{1}{K_2}\right]^{\frac{2}{3-q}} \,, \tag{4.21}$$

where

$$K_2 \equiv \frac{k_2}{2(2-q)D_q\,\beta(0)[Z_q(0)]^{q-1}} \,, \tag{4.22}$$

and

$$\tau \equiv \frac{1}{k_2(3-q)} \,. \tag{4.23}$$

To close this discussion, let us mention that it can be straightforwardly shown that

$$\int dx\, p_q(x,t) = 1 \,, \forall t \geq 0 \,. \tag{4.24}$$

(x) Equation (4.7) can be further generalized, namely to d dimensions, as follows [412]:

$$\frac{\partial p(\vec{x},t)}{\partial t} = \nabla\big[D(\vec{x})\nabla(p(\vec{x},t))^\nu\big] \quad (D \in \mathbb{R};\ \nu \in \mathbb{R};\ \vec{x} \in \mathbb{R}^d) \,. \tag{4.25}$$

If we assume isotropy and $D = D_0/r^\theta$ ($r \equiv |\vec{x}|;\ \theta \in \mathbb{R}$), this equation becomes

$$\frac{\partial p(r,t)}{\partial t} = \frac{D_0}{r^{d-1}}\frac{\partial}{\partial r}\left[r^{d-1-\theta}\frac{\partial}{\partial r}(p(r,t))^\nu\right] \,. \tag{4.26}$$

Its solution is a q-Gaussian with $q = (2 - \nu)$, and r^2 scales like t^μ with

$$\mu = \frac{2}{2 + \theta + d(1-q)} \,. \tag{4.27}$$

The $\theta = 0$ particular instance recovers (4.14) for $d = 1$, $\forall q$, and $\mu = 1$ for $q = 1$, $\forall d$ (i.e., normal diffusion).

(xi) Equation (4.1) can be generalized also as follows

$$\frac{\partial p(x,t)}{\partial t} = D_\gamma \frac{\partial^\gamma p(x,t)}{\partial |x|^\gamma} \quad (D_\gamma > 0;\ 0 < \gamma < 2)\,, \tag{4.28}$$

where we have introduced fractional derivatives (see, for instance, [413–416]), which preserves the linearity of the equation. The solution corresponding to the initial condition (4.2) is given by [217, 417]

$$p(x,t) = (D_\gamma t)^{1/\gamma}\, L_\gamma(x/(D_\gamma t)^{1/\gamma}) \quad (t \geq 0)\,, \tag{4.29}$$

where $L_\gamma(z)$ is the Lévy distribution with index γ. It follows that $\mu = 2/\gamma$, hence $\mu > 1$. Therefore this case is a superdiffusive one, in a quite strong sense in fact. Indeed, the corresponding variance, i.e., $\langle z^2 \rangle$, *diverges*, $\forall \gamma < 2$. An illustration of this fact can be seen for the Cauchy-Lorentz distribution $L_1(z) = \frac{1}{\pi(1+z^2)}$, which corresponds to $\gamma = 1$. We remind that $L_\gamma(z)$ is given by the Fourier transform of $e^{-z|k|^\gamma}$ (see, for instance, [217] and references therein). Asymptotic connections between the Lévy distributions and S_q have been discussed in the literature [407, 418–421].[3]
Lévy distributions asymptotically satisfy

$$L_\gamma(x) \propto \frac{1}{|x|^{1+\gamma}} \quad (|x| \to \infty;\ 0 < \gamma < 2)\,, \tag{4.30}$$

whereas q-Gaussians asymptotically satisfy

$$p_q(x) \propto \frac{1}{|x|^{2/(q-1)}} \quad (|x| \to \infty;\ 1 < q < 3)\,. \tag{4.31}$$

If we identify the exponents of these two asymptotic power-laws we obtain

$$\gamma = \begin{cases} 2 & \text{if } q < 5/3\,, \\ \frac{3-q}{q-1} & \text{if } 5/3 < q < 3\,, \end{cases} \tag{4.32}$$

---

[3] A long-standing confusion regretfully exists regarding this connection. There are Lévy distributions and there are q-Gaussian ones. They definitively are different, with an unique exception, namely the Cauchy-Lorentz distribution ($q = 2$ in the family of q-Gaussians, and $\gamma = 1$ in the Lévy family). However, the expression *Lévy-Tsallis distribution, which does not exist*, is repeatedly, and wrongly, used in the literature of high-energy collisions between elementary particles (e.g., in papers by the ALICE Collaboration, among others) to refer to the q-Gaussians that are used for specific interpretations of the data. This confusion has been lengthily clarified in [422] and elsewhere, but nevertheless it still persists! This is regrettable even from the physical viewpoint. Indeed, Lévy distributions are attractors in the space of probabilities under the assumption of probabilistic (strict or approximate) *independence* of the distributions that are being summed, whereas q-Gaussians are attractors under the assumption of *strong correlations of a specific wide class*. This important distinction implies in absolutely different interpretative paths for the relevant microscopic mechanisms that are in play.

where we have used the above remark (iii), i.e., that, if $q < 5/3$, there is no possible comparison between $L_\gamma(x)$ and $p_q(x)$. See Fig. 4.3.

The above $d = 1$ connection can be straightforwardly generalized to the isotropic d-dimensional case. In that case, we have, for $|\mathbf{x}| \to \infty$, $L_\gamma(\mathbf{x}) \propto 1/|\mathbf{x}|^{d+\gamma}$ (hence $L_\gamma(|\mathbf{x}|) \propto |\mathbf{x}|^{d-1}/|\mathbf{x}|^{d+\gamma} = 1/|\mathbf{x}|^{1+\gamma}$) and $p_q(\mathbf{x}) \propto 1/|\mathbf{x}|^{2/(q-1)}$, where $\mathbf{x}$ is a dimensional variable. By identifying the exponents we obtain

$$\gamma = \begin{cases} 2 & \text{if } q \leq \frac{4+d}{2+d}, \\ \frac{2}{q-1} - d & \text{if } \frac{4+d}{2+d} < q < \frac{2+d}{d}, \end{cases} \tag{4.33}$$

where we have taken into account that $p_q(\mathbf{x})$ is normalizable only if $q < \frac{2+d}{d}$, and that its variance is finite only if $q < \frac{4+d}{2+d}$. The particular instance $\gamma = 1$ corresponds to the distribution of the radial component $|\mathbf{x}|$ of the d-dimensional Cauchy-Lorentz distribution (proportional to $1/(a^2 + |\mathbf{x}|^2)$, a being a constant). The value $q = (3 + d)/(1 + d)$ precisely leads to $\gamma = 1$. Similarly, when q approaches $(2 + d)/d$ from below, γ approaches zero from above. The similarities and differences between Lévy distributions and q-Gaussians are illustrated in Fig. 4.3.

(xii) We have up to now focused on q-Gaussians with $\beta > 0$ (positive inverse temperature). We shall now pay attention to the (less frequent) case $\beta < 0$ [297]. One-dimensional q-Gaussians are defined as follows:

$$G_q(\beta; x) = C_{q,\beta} e_q^{-\beta x^2} \quad (x \in D_q), \tag{4.34}$$

where $\beta \in \mathbb{R}$, $C_{q,\beta}^{-1} = \int_{D_q} e_q^{-\beta x^2} dx$. q-Gaussians are normalizable probability distributions whenever $q < 3$ for $\beta > 0$, and $q > 2$ for $\beta < 0$. The support is $D_q = \mathbb{R}$ if $q \in [1, 3)$, $\beta > 0$, and the bounded interval $D_q = \{x \mid |x| < 1/\sqrt{\beta(1 - q)}\}$ for $q < 1$, $\beta > 0$, and $q > 2$, $\beta < 0$ (the boundary of the interval is reached only in the $\beta > 0$ case, thus having a compact support). The expression for the normalization constant is given by

$$C_{q,\beta} = \begin{cases} \frac{\sqrt{\beta(1-q)}}{B\left(\frac{2-q}{1-q}, \frac{1}{2}\right)}; & \beta > 0, \ q < 1 \quad \text{and} \quad \beta < 0, \ q > 2 \\ \frac{\sqrt{\beta(q-1)}}{B\left(\frac{3-q}{2(q-1)}, \frac{1}{2}\right)}; & \beta > 0, \ 1 < q < 3 \end{cases} \tag{4.35}$$

where $B(x, y)$ stands for the Beta function with $\lim_{q \to 1} C_{q,\beta>0} = \frac{\sqrt{\beta}}{\sqrt{\pi}}$, as expected for the Gaussian distribution.   Since β^{-1} plays the role of a generalized temperature, the $\beta < 0$ values yield q-Gaussians that are to be associated with systems at negative temperatures. As the distribution corresponding to the system is determined by the values of q and T, any allowed point in a (q, β) plane will be in correspondence with a different q-Gaussian. Figure 4.4 shows such a plane and how q-Gaussians look like depending on their location in it. To start with, any q-Gaussian with $\beta > 0$ points downward (being con-

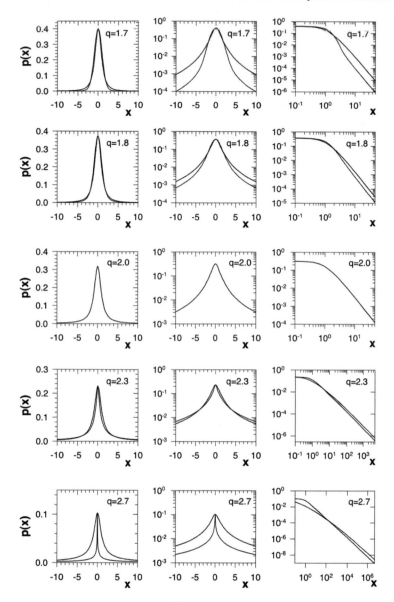

Fig. 4.3 Lévy distributions $L_\gamma(x) \equiv \frac{1}{2\pi} \int_{-\infty}^{\infty} dk \cos kx\, e^{-\alpha|k|^\gamma}$, with $0 < \gamma < 2$ and $\alpha > 0$ (black curves), and q-Gaussians $P_q(x) \equiv [1 - (1-q)\beta x^2]^{1/(1-q)}/Z_q$, with $5/3 < q < 3$, $\beta > 0$ and $Z_q = \sqrt{\frac{\pi}{\beta(q-1)}}\Gamma(\frac{3-q}{2(q-1)})/\Gamma(\frac{1}{q-1})$ (red curves). Parameters (q, γ) are related through $q = \frac{\gamma+3}{\gamma+1}$ so that the tails of both distributions decay with the same power-law exponent. Without loss of generality, we have taken $\beta = 1$ which corresponds to a simple rescaling; α was chosen such that $P_q(0) = L_\gamma(0)$. Notice that, in log–log representation, Lévy distributions may have an *inflection point*, whereas this never occurs for q-Gaussians. From [423].

cave for $q < 0$ and having two symmetric inflection points for $q > 0$) while
any q-Gaussian with $\beta < 0$ is convex. In the upper half plane, q-Gaussians
for a constant value of $\beta > 0$ are shown for decreasing values of $q < 3$ from
right to left. In the vertical axis, corresponding to $q = 1$, the Gaussian is
shown. On his right, a q-Gaussian with $1 < q < 3$ is depicted, whose support is the whole real axis. As q further increases q-Gaussians turns flatter
up to the limit $q \to 3^-$ where distributions are no longer normalizable, with
$\lim_{q \to 3^-} C_{q,\beta>0} = 0$. On the left top quadrant, q-Gaussians for $q < 1$ and a
constant value of $\beta > 0$ are shown, all of them with bounded support D_q with
$\lim_{q \to -\infty} D_q = 0$, while the area is preserved. Thus a Dirac delta function is
obtained in the $q \to -\infty$ limit provided β is kept constant. Nevertheless, the
same limit can be taken following different paths given by the graph of any function $\beta(q)$ on the plane. In [407] the choice $\beta(q) = \frac{1}{3-q}$ was made. Thus, the
support is $D_q = \{x \ / \ |x| \leqslant \sqrt{(3-q)/(1-q)}\}$ with $\lim_{q \to -\infty} D_q = [-1, 1]$
so in the limit a uniform distribution is obtained instead, as shown in Fig. 4.4.
The same limiting distribution would be obtained following any $\beta(q)$ curve
provided $\beta(q) = O((1-q)^{-1})$. In the lower plane, convex q−Gaussians for
a constant value of $\beta < 0$ are shown for increasing values of $q \geqslant 2$ from left

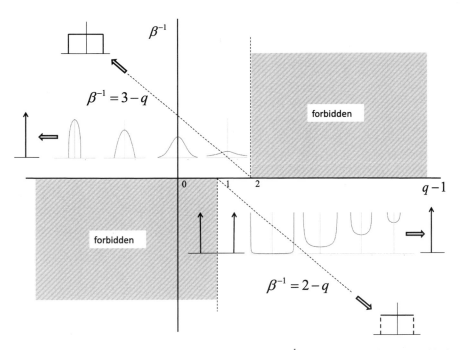

Fig. 4.4 q-Gaussians in the (q, β) plane with vertical axis β^{-1} and horizontal axis $(q - 1)$. Shad-
owed regions corresponds to forbidden values of (q, β). The support of the uniform distribution
in the upper (lower) β^{-1} semiplane is compact (open). See text here and in [297] for the limiting
distributions in the diagonal lines. From [297]

to right. All of them have a bounded support D_q with $\lim_{q\to\infty} D_q = 0$. Also, $\lim_{q\to\infty} G_q(\beta < 0; 0) = \infty$. Thus, we obtain again a Dirac delta distribution in the $q \to \infty$ limit when $\beta < 0$ is kept constant. In the opposite $q \to 2^+$ limit, with support $D_2 = [-1/\sqrt{-\beta}, 1/\sqrt{-\beta}]$ and $\lim_{q\to 2^+} C_{q,\beta<0} = 0$, a double peaked delta distribution is obtained (see [297] for further details). In analogy with the path followed in the upper half plane, taking $\beta(q) = \frac{1}{2-q}$ we obtain $D_q = \{x \ / \ |x| \leqslant \sqrt{(2-q)/(1-q)}\}$ with $\lim_{q\to\infty} D_q = [-1, 1]$. Thus, the support remains finite while the area is preserved, so we recover the uniform distribution. Again, the same limit is obtained for any path with $\beta(q) = O((1-q)^{-1})$. Let us conclude this discussion by mentioning a link to nonlinear dynamical systems: the invariant distribution of the most chaotic case of the standard logistic map, precisely $x_{t+1} = 1 - 2x_t^2$, is given by $\frac{1}{\pi\sqrt{1-x^2}}$, which is the normalized q-Gaussian with $(q, \beta) = (3, -1/2)$.

(xiii) Equation (4.1) can naturally be generalized even more, as follows:

$$\frac{\partial^\beta p(x, t)}{\partial |t|^\beta} = D_{\beta,\gamma,q} \frac{\partial^\gamma [p(x, t)]^{2-q}}{\partial |x|^\gamma} \qquad (0 < \beta \leq 1; \ 0 < \gamma \leq 2). \qquad (4.36)$$

This equation contains Eqs. (4.28) and (4.9) as particular cases. No general solution of (4.36) is yet known to the best of our knowledge. However, solutions for particular cases (e.g., for $\beta = 1$) are available [424, 425]. Being out of the present scope, we shall not discuss them here. Related extensions have been proposed in [426–428], yielding a nonlinear and inhomogeneous Fokker–Planck equation within a generalized Stratonovich (or *stochastic α, with $\alpha \in$ [0, 1])*, prescription whose values $\alpha = 0$, $1/2$ and 1, respectively, correspond to the so-called Itô, Stratonovich and anti-Itô prescriptions. Once again their basic solutions are (either precisely or related to) the q-exponential form.

Still other generalizations have been advanced in the literature along the present lines, such as diffusion-reaction equations, which have been applied to discuss a social crisis in Chile [429].

Let us finally mention that a specific restricted random walk [430] has also exhibited neat connections with q-Gaussians.

4.3 Stable Solutions of Fokker–Planck-Like Equations

For $\beta = 1$, Eq. (4.36) becomes

$$\frac{\partial p(x, t)}{\partial t} = D_{\gamma,q} \frac{\partial^\gamma [p(x, t)]^{2-q}}{\partial |x|^\gamma} \qquad (0 < \gamma \leq 2; \ q < 3). \qquad (4.37)$$

This equation has four classes of solutions (see Fig. 4.30) which provide interesting hints.

First of all, the *Gaussian class*, corresponding to $q = 1$ and $\gamma = 2$. Its basic solution is, as already shown, a Gaussian. This corresponds to the *standard Central Limit Theorem* (*G-CLT*). This theorem essentially states that, if we add (or arithmetically average) N random variables, that are probabilistically *independent* and have *finite variance*, then the distribution of the sum approaches, after appropriate centering and rescaling, a *Gaussian* when $N \to \infty$.

Second, the *Lévy class*, or α *class* (with $\alpha = \gamma$), corresponding to $q = 1$ and $0 < \gamma < 2$. Its basic solutions are, as already discussed, Lévy distributions (also called α *-stable distributions*). This corresponds to the *Lévy-Gnedenko Central Limit Theorem* (*L-CLT*). This theorem essentially states that, if we add (or arithmetically average) N random variables, that are probabilistically *independent* and have *infinite variance* (due to fat tails of the power-law class, excepting for possible logarithmic corrections; see, for instance, [217] and references therein), then the distribution of the sum approaches, after appropriate centering and rescaling, a *Lévy distribution* when $N \to \infty$.

Third, we have the class (from now on referred to as the q *-Gaussian class*) corresponding to $\gamma = 2$ and $q \neq 1$. Its basic solutions are, as already discussed, q-Gaussians. And these solutions are *stable* in the sense that, if we start with an arbitrary (symmetric) solution $p(x, 0)$, it asymptotically approaches, for increasing N, a q-Gaussian. This has been numerically verified, and analytically proved in [408–410]. As we shall see, sufficient conditions within a generalized Central Limit Theorem (noted q-*G-CLT*, or just q *-CLT*) can be established for this situation. It corresponds to the violation of the hypothesis of *independence*. Not a weak violation with correlations that gradually disappear in the $N \to \infty$ limit, but a certain class of global correlations which persist up to infinity. This makes sense since Eq. (4.37) is *nonlinear* for $q \neq 1$. The possible existence of such a theorem was first suggested in [424], and specifically conjectured in [271]. This theorem also demands the finiteness of a certain q-variance. If this q-variance diverges, then we are led to the fourth and last present class.

The fourth class (from now on referred to as the (q, α) *class*) corresponds to $\gamma \equiv \alpha < 2$ and $q \neq 1$. Its basic solutions are the so-called (q, α) *-stable distributions* that will be described later on.

The existence of such theorems is of extreme interest. Indeed, they provide a plausible mathematical basis for the ubiquity of distributions such as the q-Gaussians (generically q-exponentials) as actually observed in many natural, artificial, and even social systems (see Chap. 7). A variety of physical situations and interesting questions related with nonlinear Fokker–Planck equations are discussed in [320, 321, 431–433].

4.4   Connection Between Entropic Functionals, Fokker–Planck Equations and H-theorems

A quite general one-dimensional nonlinear Fokker-Planck equation can be written as follows [322] (see also [320, 321, 323, 324]):

$$\eta \frac{\partial P(x,t)}{\partial t} = -\frac{\partial\{A(x)\Psi[P(x,t)]\}}{\partial x} + D \frac{\partial}{\partial x}\left\{\Omega[P(x,t)]\frac{\partial P(x,t)}{\partial x}\right\}, \qquad (4.38)$$

where $\eta > 0$ is an effective friction coefficient, $D > 0$ has the dimension of an energy, $\Psi[P] > 0$ and $\Omega[P] > 0$ are smooth functionals, and $A(x)$ is an external force associated with a potential $\phi(x)$ $(A(x) = -d\phi(x)/dx)$.

Let us consider also the following quite general entropic functional:

$$S[P] = k\Lambda[Q[P]], \qquad (4.39)$$

where $k > 0$, $\Lambda[Q]$ is a smooth function of Q, where $Q[P] = \int_{-\infty}^{\infty} dx\, g[P(x,t)]$ with $g(0) = g(1) = 0$ and $\frac{d^2 g[P]}{dP^2} \le 0$; the particular case $\Lambda[Q] = Q$ corresponds to a trace-form functional for the entropy; if, in addition to that, we choose $g[P] = P \ln_q(1/P)$ we get $S = S_q$.

We can next consider the free energy functional

$$F(t) = U(t) - \theta S(t), \qquad (4.40)$$

with

$$U(t) = \int_{-\infty}^{\infty} dx\, \phi(x) P(x,t), \qquad (4.41)$$

the constant $\theta > 0$ having the dimension of a temperature. Under simple mathematical conditions (see [322] for all details), it can be proved that, assuming $D = k\theta$, the condition

$$-\frac{d\Lambda[Q]}{dQ}\frac{d^2 g[Q]}{dP^2} = \frac{\Omega[P]}{\Psi[P]}, \qquad (4.42)$$

implies a remarkable H-theorem, namely that

$$\frac{dF}{dt} \le 0. \qquad (4.43)$$

In other words, Eq. (4.42) constitutes a sufficient condition for having a sensibly general nonlinear Fokker–Planck equation whose stationary-state distribution, in the presence of a confining potential, precisely coincides with the probability distribution which optimizes, under standard constraints, a wide associated entropic functional. Many illustrations are available in the literature, e.g., [320–324, 434].

4.5 Many-Body Interacting Systems with Overdamping

4.5.1 Models

A variety of many-body interacting systems having in common overdamping are available in the literature [314, 434–445], including with connections to the Vlasov equation [446]. In all these cases, the corresponding thermostatistics analytically and/or numerically satisfies, within wide physical regimes, nonextensive statistical mechanics. As an illustration let us follow along the paradigmatic lines of [314]. We consider the equation of motion for a particle i in a system of N overdamped particles, namely

$$\mu \vec{v}_i = \sum_{j \neq i} \vec{J}(\vec{r}_i - \vec{r}_j) + \vec{F}^e(\vec{r}_i) + \eta(\vec{r}_i, t), \tag{4.44}$$

where $\vec{v}_i$ is the velocity of the ith particle, μ is the effective viscosity of the medium, the first term on the right accounts for the interactions among particles, $\vec{F}^e(\vec{r}_i)$ represents an external confining force, and η corresponds to an uncorrelated thermal noise with zero mean and variance $\langle \eta^2 \rangle = k_B T / \mu$. Here we consider a short-range repulsive particle–particle interaction in the form $\vec{J}(\vec{r}) = G(|\vec{r}|/\lambda)\hat{r}$, where $\hat{r}$ is the unit vector along the axis connecting each pair of particles and λ is a characteristic length of the short-range pairwise interaction. We may investigate the motion of N particles in a two-dimensional narrow channel of size $L_x L_y$ (with $L_y \ll L_x$) under an external restoring force in the x direction, $\vec{F}^e = -A(x)\hat{x} = -\alpha_c x \hat{x}$ ($\alpha_c > 0$); the subindex c stands for *confinement*. Under simple physical hypothesis we obtain (see details in [314]) for the concentration $\rho(x, t)$ the following Fokker–Planck-like equation:

$$\mu \frac{\partial \rho}{\partial t} = \frac{\partial}{\partial x} \left\{ \rho \left[a \frac{\partial \rho}{\partial x} - A(x) \right] \right\} + kT \frac{\partial^2 \rho}{\partial x^2}, \tag{4.45}$$

with $\int_{-L_x/2}^{L_x/2} \rho(x, t) dx = N/L_y$ and $a > 0$. We notice that μ simply rescales time t into t/μ; therefore without loss of generality we shall from now on consider $\mu = 1$. If we assume that the particles are vortices within a type II superconductor, then we typically have $J(\vec{r}) = f_0 K_1(|\vec{r}|/\lambda)\hat{r}$, where λ is the *London penetration length*, f_0 is the *repulsive interaction strength*, and $K_1(z)$ is the modified Bessel function decaying exponentially for $z > 1$. The stationary state for vanishing *kinetic temperature*, i.e., $T = 0$, is given by

$$\rho(x, \infty) = \frac{\alpha_c}{2a}(x_e^2 - x^2) = \rho(0)(1 - x^2/x_e^2) \quad (|x| \leq x_e), \tag{4.46}$$

with $\rho(0) = \alpha_c x_e^2 / 2a$, and $x_e = (Na/2\alpha_c L_y)^{1/3}$; $\rho(x, \infty)$ precisely is a q-Gaussian with $q = 0$, and x_e^2 plays the role of an *effective temperature*. The above equation can be rewritten as follows:

$$\rho(x, \infty) = \frac{1}{4\tau\lambda}\left[\left(\frac{x_e}{\lambda}\right)^2 - \left(\frac{x}{\lambda}\right)^2\right] \quad (|x| \le x_e),\qquad(4.47)$$

where we have introduced the *dimensionless temperature* $\tau \equiv (k\theta)/(\alpha\lambda^2)$.
For $T > 0$, the stationary state is given by

$$\rho(x, \infty) = \frac{k_B T}{a} W\left\{\frac{a\rho(0)}{k_B T}\exp\left[\frac{a\rho(0)}{k_B T} - \frac{ax^2}{2k_B T}\right]\right\},\qquad(4.48)$$

where the W-Lambert function is defined implicitly through the equation
$W(z)\, e^{W(z)} = z$. By using the asymptotic behavior $W(z) \sim \ln z - \ln \ln z$ in the limit
$z \to \infty$, this result recovers Eq. (4.46) in the limit $a\rho(0)/kT \to \infty$. Also, by using
the asymptotic behavior $W(z) \sim z$ in the limit $z \to 0$, this result yields, in the limit
$a\rho(0)/kT \to 0$, the BG stationary state $\rho(x, \infty) = \rho(0)e^{-\alpha_c x^2/2kT}$. See Figs. 4.5 and
4.6. Notice that for intermediate temperatures $0 < T < \infty$, the stationary state is *not*
given by a q-Gaussian with a T-dependent value of q gradually going from $q = 0$
to $q = 1$, but it is instead given by a Lambert-function. This is a consequence of the
fact that the entropy that has been used is a linear combination of the BG entropy
S_{BG} and S_q with $q = 2$. We remind that when standard *linear* constraints are used
in the entropic extremization of S_q, the q_{stat}-Gaussian which emerges has an index
$q_{stat} = 2 - q$; to have $q_{stat} = q$ we need to use *nonlinear* constraints, namely those
involving escort distributions (see [256, 447] and references therein).

We may describe as well the time evolution of the distributions of positions and
velocities.

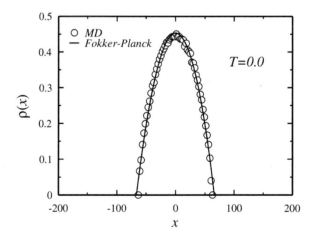

Fig. 4.5 Profile of the density of particles at stationary state and $T = 0$, obtained from molecular
dynamics by integrating Eq. (4.44) (empty circles), as compared to the theoretical estimate Eq.
(4.46). The best fit to the data (with $\alpha_c = 10^{-3} f_0/\lambda$) of a quadratic function gives the parameter
$a = 2.41 f_0\lambda^3$ (full line). The position x is measured in units of λ, whereas the steady-state density
$\rho(x) \equiv \rho(x, \infty)$ is expressed in units of λ^{-2}. From [314]

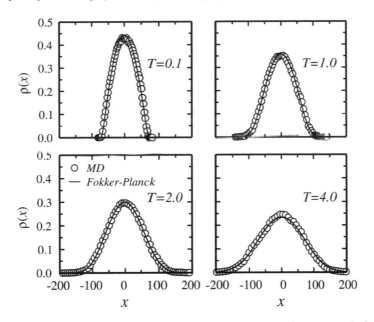

Fig. 4.6 Comparison between the density of particle profiles in the stationary state obtained from molecular dynamics (empty circles) and the theoretical prediction Eq. (4.48) (full lines), for different values of the temperature of the thermal bath T. The position x is measured in units of λ, the density is in units λ^{-2}, and T is given in units of $(f_0\lambda/k)$. The parameter $a = 2.41 f_0\lambda^3$ is the same adopted for $T = 0$, whereas $\rho(0)$ is determined by conservation of the total number of particles in the system, $\int_{-L_x/2}^{L_x/2} \rho(x)dx = N/L_y$. For low values of T, e.g., $kT/f_0\lambda = 0.1$, the density profile is approximately parabolic. At intermediate values of T, the system can be described by a W-Lambert function Eq. (4.48). At high values of T, the profile becomes typically Gaussian, as illustrated from results at $kT/f_0\lambda = 4$ [314]

$$\rho(x, t) = \frac{NB(t)}{L_y}\left[1 - \beta(t)x^2\right], \tag{4.49}$$

where

$$\frac{NB(t)\beta(t)}{L_y} = \frac{\alpha_c}{2a}\left[1 + (K_2 - 1)e^{-3\alpha_c t}\right]. \tag{4.50}$$

with $K_2 \equiv \frac{\alpha_c}{4D\beta(t_0)B(t_0)}$ and, due to normalization, $\beta(t)/\beta(t_0) = [B(t)/B(t_0)]^2$, t_0 being an arbitrary reference time. Also, it can be straightforwardly verified that, during the beginning of the time evolution, x^2 scales as $t^{2/3}$, in agreement with the general nonlinear Fokker–Planck result that x^2 scales like $t^{2/(3-q)}$ [404]. The main results are shown in Figs. 4.7, 4.8, 4.9, and 4.10 from [436].

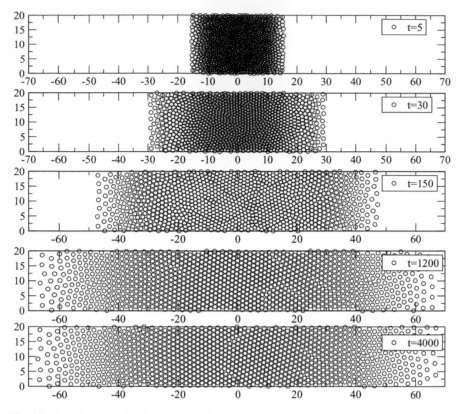

Fig. 4.7 Snapshots showing the particle positions at typical times t, for a single sample of $N = 1000$ particles. In each snapshot, the horizontal and vertical axes (x/λ and y/λ, respectively) show the relevant part of a box of size $L_x = 280\lambda$ and $L_y = 20\lambda$. At $t = 5$ the state is very close to the initial distribution, whereas the stationary state is almost reached at $t = 1200$. The time is dimensionless, measured in terms of the molecular-dynamics time step δt, as described in [436]

4.5.2 Thermodynamics, Including Clausius Relation, First and Second Principles, Equation of States, and Carnot Cycle

In this section we analytically focus on diverse thermostatistical quantities involving, among others, the entropy itself and not only the corresponding optimizing distribution. It is therefore worthy to remind the reader that, if the constraints used in the entropic optimization are written in their simplest form, the distributions corresponding to the entropy S_q are not q-exponentials and q-Gaussians but rather $(2-q)$-exponentials and $(2-q)$-Gaussians. Equivalently, if we optimize the entropy S_{2-q} with these same constraints, we directly obtain q-exponentials and q-Gaussians. This is the format that we use in what follows (see [439]). Consistently, it is the entropy S_2 which is associated with the q-Gaussian with $q = 0$ in Eq. (4.46).

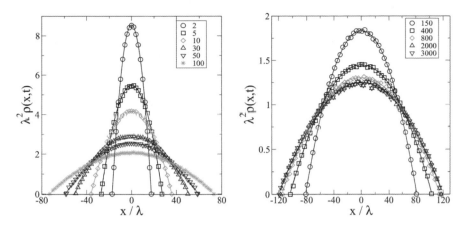

Fig. 4.8 The dimensionless density of particles $\lambda 2\rho(x, t)$ is plotted versus x/λ, for typical times t: **a** short and intermediate times; **b** in the approach to the stationary state. The symbols are results from the simulations, whereas the solid lines represent the theoretical solution of the NLFPE, as given in Eqs. (4.49) and (4.50). From [436]

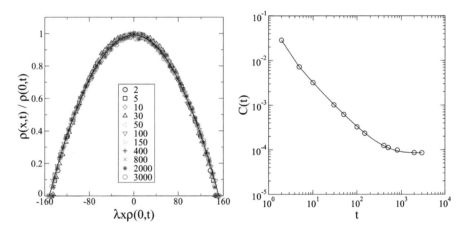

Fig. 4.9 **a** Collapse of all data of Fig. 4.8 into a universal curve, using the representation $\rho(x, t)/\rho(0, t)$ versus $\lambda x \rho(0, t)$; the solid line is a parabola, as described in the text. **b** Time evolution of the dimensionless coefficient $C(t) = \lambda 4 N B(t)\beta(t)/L_y$ which characterizes $\rho(0, t)$; the empty circles are molecular-dynamics data, whereas the solid line corresponds to the theoretical prediction obtained from Eq. (4.50) by setting $a = 5.85 f_0 \lambda^3$. The time is dimensionless, measured in terms of the molecular-dynamics time step δt. From [436]

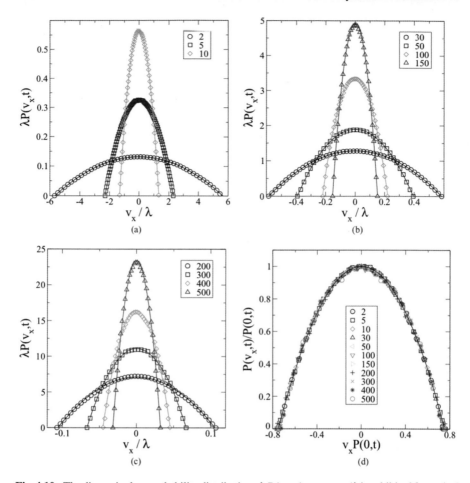

Fig. 4.10 The dimensionless probability distributions $\lambda P(v_x, t)$ versus v_x/λ is exhibited for typical increasing times: **a** at the beginning of the evolution; **b** intermediate times; **c** in the approach to the stationary state. All these data fall into a universal curve, for conveniently rescaled variables, as shown in **d**. Since time is dimensionless, velocities are presented with a dimension of length. The symbols represent simulation data, whereas solid curves are q-Gaussian distributions with $q = 0$. From [436]

Therefore we straightforwardly obtain the fundamental Clausius relation

$$\left(\frac{\partial s_2}{\partial u}\right)_\alpha = \frac{1}{\theta}. \tag{4.51}$$

Consistently we obtain the specific-heat-like quantity

$$c_\alpha = \left(\frac{\partial u}{\partial \theta}\right)_\alpha = \theta\left(\frac{\partial s_2}{\partial \theta}\right)_\alpha, \tag{4.52}$$

suggesting the definition of an infinitesimal amount of heat-like contribution to the energy change, $\delta Q = \theta ds_2$, where θ plays the role of an integrating factor.

It can be also seen that, in spite of the fact that $q \neq 1$, the First Principle of Thermodynamics is satisfied, namely that

$$du = \delta Q + \delta W = \theta ds_2 + \sigma d\alpha. \tag{4.53}$$

Let us discuss now the Carnot cycle [437]. It is verified that

$$\sigma = \frac{3^{2/3}}{10} \lambda^2 \left(\frac{k\theta}{\alpha \lambda^2} \right)^{2/3} \implies \sigma = \frac{u}{\alpha}, \tag{4.54}$$

with

$$s_2(u, \alpha) = k \left[1 - \frac{3}{5} \left(\frac{\alpha \lambda^2}{10u} \right)^{1/2} \right] \tag{4.55}$$

$$u(s_2, \alpha) = \frac{9}{250} \frac{\alpha \lambda^2}{(1 - s_2/k)^2}, \tag{4.56}$$

which yield

$$\left(\frac{\partial u}{\partial s_2} \right)_\alpha = \theta \tag{4.57}$$

$$\left(\frac{\partial u}{\partial \alpha} \right)_{s_2} = \sigma. \tag{4.58}$$

From the internal energy

$$u = \frac{3^{2/3}}{10} \alpha \lambda^2 \tau^{2/3}, \tag{4.59}$$

we obtain the specific heat

$$c = \frac{3^{2/3}}{15} k \tau^{-1/3}. \tag{4.60}$$

There should be no surprise from the fact that $\lim_{\tau \to 0} c(\tau) \neq 0$, in fact diverges. Indeed, the present calculation is on classical grounds (non-quantum), and there is therefore no expectation for the verification of the Third Principle of Thermodynamics. The present result is, in this respect, not very different from the usual $q = 1$ classical specific heats, which systematically attain a positive value in the $\tau \to 0$ limit, thus also violating the Third Principle.

It is finally obtained the celebrated Carnot efficiency (see Fig. 4.12)

$$\eta = \frac{W}{\mathcal{Q}_\infty} = \frac{\mathcal{Q}_1 - \mathcal{Q}_2}{\mathcal{Q}_1} = 1 - \frac{\theta_2}{\theta_1} \quad (0 \leq \eta \leq 1). \tag{4.61}$$

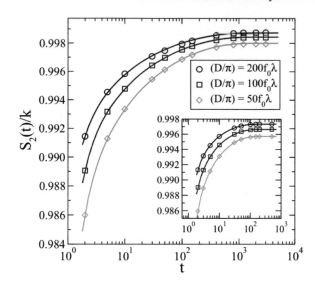

Fig. 4.11 In each case, two choices for the confining constant were employed, namely, $\alpha = 10^{-3} f_0 / \lambda$ ($\alpha = 10^{-2} f_0 / \lambda$ in the inset). Typical values for the total number of vortices are shown, namely $D = 200 \pi f_0 \lambda$, $D = 100 \pi f_0 \lambda$ and $D = 50 \pi f_0 \lambda$. At the stationary state, these choices correspond to three distinct values of the effective temperature θ. The increase of the total entropy with time shows that the second law of thermodynamics also applies, along and irreversible process, for the present $q \neq 1$ entropic form. The dimensionless time t is measured in units of the molecular-dynamics time step δt. From [440]

It is certainly remarkable that this result emerges in the realm of thermodynamics with $q \neq 1$. Although we are unaware of a general proof that the Carnot efficiency is q-invariant, i.e., valid for any value of q, it is nearly doubtless that it is indeed so.

It is worthy stressing also that the present $q = 0$ result $\langle x^2 \rangle \propto (k\theta)^{2/3}$ plays a role by all means analogous to the well known $q = 1$ result $\langle v^2 \rangle \propto kT$.

The Second Principle can be basically stated by saying that the thermodynamical entropy never decreases with time. This issue has also been discussed for the present class of systems and it has been analytically proved [440] that $dS_2(t)/dt \geq 0$, as illustrated in Fig. 4.11, thus confirming its validity for $q = 0$.

4.5.3 Zeroth Principle of Thermodynamics

Since Carathéodory's fundamental works, it is known the crucial role played in Thermodynamics by the so-called Zeroth Principle. From [439] and various related papers its validity is also verified for $q \neq 1$.

Let us consider two systems (1 and 2) in contact, the equilibrium of which is characterized by $\frac{N_1}{L_{y,1}} = \frac{N_2}{L_{y,2}}$. Now, by putting a third system in contact and in equi-

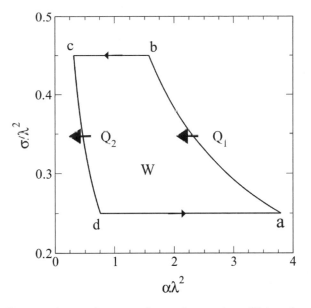

Fig. 4.12 The Carnot cycle $a \rightarrow b \rightarrow c \rightarrow d \rightarrow a$, for a system of interacting vortices under overdamped motion at $(T/\Theta) \simeq 0$. The transformations for constant σ are adiabatic, and herein they were chosen to occur for $(\sigma/\lambda^2) = 0.45$ $(b \rightarrow c)$ and $(\sigma/\lambda^2) = 0.25$ $(d \rightarrow a)$. The isothermal transformations are characterized by $\sigma \sim \alpha^{-2/3}$ [cf. Eq. (4.54)] and they occur for $k\theta_1 = 5$ (units of energy) in $a \rightarrow b$, and $k\theta_2 = 1$ (units of energy) in $c \rightarrow d$, i.e., $\theta_1 > \theta_2$. The area inside the cycle represents the total work W done on the system, which is negative, as expected from Eq. (4.53). The abscissa $\alpha\lambda^2$ presents dimensions of energy, whereas the ordinate σ/λ^2 is dimensionless; the cycle above holds for any system of units, e.g., one may consider all quantities with dimensions of energy in joules. From [437]

librium with system 1, one has $\frac{N_1}{L_{y,1}} = \frac{N_3}{L_{y,3}}$. Consequently systems 2 and 3 are also in equilibrium, i.e., $\frac{N_2}{L_{y,2}} = \frac{N_3}{L_{y,3}}$. Therefore, one can formulate the zeroth principle for the present class of systems.

Two systems of superconducting vortices for which thermal effects are negligible in comparison with those associated with their effective temperatures are said to be in thermal equilibrium if, being in contact, no heat flows in either way. The zeroth principle can be enunciated by stating that two systems in thermal equilibrium with a third one are in thermal equilibrium with each other, for an arbitrary value of q. This corresponds to the transitivity of their effective temperatures.

Before going next onto a more general model, let us stress that we have presented above, for a specific anomalous many-body system, an analytical set of explicit equations which prove that, for $q = 0$, the entire frame of thermodynamics and statistical mechanics satisfies the usual requirements of the $q = 1$ (BG) theory, namely the Clausius relation, Zero, First, and Second Principles of Thermodynamics, equations of states, distributions of positions (and of velocities), and Carnot cycle. This consti- tutes, to the best of our knowledge, the first complete analytical proof of what may be considered as a new paradigm. Indeed, *Boltzmann–Gibbs statistical mechanics is sufficient but not necessary for the validity of thermodynamics* (see also [235]). This opens wide the door for analytically proving this statement for all values of q, at least for the general class of overdamped systems.

4.5.4   An Extended Model with Repulsive Power-Law Short-Ranged Interactions

The system that we have been discussing above can be extended in the following manner [441, 443]. We consider a d-dimensional overdamped system including two-particle repulsive short-ranged interactions characterized by the potential $V_\lambda(r) = \epsilon(\sigma/r)^\lambda$ with $\lambda > d$, $\epsilon > 0$, $\sigma > 0$. Dissipative systems of repulsive particles tend to form structural lattices with (at least) local order. The effect of the local correlations is especially important for power-law interactions. [Notice that, in various other sections of this book, the notation $\lambda \equiv \alpha$ has been used. We adopt here λ to avoid confusion with the quantity α that has been discussed just above.] For the position distribution, it is obtained a q-Gaussian one with

$$q = 1 - \frac{\lambda}{d} < 1. \tag{4.62}$$

In the limit $\lambda/d \to 1$ we recover the value $q = 0$ of the above model for superconductors type II.

See Figs. 4.13, 4.14, 4.15, 4.16, and 4.17.

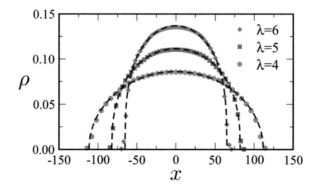

Fig. 4.13 Density profiles at the stationary state obtained from simulations (symbols). We consider two-dimensional systems of $N = 900$ particles interacting through a power-law repulsive potential $V_\lambda(r) = \epsilon(\sigma/r)^\lambda$. In the x direction, the particles are confined by a quadratic potential $U_{ext}(x) = kx^2/2$, with $k = 10^{-5}\epsilon\sigma^{-2}$. In the y direction the simulation cell has a dimension $L_y = 60\sigma$, with periodic boundary conditions. We present results for simulations of this system considering three different values of λ. The dashed lines represent the results of the continuum model via Eq. (4.63). From [441]

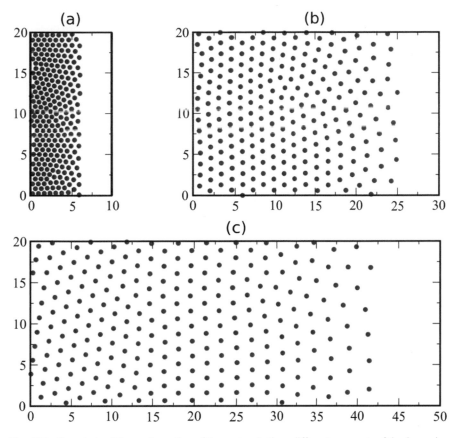

Fig. 4.14 Snapshots of the configuration of the system in three different moments of the dynamics: **a** $t = 1.0$, **b** $t = 5.88$, and **c** $t = 550$. Due to the symmetry we show only the $x > 0$ half of the system. The particles start in a very narrow region and invade the system as time goes on, eventually reaching an equilibrium position. This system comprises $N = 500$ particles, interacting through a potential $V_{ij} = \epsilon(\sigma/r_{ij})^\lambda$, with $\lambda = 4$, in a cell of lateral length $L_y = 20\sigma$, and confined by an external force $-kx$ with $k = 10^{-6}\epsilon/\sigma^2$. From [443]

From [441] we reproduce the stationary-state position distribution:

$$\rho(x) = \left[\rho_0^{\lambda/2} - \frac{\lambda}{2C_\lambda} U_{ext}(x) \right]^{2/\lambda}. \tag{4.63}$$

Let us further focus now on the time-dependent distributions of positions and also those of velocities.

The following auxiliary function emerges in [443]:

$$f(t) = \left\{ \frac{\nu C_\lambda}{k} [1 - e^{-k(\frac{\lambda}{d}+2)(t-t_0)}] \right\}^{\frac{d}{\lambda+2d}}. \tag{4.64}$$

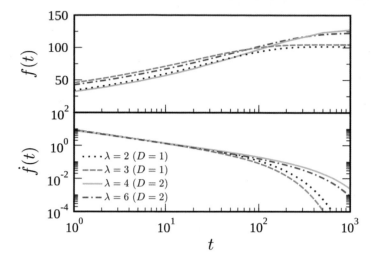

Fig. 4.15 Curves of f(t) for $d = 1$ ($\lambda = 2$ and 3) and d=2 ($\lambda = 4$ and 6) obtained from Eq. (4.64) and its derivative $\dot{f}(t)$. In our numerical simulations, we start with all particles confined in a narrow stripe, leading us to use $t_0 = 0$ in Eq. (4.64). The case d = 1 corresponds to $N = 3600$ particles with confining potential strength $k = 3.2 \times 10^{-3} \epsilon/\sigma^2$. The case $d = 2$ corresponds to $N = 4000$ particles, with confining potential strength $k = 1 \times 10^{-3} \epsilon/\sigma^2$, in a cell with transverse size $L_y = 20\sigma$. From [443]

It enters in particular in the velocity distribution

$$P\left(\frac{v_x}{\dot{f}}\right) = \frac{1}{2h} \int_{-\infty}^{\infty} d\xi \, G_q\left(\frac{v_x/\dot{f} - \xi}{q - 1}\right) e^{-\frac{|\xi|}{h}}. \tag{4.65}$$

4.5.5 Model with Direction-Depending Drag

Many physical overdamped systems exist in the presence of a drag which depends on the direction. We shall briefly present here a one-dimensional model [434] exhibiting such directional drag, the interaction between particles being a short-range repulsion. We assume a drag force given by $F_{drag} = -\alpha_r v$ ($F_{drag} = -\alpha_l v$) for $v > 0$ (for $v < 0$), with not necessarily equal $\alpha_r \geq 0$ and $\alpha_l \geq 0$. We may define the anisotropy index $\varepsilon \equiv 1 - \alpha_l/\alpha_r$, isotropy thus corresponding to $\varepsilon = 0$. Consequently the drag force can be written as follows:

$$F_{drag} = -\frac{1}{2}\left[(\alpha_r + \alpha_l)v + (\alpha_r - \alpha_l)|v|\right]. \tag{4.66}$$

The corresponding nonlinear diffusion Fokker–Planck equation is given by [434]

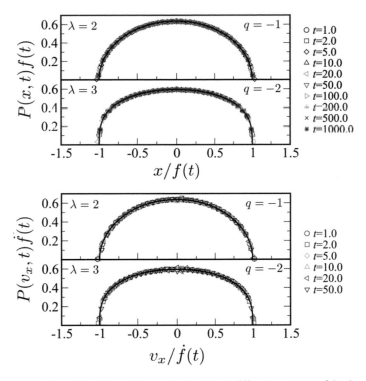

Fig. 4.16 Distributions of scaled positions and velocities at different moments of the dynamics for the one-dimensional case ($d = 1$). These results concern $N = 3600$ particles interacting through the power-law potential, $V = \epsilon(\sigma/r)^\lambda$, for $\lambda = 2$ ($q = -1$) and $\lambda = 3$ ($q = -2$). The black curves are q-Gaussians. Here we used the strength of the confining potential $k = 3.2 \times 10^{-3}\epsilon/\sigma^2$. From [443]

$$\frac{\partial \rho}{\partial t} = \frac{\partial}{\partial x}\left\{\frac{\rho}{2}\left[(2-\varepsilon)\frac{\partial}{\partial x}\left(\mathcal{D}\left(\frac{2-q}{1-q}\right)\left(\frac{\rho}{\rho_0}\right)^{1-q}\right)\right.\right.$$

$$\left.\left. +\varepsilon\left|\frac{\partial}{\partial x}\left(\mathcal{D}\left(\frac{2-q}{1-q}\right)\left(\frac{\rho}{\rho_0}\right)^{1-q}\right)\right|\right]\right\}, \qquad (4.67)$$

with $q < 3$ and $(2-q)\mathcal{D} > 0$, ρ_0 being a reference constant whose dimensionality is, like that of $\rho(x, t)$ itself, an inverse length. This equation is consistent with an H-theorem, and admits as a solution the following asymmetric Ansatz:

$$\rho(x, t) = A(t)\left[1 - \frac{1}{4}(1-q)\left(\beta_1(t)(x + |x|)^2 + \beta_2(x - |x|)^2\right)\right]_+^{\frac{1}{1-q}}, \qquad (4.68)$$

where $A(t) > 0$ is determined through normalization, and $\beta_2(0) = (1 - \varepsilon)\beta_1(0)$. See Fig. 4.18 for a $q = 0$ illustration.

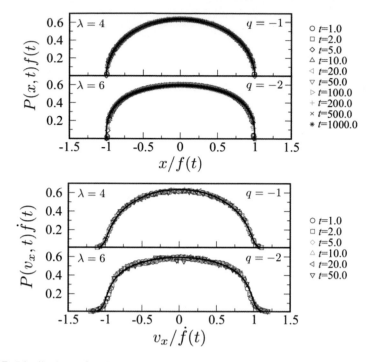

Fig. 4.17 Distributions of scaled positions and velocities at different moments of the dynamics for the case of two dimensions ($d = 2$). These results concern particles interacting through the power-law potential, $V = \epsilon(\sigma/r)^\lambda$, for $\lambda = 4$ ($q = -1$) and $\lambda = 6$ ($q = -2$). One can compare the density profiles with q-Gaussians (black curves). In the case of the velocity distributions the black curves show convolutions between q-Gaussians and Laplacian distributions, as given by Eq. (4.65), with the parameter h given by 0.045 and 0.051 for $\lambda = 4$ and 6, respectively. In these simulations we used $N = 4000$ particles in a cell of transverse size $L_y = 20\,\sigma$ with periodic boundary conditions. In the longitudinal direction (x axis) we imposed a confining force $-kx$ with $k = 1 \times 10^{-3}\epsilon/\sigma^2$. From [443]

4.5.6 Plasma in Spherical Capacitor with Overdamping

A gas of small charges free to move within a spherical capacitor has been simulated from first principles assuming overdamping [444]: see Fig. 4.19. The system clearly is electrically non-neutral, which makes impossible an important screening of the long-range Coulombian forces which, in a typical neutral system, would then effectively become short-ranged ones. A typical result is shown in Fig. 4.20. Comparison is done with the mean-field approach ($q = 1$) of Debye–Huckel for an electrolyte or, equivalently, of the Yukawa potential in nuclear physics.

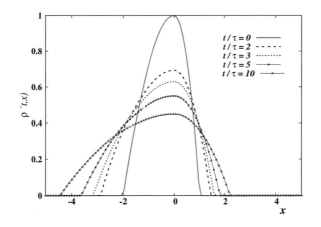

Fig. 4.18 Evolution of the asymmetric, q-Gaussian solution (4.68) to the nonlinear diffusion equation (4.67). The density $\rho(x,t)$ is depicted as a function of x for different values of t/τ, where τ is a characteristic time given by $\tau^{-1} = 12(1-\varepsilon)D\beta_1(0)A(0)$. The solution corresponds to $D = \mathcal{D}/\rho_0 = 1$, $q = 0$, and to initial conditions given by $A(0) = 1$, $\beta_1(0) = 1$, $\beta_2(0) = 1/4$. The density ρ is measured in units of $A(0)$, and the coordinate x in units of $\beta_1(0)^{-1/2}$. From [434]

Fig. 4.19 The system investigated consists of N negative charges $-q_0$ ($q_0 > 0$), inside a spherical region, between two concentrical conducting shells of radius r_1 and r_2 ($r_2 > r_1$). This region is filled with a neutral fluid that produces friction, so that the particles move under overdamped motion. The shells are charged (total charges $\pm Q$), leading to a spherically symmetric electric field $\vec{E}$ pointing outward. From [444]

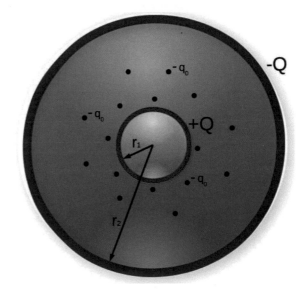

4.6 Generalizing the Langevin Equation

The standard Langevin equation is given by [448, 449]

$$\dot{x} = f(x) + \eta(t),\tag{4.69}$$

where $x(t)$ is a stochastic variable, $f(x)$ is an arbitrary function which represents some deterministic drift, and $\eta(t)$ is a Gaussian-distributed zero-mean white noise satisfying

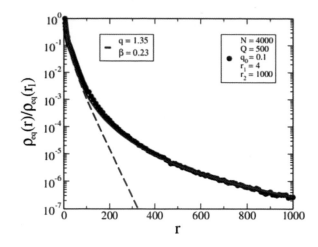

Fig. 4.20 The equilibrium density $\rho_{eq}(r)$ is plotted in log-linear representation, for typical values of the relevant parameters. The symbols (black circles) are results from simulations, whereas the full red line represents the fitting proposal $\Phi_q(r) = B_0 r^{-1} e_q^{-\beta(r-r_1)}$ with $q \simeq 4/3$. An illustrative Yukawa potential $\Phi_1(r) = B_0 r^{-1} e^{-\beta(r-r_1)}$ is represented by the dashed blue line. Lengths and charges are represented in units of a characteristic length $l(r_1, r_2)$ and a typical charge $\tilde{q}$, respectively. See details in [444]

$$\langle \eta(t)\,\eta(t')\rangle = 2\,A\,(1-q)(t-t')\,. \tag{4.70}$$

The noise amplitude $A \geq 0$ stands for *additive*. The deterministic drift $f(x)$ can be interpreted either as a damping force (whenever x is a velocity-like quantity), or as an external force (when motion is overdamped and x represents a position coordinate). Other interpretations are possible as well, depending on the particular system we are focusing on. This equation is known to lead to the standard Fokker–Planck equation (Fourier's *heat equation*), whose basic solutions are Gaussians in the variable $x/\sqrt{t}$.

This historical equation has been generalized in very many ways. Some of them yield exact solutions which are q-Gaussians. Two such examples are described in [450] (simultaneous presence of uncorrelated additive and multiplicative noises) and in [452] (dichotomous colored noise). Because of its particularly simple nature, we shall present here the first example in detail. Let us consider the following generalization of Eq. (4.69):

$$\dot{x} = f(x) + g(x)\,\xi(t) + \eta(t)\,, \tag{4.71}$$

where $g(x)$ is an arbitrary function satisfying $g(0) = 0$, and $\xi(t)$ is a Gaussian-distributed zero-mean white noise satisfying

$$\langle \xi(t)\, \xi(t') \rangle = 2\, M\, (1 - q)(t - t')\,. \tag{4.72}$$

The noise amplitude $M \geq 0$ stands for *multiplicative*. The noises $\xi(t)$ and $\eta(t)$ are assumed uncorrelated.[4] The stochastic differential equation is not completely defined and must be complemented by an additional rule. This is due to the fact that each pulse of the stochastic noise produces a jump in x, then the question arises: which is the value of x to be used in $g(x)$ This is the well known Itô-Stratonovich dilemma [449, 453]. In the Itô definition, the value before the pulse must be used, whereas in the Stratonovich definition, the values before and after the pulse contribute in a symmetric way. In the particular instance when noise is purely additive, both definitions agree. In what follows, we shall adopt the Stratonovich definition (the Itô definition leads in fact to very similar results). By using standard procedures [449, 450], Eq. (4.71) leads to

$$\frac{\partial P(u, t)}{\partial t} = -\frac{\partial j(u, t)}{\partial u}\,, \tag{4.73}$$

where the *current* is defined as follows:

$$j(u, t) \equiv J(u)P(u, t) - \frac{\partial [D(u)P(u, t)]}{\partial u}\,, \tag{4.74}$$

with

$$J(u) \equiv f(u) + Mg(u)g'(u)\,, \tag{4.75}$$

$$D(u) \equiv A + M[g(u)]^2\,. \tag{4.76}$$

Equation (4.73) can be rewritten as a Fokker–Planck equation, namely

$$\frac{\partial P(x, t)}{\partial t} = -\frac{\partial [f(x)P(x, t)]}{\partial x} + M\frac{\partial}{\partial x}\left(g(x)\frac{\partial [g(x)P(x, t)]}{\partial x}\right) + A\frac{\partial^2 P(x, t)}{\partial x^2}\,. \tag{4.77}$$

In some processes, the deterministic drift derives from a potential-like function $V(x) = (\tau/2)[g(x)]^2$, where τ is some nonnegative proportionality constant. Therefore, using $f(x) = -dV/dx$, we obtain the condition

---

[4] It is possible to combine two such noises into a single effective multiplicative one [450], but, for clarity purposes, we shall keep track of both sources separately.

$$f(x) = -\tau g(x)g'(x).$$ (4.78)

Let us note that the particular case $g(x) \propto f(x) \propto x$, which is a natural first choice for a physical system, verifies this condition. However, since no extra calculational difficulties emerge, we will discuss here the more general case of Eq. (4.78). Notice that, in the absence of deterministic forcing, condition (4.78) is satisfied for any $g(x)$ by setting $\tau = 0$.

We shall restrict here to the stationary solutions corresponding to no flux boundary conditions (i.e., $j(-\infty) = j(\infty) = j(u) = 0$), although more general conditions could in principle also be considered. If (4.78) is satisfied, the stationary solution $P(u, \infty)$ is of the q-exponential form, namely

$$P(u, \infty) \propto e_q^{-\beta[g(u)]^2},$$ (4.79)

with[5]

$$q = \frac{\tau + 3M}{\tau + M} \geq 1,$$ (4.80)

and

$$\beta \equiv \frac{1}{kT} = \frac{\tau + M}{2A}.$$ (4.81)

(T can be generically seen as the amplitude of an effective noise). For the typical case $\tau > 0$, we have that $q \geq 1$ if $M \geq 0$, the value $q = 1$ corresponds to vanishing multiplicative noise. If $|g(u)|$ grows, for $|u| \rightarrow \infty$, faster than $|u|^{1+\tau/M}$, $P(u, \infty)$ decreases faster than $1/|u|$, and is therefore normalizable. The condition (4.78) is in fact not necessary for having solutions of the q-exponential form. The interested reader can see the details in [450].

Let us mention that we have discussed here a case in which q-Gaussian distributions emerge from a *linear* Fokker–Planck equation (Eq. (4.77)). It is clear that this mechanism differs from the one focused on in Eq. (4.7), which is a *nonlinear* Fokker–Planck equation. For the Langevin discussion, i.e., from mesoscopic first principles, of this nonlinear Fokker–Planck equation, see [451] (with the notation change $q \rightarrow 2 - q$). It turns out that if we consider a mechanism involving strongly non-Markovian processes, i.e., long memory effects, the nonlinear Fokker–Planck equation (Eq. (4.7)) naturally comes out.

---

[5] Incidentally, if we had used the Itô convention, we would have obtained $q = \frac{\tau + 4M}{\tau + 2M}$. Further generalizations can be seen in [426, 427]. In the present notation, we have $q = \frac{\tau + 2(2-\alpha)M}{\tau + 2(1-\alpha)M} \geq 1$, where $\alpha \in [0, 1]$, $\alpha = 0, 1/2, 1$, respectively denoting the Itô, Stratonovich and anti-Itô conventions.

Finally, as mentioned above, another Langevin process has been studied [452] which includes a colored symmetric dichotomous noise. Although not identified in this manner by the author, the stationary state precisely has a q-Gaussian distribution of velocities with

$$q = \frac{1 - 2\gamma/\lambda}{1 - \gamma/\lambda} < 1, \tag{4.82}$$

where $\gamma > 0$ and $\lambda > 0$ are mesoscopic parameters of the model.

4.7 Other Complex Dissipative Systems

4.7.1 Coherent Noise Model

The Coherent noise model (CNM) (see [454] and references therein) has been introduced for analyzing biological extinctions. It has later on been adopted as a very simple mean-field model for earthquakes even though no geometric configuration space is introduced in the model. It is shown that the model obeys the Omori law for the temporal decay pattern of aftershocks and exhibits aging phenomena and power-law sensitivity to initial conditions.

The system consists of N agents, each one having a threshold $x_i \in [0, 1]$ against an external stress $\eta > 0$. The threshold levels and the external stress η are randomly chosen from probability distributions $p_{thresh}(x)$ and $p_{stress}(\eta)$, respectively. Throughout the simulations in [454], it has been used the exponential distribution for the external stress, namely, $p_{stress}(\eta) = (1/\sigma)e^{-\eta/\sigma}$ ($\sigma > 0$) and the uniform distribution ($0 \leq x \leq 1$) for $p_{thresh}(x)$. The dynamics of the model are very simple: (i) Generate a random stress η from $p_{stress}(\eta)$ and replace all agents with $x_i \leq \eta$ by new agents with new threshold drawn from $p_{thresh}(x)$; (ii) Choose a small fraction f of N agents and assign them new thresholds drawn again from $p_{thresh}(x)$; and (iii) Repeat the first step for the next time step. Step (ii) is necessary for the continuity of the generation of avalanches since the average value of the threshold distribution increases with the proceeding steps (up to one), and therefore any stress value cannot hit any agents anymore. The number of agents replaced in the first step of the dynamics determines the event size s for this model. The avalanche distribution has a plateau until $s = \sigma f$ followed by a power-law regime. It is possible to make the plateau practically disappear by using a proper value for f. For a fixed σ value, the behavior of the model is independent of the f values.

The corresponding avalanche size distribution $p(s)$ behaves like a $q_{avalanche}$-exponential, hence asymptotically we have $p(s) \propto s^{-\tau}$ with $\tau \equiv 1/(q_{avalanche} - 1) > 1$. The return distribution is noted $p(\Delta s)$, where

$$\Delta s = s(t+1) - s(t), \tag{4.83}$$

is the difference between two consecutive event sizes. The corresponding distributions of this model have been studied in [454–456], and also in [457, 458], and they have a q_{return}-Gaussian form with

$$q_{return} = \frac{\tau + 2}{\tau}, \tag{4.84}$$

hence

$$\frac{2}{q_{return} - 1} = \frac{1}{q_{avalanche} - 1}, \tag{4.85}$$

where the fact that the process is Markovian has enabled the straightforward establishment of the analytical relation. We verify that the intervals $q_{avalanche} \in [1, 2)$ and $q_{return} \in [1, 3)$ are perfectly consistent, as expected. See Figs. 4.21 and 4.22. There are naturally finite-size effects in the above computational approaches. Interestingly enough, they have been shown in [454] to numerically follow very well the expectations of crossover statistics, discussed in Sect. 6.1. Indeed, the finite-size distribution typically satisfies [459]

$$\frac{dy}{d(x^2)} = -a_1 y - (a_q - a_1)y^{q_{return}} \quad (a_q \geq a_1 \geq 0; y(0) = 1), \tag{4.86}$$

with $\lim_{N \to \infty} a_1(N) = 0$, $\lim_{N \to \infty} a_q(N) = a_q(\infty)$, and $\lim_{N \to \infty} q_{return}(N) = q_{return}(\infty) > 1$. The solution is given by

$$y(x; N) = \left[1 - \frac{a_q(N)}{a_1(N)} + \frac{a_q(N)}{a_1(N)} e^{[q_{return}(N)-1]a_1(N) x^2}\right]^{1/[1-q_{return}(N)]}, \tag{4.87}$$

which implies $y(x; \infty) = e^{\frac{-a_q(\infty) x^2}{q_{return}(\infty)}}$ $(a_q(\infty) \equiv \beta)$. See Fig. 4.23.

4.7.2 Ehrenfest Dog-Flea Model

The Ehrenfest dog-flea model (see [461, 462] and references therein) has N dynamical sites represented by the total number of fleas shared by two dogs, A and B. Suppose that there are N_A fleas on dog A and N_B fleas on dog B, leading to a population of fleas $N = N_A + N_B$. For convenience, N is assumed to be even. In every time step, a randomly chosen flea jumps from one dog to the other. Thus, we have $N_A \to N_A \pm 1$ and $N_B \to N_B \mp 1$. The procedure is repeated for an arbitrary number of times. In long-time run, the mean number of fleas on both dog A and dog B converges to the equilibrium value, $\langle N_A \rangle = \langle N_B \rangle = N/2$, with the fluctuations around it. A single fluctuation is described as a process that starts once the number

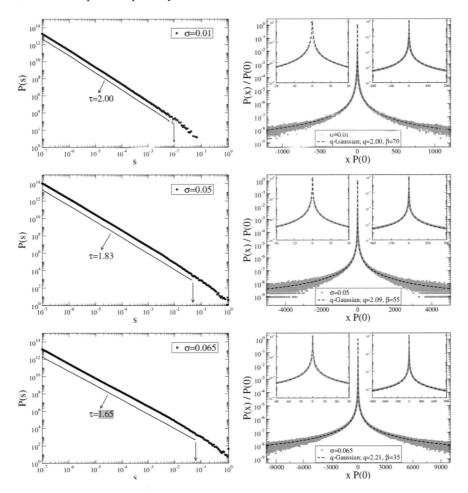

Fig. 4.21 *Left column*: Avalanche size distributions corresponding to three typical values of σ. For each case, the exponent τ is calculated using the standard regression method for that region of s. *Right column*: Return distributions for the same three examples. Two zooms of the central part are given in the insets for better visualization. For each case, $f = 10^{-7}$ and 2109 experiments are generated. From [454]

of fleas on one of the dogs becomes larger or smaller than the equilibrium value $N/2$ and stops when it gets back to it for the first time. Thus, the end of one fluctuation specifies the start of the subsequent one. The length (λ) of a fluctuation is determined by the number of time steps elapsed until the fluctuation ends. The returns are defined as $\Delta\lambda(t) \equiv \lambda(t+1) - \lambda(t)$, and the centered variable as $x \equiv \Delta\lambda - \langle\Delta\lambda\rangle$, where $\langle\ldots\rangle$ stands for the mean value of the given data set. The results for fixed N are illustrated in Figs. 4.24 and 4.25.

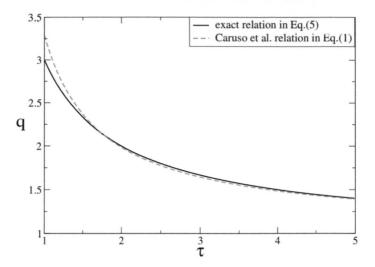

Fig. 4.22 Comparison of the exact relation (Eq. (4.85)) with a good numerical approximation previously proposed in the literature (Eq. (1) in [460]). From [454]

4.7.3 Random Walk Avalanches

It was shown in [463] that the avalanche size of a one-dimensional directed sandpile model can be mapped to the area under a Brownian curve with an absorbing boundary at the origin. This is equivalent to a random walker on $[0, \infty)$ with an absorbing boundary at the origin. If we denote the trajectory of the random walker by $x(i)$ with $i = 0, 1, ..., N$, the avalanche size can be described as

$$s = \sum_{i=1}^{N} x(i), \qquad (4.88)$$

with $x(0) = 1$. The return distribution of this model is defined as in the CNM, i.e., using Eq. (4.83), and it was numerically found a Q-Gaussian with $Q \simeq 7/3$: see Fig. 4.26 from [464].

4.7.4 Kuramoto Model

The Kuramoto model was introduced in [465] and is defined (see [466, 467] and references therein) by the following dissipative dynamics of a system of N sinusoidally coupled oscillators:

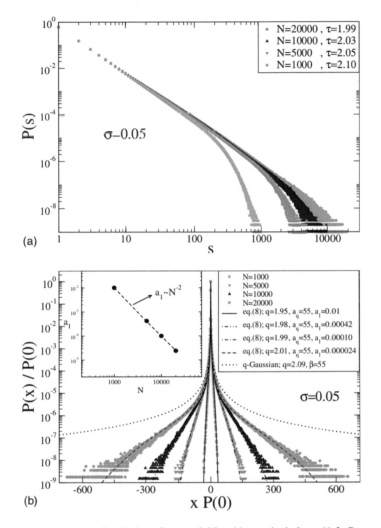

Fig. 4.23 a Avalanche size distributions for $\sigma = 0.05$ and increasingly large N. **b** Corresponding return distributions with $(q, a_1) \equiv (q_{return}, \beta)$ and N increasing from the centered distributions to the $N \to \infty$, fully tailed, one. The curves are given by Eq. (4.87) [with the convention $(x, y) \to (xP(0), P(x)/P(0))$]. *Inset: N-dependence of the parameter a_1. The gradual disappearance of the finite-size effects is found to obey $a_1 \propto 1/N^2$. From [454]

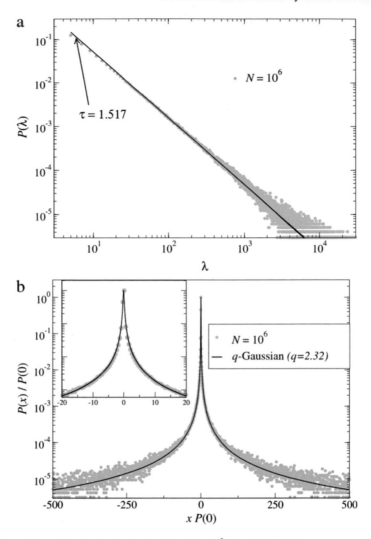

Fig. 4.24 a Fluctuation length distribution for $N = 10^6$. The full black line represents the fitting curve of the distribution with slope $\tau \simeq 1.517$. The distribution has an arbitrary normalization such that $P(\lambda = 1) = 1$. **b** Return distribution for $N = 10^6$. The full black curve represents the appropriate q-Gaussian with $q = 2.32$ determined from relation (4.84) *a priori*, and $\beta = 30$. The central part of the distribution is emphasized in the inset. From [462]

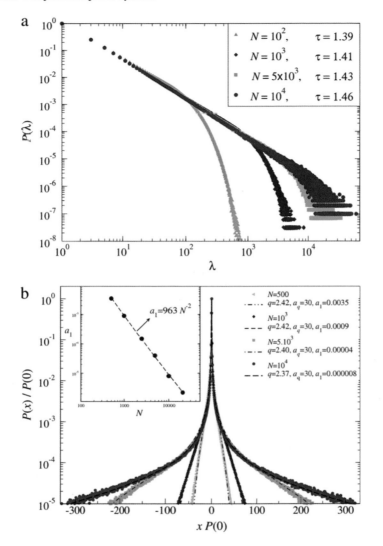

Fig. 4.25 **a** Fluctuation length distributions for typical values of N; 109 fluctuations are considered for each system size. The figure legends give the corresponding values for τ of power-law regimes. The distributions have an arbitrary normalization such that $P(\lambda = 1) = 1$. **b** Return distributions for the same values of N. The black lines on each return distribution represent the fitting curves obtained through Eq. (4.87) with values for q determined *a priori* from the relation (4.84), using the corresponding values for τ. For each curve, $a_q = 30$ which equals β of the q-Gaussian that fits the return distribution with largest system size, i.e., $N = 10^6$. In the inset, we also give the parameter a_1 as a function of N in order to better visualize how the finite-size effect disappears as N increases. It is clear that a_1 appears to vanish as $a_1 = 963\, N^{-2}$. From [462]

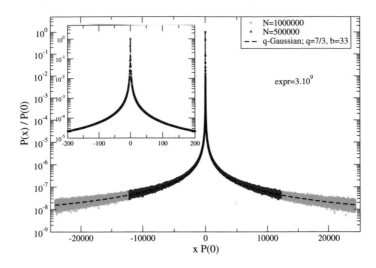

Fig. 4.26 Return distribution for the model with random walk avalanches for two typical values of N; the averaging has been done over 3×10^9 experiments. Courtesy from [464]

$$\dot{\theta}_i = \omega_i + \frac{K}{N} \sum_{j=1}^{N} \sin(\theta_j - \theta_i)\,, \qquad (4.89)$$

where θ_1 and ω_i are, respectively, the angle (phase) and the natural frequency of the i-th oscillator ($i = 1, \ldots, N$) and $K \geq 0$ is the coupling strength. The natural frequencies are time-independent and randomly chosen from a symmetric, unimodal distribution $g(\omega)$. They represent a driving term that, considering the oscillators as particles moving on the unit circle, forces each particle to rotate at its own frequency regardless of the other particles. On the other hand, the sinusoidal term, tuned by the coupling constant K, tends to synchronize the particles, so that, for values of $K < K_c$, the oscillators run independently and incoherently, while for $K > K_c$ the system undergoes a spontaneous phase transition toward a partially or fully synchronized state (depending on the value of K). For an uniform distribution $g(\omega)$ ($\omega \in [-\frac{1}{2g(0)}, \frac{1}{2g(0)}]$), the critical value of the coupling is given by $K_c = \frac{2}{\pi g(0)}$, $g(0)$ being the central value of the distribution. By following along the Central Limit Theorem lines, we define

$$y_i = \frac{1}{\sqrt{n}} \sum_{k=1}^{n} \theta_i(n\delta)\,, \qquad (4.90)$$

by picking out, for each oscillator, n values of the angle θ_1 at fixed intervals of time δ along the deterministic time evolution imposed by Eq. (4.89); the total simulation time over which the sum is calculated is given by $n\delta$. After centering and rescaling, the PDF associated with y_i appears to be given, for wide ranges of the model parameters, by q-Gaussians: a typical example is shown in Fig. 4.27.

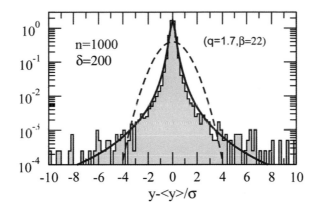

Fig. 4.27 The CLT-motivated probability distribution $p(y)$ for a Kuramoto system for uniform distribution $g(\omega)$ with $g(0) = 1/4$ (hence $K_c = 8/\pi \simeq 2.54$), $N = 20000$, $\delta = 200$ and $n = 1000$, and $K = 0.1$, hence the largest Lyapunov exponent is nearly vanishing. The numerical results are well fitted by a q-Gaussian (full line) with $q \simeq 1.7$ and $\beta \simeq 22$. A standard Gaussian with unit variance (dashed line) is shown for comparison. From [467]

4.8 Probabilistic Models with Strong Correlations—Numerical and Analytical Approaches

Before addressing the above theorems and their proofs, let us first present four interesting models that provide some degree of intuition on the type of correlations that are assumed within the present context. These models will be referred to as the *MTG model* [468], the *TMNT model* [469], the *RST1 model* [470], and finally the *RST2 model* [470]. The first three are models of the type $q_{attract} \leq 1$; also, they are strictly scale-invariant. The fourth model is defined for both $q_{attract} < 1$ and $q_{attract} \geq 1$ cases, and it is asymptotically (but not strictly) scale-invariant. All four models numerically appear to approach, when $N \to \infty$, q-Gaussian forms. The MTG and TMNT models do *not* strictly do so in fact, as analytically proved in [471]. In contrast, the RST1 model *does* approach a q-Gaussian, as analytically proved in [470]. Also *does* the RST2 model, by construction.

To establish a quite general frame let us define a system of generically correlated distinguishable binary probabilities as follows:

$$p_{N,n} = \frac{N!}{(N-n)!\,n!}\, r_{N,n}\,, \tag{4.91}$$

where the *reduced probabilities* $\{r_{N,n}\}$ are depicted in Fig. 4.28. The following normalization is satisfied:

$$\sum_{n=0}^{N} p_{N,n} = \sum_{n=0}^{N} \frac{N!}{(N-n)!\,n!}\, r_{N,n} = 1\,, \tag{4.92}$$

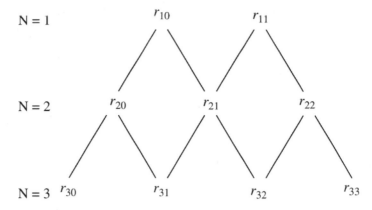

Fig. 4.28 Pascal-Leibnitz triangle configuration of reduced probabilities $\{r_{N,n}\}$ illustrated for number of composed systems $N = 1, 2, 3$. From [283]

and the corresponding q-entropy is given by

$$S_q(N) = \frac{1 - \sum_{i=1}^{2^N} p_i^q}{q - 1} = \frac{1 - \sum_{n=0}^{N} \frac{N!}{(N-n)!\,n!}\, r_{N,n}^q}{q - 1}. \tag{4.93}$$

The Leibnitz rule consists in

$$\frac{r_{N,n} + r_{N,n+1}}{r_{N-1,n}} = 1 \quad (\forall n,\ \forall N), \tag{4.94}$$

referred to as *strictly scale-invariant*. Naturally, the property

$$\lim_{N \to \infty} \frac{r_{N,n} + r_{N,n+1}}{r_{N-1,n}} = 1, \tag{4.95}$$

is referred to as *asymptotically scale-invariant*. If the system satisfies the Leibnitz rule, it is completely determined by giving one column, for example $\{r_{N,0}\}$. If $r_{N,0} = 1/2^N$, $\forall N$ (hence $r_{N,n} = 1/2^N$, $\forall(N,n)$), we have independent binary variables, whose attractor is a Gaussian ($q_{attract} = 1$). If $r_{N,0} = 1/(N+1)$, $\forall N$ (hence $p_{N,n} = 1/(N+1)$, $\forall(N,n)$), we recover the celebrated Leibnitz triangle, whose attractor is the uniform distribution, i.e., another q-Gaussian ($q_{attract} \to -\infty$), as shown in Fig. 4.2. The various stages of the independent case are illustrated in Fig. 4.29.

Let us now focus on the MTG and TMNT models and their numerical discussion, and eventually present their analytical solutions [471]. We shall later on address the RST1 and RST2 models and their corresponding results [470].

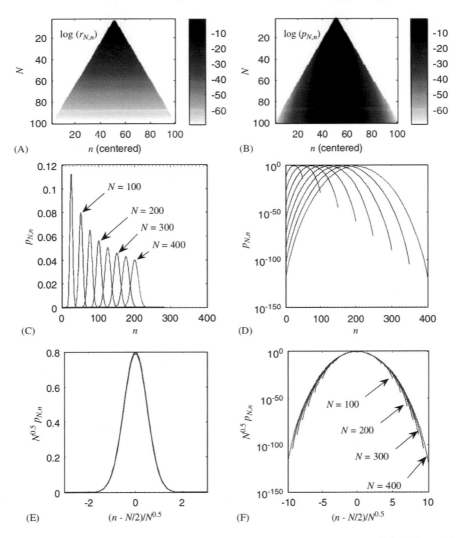

Fig. 4.29 Behavior of independent binary system composition for the case $p = 0.5$. Each model discussed in this paper has this case as a limiting case. Plots **a–b** show reduced probabilities $\log r_{N,n}$ and probabilities $\log p_{N,n}$ up to $N = 100$ as grayscale images. Plots **c–d** show uncentered and unscaled probabilities for $N = 50, 100, 400$, in both linear and logarithmic scales. Plots **e–f** show centered and scaled probabilities, illustrating the Gaussian attractor in probability space. From [283]

4.8.1 The MTG Model and Its Numerical Approach

Here we follow [468]. To avoid the Gaussian as the $N \to \infty$ attractor, we need to introduce persistent correlations. We shall do this by generalizing the case of independence (Figs. 4.30 and 4.31).

$$\frac{1}{r_{N0}} = \frac{1}{1/2} \times \frac{1}{1/2} \times \dots \times \frac{1}{1/2} \quad (N \text{ factors}). \tag{4.96}$$

in the following manner:

$$\frac{1}{r_{N0}} = \frac{1}{1/2} \otimes_q \frac{1}{1/2} \otimes_q \dots \otimes_q \frac{1}{1/2} = [2^{1-q}N - (N-1)]^{\frac{1}{1-q}} \quad (0 \le q \le 1). \tag{4.97}$$

We have generalized the product between the *inverse* probabilities, and not the probabilities themselves, in order to conform to the requirements of the q-product (see Eq. (3.81)). The Leibnitz rule is maintained, which enables us to calculate the entire set $\{r_{Nn}\}$ by assuming (4.97).

If we define $p(x) \equiv \frac{N!}{(N-n)!n!} r_{Nn}$, and $x \equiv \frac{n-(N/2)}{N/2}$, we obtain the results exhibited in Figs. 4.32 and 4.33. In other words, we verify that, in the limit $N \to \infty$, the numerical results approach

$$p(x) = \begin{cases} p(0) \, e_{q_{attract}}^{-\beta_+ x^2} & \text{if } x \ge 0, \\ p(0) \, e_{q_{attract}}^{-\beta_- x^2} & \text{if } x \le 0, \end{cases} \tag{4.98}$$

where β_+ and β_- are slightly different, i.e., the distribution is slightly asymmetric. This specific asymmetry is caused by the fact that we have imposed r_{N0}, instead of say r_{NN}, or something similar. By introducing $\beta \equiv \frac{1}{2}(\beta_+ + \beta_-)$, we obtain the dashed line of Fig. 4.32, and the results of Fig 4.33. The index $q_{attract}$ in the $q_{attract}$-Gaussian (*apparent*—but *not exact*, as we shall see!—attractor for $N \to \infty$) is a function of the index q in the q-product (which, together with Leibnitz rule, introduces the scale-invariant correlations into the probability sets). The numerics strongly suggest (see Fig. 4.34)

$$q_{attract} = 2 - \frac{1}{q} \quad (0 \le q \le 1). \tag{4.99}$$

The particular case $q = q_{attract} = 1$ recovers of course the celebrated de Moivre-Laplace theorem. This transformation is a simple combination of the *multiplicative duality*

$$\mu(q) \equiv 1/q , \tag{4.100}$$

and the *additive duality*

$$\nu(q) \equiv 2 - q . \tag{4.101}$$

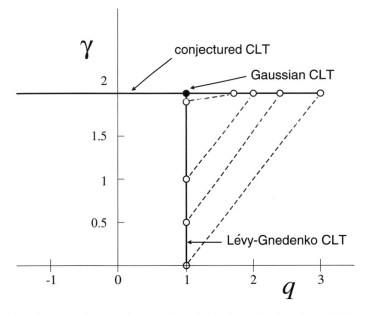

Fig. 4.30 Localization in the (q, γ)-space of the standard and Lévy-Gnedenko CLT's, as well as of the conjectural fully q-generalized CLT (based on [271], and whose sufficient conditions are established in [477]). The schematic dashed lines are curves that share the *same* exponent of the power-law behavior that emerges in the limit $|x| \to \infty$. At the $q = 1$ axis we have Lévy distributions which asymptotically decay as $1/|x|^{1+\gamma}$, and at the $\gamma = 2$ axis we have q-Gaussians which decay as $1/|x|^{2/(q-1)}$. The connection is therefore given by $q = (\gamma + 3)/(\gamma + 1)$ for $2 > \gamma > 0$, hence $5/3 < q < 3$ (see [407, 417, 420, 421, 478], based on [418, 419]). For instance, the dashed line which joins the (q, γ) points $(1, 1)$ and $(2, 2)$ schematically indicates those solutions of Eq. (4) which asymptotically decay as $1/x^2$, and the dashed line joining $(1, 1/2)$ and $(7/3, 2)$ indicates those solutions which decay as $1/|x|^{3/2}$. The dot slightly to the right of the point $(5/3, 2)$ is joint to the point slightly below $(1, 2)$

In other words, relation (4.99) can be rewritten as $q_e = \nu\mu(q) \equiv \nu(\mu(q))$. This relation as well as the two basic dualities appear again and again in the literature of nonextensive statistical mechanics, in very many contexts (see, for instance, [226, 472–476]).

Transformations (4.100) and (4.101) enable the construction of an interesting algebra.[6] Indeed, the following properties can be easily established:

$$\mu^2 = 1, \tag{4.102}$$

and

$$\nu^2 = 1, \tag{4.103}$$

---

[6] Private discussions with M. Gell-Mann in the context of a possible understanding of the numerical values determined (for the solar wind) in [480] for the q-triplet that will be discussed in Sect. 5.4. See also [274].

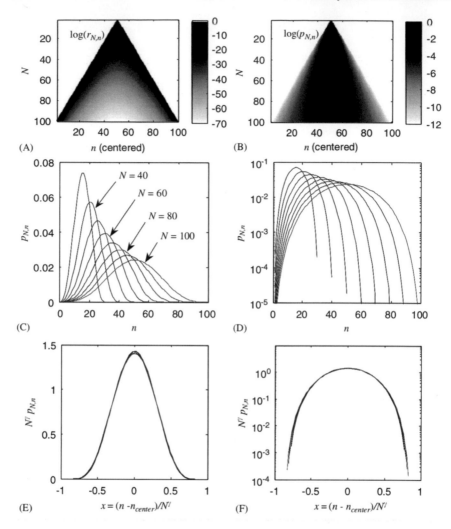

Fig. 4.31 The behavior of the MTG model with parameter $q \equiv q_{correlat} = 0.8$. Plots **a–b** show reduced probabilities $\log r_{N,n}$ and probabilities $\log p_{N,n}$ up to $N = 100$ as grayscale images. Plots **c–d** show uncentered and unscaled probabilities for $N = 30, 40, 100$, in both linear and logarithmic scales. Plots **e–f** show centered and scaled probabilities, illustrating the attractor in probability space. We find $\mu/2 \equiv \gamma = 0.88$. From [283]

where **1** represents the *identity*, i.e., $\mathbf{1}(q) = q$, $\forall q$. These properties justify the name *duality*.

We immediately verify that

$$(\mu\nu)^n (\nu\mu)^n = (\nu\mu)^n (\mu\nu)^n = \mathbf{1} \quad (n = 0, 1, 2, ...) . \tag{4.104}$$

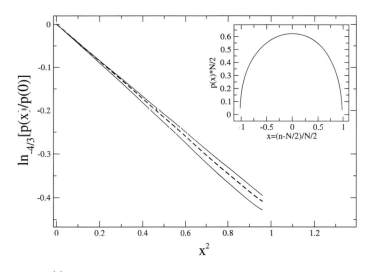

Fig. 4.32 $\ln_{-4/3} \frac{p(x)}{p(0)}$ vs x^2 for $(q, p) = (3/10, 1/2)$, and $N = 1000$. Two branches are observed due to the asymmetry emerging from the fact that we have imposed the q-product on the "left" side of the triangle; we could have done otherwise. The mean value of the two branches is indicated in a dashed line. It is through this mean line that we have numerically calculated $q_e(q)$ as indicated in Fig. 4.34. In order to minimize the tinny asymmetry, we have represented a variable x slightly displaced with regard to $\frac{n-(N/2)}{N/2}$ so that the center $x = 0$ precisely coincides with the location of the maximum of $p(x)$. INSET: Linear–linear representation of $p(x)$. From [468]

Consistently, we may define $(\mu\nu)^{-n} \equiv (\nu\mu)^n$, and $(\nu\mu)^{-n} \equiv (\mu\nu)^n$.

We verify also that, for $z = 0, \pm 1, \pm 2, \ldots$, and $\forall q$,

$$(\mu\nu)^z(q) = \frac{z - (z - 1)q}{z + 1 - zq}, \tag{4.105}$$

$$\nu(\mu\nu)^z(q) = \frac{z + 2 - (z + 1)q}{z + 1 - zq}, \tag{4.106}$$

$$(\mu\nu)^z\mu(q) = \frac{-z + 1 + zq}{-z + (z + 1)q}. \tag{4.107}$$

These three expressions have the form

$$q^* = \frac{A + Bq}{C + Dq}, \tag{4.108}$$

$q = q^* = 1$ being a fixed point, hence $A + B = C + D$. The constants A, B, C and D generically (but not necessarily) do not vanish. In the complex plane (q) and for $(AB - CD) \neq 0$, Eq. (4.108) corresponds to the conformal transformations known

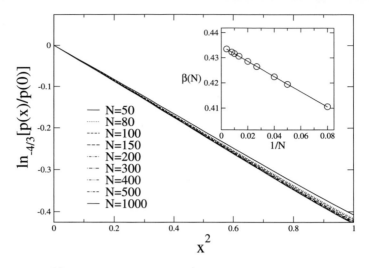

Fig. 4.33 $\ln_{-4/3} \frac{p(x)}{p(0)}$ *vs* x^2 for $(q, p) = (3/10, 1/2)$ and various system sizes N. INSET: N-dependence of the (negative) slopes of the $\ln_{q_e}$ *vs* x^2 straight lines. We find that, for $p = 1/2$ and $N \gg 1$, $\langle (n - \langle n \rangle)^2 \rangle \sim N^2/\beta(N) \sim a(q)N + b(q)N^2$. For $q = 1$ we find $a(1) = 1$ and $b(1) = 0$, consistent with *normal* diffusion as expected. For $q < 1$ we find $a(q) > 0$ and $b(q) > 0$, thus yielding *ballistic* diffusion. The linear correlation factor of the $q - log\ versus\ x^2$ curves ranges from 0.999968 up to near 0.999971 when N increases from 50 to 1000. The very slight lack of linearity that is observed is expected to vanish in the limit $N \to \infty$, but at the present stage this remains a numerically open question. From [468]

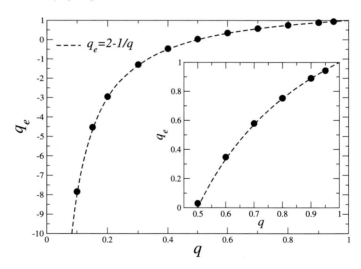

Fig. 4.34 Relation between the index q from the q-product definition, and the index q_e resulting from the numerically calculated probability distribution. The agreement with the analytical conjecture $q_e = 2 - \frac{1}{q}$ is remarkable. INSET: Detail for the range $0 < q_e < 1$. From [468]

as the *Moebius* (or *homographic* or *fractional linear*) *transformations*, to be focused on in Sect. 5.4, and more specifically in [479].

The numerical discussion that we have provided in this subsection is restricted to $q \leq 1$, hence to $q_{attract}$-Gaussians with $q_{attract} \leq 1$. As we shall see later on, similar arguments exist for $q_{attract} > 1$ (in this case, one should of course *avoid* to scale, after centering, the variable n in such a way that it yields a compact support, as it occurs in Figs. 4.31, 4.32 and 4.33). Further related analytical and numerical results can be found in [241, 481]. These results provide a preliminary basis reinforcing the conjecture of the existence of a q-CLT [271].

Let us remind that, in the (q, γ) plane of Fig. 4.30, we have addressed four classes of CLT's. These various CLT's will be shown to correspond to classes of global correlations or absence of correlations. Two of those theorems violate the traditional hypothesis of independence of the random variables that are being summed or arithmetically averaged. In what concerns the region that simultaneously has $q > 1$ and $\gamma < 2$, the attractors will be shown to be different from q-Gaussians. However, they all share asymptotic power-law behaviors for large values of $|x|$ (see the dashed lines in Fig. 4.30). This is so for large t if we are addressing the corresponding Fokker–Planck-like equation, or large N if we are addressing the corresponding CLT. The situation is of course expected to be even richer in the (q, γ, β) space characterizing Eq. (4.36), or for different types of strong correlations [283].

4.8.2 The TMNT Model and Its Numerical Approach

Here we follow [469]. This model, in contrast with the *MTG* one, concerns *continuous* random variables. Let us consider N correlated uniform random variables

$$f(x) = \begin{cases} 1 & \text{if } -1/2 \leq x \leq 1/2, \\ 0 & \text{otherwise.} \end{cases} \tag{4.109}$$

The correlation is introduced through the following multivariate Gaussian $N \times N$ covariance matrix, using probability integral transform (component by component):

$$\begin{pmatrix} 1 & \rho & \rho & \dots & \rho \\ \rho & 1 & \rho & \dots & \rho \\ \rho & \rho & 1 & \dots & \rho \\ \dots & \dots & \dots & \dots & \dots \\ \rho & \rho & \rho & \dots & 1 \end{pmatrix} \tag{4.110}$$

with $-1 \leq \rho \leq 1$ ($\rho = 0$ means *independence*; $\rho = 1$ means *full correlation*). See Fig. 4.35 for the influence of ρ for fixed N, and Figs. 4.36 and 4.37 for the influence of N for fixed ρ. The $N \to \infty$ limiting distribution of the sum of N random variables appears to be very well fitted by q-Gaussians with

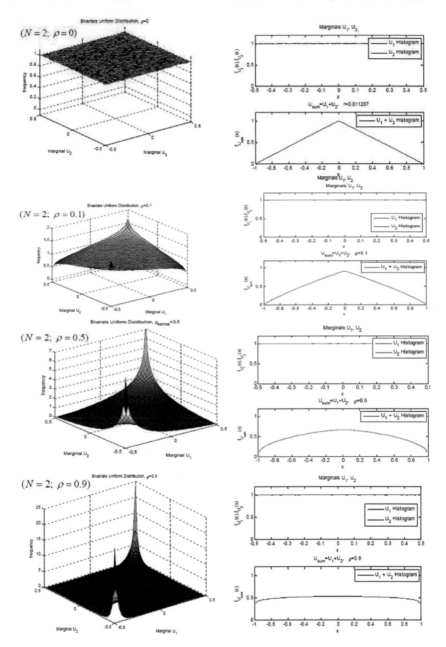

Fig. 4.35 TMNT model for $N = 2$ random variables with increasingly large correlation ρ ($\rho = 0$ corresponds to independence). *Left*: Joint distribution of the two variables. *Right up*: Marginal distribution of each of the two variables. *Right bottom*: Distribution of the sum of the two variables. Notice that, whereas for $\rho = 0$ phase space is equally probable, ρ approaching unity concentrates the probability on only two of the four corners. Notice also that the marginal distribution does *not* seemingly depend on ρ. From [469] and related unpublished results

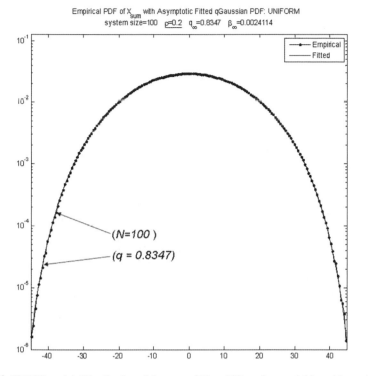

Fig. 4.36 TMNT model: Distribution of the sum of $N = 100$ random variables with $\rho = 0.2$. It is remarkably well fitted by a q-Gaussian with $q = 0.8347$ (continuous curve). From [469]

$$q(\rho, N) = q_\infty(\rho) - \frac{A(\rho)}{N^{\delta(\rho)}} . \tag{4.111}$$

For example, $q_\infty(0.5) \simeq 0.3545$, $A(0.5) \simeq 0.5338$ and $\delta(0.5) \simeq 1.9535$. We present $q_\infty(\rho)$ in Fig. 4.38. It is well fitted by the heuristic relation

$$q_\infty(\rho) = \frac{1 - (5/3)\rho}{1 - \rho} . \tag{4.112}$$

Let us now apply the present numerical approach to a model which generalizes that of matrix (4.110). We assume the following covariance matrix:

$$\begin{pmatrix} 1 & \rho(2) & \rho(3) & \cdots & \rho(N) \\ \rho(2) & 1 & \rho(2) & \cdots & \rho(N-1) \\ \rho(3) & \rho(2) & 1 & \cdots & \rho(N-2) \\ \cdots & \cdots & \cdots & \cdots & \cdots \\ \rho(N) & \rho(N-1) & \rho(N-2) & \cdots & 1 \end{pmatrix} \tag{4.113}$$

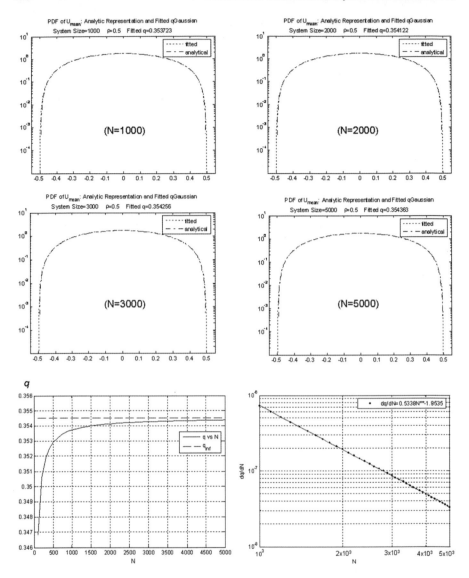

Fig. 4.37 TMNT model with $\rho = 0.5$. *Top*: Distributions of the sum of N random variables for increasingly large values of N, and their fittings with q-Gaussians (continuous curves). *Bottom*: Influence of N on the fitting value of q. These results provide numerical support to the relation (4.111). From [469]

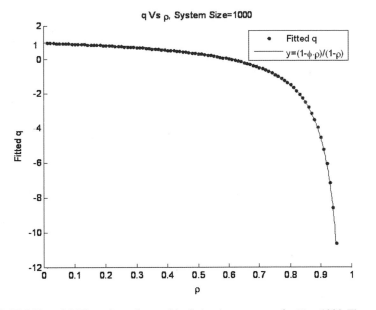

Fig. 4.38 TMNT model: The ρ-dependence of the fitting parameter q_∞ for $N = 1000$. These results provide numerical support to the heuristic relation (4.112) (continuous curve, with $\phi = 5/3$). Notice that $\rho = 3/5$ implies $q = 0$. From [469]

with

$$\rho(r) = \frac{\rho}{r^\alpha} \quad (-1 \le \rho \le 1; \ \alpha \ge 0; \ r = 2, 3, 4, ..., N).$$ (4.114)

As in the $\alpha = 0$ case (i.e., matrix (4.110)), q-Gaussians provide an excellent fitting. The dependence of q on (ρ, α) is depicted in Fig. 4.39. This numerical result is totally consistent with what is expected in terms of the motivations of nonextensive statistical mechanics. Indeed, for $\rho = 0$ (independent variables) we obtain $q = 1$, whereas, for $\rho \ne 0$ ($d = 1$ system of correlated variables with periodic boundary conditions), we obtain $q = 1$ for $\alpha > 1$ (*short-range* correlations) and $q \ne 1$ for $0 < \alpha < 1$ (*long-range* correlations). This is the scenario conjectured for many-body Hamiltonian systems in the $t \to \infty$ limit *after* the $N \to \infty$ limit has been taken (or both limits are simultaneously taken, with t diverging slower than a critical power of N [482]). In other words, BG statistical mechanics for no interactions or short-range interactions, and nonextensive statistical mechanics for long-range interactions. All this would be just perfect, but—there is a *but!*—, as we shall show in the next subsection, the $N \to \infty$ distributions of the MTG and $TMNT$ probabilistic models are *not exactly* q-Gaussians, even if numerically extremely close to them.[7]

---

[7] Anticipating the notion of q-*independence* that will soon be introduced in the context of the q-generalization of the Central Limit Theorem, this means that the N random variables introduced in the present two models are *not exactly*, but *only approximatively* q-independent. If they were *exactly* q-independent, the attractors ought to be *exactly* q-Gaussians.

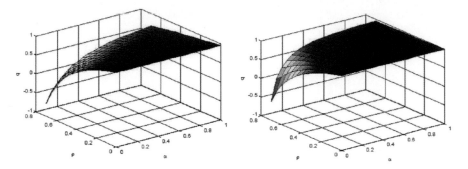

Fig. 4.39 TMNT model: The (ρ, α)-dependence of the fitting parameter q for $N = 10$ (left) and $N = 100$ (right). The $\alpha = 0$ particular case corresponds to what is presented in Fig. 4.38. These results suggest that, in the $N \to \infty$ limit, $q = 1$ for $\rho = 0$ ($\forall \alpha$) as well as for $\rho \neq 0$ and $\alpha > 1$; and $q < 1$ for $\rho \neq 0$ and $0 \leq \alpha < 1$. From [469]

4.8.3 Analytical Approach of the MTG and TMNT Models

Here we follow [471], where the MTG and the $\alpha = 0$ $TMNT$ models are analytically discussed. As we shall see, the $N \to \infty$ limiting distributions are *not q-Gaussians*, but distributions instead which numerically are *amazingly close to q-Gaussians*, although *distinctively differing from them*. It is of course trivial,—and ubiquitous in experimental, observational, and computational sciences—, the fact that a *finite* number of *finite-precision* values can *never* guarantee analytical results. The history of science is full of such illustrations. Nevertheless, the present two examples are particularly instructive. Indeed, the numerics are strongly consistent with q-Gaussians. Nevertheless, the exact distributions conspire in such a way as to be numerically extremely close to q-Gaussians, and still differing from them! Let us briefly report now the analytical results in [471] (see details therein), and then discuss the possible reason for which the $N \to \infty$ distributions do not strictly coincide, for these two models, with q-Gaussians.

Let us start with the $\alpha = 0$ *TMNT* model. The $N \to \infty$ distribution is given by [471]

$$P(U) = \left(\frac{2 - \rho}{\rho}\right)^{1/2} \exp\left(-\frac{2(1 - \rho)}{\rho}[erf^{-1}(2U)]^2\right) \quad \left(-\frac{1}{2} \leq U \leq \frac{1}{2}\right). \quad (4.115)$$

Clearly, this distribution is not a q-Gaussian, even if numerically it is amazingly close to it: see Fig. 4.40. If we approximate it by the best q-Gaussian (by imposing the matching of the second and fourth moments), we obtain for q precisely the conjectural Eq. (4.112)!

Fig. 4.40 Exact distribution (dots) for $\rho = 7/10$ and its best q-Gaussian approximant with $q = -5/9$ (continuous curve). From [471]

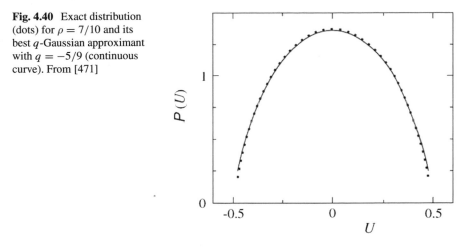

Let us address now the *MTG* model. The $N \rightarrow \infty$ distribution is given by [471]

$$R(y) = A_\rho^{-1}(1 - y)^{a_\rho}[-\ln(1 - y)]^{(1-\rho)/\rho} \quad (0 \leq y \leq 1), \qquad (4.116)$$

$$a_\rho \equiv \frac{2 - 2^\rho}{2^\rho - 1}, \qquad (4.117)$$

A_ρ being a normalizing constant. Once again, this distribution is not a q-Gaussian, even if numerically it is very close to it: see Fig. 4.41. If we approximate it by the best q-Gaussian (by imposing the matching of the second and fourth moments), we obtain

$$q \simeq \frac{1 - 2\rho}{1 - \rho}. \qquad (4.118)$$

Through the identification $\rho \equiv 1 - q_{correlat}$, this relation becomes

$$q \simeq 2 - \frac{1}{q_{correlat}}, \qquad (4.119)$$

which, with the notation change $(q_{correlat}, q) \rightarrow (q, q_e)$, recovers the conjectural Eq. (4.99)!

Further understanding is obviously needed. Why these two strictly scale-invariant models (*MTG* and *TMNT*) are *so close* to q-Gaussians?, and why they do *not* precisely coincide with them? At the present stage, this remains as an open problem.

4.8.4 The MFMT Models and Their Numerical Approach

In order to further illustrate the complexity that purely probabilistic models can have, we briefly review here the MFMT models (MFMT1, MFMT2, and MFMT3)

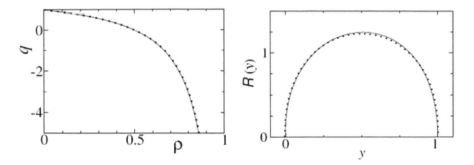

Fig. 4.41 *MTG model. Left*: The ρ-dependence of the index q of the best q-Gaussian approximant (dots), compared to Eq. (4.118). *Right*: Exact limiting distribution for $\rho = 7/10$ (hence $q_{correlat} = 3/10$ (continuous curve), and its best q-Gaussian approximant with $q = -4/3$ (dots). From [471]

introduced in [283]. All of them include the model of independent binary variables as a particular instance. Otherwise, in all of them, the correlations are sufficiently strong to sensibly depart from the q-Gaussian attractor and from the value $\mu = 1/2$ characterizing normal diffusion, but still remaining in the frame of S_q entropic functionals yielding extensivity for a special value $q_{ent} < 1$.

The MFMT1 model is defined through

$$r_{N,n} = \frac{N^{\alpha n}}{\sum_{k=0}^{N} \frac{N!}{(N-k)!k!} N^{\alpha k}} \quad (\alpha \in \mathbb{R}) . \tag{4.120}$$

We verify that $S_{q_{ent}}$ is extensive for $q_{ent} = e_Q^{-\alpha^4}$ with $Q \simeq 4.95$: see Fig. 4.42. Also, we verify in Fig. 4.43 that the attractor sensibly differs from a q-Gaussian.

The MFMT2 model is defined through

$$r_{N,n} = \frac{N^{\alpha n'}}{Z} \quad (\alpha \in \mathbb{R}) , \tag{4.121}$$

where Z is a constant chosen to satisfy normalization, and $n' = n$ for the first half of the interval $[0, N]$ and $n' = (N - n)$ for its second half. We verify that $S_{q_{ent}}$ is extensive for q_{ent} depicted in Fig. 4.44. Also, we verify in Fig. 4.45 that the attractor sensibly differs from a q-Gaussian.

Finally, the MFMT3 model is defined through a cutoff d $(d = 1, 2, 3, \dots)$ similar to that of the TGS cutoff model, more precisely through

$$r_{N,n} = \frac{N^{\alpha n}}{Z} \quad (\alpha \in \mathbb{R}) , \tag{4.122}$$

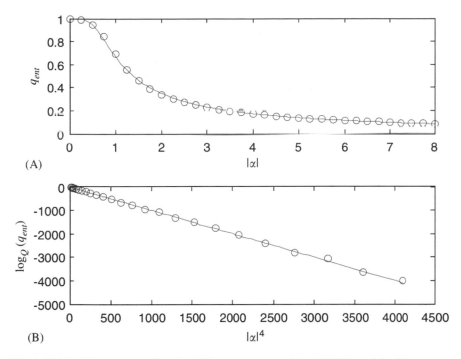

Fig. 4.42 The index q_{ent} as a function of the parameter α of the MFMT1 model. **a** Linear axes, and **b** scaled axes. The open circles represent data obtained numerically, and the solid line is a fit to the stretched Q-Gaussian $q_{ent} = e_Q^{-\alpha^4}$ with $Q \simeq 4.95$, which was determined by analyzing the linearity of $S_q(N)$ with N ranging up to $N = 25$. From [283]

for $n \leq d$ and zero otherwise; Z is a N-dependent quantity chosen to satisfy normalization. We verify that $S_{q_{ent}}$ is extensive for q_{ent} depicted in Fig. 4.46. The corresponding attractor sensibly differs from a q-Gaussian.

4.8.5 The RST1 Model and Its Analytical Approach

In Table 3.7 we have the celebrated Leibnitz triangle (merged in fact with the Pascal triangle). It satisfies the recursive relation (3.142) and, as previously mentioned, it is completely determined by the marginal coefficient

$$r_{N,0}^{(1)} = \frac{1}{N+1} \quad (N = 1, 2, 3, \ldots).$$
(4.123)

Let us now generalize this triangle by still imposing (3.142), and nevertheless generalizing (4.123) as follows [470]:

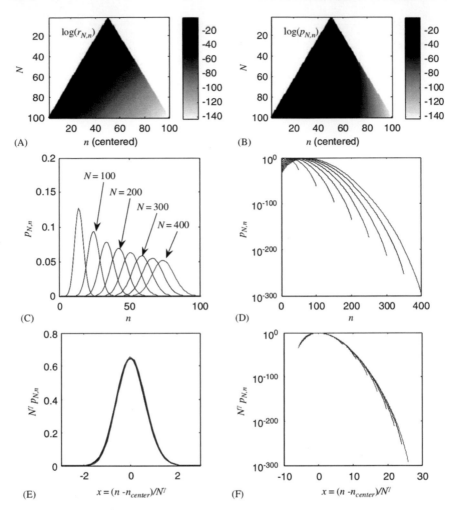

Fig. 4.43 Details of the MFMT1 attractor with $\alpha = -0.25$. Plots **a–b** show reduced probabilities $\log r_{N,n}$ and probabilities $\log p_{N,n}$ up to $N = 400$ as grayscale images. Plots **c–d** show uncentered and unscaled probabilities for $N = 50, 100, \ldots, 400$, in both linear and logarithmic scales. Plots **e–f** show centered and scaled probabilities, illustrating the attractor in probability space. We find $\mu/2 \equiv \gamma = 0.42$. From [283]

$$
\begin{aligned}
r_{N,0}^{(1)} &= \frac{1}{N+1}, \\
r_{N,0}^{(2)} &= \frac{2 \cdot 3}{(N+2)(N+3)}, \\
r_{N,0}^{(3)} &= \frac{3 \cdot 4 \cdot 5}{(N+3)(N+4)(N+5)}, \\
r_{N,0}^{(\nu)} &= \frac{\nu \cdots (2\nu - 1)}{(N+\nu) \cdots (N + 2\nu - 1)} = \frac{(2\nu - 1)!(N + \nu - 1)!}{(\nu - 1)!(N + 2\nu - 1)!}.
\end{aligned}
\tag{4.124}
$$

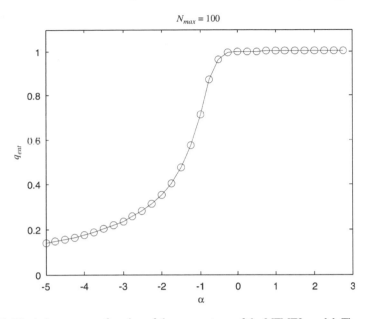

Fig. 4.44 The index q_{ent} as a function of the parameter α of the MFMT2 model. The symmetry of the MFMT1 model has been broken, and $q_{ent} = 1$ for all $\alpha \geq 0$. The solid line is a guide for the eye. The values q_{ent} for $\alpha < 0$ coincide with those of Fig. 4.42. From [283]

We verify that, $\forall \nu$, $\lim_{N \to 0} r_{N,0}^{(\nu)} = 1$, and that $r_{N,0}^{(\nu)} \sim \frac{(2\nu-1)!}{(\nu-1)!\, N^\nu}$ $(N \to \infty)$. Also, $\lim_{\nu \to \infty} r_{N,0}^{(\nu)} = \frac{1}{2^N}$, hence independence. As an example, the $\nu = 2$ triangle (merged with the Pascal triangle) is indicated in Table 4.1.

It has been analytically shown [470] that, after appropriate centering and scaling, the $N \to \infty$ limit of these distributions is exactly a q-Gaussian with

$$q = \frac{\nu - 2}{\nu - 1} = 1 - \frac{1}{\nu - 1}. \tag{4.125}$$

Also, if we associate $\sigma_1 = \pm 1$ $(i = 1, 2, ..., N)$ with the N random variables, we can easily obtain (in addition to $\langle \sigma_i \rangle = 0$, $\forall i$) the following interesting result:

$$\langle \sigma_i \sigma_j \rangle = \frac{1}{2\nu + 1} \quad (\forall\, i \neq j;\ \forall N). \tag{4.126}$$

As expected, for the case of independence, i.e., when $\nu \to \infty$, the correlation vanishes.

This model, like the MTG and TMNT ones, is strictly scale-invariant. But, in variance with those two, it asymptotically approaches a q-Gaussian. Moreover, using the Laplace-de Finetti representation, this model has been extended to *real* values of q, both above and below unity [483].

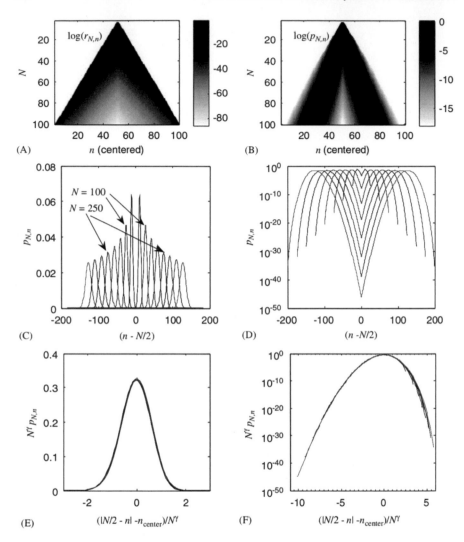

Fig. 4.45 Details of the MFMT2 attractor with $\alpha = -0.25$. Plots **a–b** show reduced probabilities $\log r_{N,n}$ and probabilities $\log p_{N,n}$ up to $N = 400$ as grayscale images. Plots **c–d** show uncentered and unscaled probabilities for $N = 50, 100, \ldots, 400$, in both linear and logarithmic scales. Plots **e–f** show centered and scaled probabilities, illustrating the attractor in probability space. We find $\mu/2 \equiv \gamma = 0.42$. From [283]

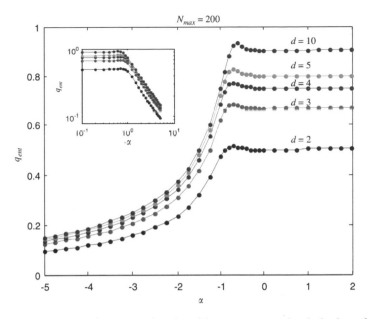

Fig. 4.46 Numerical results for q_{ent} as a function of the parameter α and typical values of d for the MFMT3 model. For $\alpha \geq 0$, we observe $q_{ent} = 1 - 1/d$, as predicted based on reduced occupancies, like in the TGS cutoff model. For $\alpha < 0$, rapidly decreasing probabilities, coupled with explicitly reduced occupancy, leads to a behavior much like that in the MFMT1 model, where $q_{ent} \sim 1/|\alpha|$ as $\alpha \to -\infty$. In the region $-1 < \alpha < 0$, we find the two effects in competition, leading to more complex behavior. The inset contains a log–log plot of the negative α portion of the data, showing explicitly that $q_{ent} \sim 1/|\alpha|$, and the sharp change in behavior near $\alpha = -1$. From [283]

Table 4.1 Merging of the Pascal triangle (the set of all *left* members) with the $\nu = 2$ triangle (the set of all *right* members) associated with N equal subsystems

$(N = 0)$	$(1, 1)$
$(N = 1)$	$(1, \frac{1}{2}) (1, \frac{1}{2})$
$(N = 2)$	$(1, \frac{3}{10}) (2, \frac{1}{5}) (1, \frac{3}{10})$
$(N = 3)$	$(1, \frac{1}{5}) (3, \frac{1}{10}) (3, \frac{1}{10}) (1, \frac{1}{5})$
$(N = 4)$	$(1, \frac{1}{7}) (4, \frac{2}{35}) (6, \frac{3}{70}) (4, \frac{2}{35}) (1, \frac{1}{7})$

4.8.6 The RST2 Model and Its Numerical Approach

We shall now define a model by discretizing (symmetrically) a q-Gaussian into $(N + 1)$ values (identified by $n = 0, 1, 2, ..., N$) [470]. These values can be interpreted as the probabilities corresponding to N equal and distinguishable binary random variables. This model, referred to as the RST2 one, will approach *by construction* the q-Gaussian that has been discretized (in fact, two slightly different discretizations have been used). The interest of such a model is of course not its limit (since this is

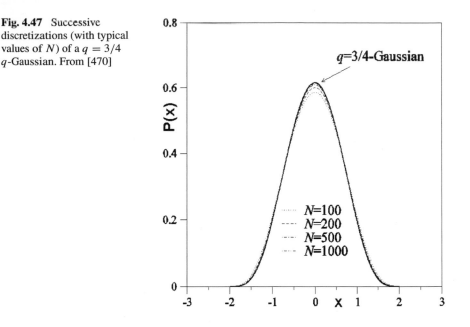

Fig. 4.47 Successive discretizations (with typical values of N) of a $q = 3/4$ q-Gaussian. From [470]

imposed), but *how* the limit is approached for increasingly large values of N. The relation (3.142) corresponds to strict scale-invariance. We can numerically (and in some cases analytically) follow the ratio

$$Q_{N,n} \equiv \frac{r_{N,n}}{r_{N+1,n} + r_{N+1,n+1}}. \tag{4.127}$$

We verify that $Q_{N,n}$ tends to 1 (or equivalently $(Q_{N,n} - 1) \to 0$) as N increases, i.e., the model is asymptotically scale-invariant. Note that $Q_{0,0} = Q_{1,0} = Q_{1,1} = 1$ for arbitrary values of $r_{0,0}$, $r_{1,n}$ and $r_{2,n}$. See Figs. 4.47, 4.48 and 4.49.

4.8.7 The HTT Model and Its Analytical Approach

A strictly scale-invariant model which generalizes the RST1 one to continuous values of q for $q \geq 1$, and also for $q < 1$, has been introduced in [483]. It is defined, for $q_{attract} \in (-\infty, 3)$, as follows:

$$r_{N,n}^{(q_{attract})} = \begin{cases} \dfrac{B\left(\frac{2-q_{attract}}{1-q_{attract}}+n, \frac{2-q_{attract}}{1-q_{attract}}+N-n\right)}{B\left(\frac{2-q_{attract}}{1-q_{attract}}, \frac{2-q_{attract}}{1-q_{attract}}\right)} & \text{if } q_{attract} < 1, \\[4mm] \dfrac{1}{2^N} & \text{if } q_{attract} = 1, \\[4mm] \dfrac{B\left(\frac{3-q_{attract}}{2q_{attract}-2}+n, \frac{3-q_{attract}}{2q_{attract}-2}+N-n\right)}{B\left(\frac{3-q_{attract}}{2q_{attract}-2}, \frac{3-q_{attract}}{2q_{attract}-2}\right)} & \text{if } 1 < q_{attract} < 3, \end{cases} \tag{4.128}$$

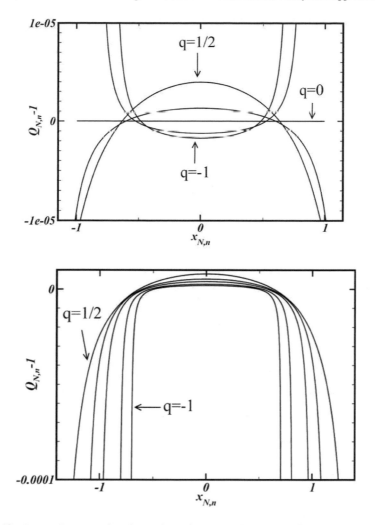

Fig. 4.48 $Q_{N,n} - 1$ as a function of n for $N = 500$ and different values of $q = -1, -1/2, 0, 1/4, 1/2$ for discretizations D1 (top) and D2 (bottom). Strict scale-invariance is observed for $q = 0$ and discretization D1. From [470]

where $B(x, y)$ is the Beta function. In the $N \to \infty$ limit, it yields a $q_{attract}$-Gaussian and, being its binary variables exchangeable, it implies $q_{ent} = 1$. For $q_{attract} = \frac{\nu-2}{\nu-1}$ with $\nu = 1, 2, 3, \ldots$ (i.e., for $q_{attract} = -\infty, 0, 1/2, 2/3, 3/4, \ldots, 1$) it recovers the RST1 model (see Sect. 4.8.5).

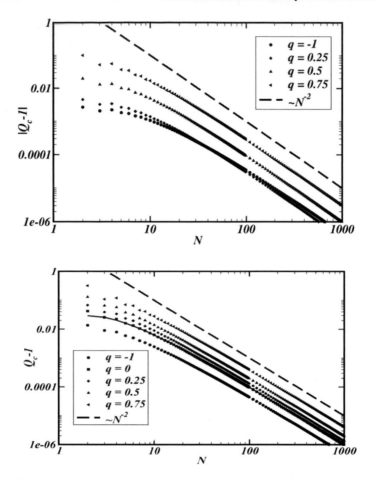

Fig. 4.49 $Q_c - 1 = Q_{N,N/2} - 1$ as a function of N for different values of q for discretizations D1 (top) and D2 (bottom). The power-law has the exponent -2. From [470]

4.8.8 Comparative Considerations

For purely probabilistic models, it is enlightening to consider various aspects in the $N \to \infty$ limit, namely,

(i) For what entropic functional $S(\{p_{N,n}\})$ (with $\sum_{n=0}^{N} p_{N,n} = 1$) is the entropy extensive, i.e., $S(N) \propto N$? If it is S_q, for what value of the index $q_{ent} \leq 1$?

(ii) What is the "diffusion" class of the system? If it is the usual asymptotic power-law class where $(n - \langle n \rangle)^2$ scales as N^μ (hence $\langle (n - \langle n \rangle)^2 \rangle = \langle n^2 \rangle - \langle n \rangle^2 \propto N^\mu$ if the variance is finite), what is the value of the exponent $\mu \geq 0$? (we remind that $\mu = 0$, $0 < \mu < 1$, $\mu = 1$, and $\mu > 1$, respectively, correspond to localization, subdiffusion, normal diffusion, and superdiffusion; $\mu = 2$ is currently referred to as *ballistic*).

(iii) What is, after appropriate centering on $\langle n \rangle$ and scaling (currently through N^μ), the attractor[8] in the space of probability distributions? If it is a q-Gaussian, for what value of the index $q_{attract}$?

(iv) How the above entropic functional S associated with the above centered and scaled attractor increases with N?

To a certain degree, the answers to the above questions do not follow trivial paths and connections. Typically, but not necessarily, discrepancy of the attractor from the Gaussian form (and from the improbable *full* Lévy form) characterizes space correlations. Also, typically, but not necessarily, $\mu \neq 1$ characterizes time correlations (memory-like). Consistently, as we shall illustrate, a considerable diversity exists in such answers.

The basic canonical reference is N coins, i.e., a system of N independent non-biased binary distinguishable random variables, hence $p_{N,n} = \frac{1}{2^N} \frac{N!}{n!(N-n)!}$ ($N = 1, 2, 3, \ldots$; $n = 0, 1, 2, \ldots, N$). The answer to (i) is $q_{ent} = 1$. The answer to (ii) is $\mu = 1$. The answer to (iii) is $q_{attract} = 1$, more precisely, the Gaussian distribution $P(x, N) = \frac{1}{\sqrt{2\pi N}} e^{-x^2/2N}$. The answer to (iv) is given by $S_{BG}[P(x, N)] = -k \int_{-\infty}^{\infty} dx \, P(x, N) \ln P(x, N) \propto \ln N$. It is worth mentioning here that if we consider an attractor isomorphic to fractal Brownian motion [484], e.g., $P(x, N) \propto e^{-x^2/DN^\mu}$ (with $D > 0$ and $\mu = 2H$, where H denotes the Hurst exponent: $0 \leq H < 1/2$, $H = 1/2$, and $1/2 < H \leq 1$ characterize, respectively, anti-persistent, uncorrelated, and persistent evolution), we still have $S_{BG}[P(x, N)] \propto \ln N$, $\forall \mu$. If we have instead $P_q(x, N) \propto \frac{1}{N^{\mu/2}} e_q^{-x^2/DN^\mu}$ with $q \geq 1$ and $D(2 - q) > 0$,[9] we verify that, for $Q \leq 1$, $S_Q[P_q(x, N)] \propto \ln_Q N^{\mu/2} = \frac{N^{(1-Q)\mu/2}}{1-Q}$. In particular, $Q = (2 - q)$ yields $S_{2-q}[P_q(x, N)] \propto \ln_{2-q} N^{\mu/2}$, i.e., $S_{2-q} \propto N^{(q-1)\mu/2}$ for $q > 1$, and recovers $S_{BG} \propto \ln N$ for $q = 1$.

Let us make now some instructive comparisons through the various probabilistic models discussed until this point.

In Sect. 3.3.6 we have reviewed three models with correlated probabilities, namely the Leibnitz (TGS1), the stretched exponential (TGS2), and the cut-off triangles (TGS3) [274, 485]. All three are scale-invariant (i.e., satisfy Leibnitz rule), either strictly or asymptotically. The first two have $q_{ent} = 1$, the third one has $q_{ent} = 1 - 1/d \leq 1$ where d denotes the width of the cut-off, out of which the probabilities vanish: see Figs. 3.14 and 3.15.

The MTG model, reviewed in Sect. 4.8.1, introduces [468] strong correlations through the parameter $q \in [0, 1]$, noted here $q_{correlat}$, where the limit $q_{correlat} = 1$

---

[8] We are using here the term *attractor* in a loose sense. Indeed, in most cases, we have not proved the stability which would fully legitimate the use of the word. However, we have no reasons to believe that, for the present models, those limits would not be stable with regard to small perturbations. Indeed, the approach to those limits systematically appears to be numerically robust.

[9] The condition $D(2 - q) > 0$ is clarified in Sect. 4.2.2.

corresponds to uncorrelated. Its numerical study suggested that the attractor would be a $q_{attract}$-Gaussian with $q_{attract} = 2 - 1/q_{correlat} \in (-\infty, 1]$. But its analytical study [471] (see Sect. 4.8.3) showed that it is strictly *not* so, even being an excellent approximation everywhere except in the neighborhood of vanishing probabilities. For all values of $q_{correlat}$ we verify $q_{ent} = 1$, which illustrates a quite generic tendency, namely relatively weak correlations leave invariant both $q_{ent} = 1$ and $q_{attract} = 1$, stronger correlations modify the attractor being thus able to yield $q_{attract} \neq 1$ while it still remains $q_{ent} = 1$, and, finally, even stronger correlations can yield both $q_{attract} \neq 1$ and $q_{ent} \neq 1$.[10] In what concerns the diffusion exponent μ, it depends on $q_{correlat}$, and only for $q_{correlat} = 1$ we have $\mu = 1$.

Very similar is the case of the TMNT model [469], reviewed in Sect. 4.8.2. The $d = 1$ correlations are characterized by the covariance matrix with ρ/r^{α} where ρ is the correlation strength and α determines the range of the correlation ($\rho = 0$ corresponds to the uncorrelated case). The numerics and analytics indicated a $N \to \infty$ attractor very close to a q-Gaussian with $q_{attract} = [1 - \Phi \rho]/[1 - \rho]$ with $\Phi = 2$ for $\alpha = 0$ and $\Phi = 1$ for $\alpha \geq 1$. The analytical study for $\alpha = 0$ showed [471] that the attractor is strictly *not* a q-Gaussian. However, the numerical discrepancy is very slight. If we identify $\rho \equiv 1 - q_{correlat}$ we obtain, for $\alpha = 0$, an approximate attractor with $q_{attract} = 2 - 1/q_{correlat} \in (-\infty, 1]$, in curious coincidence with the approximate attractor for the MTG model.

Let us focus now on the MFMT models [283] (MFMT1, MFMT2, and MFMT3), reviewed in Sect. 4.8.4. These models profusely exhibit that, rather easily, attractors can differ from q-Gaussians and the diffusion exponent differ from the normal diffusion value. In contrast, the entropic functional is very robust. Indeed, $S_q(N)$ remains capable of yielding the expected thermodynamical extensivity for a special value of q_{ent}. In what concerns the influence of correlations on the diffusion exponent, see Fig. 4.50.

Let us make, at this stage, a few comments about the RST models introduced in Sects. 4.8.5 and 4.8.6 [470] (RST1 and RST2) and the HTT model introduced in Sect. 4.8.7. They are either strictly (RST1 and HTT) or asymptotically (RST2) scale-invariant. All three yield $q_{ent} = 1$, and have q-Gaussian attractors. The fact that $q_{ent} = 1$ is consistent with the exchangeable nature of the binary random variables of those models [483].

A question emerges at this point: is it possible to have $q_{ent} \neq 1$, and having simultaneously a $q_{attract}$-Gaussian attractor with $q_{attract} \neq 1$? The answer is *yes*, as illustrated in [486]. Correlations between N binary variables are there introduced by imposing that the N sequence contains $(d_R + 1)$ domains, each of them having parallel "spins" (the subindex R stands for *Ruseckas*): the combinatorics is performed within the windows $\{n_1, n_2, \ldots, n_{d_R+1}\}$ such that $\sum_{i=1}^{d_R+1} n_i = N$. The number of microscopic possibilities is given, for $N >> 1$, by $W(N) \sim \frac{2}{d_R!} N^{d_R}$. Consequently

---

[10] The typical scenario within which BG statistical mechanics is expected to be legitimately applicable is that where nonvanishing correlations are sufficiently weak so that both $q_{attract}$ and q_{ent} remain equal to unity, e.g., classical d-dimensional short-range-interacting fluid or inertial magnetic models.

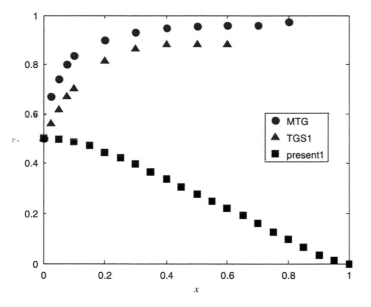

Fig. 4.50 Diffusion exponent $\gamma \equiv \mu/2$ versus model parameters x, for the MTG, TGS1, and MFMT1 ("present1") models. For all three models, $x = 0$ corresponds to the independent case, with normal diffusion exponent $\gamma = 1/2$. In the model MTG $x = 1 - q_{correlat}$, and in the model TGS1 $x = 1 - \alpha$, both models exhibit superdiffusion. In the model MFMT1, $x = 1 - |\alpha|$ and it exhibits subdiffusion. The values $\gamma = 0, 1$, respectively, correspond to localization and ballistic superdiffusion. From [283]

$q_{ent} = 1 - 1/d_R$, which formally coincides with the result (3.155) obtained for the TGS2 cutoff model [274]. Therefore d_R plays the same role as the cutoff d in the TGS2 model. However and in spite of this similarity, a great difference exists between these two models in what concerns the $N \to \infty$ attractors. Indeed, the TGS2 attractor has no particularly interesting functional form (see Fig 3.17), whereas the Ruseckas model has a $q_{attract}$-Gaussian attractor, with $q_{attract} = 1 - \frac{1}{\lfloor (d_R-1)/2 \rfloor}$, where $\lfloor x \rfloor$ denotes the integer part of x. A simple relation emerges between q_{ent} and $q_{attract}$, namely

$$\frac{2}{1 - q_{ent}} = \frac{1}{1 - q_{attract}} - 1, \qquad (4.129)$$

for odd d_R, and

$$\frac{2}{1 - q_{ent}} = \frac{1}{1 - q_{attract}} - 2, \qquad (4.130)$$

for even d_R.

To conclude the review of correlated probabilistic models having q-statistical features, let us mention the existence of another interesting symmetric model related to Pólya's urns. It is introduced and discussed in [487, 488, 490]. It satisfies Leibnitz triangle rule and presents a rich q-statistical structure. We will delay its detailed review until Sect. 4.10 because it constitutes an interesting example within the q-generalized Large Deviation Theory.

4.9 Central Limit Theorems

The standard and Lévy-Gnedenko central limit theorems (CLT) play a central role in the theory of probabilities. The standard CLT, with its Gaussian attractor in the space of probabilities, constitutes one of the mathematical foundations of BG statistical mechanics, reflected in particular in the Maxwellian distribution of velocities of inertial Hamiltonian systems. This, as well as the Lévy-Gnedenko CLT, have been q-generalized in [477] (see also [240, 491–495]). Let us review now this generalization.

We start with a *definition*. The q-*Fourier transform* (q-FT) of a function $f(x)$ is defined as follows:

$$F_q[f](\xi) \equiv \int dx \, e_q^{i\xi x} \otimes_q f(x) . \tag{4.131}$$

This definition holds for any real value of q. However, its implementation is quite simple for $q \geq 1$. We shall therefore restrict to this interval from now on. It can be shown [477] that

$$F_q[f](\xi) = \int_{-\infty}^{\infty} dx \, e_q^{i\,\xi\,x[f(x)]^{q-1}} f(x) \quad (q \geq 1) . \tag{4.132}$$

It is transparent that this transformation is, for $q \neq 1$, *nonlinear*. Indeed, if we do $f(x) \rightarrow \lambda f(x)$, λ being any constant, we verify that $F_q[\lambda f](\xi) \neq \lambda F_q[f](\xi)$ ($q \neq 1$).

This generalization of the standard Fourier transform ($F_1[f](\xi)$) has a remarkable *property*: it transforms q-Gaussians into q-Gaussians.[11] Indeed, we verify

$$F_q\left[B_q \sqrt{\beta} \, e_q^{-\beta x^2}\right](\xi) = e_{q_1}^{-\beta_1 \xi^2} , \tag{4.133}$$

where

$$q_1 = \frac{1+q}{3-q} , \tag{4.134}$$

and

$$\beta_1 = \frac{3-q}{8 \, \beta^{2-q} B_q^{2(q-1)}} , \tag{4.135}$$

---

[11] The definition (4.131) was in fact introduced with the specific aim of satisfying this property

with $B_q = A_q/\sqrt{\pi}$, where A_q is given in Eqs. (4.12).[12]

It follows that, within a subtle procedure involving extra information that is unnecessary for $q = 1$, the q-Fourier transform has a unique inverse transform [496–499, 501] which transforms, in particular, q-Gaussians into q-Gaussians (see [502]) (the (q, α)-stable distributions will be soon defined).[13]

See Fig. 4.51 for illustrations of the interesting *closure* property (4.133), which does *not* exist for *any* other of the presently known linear or nonlinear integral transforms.

Equation (4.135) can be rewritten as follows

$$\beta^{\sqrt{2-q}} \beta_1^{\frac{1}{\sqrt{2-q}}} = K(q), \tag{4.136}$$

$$K(q) \equiv \left[\frac{3-q}{8 \, B_q^{2(q-1)}} \right]^{\frac{1}{\sqrt{2-q}}}, \tag{4.137}$$

depicted in Fig. 4.52.

Through direct derivation we can easily verify another interesting *property* of the q-Fourier transform, namely the following set of relations (for $q \geq 1$) [505]:

---

[12] If we apply twice (4.134) we obtain $q \to \frac{1}{2-q}$, which is the inverse of the transformation $q \to 2 - 1/q$, already appeared in various occasions.

[13] Hilhorst [503, 504] has recently produced an interesting example which is noninvertible unless we use the extra information mentioned above. Consider $f(x) = (\lambda/x)^{\frac{1}{q-1}}$ if $a < x < b$, and zero otherwise; $q > 1$, $0 < a < b$, and $\lambda > 0$. Imposition of normalization straightforwardly yields $\lambda = \left[\frac{q-1}{q-2} \left(b^{\frac{q-2}{q-1}} - a^{\frac{q-2}{q-1}} \right) \right]^{-(q-1)}$. It immediately follows that $F_q[f](\xi) = [1 + (1-q)i\xi\lambda]^{\frac{1}{1-q}}$. Therefore this solution is, for fixed q, *one and the same* for a one-parameter family of normalized functions $f(x)$. Indeed, for all (a, b) having the same λ, the q-Fourier transform is the same. Therefore, for this example, the inverse q-Fourier/indexFourier transform does not exist in the sense that it does not yield a single function, but rather a family of them. In other words, the q-Fourier transform is not invertible in the image of *all* probability density functions. To further understand the domain of impact of this example, let us consider a more general situation, namely $F_{q'}[f](\xi) = \int_a^b dx \left[1 + (1-q')i\,\xi x[f(x)]^{q'-1} \right]^{\frac{1}{1-q'}} f(x)$ $(q' > 1)$, i.e., $F_{q'}[f](\xi) = \int_a^b dx \left[1 + (1-q')i\,\xi \lambda^{\frac{q'-1}{q-1}} x^{\frac{q-q'}{q-1}} \right]^{\frac{1}{1-q'}} f(x)$. This integral can be expressed as a combination of two hypergeometric functions, where we see through inspection that, due to the presence of $x^{\frac{q-q'}{q-1}}$, the one-parameter invariance has disappeared for all (q, q') such that $q \neq q'$. In other words, for fixed q, all q'-Fourier transforms are invertible, except if $q' = q$. Equivalently, for fixed q', all the above functions $f(x) = (\lambda/x)^{\frac{1}{q-1}}$ are invertible excepting if $q = q'$. The generic inversion problem has a unique answer only within the extra information described in [496–499, 501].

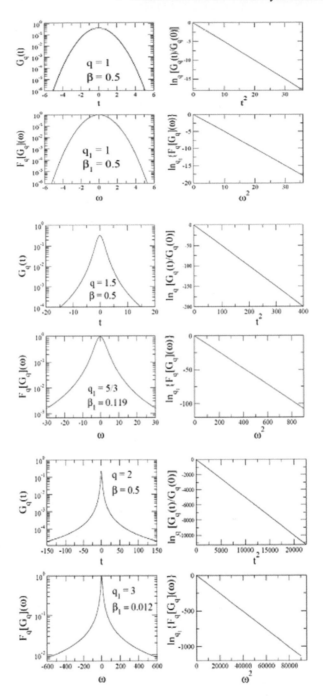

Fig. 4.51 q-Gaussians (in log-linear and q-log-quadratic scales) and their q-Fourier transforms (in log-linear and q-log-quadratic scales) for $q = 1$ (*top*), $q = 3/2$ (*middle*), and $q = 2$ (*bottom*)

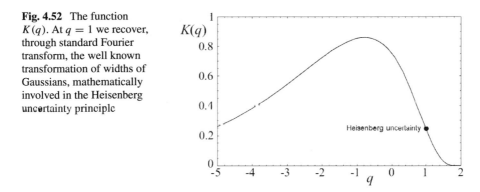

Fig. 4.52 The function $K(q)$. At $q = 1$ we recover, through standard Fourier transform, the well known transformation of widths of Gaussians, mathematically involved in the Heisenberg uncertainty principle

$$F_q[f](0) = \int_{-\infty}^{\infty} dx \, f(x), \tag{4.138}$$

$$\left. \frac{d F_q[f](\xi)}{d\xi} \right|_{\xi=0} = i \int_{-\infty}^{\infty} dx \, x \, [f(x)]^q, \tag{4.139}$$

$$\left. \frac{d^2 F_q[f](\xi)}{d\xi^2} \right|_{\xi=0} = -q \int_{-\infty}^{\infty} dx \, x^2 \, [f(x)]^{2q-1}, \tag{4.140}$$

$$\left. \frac{d^n F_q[f](\xi)}{d\xi^n} \right|_{\xi=0} = (i)^n \left\{ \prod_{m=0}^{n-1} [1 + m(q-1)] \right\}$$

$$\times \int_{-\infty}^{\infty} dx \, x^n \, [f(x)]^{1+n(q-1)} \quad (n = 1, 2, 3, ...). \tag{4.141}$$

If $f(x)$ is a real, nonnegative, integrable function, we can define a probability distribution, namely $p(x) \equiv f(x)/\int_{-\infty}^{\infty} dx \, f(x)$. We can also define a family of escort distributions, namely [505]

$$P^{(n)}(x) \equiv \frac{[f(x)]^{1+n(q-1)}}{\int_{-\infty}^{\infty} dx \, [f(x)]^{1+n(q-1)}} \quad [n = 0, 1, 2, ...; P^{(0)}(x) = p(x); P^{(1)}(x) = P(x)]. \tag{4.142}$$

With the following definition of associated q-expectation values

$$\langle (...) \rangle_n \equiv \int_{-\infty}^{\infty} dx \, (...) \, P^{(n)}(x) = \frac{\int_{-\infty}^{\infty} dx \, (...)[f(x)]^{1+n(q-1)}}{\int_{-\infty}^{\infty} dx \, [f(x)]^{1+n(q-1)}} \quad (n = 0, 1, 2, ...), \tag{4.143}$$

we can rewrite the set of Eqs. (4.141) as follows:

$$\left. \frac{1}{\nu_{q_n}} \frac{d^n F_q[f](\xi)}{d\xi^n} \right|_{\xi=0} = (i)^n \left\{ \prod_{m=0}^{n-1} [1 + m(q-1)] \right\} \langle x^n \rangle_n \quad (n = 1, 2, 3, ...), \tag{4.144}$$

where

$$\nu_{q_n} \equiv \int_{-\infty}^{\infty} dx \, [f(x)]^{q_n} \quad (n = 0, 1, 2, ...), \tag{4.145}$$

with

$$q_n = 1 + (q - 1)n \quad (n = 0, 1, 2, ...). \tag{4.146}$$

Notice that

(i) For $q = 1$, we recover the well known relations involving the *generating func-tion* in theory of probabilities;

(ii) For $n = 1$, we obtain $q_1 = q$, hence the usual escort distribution (used to define the energy-related constraint under which S_q is to be extremized) emerges naturally;

(iii) *All* q-expectation values in nonextensive statistical mechanics are well defined (i.e., *finite*) up to one and the same value of q (more precisely, for $q < 2$ for a discrete energy spectrum);

(iv) If we consider that $F_q[f](\xi) = 1 + \left[\frac{dF_q[f](\xi)}{d\xi}\right]_{\xi=0} \xi + \frac{1}{2}\left[\frac{d^2 F_q[f](\xi)}{d\xi^2}\right]_{\xi=0} \xi^2 + \frac{1}{3!}\left[\frac{d^3 F_q[f](\xi)}{d\xi^3}\right]_{\xi=0} \xi^3 + ...$, then $F_q[f](\xi)$ is uniquely determined by the knowl-edge of the sets $\{\langle x^n \rangle_n\}$ and $\{\nu_{q_n}\}$ ($n = 0, 1, 2, 3, ...$). Finally, since the inverse q-Fourier transform can exist and be unique (see details in [477]), the same knowledge determines in principle $f(x)$ itself [505].

(v) If $f(x) \sim |x|^\gamma$ ($|x| \to \infty$; $\gamma > 0$), then we define $q \equiv 1 + \frac{1}{\gamma}$. This determines $q_n = 1 + (q - 1)n$, hence all the moments $\langle x^n \rangle_n$. For example, if $f(x)$ is a Q-Gaussian, we have that $\gamma = 2/(Q - 1)$, hence $1/(q - 1) = 2/(Q - 1)$. Therefore, the upper admissible limit $q = 2$ precisely corresponds to the well known upper admissible value $Q = 3$.

Let us now introduce another *definition*. A random variable is said to have a (q, α)-stable distribution $L_{q,\alpha}(x)^{14}$ if its q-Fourier transform has the form

$$a \, e_{q_{\alpha,1}}^{-b|\xi|^\alpha} \quad (a > 0, \, b > 0, \, 0 < \alpha \le 2), \tag{4.147}$$

with

$$q_{\alpha,1} \equiv \frac{\alpha q + 1 - q}{\alpha + 1 - q}, \tag{4.148}$$

i.e., if

$$F_q[L_{q,\alpha}](\xi) = a \, e_{q_{1,\alpha}}^{-b|\xi|^\alpha} \quad (a > 0, \, b > 0, \, 0 < \alpha \le 2). \tag{4.149}$$

---

[14] The reason for the word *stable* will become clear soon.

Therefore, $L_{1,2}(x)$ corresponds to Gaussians, $L_{1,\alpha}(x)$ corresponds to α-stable Lévy distributions, and $L_{q,2}(x)$ corresponds to q-Gaussians. Notice that $q_{2,1} = q_1 = (1 + q)/(3 - q)$, as given by Eq. (4.134).

If we successively apply n times the q-Fourier transform onto $L_{q,\alpha}(x)$, we obtain the following algebra:

$$\frac{\alpha}{1 - q_{\alpha,n}} = \frac{\alpha}{1 - q} + n \quad (n = 0, \pm 1, \pm 2, \ldots). \tag{4.150}$$

See Fig. 4.53. The $n = 1$ case recovers relation (4.148). From Eq. (4.150) we immediately obtain

$$q_{\alpha,n} = \frac{(2 + \alpha)q_{\alpha,n+2} - 2}{2q_{\alpha,n+2} + \alpha - 2} \quad (n = 0, \pm 1, \pm 2, \ldots). \tag{4.151}$$

If $\alpha = 2$, this recursion becomes

$$q_{\alpha,n} = 2 - \frac{1}{q_{\alpha,n+2}} \quad (n = 0, \pm 1, \pm 2, \ldots), \tag{4.152}$$

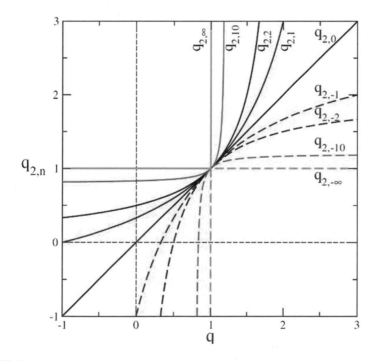

Fig. 4.53 The index $q_{2,n}$ versus q for typical values of n. The countable infinite family merges on a single point only for $q = 1$. This reflects the fact that the structure of BG statistical mechanics is considerably simpler than that of the nonextensive one

which, interestingly enough, coincides with Eq. (4.99).

Let us finally introduce one more *definition*. Two random variables X (with distribution $f_X(x)$) and Y (with distribution $f_Y(y)$) having zero q-mean values are said q-*independent* if

$$F_q[X + Y](\xi) = F_q[X](\xi) \otimes_{\frac{1+q}{3-q}} F_q[Y](\xi),\qquad(4.153)$$

i.e., if

$$\int dz\, e_q^{i\xi z} f_{X+Y}(z) = \left[\int dx\, e_q^{i\xi x} f_X(x)\right] \otimes_{\frac{1+q}{3-q}} \left[\int dy\, e_q^{i\xi y} f_Y(y)\right],\qquad(4.154)$$

with

$$f_{X+Y}(z) = \int dx \int dy\, h(x, y)\,(1 - q)(x + y - z) = \int dx\, h(x, z - x) = \int dy\, h(z - y, y),$$
$$(4.155)$$

where $h(x, y)$ is the joint distribution. Therefore, q -*independence* means *independence* for $q = 1$ (i.e., $h(x, y) = f_X(x) f_Y(y)$), and it means *strong correlation* (of a certain class) for $q \neq 1$ (i.e., $h(x, y) \neq f_X(x) f_Y(y)$).

We can now present the structure of the q-generalization of the CLTs: see Fig. 4.54. To better understand the four theorems therein, we shall illustrate some crucial aspects of them.

Let us start with the $q = 1$ cases, i.e., the standard and the Lévy-Gnedenko CLTs. The $\alpha = 2$ attractor is a Gaussian. The asymptotic behavior of the $\alpha < 2$ attractor is proportional to $1/|x|^{1+\alpha}$. Consequently the $\alpha \to 2$ limit yields a $1/|x|^3$ tail, which is definitively different from a Gaussian tail. How can this occur? Through a *crossover*! (which corresponds to an inflection point in log–log plots). The situation is depicted in Fig. 4.55.

We can also see the attractive effect in the space of distributions as N increases. If the distributions that we compose are q-Gaussians assumed to be independent, the nature of the attractor will depend on whether the variance is finite (which occurs for $q < 5/3$) or infinite (which occurs for $q \geq 5/3$). Both cases are illustrated in Figs. 4.56 and 4.57, respectively.

Let us see now the $q > 1$ cases. The attractors are now q-Gaussians when the $(2q - 1)$-variance is finite (i.e., $\alpha = 2$), and (q, α)-stable distributions when it diverges (i.e., $0 < \alpha < 2$). These distributions must somehow match with q-Gaussians when α approaches 2, and must match with α-stable Lévy distributions when q approaches 1. This happens through a *double crossover*! See Fig. 4.58. We see that, while $|x|$ increases, the distribution goes essentially through *two* different power-law regimes, a *distant* one, which will match with α-distributions when q approaches unity, and an *intermediate one*, which will match with q-Gaussians when α approaches 2.

	$q=1$ [*independent*]	$q \neq 1$ (i.e., $Q \equiv 2q-1 \neq 1$) [*globally correlated*]
$\sigma_Q < \infty$ ($\alpha = 2$)	$\mathbb{F}(x) = $ *Gaussian* $G(x)$, with same σ_1 of $f(x)$	$\mathbb{F}(x) = G_q(x) \equiv G_{(3q_1-1)/(1+q_1)}(x)$, *with same* σ_Q *of* $f(x)$ $G_q(x) \sim \begin{cases} G(x) & \text{if } \|x\| << x_c(q,2) \\ f(x) \sim C_q / \|x\|^{2/(q-1)} & \text{if } \|x\| >> x_c(q,2) \end{cases}$ *with* $\lim_{q \to 1} x_c(q,2) = \omega$
$\sigma_Q \to \infty$ ($0 < \alpha < 2$)	$\mathbb{F}(x) = $ *Levy distribution* $L_\alpha(x)$, *with same* $\|x\| \to \infty$ *behavior* $L_\alpha(x) \sim \begin{cases} G(x) & \text{if } \|x\| << x_c(1,\alpha) \\ f(x) \sim C_\alpha / \|x\|^{1+\alpha} & \text{if } \|x\| >> x_c(1,\alpha) \end{cases}$ *with* $\lim_{\alpha \to 2} x_c(1,\alpha) = \infty$	$\mathbb{F}(x) = L_{q,\alpha}$, *with same* $\|x\| \to \infty$ *asymptotic behavior* $L_{q,\alpha} \sim \begin{cases} G_{\frac{2(1-q)-\alpha(1+q)}{2(1-q)-\alpha(3-q)}, \alpha}(x) \sim C^*_{q,\alpha} / \|x\|^{\frac{2(1-q)-\alpha(3-q)}{2(1-q)}} \\ \qquad\qquad\qquad\qquad \textit{(intermediate regime)} \\ G_{\frac{2\alpha q - \alpha + 3}{\alpha+1}, 2}(x) \sim C^L_{q,\alpha} / \|x\|^{(1+\alpha)/(1+\alpha q - \alpha)} \\ \qquad\qquad\qquad\qquad \textit{(distant regime)} \end{cases}$

Fig. 4.54 $N^{1/[\alpha)2-q)]}$ -scaled attractors $\mathcal{F}$ (x) when summing $N \to \infty$ q-independent identical random variables with symmetric distribution $f(x)$ with Q-variance $\sigma_Q \equiv \int_{-\infty}^{\infty} dx\, x^2 [f(x)]^Q / \int_{-\infty}^{\infty} dx\, [f(x)]^Q$ ($Q \equiv 2q-1$; $q_1 = (1+q)/(3-q)$; $q \geq 1$). *Top left:* The attractor is the Gaussian sharing with $f(x)$ the same variance σ_1 (standard CLT). *Bottom left:* The attractor is the α-stable Lévy distribution which shares with $f(x)$ the same asymptotic behavior, i.e., the coefficient C_α (Lévy-Gnedenko CLT, or α-generalization of the standard CLT). *Top right:* The attractor is the q-Gaussian which shares with $f(x)$ the same $(2q-1)$-variance, i.e., the coefficient C_q (q-generalization of the standard CLT, or q-CLT); see [477]. *Bottom right:* The attractor is the (q, α)-stable distribution which shares with $f(x)$ the same asymptotic behavior, i.e., the coefficient $C^L_{q,\alpha}$ (q-generalization of the Lévy-Gnedenko CLT and α-generalization of the q-CLT); see [495]. The case $\alpha < 2$, for both $q = 1$ and $q \neq 1$ (more precisely $q > 1$), further demands specific asymptotics for the attractors to be those indicated; essentially the divergent q-variance must be due to fat tails of the power-law class, excepting for possible logarithmic corrections (for the $q = 1$ case see, for instance, [217] and references therein)

See Figs. 4.59 and 4.60, which illustrate the steps of the q-Fourier transform of q-Gaussians. Also, see in [506] a method for solving a nonlinear Fokker–Planck equation through the use of the q-Fourier transform.

For the q-Fourier transform for $q < 1$, one may use the transformation $q \to \frac{5-3q}{3-q}$ which biunivocally transforms the interval $[1, 3)$ into the interval $[1, -\infty)$: see [507], where this transformation is noted $Q \to \frac{-2Q}{2+Q}$ with $Q \equiv 1 - q$.

The q-CLT that we have described above constitutes a mathematical extension of one of the fingerprints of BG statistical mechanics, namely the Maxwellian distribution of velocities. In the next section we focus on the possible q-generalization of the Large Deviation Theory (LDT), which would constitute (see [508] and references therein) a mathematical extension of the most celebrated fingerprint of the BG theory, namely the BG factor.

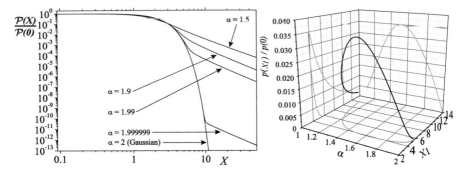

Fig. 4.55 *Left*: Gaussian and α-stable Lévy distributions for α approaching 2 in the inverse Fourier transform of $e^{-|\xi|^\alpha}$. For values of α closer to 2, the Lévy distribution becomes almost equal to a Gaussian up to some critical value above which the power-law behavior emerges. *Right*: Locus of the inflection point of the same α-stable Lévy distributions. Contrarily to what happens with q-Gaussians, when Lévy distributions are represented in log-log scales, they exhibit an inflection point which disappears both as $\alpha \to 1$ (Cauchy-Lorentz distribution, i.e., $q = 2$) and as $\alpha \to 2$ (Gaussian distribution). We also show the projections onto the planes $\left[\frac{p(X_I)}{p(0)}, X_I\right]$, $\left[\frac{p(X_I)}{p(0)}, \alpha\right]$, and $\left[\alpha, X_I\right]$. From [240]

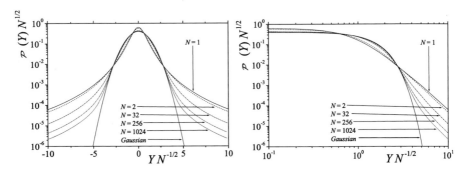

Fig. 4.56 Both panels represent probability density function $\mathcal{P}(Y)$ versus Y (properly scaled) in log-linear (left) and log-log (right) scales, where Y represents the sum of N independent variables X each of them having a q-Gaussian distribution with $q = 3/2$ ($< 5/3$). Since the variables are independent and the variance is finite, $\mathcal{P}(Y)$ converges to a Gaussian as it is visible. It is also visible in the log-linear representation that, although the central part of the distribution approaches a Gaussian, the power-law decay subsists even for relatively large N as depicted in log-log representation. From [240]

4.10   Large Deviation Theory

4.10.1   Purely Probabilistic Models

Independent coins

If we toss $N = 10$ coins, what is the probability to have up to $n = 2$ heads? This is an easy exercise to solve [509]. If we toss now $N = 100$ coins, what is the probability

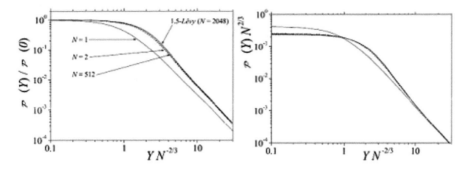

Fig. 4.57 Both panels represent probability density function $\mathcal{P}(Y)$ versus Y (properly scaled) in two different log–log scales, where Y represents the sum of N independent variables X each of them having a q-Gaussian distribution with $q = 9/5$ ($> 5/3$). Since the variables are independent and the variance diverges, $\mathcal{P}(Y)$ converges to a Lévy distribution as it is visible. From [240]

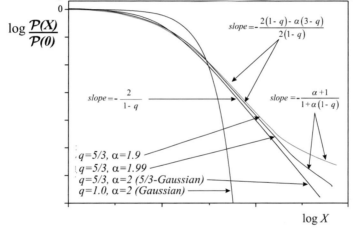

Fig. 4.58 Outline of (q, α)-stable distributions (inverse q-Fourier transforms of $a\, e_q^{-b\,|\xi|^\alpha}$) for the case in which the correlation is given by $q_1 = 2$. As α approaches 2, the (q, α)-stable distributions become closer and closer to a q-Gaussian with $= 5/3$, with an exponent $[2(q-1) + \alpha(3-q)] / [2(q-1)]$. However, since $\alpha \neq 2$, for some value X^*, a crossover occurs through which the distribution changes from the intermediate regime toward the distant regime with a tail exponent $(\alpha+1)/(1+\alpha q - \alpha)$. The inequalities $2/(q-1) \geq [2(q-1) + \alpha(3-q)]/[2(q-1)] > (1+\alpha)/[1+\alpha(q-1)]$ are satisfied. From [492]

to have up to $n = 20$ heads? Easy again. Generically, if we toss N coins, what is the probability $P(N; n/N < x)$ of getting up to $x = 20\%$ heads? It is intuitive that this probability $P(N; n/N < x)$ decreases, for $0 \leq x < 1/2$, to zero in the $N \to \infty$ limit. And it does so *exponentially* with N. More precisely [509],

$$P(N; n/N < x) \sim e^{-r_1(x)N}, \qquad (4.156)$$

where the *rate function* is given by the relative BG entropy per particle

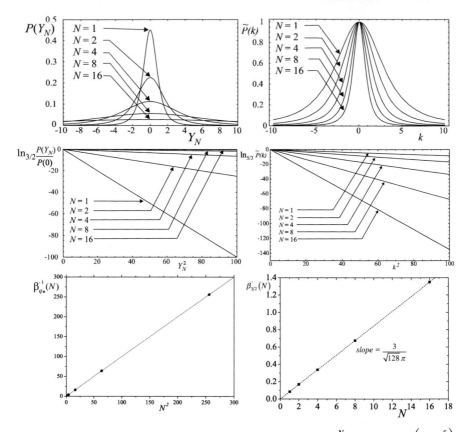

Fig. 4.59 *Top*: Probability distribution $P(Y_N)$ versus Y_N, with $Y_N \equiv \sum_{i=1}^{N} X_i$, X_i being $\left(q = \frac{5}{3}\right)$-independent random variables associated with a $\mathcal{G}_{\frac{3}{2}}(X)$ distribution with $\beta = 1$ (*left*), and the respective $\left(q = \frac{3}{2}\right)$-Fourier Transform, $\tilde{P}(k)$, *vs.* k (*right*). *Middle*: Same as above, in $\ln_{\frac{3}{2}}$-squared scale (*left*), and $\ln_{\frac{5}{3}}$-squared scale (*right*). The straight lines indicate that $P(Y_N)$ and $\tilde{P}(k)$ are q-Gaussians with $q = \frac{3}{2}$ and $q = \frac{5}{3}$, respectively. Their slopes are $\beta_{q_*=3/2}^{-1}(N)$ for left panel curves and $\beta'_{q_*}(N)$ for right panel curves. *Bottom*: $\beta_{q_*=3/2}^{-1}(N)$ versus N^2, which is a straight line with slope 1 (*left*); $\beta'_{q_*=3/2}(N)$ versus N which is also a straight line but with slope $\left.\frac{3-q_*}{8\,C_{q_*}^{2(q_*-1)}}\right|_{q_*=3/2} = 0.088844\ldots$ (*right*). From [492]

$$r_1(x) = \ln 2 + x \ln x + (1-x)\ln(1-x), \qquad (4.157)$$

as shown in Fig. 4.61. The fact that $r_1 N$ is extensive is directly connected, for say many-body classical Hamiltonian systems, with the classical thermodynamical requirement of extensivity of the total entropy. Indeed, Eq. (4.156) is a mathematical extension [508], of the the BG factor $p_{BG} \propto e^{-\beta \mathcal{H}_N} = e^{-[\beta \mathcal{H}_N/N]N}$, where $\mathcal{H}_N$

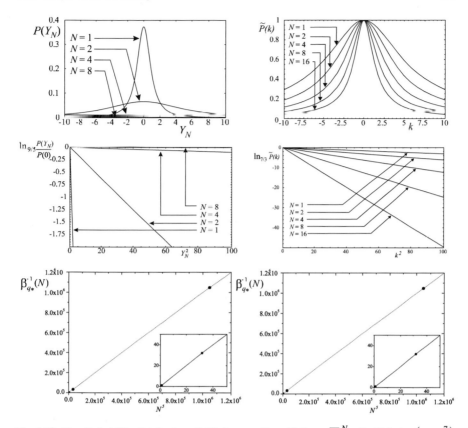

Fig. 4.60 *Top*: Probability distributions $P(Y_N)$ versus Y_N, with $Y_N \equiv \sum_{i=1}^{N} X_i$, X_i being $\left(q = \frac{7}{3}\right)$-independent random variables associated with a $\mathcal{G}_{\frac{9}{5}}(X)$ distribution with $\beta = 1$ (*left*), and the respective $\left(q = \frac{9}{5}\right)$-Fourier Transform, $\tilde{P}(k)$, versus k (*right*). *Middle*: Same as above, in $\ln_{\frac{9}{5}}$-squared scale (*left*), and $\ln_{\frac{7}{3}}$-squared scale (*right*). The straight lines indicate that $P(Y_N)$ and $\tilde{P}(k)$ are q-Gaussians with $q = \frac{9}{5}$ and $q = \frac{7}{3}$, respectively. Their slopes are $\beta_{q_*=9/5}^{-1}(N)$ for left panel curves and $\beta'_{q_*=9/5}(N)$ for right panel curves. *Bottom*: $\beta_{q_*=9/5}^{-1}(N)$ versus N^5, which is a straight line with slope 1 (*left*); $\beta'_{q_*=9/5}(N)$ versus N, which is also a straight line, but with slope

$$\left. \frac{3-q_*}{8\, C_{q_*}^{2(q_*-1)}} \right|_{q_*=9/5} = 0.030995 \ldots (right).$$ From [492]

denotes the N-body Hamiltonian with *short*-range interactions, $[\beta \mathcal{H}_N / N]$ being an intensive quantity.

The situation is quite different if $\mathcal{H}_N$ includes *long*-range interactions. Indeed we expect (see Sect. 3.3.1, and the classical systems focused on in Sect. 5.3) p_{BG} to be typically generalized into

$$p_q \propto e_q^{-\beta_q \mathcal{H}_N} = e_q^{-(\beta_q \tilde{N}) \frac{\mathcal{H}_N}{N\tilde{N}} N}, \tag{4.158}$$

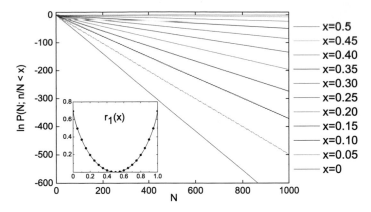

Fig. 4.61 Tossing N independent coins: the large deviation probability $P(N; n/N < x)$ decays exponentially with N, and the slopes of the logarithms, $\ln P(N; n/N < x)$, provide the rate function numerical values $r_1(x)$, represented in the inset as dots. Continuous curve in the inset corresponds to Eq. (4.157). From [509]

$[(\beta_q \tilde{N})(\mathcal{H}_N / N \tilde{N})]$ being an intensive quantity. As a mathematical extension of such a wide class of strongly correlated systems, we consistently expect Eq. (4.156) to be generalized as follows:

$$P(N; n/N < x) \sim e_{q_{LDT}}^{-r_{q_{LDT}}(x)N} , \qquad (4.159)$$

where the q_{LDT}-generalized *rate function* is expected to be given by an appropriate relative q-entropy per particle. This expression directly reflects the fact that, even for these long-range-interacting systems, the total entropy should remain extensive (thus corresponding to $r_{q_{LDT}}(x)N$). In other words, the functional form of $P(N; n/N < x)$ might change (from exponential to say q-exponential), but, in all cases, its argument is expected to *invariably remain extensive!*

Such is the scenario that we present as highly plausible in what follows: in this section for purely probabilistic systems [488, 509–511] (see also [489]), and, in the following section, for dynamical systems as well. In other words, we shall provide concrete nontrivial examples toward the highly desirable q-generalization of the standard LDT, up to now restricted [508] to the mathematical hypotheses associated with the BG systems.

Correlated coins

The first model we focus on is a discretized version of a $q_{attract}$-Gaussian ($q_{attract} \geq 1$) that has been proposed in [470] (the notation identification is $(Q, q) \equiv (q_{attract}, q_{LDT})$), and whose LDT has been studied in [509, 511]. When N increases, the continuous $q_{attract}$-Gaussian attractor is gradually approached, and the LDT consists in calculating $P(N; n/N < x)$, which characterizes the speed at which the attractor is attained. The discretization is determined by the quantity $\Delta n =$

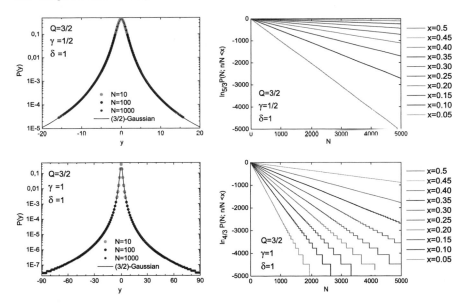

Fig. 4.62 Illustrative histograms of the discretized model (left column) and their corresponding distributions $P(N; n/N < x)$, in semi q-log representation (right column). From [509]

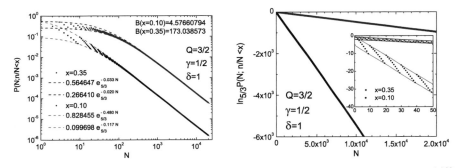

Fig. 4.63 Comparison of the numerical data (dots) with upper and lower bounds $a(x)e_q^{-r_q(x)N}$ for $x = 0.10$ and $x = 0.35$. *Left*: log–log representation. *Right*: q-log-linear representation. Notice that, quite remarkably, the straight-line structure is observed for all values of N, starting at $N = 1$. From [511]

$\delta(N + 1)^\gamma$ ($\delta > 0$; $0 < \gamma < 1$); the model is fully determined by $(q_{attract}, \gamma, \delta)$. The numerical study of this system provides Eq. (4.159), where

$$q_{LDT} = \frac{q_{attract} - 1}{\gamma(3 - q_{attract})} + 1 \quad (\forall \delta). \tag{4.160}$$

Illustrations are shown in Figs. 4.62 and 4.63, which validate the behavior conjectured in Eq. (4.159).

The second probabilistic model we focus on is introduced in [487, 490] and numerically studied in [488]. It is a symmetric scale-invariant system of N strongly correlated exchangeable (hence $q_{ent} = 1$) binary random variables which is defined by adopting $e_{q_{correlat}}^z$ as generating function, with $q_{correlat} \geq 1$, where $q_{correlat} = 1$ corresponds to the independent binomial distribution (in [488] the notation $q_{correlat} \equiv q^{gen} \equiv 1 + 1/\alpha$ is used, hence $\alpha > 0$). The corresponding $N \to \infty$ attractor is a $q_{attract}$-Gaussian which is obtained as follows. Given $q_{correlat}$ we obtain the most general normalized symmetric distribution satisfying the Leibnitz rule, namely

$$p_{N,n} = \frac{x_N!}{x_{N-n}! \, x_n!} q_n q_{N-n} \,, \tag{4.161}$$

where q_n is a polynomial of degree n, and $x_N = x_1 x_2 \ldots x_N$ $(x_0! = 1)$, $\{x_N\}$ being any strictly increasing infinite sequence of nonnegative real numbers. This structure eventually yields

$$p_{N,n} = \frac{N!}{(N-n)! \, n!} \frac{(\alpha/2)_n \, (\alpha/2)_{N-n}}{(\alpha/2)_N} \left(\sum_{n=0}^{N} p_{N,n} = 1 \right), \tag{4.162}$$

where the *Pochhammer symbol* is defined $(a)_b \equiv a(a+1) \ldots (a+b-1)$. The $N \to \infty$ attractor is a $q_{attract}$-Gaussian with

$$q_{attract} = \frac{\alpha - 4}{\alpha - 2} \leq 1 \,. \tag{4.163}$$

We straightforwardly verify that

$$\frac{1}{q_{correlat} - 1} - \frac{2}{1 - q_{attract}} = 2 \ (\forall \alpha) \,. \tag{4.164}$$

By extending Eq. (4.159) into

$$P(N; n/N^{\gamma(\alpha)} < x) \sim e_{q_{LDT}}^{-r_{q_{LDT}}(x)N} \ (\gamma(\alpha) > 0) \,, \tag{4.165}$$

the corresponding numerical approach indicates that large deviations are closely described by Eq. (4.165) with

$$q_{LDT} = 1 + 2/\alpha \,, \tag{4.166}$$

hence

$$\frac{1}{q_{LDT} - 1} - \frac{1}{1 - q_{attract}} = 1 \quad (\forall \alpha), \tag{4.167}$$

and

$$\frac{1}{q_{LDT} - 1} = \frac{1}{2 \left(q_{correlat} - 1 \right)} \quad (\forall \alpha). \tag{4.168}$$

Other strongly correlated probabilistic models have also been numerically studied [1198, 1199], and once again the q-LDT scenario is confirmed.

4.10.2   Dynamical Models

Let us know to turn our attention onto large deviations in four dynamical systems [512] whose CLT attractors are $q_{attract}$-Gaussians. The first of these dynamical systems is a prototypical two-dimensional conservative map, namely the standard map; the other three are many-body nonconservative systems, namely the Coherent Noise Model (CNM), the Ehrenfest Dog Flea Model (EDFM), and a Random Walk Avalanches Model (RWAM).

Standard map
The standard map is a low-dimensional conservative deterministic map and as such it belongs to Sect. 5.2. But in order to jointly present the q-generalized LDT available evidences, we shall review it here, following [513]. It was introduced by B.V. Chirikov in 1979 (see [514] and references therein) and is defined through

$$
\begin{aligned}
p_{i+1} &= p_i - K \sin x_i \quad (mod \, 2\pi) \quad (K \geq 0) \\
x_{i+1} &= x_i + p_{i+1} \quad\quad (mod \, 2\pi).
\end{aligned}
\tag{4.169}
$$

Typical phase portraits are shown in Fig. 4.64. Each (x, p) point yields a Lyapunov exponent $\lambda^{(1)} = -\lambda^{(2)} \geq 0$: see Fig. 4.65. Next, along the CLT lines, we define the following quantity

$$\bar{y} \equiv \sum_{i=1}^{T} (x_i^{(j)} - \langle x \rangle), \tag{4.170}$$

with

$$\langle x \rangle \equiv \frac{1}{M} \frac{1}{T} \sum_{j=1}^{M} \sum_{i=1}^{T} x_i^{(j)}, \tag{4.171}$$

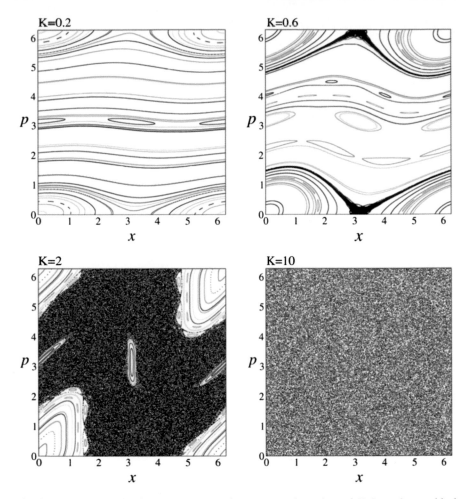

Fig. 4.64 Phase portrait of the standard map for representative values of K. In each case, black dots represent the region of chaotic sea in the available phase space and all other colors represent different stability islands. From [513]

where $M >> 1$ (typically $M \geq 10^7$) is the number of initial conditions and $T >> 1$ (typically $T \geq 2^{23} \simeq 10^7$) is the number of iterations for each of those M initial conditions. Finally, by constructing the histogram for $\bar{y}$, we obtain the results indicated in Figs. 4.66, 4.67, and 4.68. The $K = 0$ case (hence a linear map) has been revisited, along the same lines, with higher precision [515] and the value $q_{attract} \simeq 1.935$ has been re-obtained. Nevertheless, since some numerical error is naturally unavoidable, an analytical effort has been accomplished [516] and the exact result turns out to be $q_{attract} = 2$.

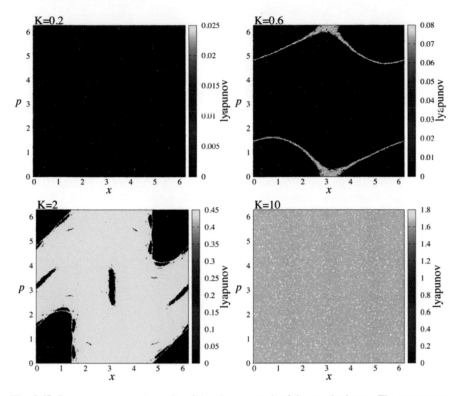

Fig. 4.65 Lyapunov exponent results of the phase portrait of the standard map. The same representative K values are used. For each case, Lyapunov exponents are calculated for 200000 initial conditions. In the calculation, each initial condition is iterated 107 times. From [513]

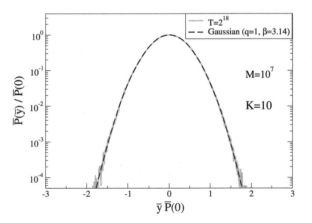

Fig. 4.66 Normalized probability distribution function for $K = 10$ with $T = 2^{18}$. From [513]

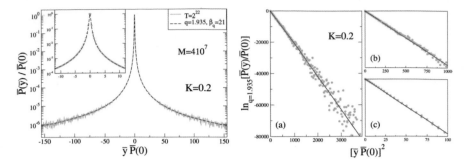

Fig. 4.67 Normalized probability distribution function for $K = 0.2$ with $T = 2^{22}$. *Left:* log-linear representation. *Right:* q-log - quadratic representation in three different scales. From [513]

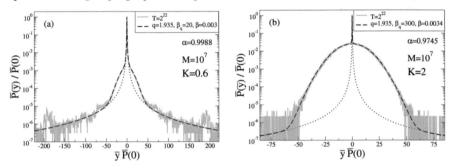

Fig. 4.68 Normalized probability distribution function for $K = 0.6$ (left) and $K = 2$ (right). The initial conditions are randomly taken from the entire available phase space. The fittings (dashed black lines) have been done with $\bar{P}(\bar{y})/\bar{P}(0) = a\, e_q^{-\beta_q[\bar{y}\bar{P}(0)]^2} + (1-a)\, e^{-\beta[\bar{y}\bar{P}(0)]^2}$. In both cases, dotted red lines are the related $q_{attract}$-Gaussian distributions. From [513]. The value of a is directly related to the respective sizes of the weakly and strongly chaotic phase space regions. Consequently, there is in fact no need to consider it as a fitting parameter and can be calculated from the available data, as detailed in [517]

Now that we have described the CLT features of the standard map, let us focus on its LDT aspects: see Fig. 4.69.

Many-body models
For the LDT aspects of CNM, EDFM, and RWAM: see Fig. 4.70.

4.11 Time-Dependent Ginzburg–Landau d-Dimensional $O(n)$ Ferromagnet with $n = d$

The standard Langevin and associated Fokker–Planck equations are by no means equivalent to BG statistical mechanics, but they surely are consistent with it if the system satisfies assumptions such as mixing/ergodicity. However, if they are violated,

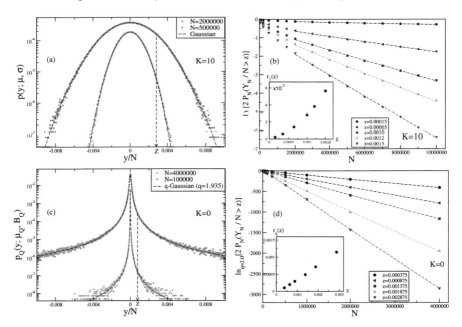

Fig. 4.69 (Color online) The probability density function of the standard map for two representative values of N for **a** $K = 10$ and **c** $K = 0$. If we multiply both the ordinate and the abscissa by $\sqrt{N}$ in (**a**) and by N^γ with $\gamma \simeq 0.65$ in (**c**), the present data collapse onto a single Gaussian (Q-Gaussian with $Q \simeq 1.935$; the analytical results is $Q = 2$) for $K = 10$ ($K = 0$). A typical value $z > 0$ is indicated as well. In **b** and **d**, we see, respectively, that the large deviation probability $P_N(Y_N/N > z)$ decays with N exponentially for $K = 10$ and q-exponentially with $q = 2.0$ for $K = 0$. The slopes provide the rate function $r_q(z)$ as shown in the Insets. The conjectural relation $q = 2/(3 - Q)$ has only two fixed points, namely $q = Q = 1$ and $q = Q = 2$, which coincide with the results corresponding to $K = 10$ and $K = 0$, respectively. From [512]

discrepancies might emerge, for example in the distribution of velocities, which reminds results that are typical within nonextensive statistical mechanics. It is our purpose here to exhibit one such example, even if the authors did not make the connection in their original papers [518, 519].

An interesting short-range-interacting d-dimensional ferromagnetic system is that whose symmetry is dictated by rotations in n dimensions, i.e., the so-called $O(n)$ symmetry ($n = 2$ corresponds to the XY model, $n = 3$ corresponds to the Heisenberg model, and so on; $n \to \infty$ corresponds to the spherical model; the analytic limit $n \to 1$ would yield the Ising model). We specifically address the kinetics of point defects (see the vortices in Fig. 4.71) during quenching from high temperature to zero temperature for the $d = n$ model. The theoretical description is done in terms of a time-dependent Ginzburg–Landau equation (similar to a Langevin equation). As the main outcome, one obtains that the distribution of the vortex velocity **v** is, although not written in this manner by the authors [518, 519], given by

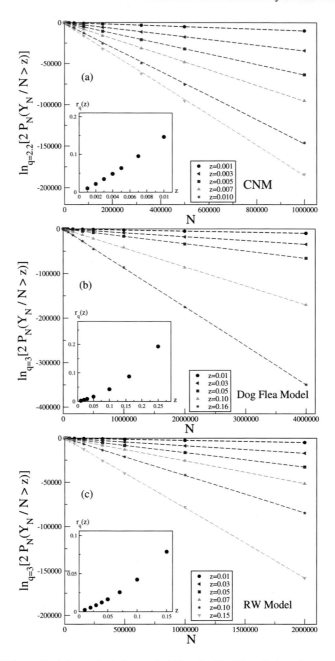

Fig. 4.70 (Color online) Large deviation probability $P_N(Y_N/N > z)$ decaying q-exponentially with N for (a) the CNM with $\sigma = 0.05$ ($Q \simeq 2.1$ and $q \simeq 2.2$), (b) the Ehrenfest dog flea model ($Q \simeq 2.3$ and $q \simeq 3$), and (c) the random walk avalanche model ($Q \simeq 2$ and $q \simeq 3$). The behavior of $r_q(z)$ can be seen in the insets of each figure. The conjectural relation $q(3 - Q)/2 = 1$ is satisfied within less than 1% error for the CNM and EDFM, but, intriguingly enough, it completely fails for the RWAM. From [512]

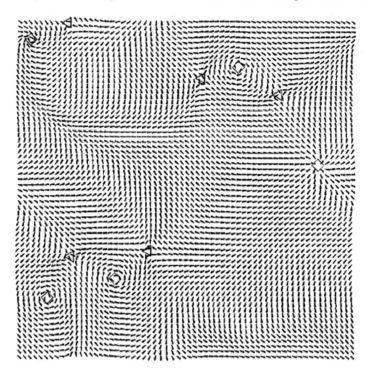

Fig. 4.71 A typical vortex configuration in a 256×256 $n = d = 2$ system. The arrow on each site represents the order parameter at that point. Not all the lattice sites are shown. The squares and triangles are in the core regions of $+1$ and -1 vortices, respectively, where the magnitude of the order parameter is near zero. From [519]

$$P(\mathbf{v}) \propto e_q^{-|\mathbf{v}|^2/v_0^2},\tag{4.172}$$

with

$$q = \frac{d+4}{d+2},\tag{4.173}$$

v_0 being a reference velocity which approaches zero for time increasing after the moment at which the quenching was done. The index q decreases from 2 to 1 when d increases from zero to infinity. It is certainly intriguing, although yet unexplained, to notice that this value of q precisely is the one which separates the finite from the infinite variance regions of q at d dimensions (see Eq. (4.33)).

Chapter 5
Deterministic Dynamical Foundations of Nonextensive Statistical Mechanics

Il dépend de celui qui passe, Que je sois tombe ou trésor, Que je parle ou me taise, Ceci ne tient qu'à toi, Ami n'entre pas sans désir.

Paul Valéry, *Palais de Chaillot*

In this Chapter we focus on *microscopic*-like nonlinear dynamical systems, in the sense that the time evolution is expressed exclusively with *deterministic* ingredients. Such approach is currently referred to in the literature as from *first principles*. We will first discuss, analytically and/or numerically, low-dimensional dissipative maps, and then low-dimensional conservative maps. We address next, numerically, many-body problems, first symplectic systems constituted by coupled simple low-dimensional conservative maps, and finally classical Hamiltonian systems. Our intention is to focus, in an unified manner, on those common aspects which relate to nonextensive statistical mechanical concepts. We shall see that, every time we have nonlinear dynamics which is only weakly chaotic (typically at the frontier between regular motion and strong chaos), the need typically emerges to q-generalize various concepts and functions, and very especially the entropy.

5.1 Dissipative Maps

5.1.1 One-Dimensional Dissipative Maps

Let us start by defining the maps we are going to deal with. We focus on unimodal one-dimensional maps. Some of them are since long well known in the literature; others have been recently introduced with the purpose of illustrating specific features that we are interested in.

© Springer Nature Switzerland AG 2023
C. Tsallis, *Introduction to Nonextensive Statistical Mechanics*,
https://doi.org/10.1007/978-3-030-79569-6_5

The *z-logistic map* is defined as follows (see, for instance, [187]):

$$x_{t+1} = 1 - a|x_t|^z \quad (z > 1; 0 \le a \le 2; |x_t| \le 1). \tag{5.1}$$

The standard case is recovered for $z = 2$, and its primary edge of chaos occurs at $a_c(2) = 1.40115518909\ldots$ For this simple $z = 2$ case, and with $y \equiv x + 1/2$, we obtain the traditional form

$$y_{t+1} = \mu\, y_t (1 - y_t) \quad (0 \le \mu \le 4; 0 \le y_t \le 1. \tag{5.2}$$

The *z-periodic map* is defined as follows [188]:

$$x_{t+1} = d\cos(\pi|x_t - 1/2|^{z/2}) \quad (z > 1; d > 0; |x_t| \le d) \tag{5.3}$$

It belongs to the same universality class of the z-logistic map since they both share an extremum with inflection of order z. The standard case is recovered for $z = 2$, and its primary edge to chaos occurs at $d_c(2) = 0.8655\ldots$

The *z-circular map* is defined as follows [190, 520]:

$$\theta_{t+1} = \Omega + [\theta_t - \frac{1}{2\pi}\sin(2\pi\theta_t)]^{z/3} \quad (z > 0) \tag{5.4}$$

The case $z = 3$ recovers the standard case, and its primary edge to chaos occurs at $\Omega_c(3) = 0.6066\ldots$. Various interesting properties and analytical results can be seen in [521].

The *z-exponential map* is defined as follows [197]:

$$x_{t+1} = 1 - a\, e^{-1/|x_t|^z}\, (z > 0; a \in [0, a^*(z)]; |x_t| \le 1) \tag{5.5}$$

where $a^*(z)$ depends slowly from z (e.g., $a^*(1/2) \simeq 5.43$). This map was introduced [197] in order to have an extremum *flatter than any power*, which is the case of the z-logistic and z-periodic ones. It shares with the z-logistic and z-periodic maps the same topological properties, although they differ in the metric ones. The case $z = 1/2$ is a typical one, and its primary edge to chaos occurs at $a_c(1/2) = 3.32169594\ldots$

Sensitivity to the initial conditions

The sensitivity to the initial conditions ξ for a one-dimensional dynamical system is, as previously addressed, defined as follows:

$$\xi \equiv \lim_{\Delta x(0) \to 0} \frac{\Delta x(t)}{\Delta x(0)}, \tag{5.6}$$

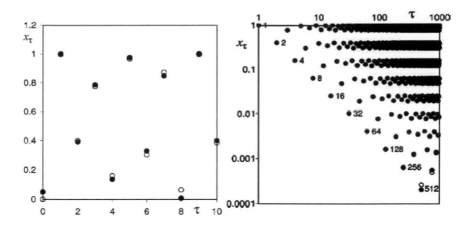

Fig. 5.1 *Left*: Absolute values of positions of the first 10 iterations τ for two trajectories of the logistic map at the edge of chaos, with initial conditions $x_0 = 0$ (empty circles) and $x_0 \simeq 5 \times 10^{-2}$ (full circles). *Right*: The same (in log-log plot) for the first 1000 iterations, with $\Delta x_0 = 10^{-4}$. From [195]

where x denotes the phase space variable. For wide classes of systems, the sensitivity ξ is expected to satisfy

$$\frac{d\xi}{dt} = \lambda_{q_{sen}} \xi^{q_{sen}} \quad (\xi(0) = 1), \tag{5.7}$$

hence [186, 194, 195, 200]

$$\xi = e_{q_{sen}}^{\lambda_{q_{sen}} t}, \tag{5.8}$$

where $q_{sen} = 1$ if the Lyapunov exponent $\lambda_1 \neq 0$ (*strongly sensitive* if $\lambda_1 > 0$, and *strongly insensitive* if $\lambda_1 < 0$), and $q_{sen} \neq 1$ otherwise; *sen* stands for *sensitivity*. At the edge of chaos, $q_{sen} < 1$ (*weakly sensitive*), and at both the period-doubling and tangent bifurcations, $q_{sen} > 1$ (*weakly insensitive*). The case $q_{sen} < 1$ yields, in (5.8), a power-law behavior $\xi \propto t^{1/(1-q_{sen})}$ in the limit $t \to \infty$. This power-law asymptotics were since long known in the literature [181–185]. The case $q_{sen} < 1$ is in fact more complex than indicated in Eq. (5.8). This equation only reflects the maximal values of an entire family, fully (and not only asymptotically) described in [200, 306]. See Figs. 5.1 and 5.2 from [195].

Multifractality

Multifractals are conveniently characterized by the multifractal function $f(\alpha)$ [163]. Typically, this function is concave, defined in the interval $[\alpha_{min}, \alpha_{max}]$ with $f(\alpha_{min}) = f(\alpha_{max}) = 0$, within this interval it attains its maximum d_H, d_H being the *Hausdorff* or *fractal dimension*.

Fig. 5.2 Numerical corroboration (full circles) of the q-generalized Pesin-like identity $K_q^{(k)} = \lambda_q^{(k)}$ at the edge of chaos the logistic map. On the ordinate we plot the q-logarithm of ξ_{t_k} (equal to $\lambda_q^{(k)}\, t$), and in the abscissa S_q (equal to $K_q^{(q)}\, t$), both for $q = 0.2445\ldots$ The dashed line is a linear fit. *Inset*: The full lines are from the analytic result. From [198]

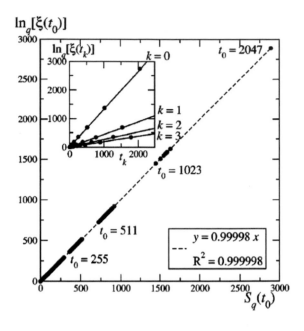

It has been proved [188, 195], that, at the edge of chaos, we have[1]

$$\frac{1}{1 - q_{sen}} = \frac{1}{\alpha_{min}} - \frac{1}{\alpha_{max}} \qquad (q_{sen} < 1) \tag{5.9}$$

For unimodal maps with inflection z, negative Schwarzian derivative in the bounded interval, and partition scale b we have

$$\alpha_{max}(z) = \frac{\ln b}{\ln \alpha_F(z)}$$

$$\alpha_{min}(z) = \frac{\ln b}{z \ln \alpha_F(z)} \tag{5.10}$$

where α_F is the so called Feigenbaum constant. Hence

---

[1] Virtually all the q-formulae of the present book admit the limit $q \to 1$. This is not the case of Eq. (5.9), since the left member diverges whereas the right member vanishes. Indeed, $q = 1$ typically corresponds to the case of dynamics with positive Lyapunov exponent, hence mixing, hence ergodic, hence leading to an Euclidean, nonfractal, geometry. For such a standard one-dimensional geometry, it should be $\alpha_{min} = \alpha_{max} = f(\alpha_{min}) = f(\alpha_{max}) = 1$, which clearly makes the right member of (5.9) to diverge. A relation more general than (5.9) is therefore needed before taking the $q \to 1$ limit. A relation such as $\frac{1}{1-q_{sen}} = \frac{1}{\alpha_{min}-f(\alpha_{min})} - \frac{1}{\alpha_{max}-f(\alpha_{mx})}$ for instance. Indeed, it recovers Eq. (5.9) for $f(\alpha_{min}) = f(\alpha_{max}) = 0$, and also admits $q \to 1$, being now possible for both members to diverge. It should be however noticed that this more general relation is totally heuristic: we do not yet dispose of neat numerical indications, and even less of a proof.

$$\frac{1}{1 - q_{sen}(z)} = (z - 1)\frac{\ln \alpha_F(z)}{\ln b} \tag{5.11}$$

For the universality class of the z-logistic map, we have $b = 2$ hence

$$\frac{1}{1 - q_{sen}(z)} = (z - 1)\frac{\ln \alpha_F(z)}{\ln 2} \tag{5.12}$$

Broadhurst calculated the $z = 2$ Feigenbaum constant α_F with 1,018 digits [522]. Through Eq. (5.12), it straighforwardly follows that

$$q_{sen}(2) = 0.244487701341282066198\ldots \tag{5.13}$$

See [187] for $q_{sen}(z)$.

The same type of information is available for the edge of chaos of other unimodal maps. For example, for the universality class of the z-circular map, we must use [190] $b = (\sqrt{5} + 1)/2 = 1.6180\ldots$ into Eq. (5.11). We then obtain [190] $q_{sen}(3) = 0.05 \pm 0.01$. Similar results are available for the universality class of the z-exponential map [197].

Entropy production and the Pesin theorem

There are quite generic circumstances under which the entropy increases with time, typically while dynamically exploring the phase space of the system. If this increase is (asymptotically) linear with time we may define an *entropy production per unit time*, which is the rate of increase of the entropy. One such concept, based on single trajectories as already mentioned, is the so called *Kolmogorov-Sinai entropy rate* or just *KS entropy* [111]. It satisfies, under quite general conditions, an identity, namely that it is equal to the sum of all positive Lyapunov exponents (which reduces to a single Lyapunov exponent if the system is one-dimensional). This equality is frequently referred in the literature as the *Pesin identity*, or the *Pesin theorem* [114]. Here, instead of the KS entropy (computationally very inconvenient), we shall use K_q, the ensemble-based entropy production rate that we defined in Sect. 3.2. We refer to Eq. (3.59). A special value of q, noted q_{ent}, generically exists such that $K_{q_{ent}}$ is finite, whereas K_q vanishes (diverges) for any $q > q_{ent}$ ($q < q_{ent}$). For systems strongly chaotic (i.e., whose single Lyapunov exponent is positive), we have $q_{ent} = 1$, thus recovering the usual case of ergodic systems and others. For systems weakly chaotic (i.e., whose single Lyapunov exponent vanishes, such as in the case of an edge of chaos), we have $q_{ent} < 1$. Many nonergodic (but certainly not all) systems belong to this class.

For quite generic low-dimensional systems we expect [186] (see 5.2)

$$q_{ent} = q_{sen}, \tag{5.14}$$

and

$$K_{q_{ent}} = \lambda_{q_{sen}}. \tag{5.15}$$

For $q_{ent} = 1$, this entropy production is expected to concide, quite generically, with the KS entropy rate. Although a rigorous proof is, to the best of our knowledge, still lacking, examples can be seen in [191, 192, 198]. For many $K_1 = \lambda_1 = 0$ systems, we expect the straightforwardly q-generalized Kolmogorov-Sinai entropy rate to coincide with $K_{q_{ent}}$.

The properties that have been exhibited here for the sensitivity to the initial conditions and the entropy production have also been checked [523, 524] for other entropies directly related to S_q. The scheme remains the same, excepting for the slope $K_{q_{ent}}$, which does depend on the particular entropy. The slope for S_q turns out to be the maximale one among those that have been analyzed. For all these $q \neq 1$ examples, the Renyi entropy S_q^R fails in providing a *linear* time dependence: it provides instead a *logarithmic* time dependence.[2]

Relaxation

In the previous paragraphs, you were dealing with the value of q, q_{sen}, associated with the sensitivity to the initial conditions, and also with multifractality and the entropy production. We address now a different property, namely relaxation. As we shall see, a new value of q, denoted q_{rel} (where *rel* stands for *relaxation*), emerges. Typically $q_{rel} \geq 1$, the equality holding for strongly chaotic systems (i.e., when $q_{sen} = 1$). Relaxation was systematically studied for the z-logistic map in [526]. The procedure consists in starting, at the edge of chaos, with a distribution of $M >> 1$ initial conditions which is uniform in phase space ($x_0 \in [-1, 1]$ for the z-logistic map), and let evolve the ensemble towards the multifractal attractor. A partition of the phase space is established with $W(0) >> 1$ little equal cells, and then the covering is followed along time by only counting those cells which have at least one point at time t. This determines $W(t)$, which gradually decreases since the measure of the multifractal attractor is zero. In the $M \to \infty$ and $W(0) \to \infty$ limits, and disregarding small oscillations, it is verified

$$\frac{W(t)}{W(0)} \simeq e_{q_{rel}}^{-t/\tau_{q_{rel}}}, \tag{5.16}$$

with $q_{rel}(z) \geq 1$ and $\tau_{q_{rel}}(z) > 0$. If it is taken into account the fact that, for the z-logistic map, also the Hausdorff dimension depends on z, it can be numerically verified the following quite intriguing, and yet unexplained, relation:

---

[2] In this Section we are focusing on maps (i.e., discrete-time systems). It is worthy however to mention that systems of ordinary differential equations (i.e., continuous-time systems), e.g., the $d = 3$ Rossler non-conservative system, can follow [525], in what concerns the Pesin-like identity, essentially the same path as the maps presently discussed.

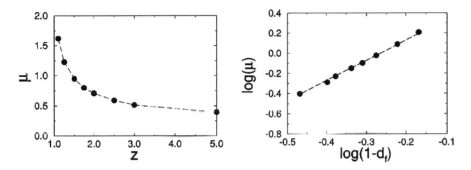

Fig. 5.3 The exponent $\mu \equiv 1/(q_{rel} - 1)$ for the z-logistic map, as a function of z (*left*), and of the fractal dimension $d_H \equiv d_f$ (*right*). From [526]

$$\frac{1}{q_{rel}(z) - 1} \simeq a\,[1 - d_H(z)]^2 \quad (z \in [1.1, 5.0]), \tag{5.17}$$

with $a = 3.3 \pm 0.3$. See Fig. 5.3. Higher precision calculations are available for $z = 2$, namely $1/[q_{rel}(2) - 1] = 0.800138194...$, hence $q_{rel}(2) = 2.249784109...$ [527, 528].[3]

For the z-circular map, it is numerically found [526] $q_{rel}(z) \to \infty$ and $d_H(z) = 1$, $\forall z$, which also is consistent with a relation like (5.17) ($q_{rel} \to \infty$ suggests a logarithmic-like behavior instead of the asymptotic power-law in (5.16)).[4]

An alternative way for studying q_{rel} has been proposed in [193]. We consider $S_1(t)$ for a map which is strongly chaotic (or $S_{q_{ent}}(t)$ for a map which is weakly chaotic) for a given number $W(0)$ of little cells within which the phase space has been partitioned, and then we initially populate one of those windows, we typically observe the following behavior. For small values of t there is a transient; for intermediate values of t there is a *linear* regime (which enables the calculation of the entropy production per unit time, and becomes longer and longer with increasing $W(0)$); finally, for larger values of t the entropy approaches (typically from above!) its saturation value $S_{q_{ent}}(\infty)$. Therefore, $S_{q_{ent}}(t) - S_{q_{ent}}(\infty)$ vanishes with diverging t, and it does so as follows:

$$S_{q_{ent}}(t) - S_{q_{ent}}(\infty) \propto e_{q_{rel}}^{-t/\tau_{q_{rel}}}, \tag{5.18}$$

which enables the determination of q_{rel}, as well as that of $\tau_{q_{rel}}$. See Fig. 5.4.

---

[3] These two references concern the *approach* to the multifractal attractor as a function of time. However, [527] claims a general criticism concerning also the time evolution *within* the attractor. This claim is rebutted in [200]. In fact, [527] contains, side by side, correct results (most of them long known), and severely incorrect statements, rebutted in appropriate places along the content of the present book).

[4] The d-dimensional generalization of Eq. (5.17) might well be $1/(q_{rel} - 1) \propto (d - d_H)^2$. Therefore, if this conjecture is correct, all the so-called *fat-fractal* dynamical attractors (i.e., $d_H = d$) would yield $q_{rel} \to \infty$.

Fig. 5.4 Time evolution of $S_{q_{ent}}$ for the $z = 2$ logistic map, for strongly chaotic (**a**) and weakly chaotic (**b**) cases. In all cases, $S_{q_{ent}}(t) < \ln_{q_{ent}} W(0)$; however, for fixed $W(0)$, the maximal value attained by $S_{q_{ent}}(t)$ is close to $\ln_{q_{ent}} W(0)$. From [193]

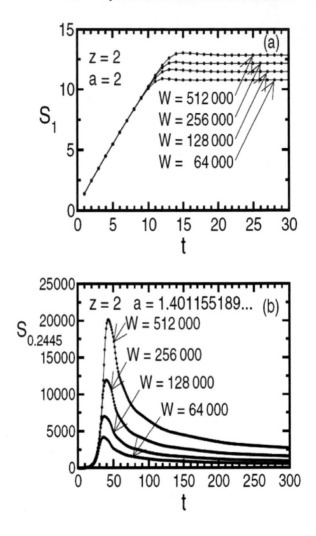

Influence of averaging

We briefly present here how results are modified [197, 199] when averaging is done over the initial conditions. Depending on the "experimental" setup of computational or real experiments, we might be interested in the dynamics related to essentially one or many initial conditions. To illustrate these effects, we focus on averages done over initial conditions within a single window, but those initial windows are uniformly distributed within the phase space of the system. We numerically verify the following behaviors:

$$\langle \ln_{q_{sen}^{av}} \xi \rangle (t) = \lambda_{q_{sen}^{av}}\, t, \tag{5.19}$$

and

$$\langle S_{q_{ent}^{av}} \xi \rangle (t) = K_{q_{ent}^{av}} t, \tag{5.20}$$

with

$$q_{ent}^{av} = q_{sen}^{av}, \tag{5.21}$$

and

$$K_{q_{ent}^{av}} = \lambda_{q_{ent}^{av}}, \tag{5.22}$$

where *av* stands for *average*. See Fig. 5.5. Notice however that, although the structure and properties remain the same, the values of $(q_{sen}^{av}, \lambda_{q_{sen}^{av}})$ differ from $(q_{sen}, \lambda_{q_{sen}})$, being $q_{sen} < q_{sen}^{av} < 1$ (see Fig. 5.6). The analytical discussion of these facts is by no means trivial and has not yet been undertaken. Indeed, it involves the simultaneous consequences of the gradual approach to the multifractal attractor and the time evolution on the attractor itself.

Averaging introduces a further complication. Let us illustrate with the z-logistic map. The edge of chaos that we have been primarily focusing on is that which emerges as an accumulation point of successive bifurcations (noted *cycle 2*). But there are edges of chaos corresponding to the accumulation points of trifurcations (noted *cycle 3*), or the various penta-furcations (noted *cycle 5*) and so on. They all belong to the same universality class in the sense that $q_{sen}(z)$ is one and the same for all of them. But it is not so for $q_{sen}^{av}(z)$. The situation is depicted in Figs. 5.7, 5.8 and 5.9. Also, we numerically verify an intriguing relation between $q_{sen}^{av}(cycle\,n;\,z)$ and $q_{rel}(cycle\,n;\,z)$, namely (see Fig. 5.10).

$$q_{rel}(cycle\,n;\,z) - 1 \simeq A_n[1 - q_{sen}^{av}(cycle\,n;\,z)]^{\alpha_n}$$
$$(A_n > 0;\; \alpha_n > 0;\; n = 2, 3, 5, ...). \tag{5.23}$$

The limit $q_{rel}(cycle\,n;\,z) = q_{sen}^{av}(cycle\,n;\,z) = 1$ corresponds to the BG case.

Finally, we verify (see Fig. 5.11) that

$$q_{sen}^{av}(cycle\,3;\,z) \simeq 2.5\,q_{sen}^{av}(cycle\,2;\,z) - 0.03, \tag{5.24}$$

$$q_{sen}^{av}(cycle\,5;\,z) \simeq 2.5\,q_{sen}^{av}(cycle\,2;\,z) + 0.03, \tag{5.25}$$

hence

$$q_{sen}^{av}(cycle\,5;\,z) - q_{sen}^{av}(cycle\,3;\,z) \simeq 0.06. \tag{5.26}$$

The full understanding of all these empirical relations remains an open problem.

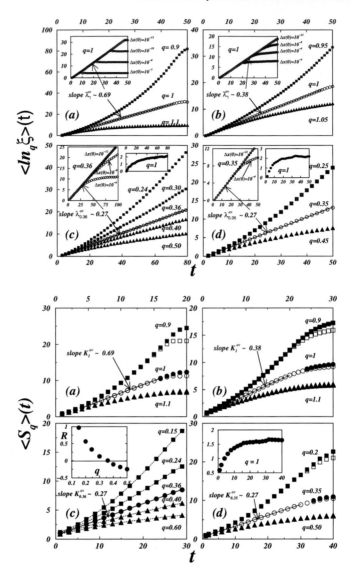

Fig. 5.5 Time dependence of $\langle \ln_q \xi \rangle$ and $\langle S_q \rangle$: $z = 2$ logistic map for strong [(**a**) $a = 2$] and weak [(**c**) $a = 1.401155189$] chaos, and $z = 0.5$ exponential map for strong [(**b**) $a = 4$] and weak [(**d**) $a = 3.32169594$] chaos. **Sensitivity function** $\langle \ln_q \xi \rangle(t)$: averages over 10^5 (10^7) runs for (**a**) and (**b**) ((**c**) and (**d**)); we use $\Delta x(0) = 10^{-12}$ as the initial discrepancy unless otherwise indicated; in the insets, we show the *linear* tendency of the sensitivity function for q_{sen}^{av} with various values of $\Delta x(0)$; at the edge of chaos ((**c**) and (**d**)) we exhibit the $q = 1$ curve *nonlinearity*. **Entropy** $\langle S_q \rangle(t)$: (**a**, **b**) 3000 runs with $N = 10W$ with $W = 10^5$ and $W = 3.10^5$ (empty and filled symbols respectively); 50000 runs with $N = 10W$ with $W = 10^5$ (for (**c**)) and $W = 5.10^4$ and $W = 10^5$ (for (**d**)). (**c**) inset: determination of q_{sen}^{av}. (**d**) inset: we exhibit the $q = 1$ curve *nonlinearity*. From [197]

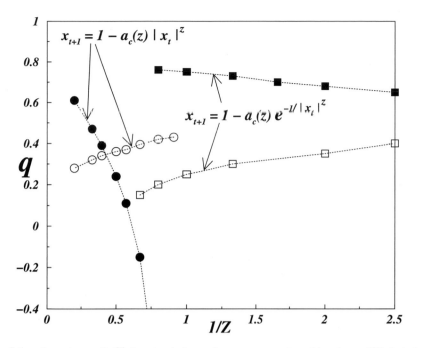

Fig. 5.6 z-dependence of q_{sen}^{av} (empty circles and squares: present work) and q_{sen} (filled circles: from [187, 188]; filled squares: from [197]). Dotted lines are guides to the eye

Fig. 5.7 The volume occupied by the ensemble as a function of discrete time. After a transient period, which is nearly the same for all N_{box} values, the power-law behavior (as well as a small log-periodic oscillation) becomes evident. For each case, a set of $10\,N_{box}$ copies of the system is followed. From [199]

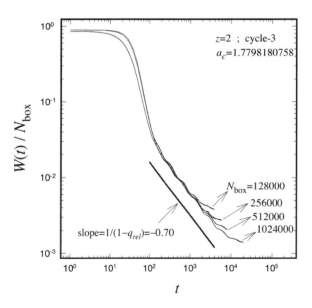

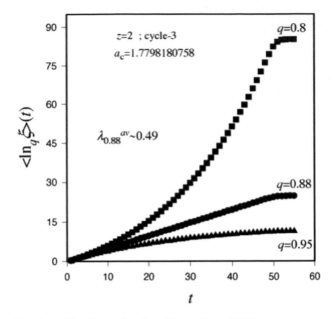

Fig. 5.8 The behavior of $\langle \ln_q \xi \rangle$ as a function of time. From [199]

Central Limit Theorem Attractor

Let us now focus on an important limiting property, directly related to the Central Limit Theorem (CLT). It is in fact a dynamical version of the CLT. As an example of unimodal one-dimensional map, let us consider the z-logistic one for values of the control parameter a such that the Lyapunov exponent λ_1 is *positive* (i.e., a strongly chaotic map)), and start from a given initial condition x_0. The successive N iterates x_1, x_2, x_3 ..., constitute a time series which associates, with each value of x_0, the uniquely defined sum of the first N terms. For fixed N, we may consider a large set of initial conditions uniformly distributed within the allowed phase space. The distribution of the sums, appropriately centered and scaled, approaches, for $N \to \infty$, a Gaussian [529]. See Figs. 5.12, 5.13 and 5.14 (from [530]).

The situation changes dramatically if we are at the edge of chaos, where $\lambda_1 = 0$ (i.e., a weakly chaotic map). The limiting distribution appears to be a q-Gaussian with $q = q_{stat} \simeq 1.65$ ($stat$ stands for *stationary state*; this qualification will become more transparent later on) [530–532]. See Fig. 5.15 as well as Fig. 3 of [532], where the approach of $(1/N, (a - a_c)^s)$, with $s \simeq 0.9$, to $(0, 0)$ is depicted.

Before proceeding, we may summarize the case of simple one-dimensional dissipative maps at the edge of chaos by reminding that we have strong numerical indications of the existence of a basic q-triplet. In particular, for the $z = 2$ logistic map we have $(q_{sen}, q_{rel}, q_{stat}) \simeq (0.2445, 2.2497, 1.65)$. Later on [479], we turn back onto q-triplets (as well as other values of q; see, for instance, [200, 306]).

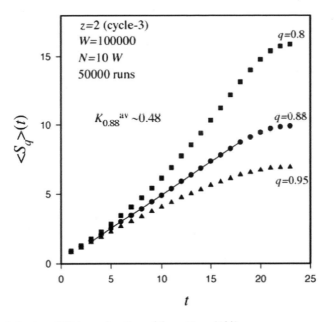

Fig. 5.9 The behavior of $\langle S_q \rangle$ as a function of time. From [199]

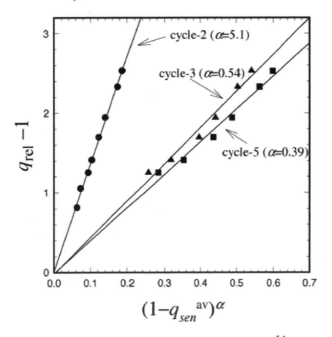

Fig. 5.10 Straight lines: $q_{rel}(cycle\,2) - 1 = 13.5\,[1 - q_{sen}^{av}(cycle\,2)]^{5.1}$, $q_{rel}(cycle\,3) - 1 = 4.6\,[1 - q_{sen}^{av}(cycle\,3)]^{0.54}$, and $q_{rel}(cycle\,5) - 1 = 4.1\,[1 - q_{sen}^{av}(cycle\,5)]^{0.39}$. The $q_{rel} = q_{sen} = 1$ corner corresponds to the Boltzmann-Gibbs limit. From [199]

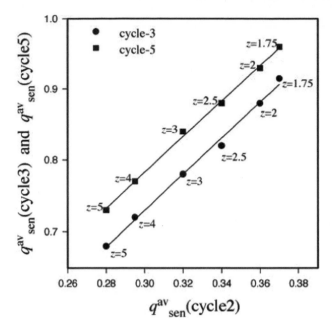

Fig. 5.11 Straight lines: $q_{sen}^{av}(cycle\,3) = 2.5\,q_{sen}^{av}(cycle\,2) - 0.03$, and $q_{sen}^{av}(cycle\,5) = 2.5\,q_{sen}^{av}(cycle\,2) + 0.03$, which suggests $q_{sen}^{av}(cycle\,5) - q_{sen}^{av}(cycle\,3) \simeq 0.06$. From [199]

Fig. 5.12 Probability
density of rescaled sums of
iterates of the logistic map
with $a = 2$; $N = 2 \cdot 10^6$ and
$N = 100$. The number of
initial values contributing to
the histogram is
$n_{ini} = 2 \cdot 10^6$, respectively
$n_{ini} = 10^7$. The solid lines
correspond to the analytical
expressions for finite N.
From [530]

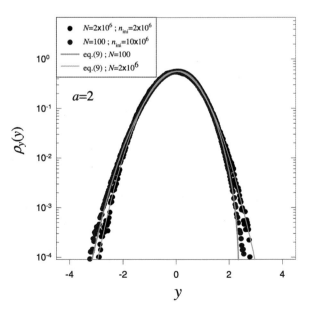

Fig. 5.13 Probability density of rescaled sums of iterates of the cubic map (which belongs to the same universality class of the logistic map) for $N = 10^7$ and $N = 10$. The number of initial values is $n_{ini} = 10^6$, respectively $n_{ini} = 5 \cdot 10^6$. The solid lines correspond the analytical expressions for finite N. From [530]

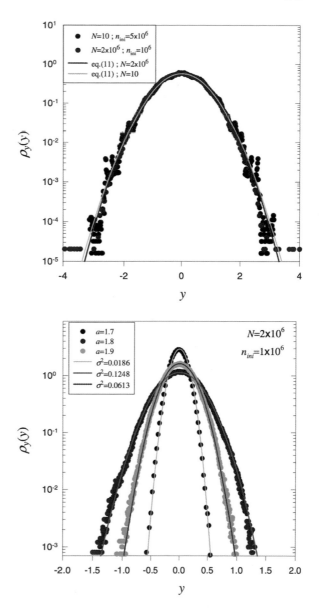

Fig. 5.14 Probability density of rescaled sums of iterates of the logistic map for $a = 1.7, 1.8, 1.9$ and $N = 2 \cdot 10^6, n_{ini} = 10^6$. The solid lines show Gaussians $e^{-y^2/(2\sigma^2)}/\sqrt{2\pi\sigma^2}$ with variance parameter σ^2 (i.e., $q_{stat} = 1$). From [530]

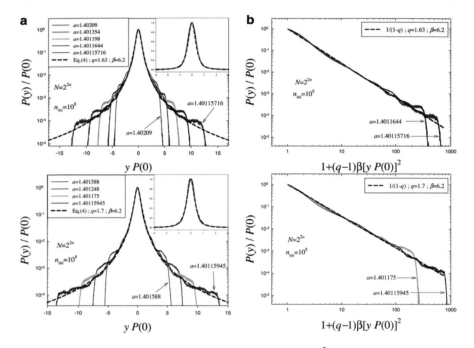

Fig. 5.15 Data collapse of probability density functions for $N = 2^{2n}$, where $2n$ is (a) odd and (b) even. As n increases, a good fit using a q_{stat}-Gaussian with (a) $(q_{stat}, \beta) = (1.63, 6.2)$, and (b) $(q_{stat}, \beta) = (1.70, 6.2)$ is obtained for increasingly large regions. *Left*: log-linear plots [*Inset*: The linear-linear plot of the data for a better visualization of the central part]. *Right*: log-log plots of the same data, having as abscissa $\{1 + (q_{stat} - 1)\beta[yP(0)]^2\}$. A straight line is expected with a slope $1/(1 - q_{stat})$ if the curve is a q_{stat}-Gaussian. It is clearly seen that the straight line is surrounded by log-periodically modulated curves. From [531]

Horizontal Visibility

The edge of chaos for the z-logistic map (see Eq. (5.1)) has also been studied by using the horizontal visibility algorithm [533]. It has been obtained, for the so called Feigenbaum graph, $q_{sen} = 1 - \frac{\ln 2}{2}$, and an upper bound for λ_q at $\lambda_{q_{sen}} = \frac{1}{1 - q_{sen}} = \frac{2}{\ln 2}$. The validity of the corresponding q-generalized Pesin identity has been verified as well.

The result $q_{sen} = 1 - \frac{\ln 2}{2}$ is derived from topological arguments, and therefore does not depend on z. This result is to be distinguished from the z-dependent value $q_{sen}(z)$ given in Eq. (5.12), which naturally includes metric issues as well.

5.1.2 Two-Dimensional Dissipative Maps

Although not with the same detail as for the one-dimensional ones, some two-dimensional dissipative maps have been studied as well [534–536]. More specifically, the Henon and the Lozi maps. Let us illustrate with the Henon map. It is defined as follows:

$$x_{t+1} = 1 - ax_t^2 + y_t \qquad\qquad (5.27)$$

$$y_{t+1} = bx_t. \qquad\qquad (5.28)$$

The $b = 0$ particular case corresponds precisely to the $z = 2$ logistic map. For $b \geq 0$, a line (an infinite number of them, in fact) exists in the (a, b) space on which the system is at the edge of chaos, with vanishing Lyapunov exponents. It is since long well known (see [537] and references therein) that two universality classes exist along this line, namely the dissipative (logistic map) universality class for any $b \neq 1$, and the conservative universality class for $b = 1$. One consistently expects that the values of q should be consistent with the same two classes. In particular, the value of q_{sen} for $0 < b < 1$ should be the same as that for $b = 0$, i.e., $q_{sen} = 0.2445....$ Indeed, it is precisely this what, within some numerical precision, has been verified in [534–536].

5.1.3 High-Dimensional Dissipative Maps

Let us discuss here many coupled dissipative maps [538]. We focus on the following model of a linear chain of N coupled logistic maps

$$x_{t+1}^{(i)} = (1 - \epsilon)f\left(x_t^{(i)}\right) + \frac{\epsilon}{2}\left[f_t^{(i-1)} + f_t^{(i+1)}\right] + \sigma(t) \qquad\qquad (5.29)$$

where $\epsilon \in [0, 1]$ is the strength of the local coupling of each map with its first-neighboring sites on the chain with periodic boundary conditions (i.e., on a circle), and the additive noise $\sigma(t)$ is a random variable, fluctuating in time but equal for all the maps, uniformly extracted from an interval whose width is σ_{max}. In our case, the ith logistic map at time t is in the form $f\left(x_t^{(i)}\right) = 1 - \mu\left(x_t^{(i)}\right)^2$, with $\mu \in [0, 2]$ and $-1 \leq x_t^{(i)} \leq 1$ (whenever the noise takes $x_t^{(i)}$ out from this interval it is refolded inside through module 1). The quantity $d_t = \frac{1}{N}\sum_{i=1}^{N}|x_t^{(i)} - \langle x_t^{(i)}\rangle|$ measures the distance from the synchronization regime at time t. If all maps are trapped in some synchronized pattern then this quantity remains close to zero, otherwise oscillations are found. As commonly used in turbulence or in finance, we analyze these oscillations by considering the two-time returns Δd_t with an interval of τ time steps, defined as $\Delta d_t = d_{t+\tau} - d_t$. The results are presented in Figs. 5.16 and 5.17.

The interoccurrence times τ_{inter} for a fixed threshold $Q > 0$ have been numerically studied as well [538] and, for the illustrative case $(N, \epsilon, \sigma_{max}, \tau) =$

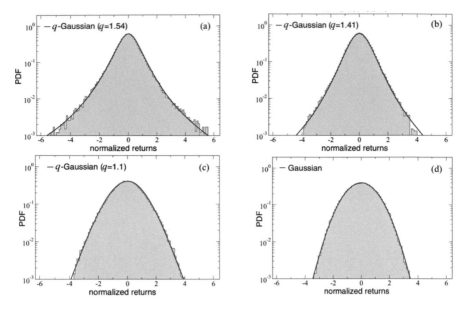

Fig. 5.16 Asymptotic pdfs of the returns for $N = 100$ maps at the edge of chaos (i.e., $\mu_c = 1.4011551\ldots$), with $\epsilon = 0.8$, $\tau = 32$ (after 100000 iterations) and for increasing values of the noise: **a** $\sigma_{max} = 0.002$, **b** $\sigma_{max} = 0.01$, **c** $\sigma_{max} = 0.05$, and (**d**) $\sigma_{max} = 0.3$. Fat tails are more evident for weak noise and tend to diminish by increasing noise. Full curves correspond to q-Gaussian fits with values $q = 1.54$, $q = 1.41$, $q = 1.1$, and $q = 1$, respectively. Returns are also normalized to the standard deviation to have a pdf with unit variance. From [538]

$(100, 0.8, 0.002, 32)$, their distribution appears to be given by the q-exponential form $p(\tau_{inter}) \propto e_{q_{inter}}^{-\tau_{inter}/\tau_{q_{inter}}}$ with $q_{inter} \simeq 1 + 0.204Q$.

5.2 Conservative Maps

We remind that a d-dimensional map has d Lyapunov exponents $\lambda_1^{(1)}$, $\lambda_1^{(2)}$, ..., $\lambda_1^{(d)}$ ([112] and references therein). If it is *conservative*, it satisfies

$$\sum_{i=1}^{d} \lambda_1^{(i)} = 0. \tag{5.30}$$

If, in addition to that, it is symplectic, d is an even integer, and we can therefore conveniently define $d = 2N$ ($N = 1, 2, \ldots$). Furthermore, the Lyapunov exponents are in pairs which differ only in the sign. Obviously, two-dimensional conservative maps are necessarily symplectic.

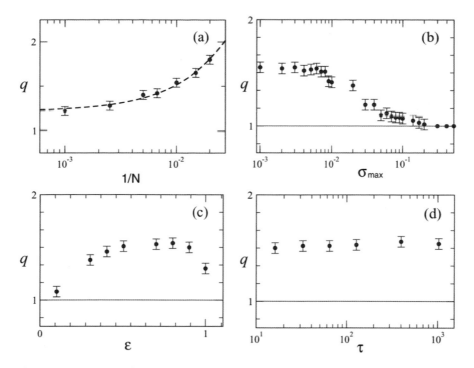

Fig. 5.17 Dependence of q in Fig. 5.16 on **a** the size of the system N, **b** the level of noise σ_{max}, **c** the coupling ϵ, and **d** the returns' interval τ. All maps are at the edge of chaos (where the system has a multifractal structure which is consistent with the τ independence of q). From [538]

Entropic properties in low-dimensional maps have already been addressed for $d = 2$ ([113, 333, 539–541], among others) and $d = 4$ ([539, 540], among others). The review of some of their peculiarities will pave the understanding of many-body Hamiltonian systems, the primary object of study in statistical mechanics.

5.2.1 Strongly Chaotic Two-Dimensional Conservative Maps

In order to illustrate relevant properties, we shall focus on three paradigmatic (strongly chaotic, area-preserving, and transforming the unit square into itself), two-dimensional conservative maps, first the so called *baker map*, second the *generalized cat map* [113], and third the *standard map* [113, 513, 539].

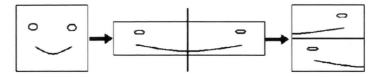

Fig. 5.18 The nondissipative baker map. From [541]

Baker map

The *baker map* is defined as follows [541]:

$$(x_{t+1}, y_{t+1}) = \begin{cases} (2x_t, y_t/2) & (0 \le x_t < 1/2) \\ (2x_t - 1, (y_t + 1)/2) & (1/2 \le x_t \le 1). \end{cases} \quad (5.31)$$

We verify that $|\partial(x_{t+1}, y_{t+1})/\partial(x_t, y_t)| = 1$, and $\lambda_1^{(1)} = -\lambda_1^{(2)} = \ln 2$. See Fig. 5.18. The time dependence of the entropy $S_q(t)$ is depicted in Figs. 5.19 and 5.20.

Generalized cat map

The *generalized cat map* is defined as follows [113]

$$\begin{aligned} p_{t+1} &= p_t + k\,x_t & (mod\ 1) \\ x_{t+1} &= p_t + (1+k)\,x_t & (mod\ 1) \quad (k \ge 0). \end{aligned} \quad (5.32)$$

We verify that $|\partial(p_{t+1}, x_{t+1})/\partial(p_t, x_t)| = 1$, and $\lambda_1^{(1)} = -\lambda_1^{(2)} = \ln \frac{2+k+\sqrt{k^2+4k}}{2}$. For typical values of k, it has been numerically verified in [113] that $\lim_{t\to\infty} \lim_{W\to\infty} \lim_{M\to\infty} S_{BG}(t)/t = \lambda_1^{(1)}$. An analytic proof would naturally be very welcome.

Standard map

The *standard map* (or *kicked rotor map*) has already been defined in Eq. (4.169) but, to accommodate with other authors, it can equivalently[5] be defined as follows [113, 539]:

$$\begin{aligned} p_{t+1} &= p_t + \frac{a}{2\pi}\sin(2\pi\theta_t) & (mod\ 1) \\ \theta_{t+1} &= \theta_t + p_{t+1} & (mod\ 1) \quad (a \ge 0). \end{aligned} \quad (5.33)$$

---

[5] By doing $(p, \theta, a) \to (p/2\pi, \theta/2\pi, -K)$ in Eqs. (5.33), we precisely recover Eqs. (4.169).

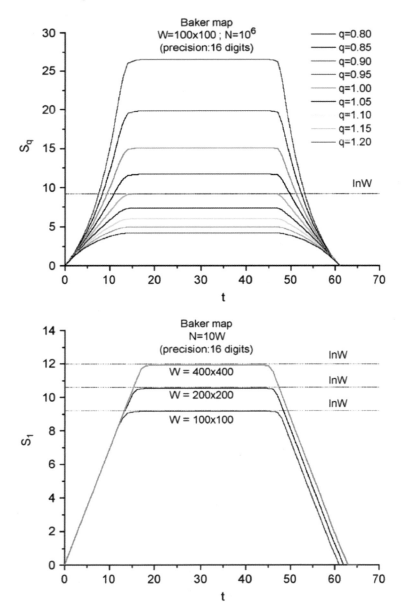

Fig. 5.19 Time evolution of the q-entropy for the non-dissipative baker map, using 16 digit calculations. *Top*: $S_q(t)$ for entropic indices $q = 0.80, 0.85, 0.90, 0.95, 1.00, 1.05, 1.10, 1.15, 1.20$ (from top to bottom) when $W = 10^4$ and $N = 10^6$. *Bottom*: $S_1(t)$ for typical values of W ($W = 10^4$, 4×10^4, 16×10^4; $N = 10W$). Notice that the bounding value for the $q_{ent} = 1$ entropy corresponds, in all cases, to equiprobability, i.e., $\ln W$. The slope $dS_1(t)/dt$ (on the left side) recovers the well known values for the Lyapunov exponents $\lambda_1^{(1)} = -\lambda_1^{(2)} = \ln 2$. From [541]

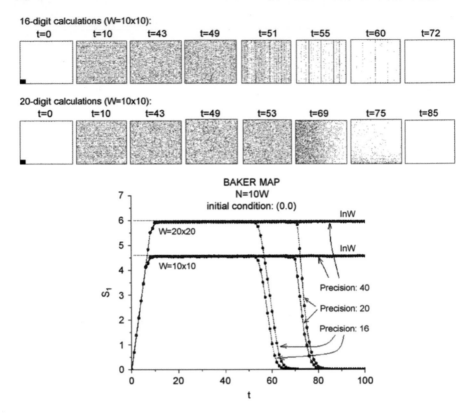

Fig. 5.20 Numerical study of the baker map, with controlled fixed precision. The sequences of the top figure exhibit the evolution in phase space with a fixed precision. The corresponding curves for $S_1(t)$ are shown in the bottom figure. The evolution of $S_1(t)$ corresponding to a higher fixed precision experiment (40 digits) is shown as well; the time reversal of the entropy is not observed before $t = 100$. From [541]

This map is only partially chaotic (i.e., the size of the "chaotic sea" is smaller than the unit square), and the percentage of chaos tends to increase for increasing a. Inside the chaotic sea, it has been numerically verified [113] that $\lim_{t \to \infty} \lim_{W \to \infty} \lim_{M \to \infty} S_{BG}(t)/t = \lambda_1^{(1)}$ equals 0.98, 1.62 and 2.30 for $a = 5$, 10 and 20 respectively. It is instructive to define [539] a *dynamical "temperature"* T as the variance of the angular momentum, i.e., $T \equiv \langle (p - \langle p \rangle)^2 \rangle = \langle p^2 \rangle - \langle p \rangle^2$, where $\langle \rangle$ denotes ensemble average. The temperature associated with the uniform ensemble (that we will call *BG temperature* T_{BG} because of its similarity with the equal-a-priori-probability postulate) is given by $T_{BG} = \int_0^1 dp\, p^2 - (\int_0^1 dp\, p^2) = 1/12$. Notice that, in the present conservative model, the "temperature" T is necessarily bounded since p itself is bounded, in contrast with a true thermodynamical temperature, which is of course unbounded. The time evolution of the system and of T, for typical values of a are depicted in Fig. 5.21.

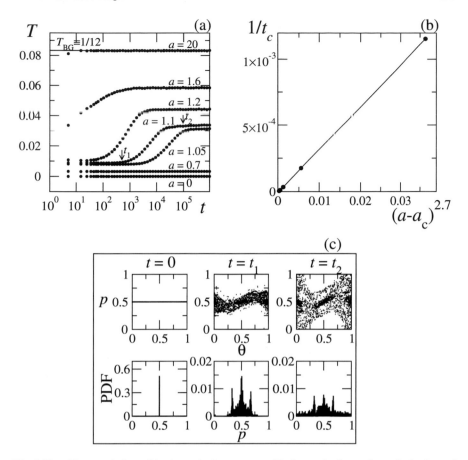

Fig. 5.21 **a** Time evolution of the dynamical temperature T of a standard map, for typical values of a. We start with 'water bag' initial conditions ($M = 2500$ points in $0 \leq \theta \leq 1$, $p = 0.5 \pm 5\ 10^{-4}$). In order to eliminate cyclical fluctuations, the dots represent average of 10 iteration steps; moreover, each curve is the average of 50 realizations. **b** Inverse crossover time t_c (inflection point between the QSS and the BG-like regimes) versus $1/(a - a_c)^{2.7}$. The critical value $a_c = 0.971635406\ldots$ is defined as the value at which the strongest cantori close and the relaxation to a higher temperature is prevented. No inflection points subsist if t is linearly represented. **c** Time evolution of the ensemble in (**a**) for $a = 1.1$ (first row) and PDF of its angular momentum (second row). $t = 0$: 'water bag' initial conditions; $t = t_1 = 500$: the ensemble is mostly restricted by cantori; $t = t_2 = 10^5$: the ensemble is confined inside KAM-tori. From [539]

5.2.2 Weakly Chaotic Two-Dimensional Conservative Maps

In the previous Subsections we have analyzed low-dimensional systems that are strongly chaotic. We shall dedicate the present Subsection to weakly chaotic two-dimensional conservative systems. We may list among those the *Casati-Prosen map* (or *triangle map*) [12, 542, 543], the *Macmillan map* [544], the Chirikov standard map [513, 515–517] (focused on also in Sects. 4.10.2 and 5.2.1), the *Web* and *Harper* maps [545, 546], and the *Moore map* [541, 547–549]. We briefly focus on them here.

Casati-Prosen map

We follow here [333]. The *Casati-Prosen map* $z_{n+1} = T(z_n)$ is defined on a torus $z = (x, y) \in [-1, 1) \times [-1, 1)$

$$
\begin{aligned}
y_{n+1} &= y_n + \alpha\, \mathrm{sgn} x_n + \beta \quad (\mathrm{mod}\ 2), \\
x_{n+1} &= x_n + y_{n+1} \quad (\mathrm{mod}\ 2),
\end{aligned}
\tag{5.34}
$$

where $\mathrm{sgn} x = \pm 1$ is the sign of x, and α, β are two parameters ($n = 0, 1. \ldots$). This map is linearly unstable. For rational values of (α, β) the system is in principle integrable, as the dynamics is confined on invariant curves. If $\beta = 0$ and α is irrational, the dynamics is ergodic but the phase-space is filled very slowly, while for incommensurate irrational values of α, β the dynamics is ergodic and mixing with dynamical correlation function decaying as $t^{-3/2}$ (i.e., $q_{rel} = 5/3$, according to a notation that will be discussed later on). This map does not have any secondary time scales, and the exploration of the phase phase by a given orbit is arbitrarily close to that of a random model.

For the sake of definiteness, in the following we will fix (see [333]) the parameter values $\alpha = [\frac{1}{2}(\sqrt{5} - 1) - e^{-1}]/2$, $\beta = [\frac{1}{2}(\sqrt{5} - 1) + e^{-1}]/2$ although it should be noticed that qualitatively identical results are obtained for other irrational parameter values. Figure 5.22 shows the mixing process of an ensemble of points initially localized inside a small square. The action of the map (5.34) initially divides the area covered by the ensemble into different unconnected portions, each essentially stretched along a straight line. After a certain amount of time these portions overlap until a slow relaxation process to a complete mixing is observed. We can verify in Fig. 5.23 that $q_{ent} = 0$. The linear time-dependence [12, 542] of the sensitivity ξ implies $q_{sen} = 0$, which, as usual, coincides with q_{ent}. Furthermore, we can verify that the q-generalized Pesin-like identity is once again satisfied.

In conclusion, while positivity of Lyapunov exponents is *sufficient* for a meaningful statistical description, basically the BG statistical mechanics, it appears be *not necessary*. Indeed, we have illustrated, for a conservative, mixing and ergodic nonlinear dynamical system, that the use of the more general entropy S_q (with the value $q = 0$ for this case) provides a satisfactory frame for handling nonlinear dynamical systems whose maximal Lyapunov exponent vanishes. In particular, we have shown

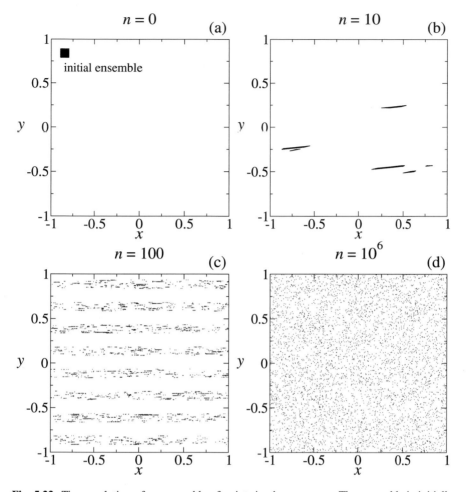

Fig. 5.22 Time evolution of an ensemble of points in phase space. **a** The ensemble is initially located inside a single cell. **b–d** Phase space distribution after $n = 10, 10^2, 10^6$ map iterations. From [333]

that (the upper bound of) the coefficient λ_q of the sensitivity to the initial conditions *coincides* with the entropy production per unit time, in total analogy with the Pesin theorem for standard chaotic systems. These results suggest that a thermostatistical approach of such systems is possible. Indeed, the structure that we have exhibited here for the time dependence of S_q, is totally analogous to the one that has been recently exhibited [274] for the N-dependence of S_q at the stationary state, where N is the number of elements of a many-body system. When the number of nonzero-probability states of the system increases as a power of N (instead of exponentially with N as usually), a special value of q below unity might exist such that S_q is

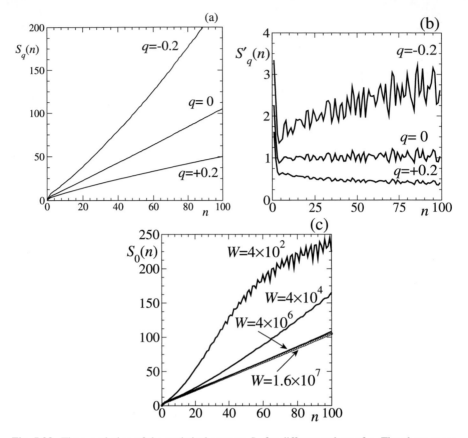

Fig. 5.23 Time-evolution of the statistical entropy S_q for different values of q. The phase space has been divided into $W = 4000 \times 4000$ equal cells of size $l = 5 \times 10^{-4}$ and the initial ensemble is characterized by $N = 10^3$ points randomly distributed inside a partition-square. Curves are the result of an average over 100 different initial squares randomly chosen in phase space. The analysis of the derivative of S_q in **b** shows that only for $q = 0$ a linear behaviour is obtained. In fact, a linear regression provides $S_0(n) = 1.029\, n + 1.997$ with a correlation coefficient $R = 0.99993$. **c** Shows that the linear growth for S_0 is reached from above, in the limit $W \to \infty$. From [333]

extensive. In other words, S_q asymptotically increases *linearly* with N, whereas S_{BG} does not.

Macmillan map

We follow here [544]. The conservative MacMillan map is defined as follows (see [544] and references therein):

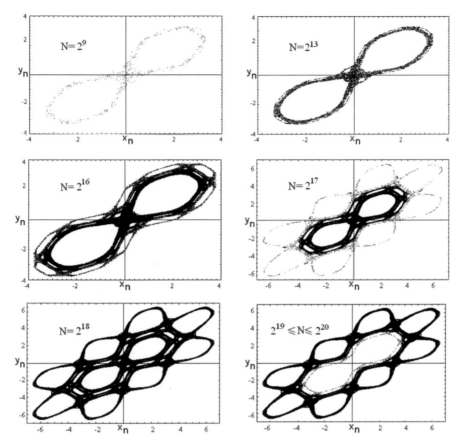

Fig. 5.24 Structure of phase space plots of the MacMillan map for parameter values $(\epsilon, \mu) =$ (1.2, 1.6), starting from a randomly chosen initial condition in the square $(0, 10^{-6}) \times (0, 10^{-6})$, and for N iterates. From [544]

$$
\begin{aligned}
x_{n+1} &= y_n \\
y_{n+1} &= -x_n + 2\mu \frac{y_n}{1 + y_n^2} + \epsilon\, y_n,
\end{aligned} \tag{5.35}
$$

where $(\epsilon, \mu) \in \mathcal{R}^2$. We illustrate its dynamical behaviour for $(\epsilon, \mu) = (1.2, 1.6)$, and the corresponding distributions when summing up $N \gg 1$ successive iterations of x_n: see Figs. 5.24 and 5.25.

Web map

We follow here [545]. The so-called conservative web map of the Q-fold symmetry is generated as follows (see [545] and references therein)

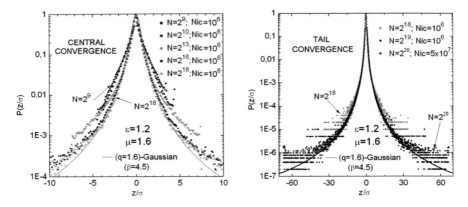

Fig. 5.25 Probability distributions of the rescaled sums of iterates of the MacMillan map for $(\epsilon, \mu) = (1.2, 1.6)$ appear to converge to converge to a q-Gaussian with $q \simeq 1.6$ during a long-lived quasi-stationary state. This is shown for the central part of the pdf (for $N < 2^{18}$) (*left*), and for the tails ($N > 2^{18}$) (*right*). N_{ic} is the number of initial conditions that have been randomly chosen from the square $(0, 10^{-6}) \times (0, 10^{-6})$. From [544]

$$u_{n+1} = (u_n + K \sin v_n) \cos \alpha + v_n \sin \alpha \qquad (mod\ 2\pi)$$
$$v_{n+1} = -(u_n + K \sin v_n) \sin \alpha + v_n \cos \alpha \qquad (mod\ 2\pi) \qquad (5.36)$$

with $\alpha = 2\pi/Q$, Q being an integer, and $K \in R$. Consequently, the 4-fold particular case corresponds to the following simple map

$$u_{n+1} = v_n \qquad\qquad (mod\ 2\pi)$$
$$v_{n+1} = -u_n - K \sin v_n \qquad (mod\ 2\pi) \qquad (5.37)$$

By summing a large number of successive iterates we obtain, after appropriate centring and scaling, a Gaussian for $K = 5$ (strong chaos) and a q-Gaussian with $q \simeq 1.935$ for $K = 0.1$ (weak chaos), as respectively shown in Figs. 5.26 and 5.27. The value $q \simeq 1.935$ numerically obtained here coincides with that obtained for the standard map with low values of K. Since the analytical result for that case is now known to be $q = 2$ [516], it is allowed to suspect that possibly the same happens for the above web map.

Further illustrations of this type of contrasting behaviors (strong versus weak chaos) can be found in [546], where the Harper and generalized standard maps are studied.

Moore map

The *Moore map* is a paradigmatic one belonging to the generalized shift family of maps proposed by Moore [547]. This class of dynamical systems posses some sort

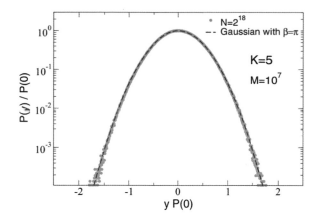

Fig. 5.26 The probability distribution function of the sum of the iterates for the $K = 5$ web map (green dots) neatly fits the Gaussian (dashed line), $P(y) = P(0)e^{-\beta y^2}$, where $\beta = \pi$; N and M are the numbers of iterates and of initial conditions respectively. From [545]

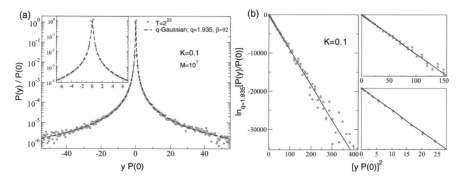

Fig. 5.27 a The $K = 0.1$ probability distribution function $P(y)$ (green dots) is well fitted with a q-Gaussian (dashed line), namely $P(y) = P(0) e_q^{-\beta y^2}$ with $q = 1.935$ and $\beta = 92$; $N = 2^{23}$ and $M = 10^7$ are the numbers of iterates and of initial conditions respectively. **b** q-logarithmic representation of the same distribution for the tail, intermediate, and central regions. From [545]

of undecidability, as compared with other low—dimensional chaotic systems [547, 548]. It is equivalent to the piecewise linear map shown in Fig. 5.28 . When this map is recurrently applied, the area in phase space is conserved, while the corresponding shape keeps changing in time, becoming increasingly complicated. This map appears to be ergodic, possibly exhibits a Lyapunov exponent $\lambda_1 = 0$, and, presumably, the divergence of close initial conditions follows a power-law behavior [550]. When we consider a partition of W equal cells and select N random initial conditions inside one random cell, the points spread much slower than they do on the baker map. More precisely, they spread, through a slow relaxation process, all over the phase space, each orbit appearing to gradually fill up the entire square. See Figs. 5.28 and 5.29 from [541].

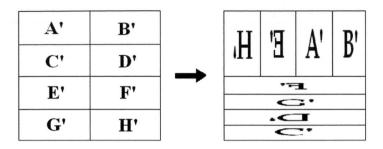

Fig. 5.28 The Moore map. Alphabetic symbols are written on the cells to show how local dynamics evolves. From [541]

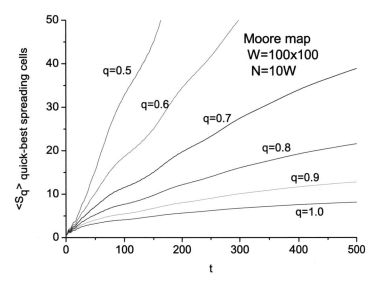

Fig. 5.29 Time dependence of S_q averaged over the 10% quick-best spreading cells, on the far-from-equilibrium regime, for typical values of q. From [541]

In order to have a *finite* entropy production for S_q we need a value of q which is definitively smaller than unity, i.e., the Boltzmann-Gibbs entropy does not appear as the most adequate tool. Deeper studies are needed in order to establish whether another value of q can solve this problem.

5.2.3 Strongly Chaotic Four-Dimensional Conservative Maps

In the previous subsection we considered $N = 1$ particle. Let us consider here $N = 2$, on the road to the thermodynamic limit $N \to \infty$ [113, 539, 540]. We shall focus on a simple symplectic system of two coupled standard maps, defined as follows:

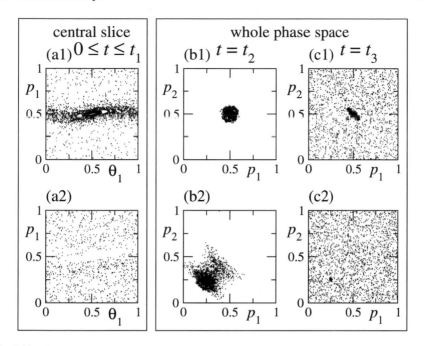

Fig. 5.30 Phase space analysis of the evolution of 'water bag' ensembles for two coupled standard maps for $(\tilde{a}, b) = (0.4, 2)$. First row: 'Water bag' initial conditions $0 \le \theta_1, \theta_2 \le 1$, $p_1, p_2 = 0.5 \pm 5 \, 10^{-3}$. Second row: 'Water bag' initial conditions $0 \le \theta_1, \theta_2 \le 1$, $p_1, p_2 = 0.25 \pm 5 \, 10^{-3}$. **a** Projection on the (θ_1, p_1)-plane of the central slice of the phase space $(\theta_2, p_2 = 0.5 \pm 10^{-2})$, for the orbit $0 \le t \le t_1 = 10^4$. **b, c** Projection on the (p_1, p_2)-plane of whole phase space for the iterate at time $t_2 = 15$ and $t_3 = 2 \, 10^4$. From [539]

$$\theta_1(t+1) = p_1(t+1) + \theta_1(t) + b \, p_2(t+1),$$
$$p_1(t+1) = p_1(t) + \frac{a_1}{2\pi} \sin[2\pi\theta_1(t)], \qquad\qquad (5.38)$$
$$\theta_2(t+1) = p_2(t+1) + \theta_2(t) + b \, p_1(t+1),$$
$$p_2(t+1) = p_2(t) + \frac{a_2}{2\pi} \sin[2\pi\theta_2(t)],$$

where $a_1, a_2, b \in \mathbb{R}$, $t = 0, 1, \ldots$, and all variables are defined mod 1. If the coupling constant b vanishes the two standard maps decouple; if $b = 2$ the points $(0, 1/2, 0, 1/2)$ and $(1/2, 1/2, 1/2, 1/2)$ are a 2-cycle for all (a_1, a_2), hence we preserve in phase space the same referential that we had for a single standard map. For a generic value of b, all relevant present results remain qualitatively the same. Also, we set $a_1 = a_2 \equiv \tilde{a}$ so that the system is invariant under permutation $1 \leftrightarrow 2$. Since we have two rotors now, the *dynamical temperature* is naturally given by $T \equiv \frac{1}{2} \left(< p_1^2 > + < p_2^2 > - < p_1 >^2 - < p_2 >^2 \right)$, hence the BG temperature remains $T_{BG} = 1/12$. The time evolution of the system is depicted in Figs. 5.30, 5.31 and 5.32.

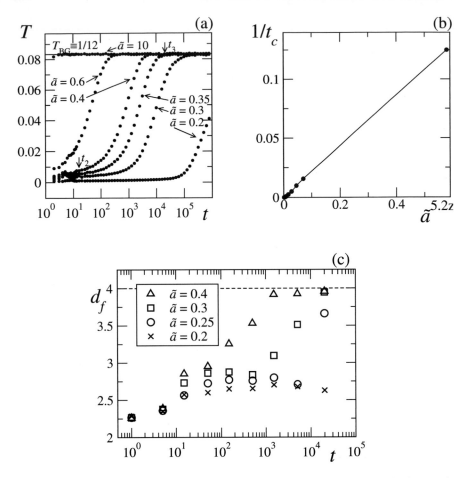

Fig. 5.31 a Time evolution of the dynamical temperature T of two coupled standard maps, for $b = 2$ and typical values of $\tilde{a}$. We start with water bag initial conditions ($M = 1296$ points with $0 \leq \theta_1, \theta_2 \leq 1$, and $p_1, p_2 = 0.25 \pm 5 \; 10^{-3}$); moreover, an average was taken over 35 realizations. See Fig. 5.30 for t_2 and t_3. **b** Inverse crossover time t_c versus $1/\tilde{a}^{5.2}$. **c** Time evolution of the fractal dimension of a single initial ensemble in the same setup of (**a**). From [539]

5.2.4 High-Dimensional Conservative Maps

Let us discuss now conservative systems of many coupled maps. The model we focus on here [551] is a set of N symplectically coupled (hence conservative) standard maps, where the coupling is made through the *coordinates* as follows:

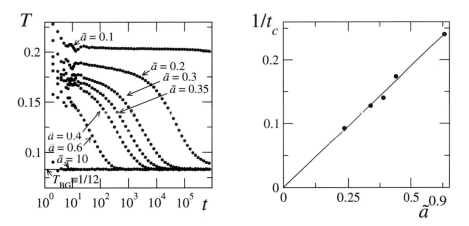

Fig. 5.32 Same as Fig. 5.31a, b but with 'double water bag' initial conditions: $0 \leq \theta_1, \theta_2 \leq 1$; p_1, p_2 randomly distributed inside one of the two regions $p_1, p_2 = 0 + 10^{-2}$, $p_1, p_2 = 1 - 10^{-2}$. From [539]

$$\theta_i(t+1) = \theta_i(t) + p_i(t+1) \qquad \text{(mod 1)},$$

$$p_i(t+1) = p_i(t) + \frac{a}{2\pi} \sin[2\pi\theta_i(t)]$$

$$\qquad\qquad + \frac{b}{2\pi\tilde{N}} \sum_{\substack{j=1 \\ j\neq i}}^{N} \frac{\sin[2\pi(\theta_i(t) - \theta_j(t))]}{r_{ij}^{\alpha}} \qquad \text{(mod 1)},$$

$$(5.39)$$

where $t = 1, 2, \ldots$, and $\alpha \geq 0$. The a parameter is the usual nonlinear constant of the single standard map, whereas the b parameter modulates the overall strength of the coupling. Both parameters contribute to the nonlinearity of the system, which becomes integrable when $a = b = 0$. For simplicity, we have studied only the cases $a > 0$, $b > 0$, but we expect similar results when one of these parameters is negative; if both are negative, the dynamics is isomorphic through the transformation $(\theta_i, p_i) \rightarrow (-\theta_i, -p_i)$. The systematic study of the whole parameter space is certainly welcome. Notice that, in order to describe a system whose phase space is bounded, we are considering, as usual, only the torus (mod 1). Additionally, the maps are placed in a one dimensional ($d = 1$) regular lattice with periodic boundary conditions. The distance r_{ij} is the minimum distance between maps i and j, hence it can take values from unity (with a conventional unit) to $\frac{N}{2}$ ($\frac{N-1}{2}$) for even (odd) number N of maps. Note that $b\, r_{ij}^{-\alpha}$ is a fixed quantity that, modulated through the power α, enters Eq. (5.39) as an effective time-independent coupling constant between sites i and j. In other words, α regulates the range of the interaction between maps. The sum is global (i.e., it includes every pair of sites), so the limiting cases $\alpha = 0$ and $\alpha = \infty$ correspond respectively to infinitely long range and nearest neighbours. In our case $d=1$, thus $0 \leq \alpha \leq 1$ ($\alpha > 1$) means *long-range* (*short-range*) coupling. Moreover, the coupling term is normalised by the sum [219, 552] $\tilde{N} \equiv d \int_1^{N^{1/d}} dr \, r^{d-1} \, r^{-\alpha} \to \frac{N^{1-\alpha/d}-\alpha/d}{1-\alpha/d}$, to yield a non-diverging quantity as the system size grows (for simplicity, we have replaced here $(d = 1)$ the discrete sum $\sum_{r=1}^{N} r^{-\alpha}$ over integer r by its continuous approximation).

If $G(\bar{x})$ denotes a map system, then G is symplectic when its Jacobian $\partial G / \partial \bar{x}$ satisfies the relation [110]:

$$\left(\frac{\partial G}{\partial \bar{x}}\right)^T J \left(\frac{\partial G}{\partial \bar{x}}\right) = J, \tag{5.40}$$

where the superindex T indicates the transposed matrix, and J is the Poisson matrix, defined by

$$J \equiv \begin{pmatrix} 0 & I \\ -I & 0 \end{pmatrix}, \tag{5.41}$$

I being the $N \times N$ identity matrix. A consequence of Eq. (5.40) is that the Jacobian determinant $|\partial G / \partial \bar{x}| = 1$, indicating that the map G is *(hyper)volume-preserving*. In particular, for our model

$$\frac{\partial G}{\partial \bar{x}} = \begin{pmatrix} I & I \\ B & (I + B) \end{pmatrix}, \tag{5.42}$$

where $\bar{x}$ is the $2N$-dimensional vector $\bar{x} \equiv (\bar{p}, \bar{\theta})$, and

$$B = \begin{pmatrix} K_{\theta_1} & c_{21} & \dots & c_{N1} \\ c_{12} & K_{\theta_2} & \dots & c_{N2} \\ \vdots & \vdots & \vdots & \vdots \\ c_{1N} & c_{2N} & \dots & K_{\theta_N} \end{pmatrix}, \tag{5.43}$$

with

$$K_{\theta_i} \equiv a \cos[2\pi\theta_i(t)] + \frac{b}{\tilde{N}} \sum_{j \neq i} \frac{cos[2\pi(\theta_i(t) - \theta_j(t))]}{r_{ij}^{\alpha}}, \tag{5.44}$$

and

$$c_{ij} = c_{ji} \equiv -\frac{b}{\tilde{N}} \frac{cos[2\pi(\theta_i(t) - \theta_j(t))]}{r_{ij}^{\alpha}}, \tag{5.45}$$

where $i, j = 1, \dots, N$. It can be seen that,

$$\left(\frac{\partial G}{\partial \bar{x}}\right)^T = \begin{pmatrix} I & B \\ I & (I + B) \end{pmatrix}, \tag{5.46}$$

hence

$$\left(\frac{\partial G}{\partial \bar{x}}\right)^T J = \begin{pmatrix} -B & I \\ -(I+B) & I \end{pmatrix}. \tag{5.47}$$

This quantity, multiplied (from the right side) by the matrix (5.42) yields J. Therefore our system is indeed symplectic. Consequently, the $2N$ Lyapunov exponents $\lambda_1 \equiv \lambda_M, \lambda_2, \lambda_3, ..., \lambda_{2N}$ are coupled two by two as follows: $\lambda_1 = -\lambda_{2N} \geq \lambda_2 = -\lambda_{2N-1} \geq ... \geq \lambda_N = -\lambda_{N+1} \geq 0$. In other words, as a function of time, an infinitely small *length* typically diverges as $e^{\lambda_1 t}$, an infinitely small *area* diverges as $e^{(\lambda_1+\lambda_2)t}$, an infinitely small *volume* diverges as $e^{(\lambda_1+\lambda_2+\lambda_3)t}$, an infinitely small N-dimensional *hypervolume* diverges as $e^{(\sum_{i=1}^N \lambda_i)t}$ ($\sum_{i=1}^N \lambda_i$ being in fact equal to the Kolmogorov-Sinai entropy rate, in agreement with the Pesin identity), an infinitely small $(N+1)$-hypervolume diverges as $e^{(\sum_{i=1}^{N-1} \lambda_i)t}$, and so on. For example, a $(2N-1)$-hypervolume diverges as $e^{\lambda_1 t}$, and finally a $2N$-hypervolume remains constant, thus recovering the conservative nature of the system (of course, this corresponds to the Liouville theorem in classical Hamiltonian dynamics).

Typical results are depicted in Figs. 5.33, 5.34, 5.35, 5.36 and 5.37.

5.3 Many-Body Long-Range-Interacting Hamiltonian Systems

In this Section we focus on a central question, namely inertial many-body Hamiltonian systems with interactions that can have a long-range character (i.e., $0 \leq \alpha/d \leq 1$ for classical systems). To isolate the role of the range of the interaction from any other influence, we shall consider interactions which present no particular difficulty at the origin (consequently, Newtonian gravitation is excluded since it has a divergent attraction at the origin). More precisely, either we shall assume that the elements of the system (e.g., classical rotors) are localized on a lattice, and the long-range manifests itself through a slowly decaying coupling constant, or the elements of the system (e.g., point atoms of a gas) are free to move translationally but then a short-distance strong repulsion (such as the $1/r^{12}$ potential term in the Lennard-Jones model for a real gas) inhibits them from being too close to each other.

5.3.1 Planar Rotators (Inertial XY-Like Ferromagnetic Model)

The model

As a paradigmatic system along the above lines, we shall focus on the following model of classical planar rotors [219]. The Hamiltonian is assumed to be

$$\mathcal{H} = \frac{1}{2}\sum_{i=1}^{N} p_i^2 + \frac{1}{2}\sum_{i \neq j}\frac{1 - \cos(\theta_i - \theta_j)}{r_{ij}^\alpha} \equiv K + V \quad (\alpha \geq 0), \qquad (5.48)$$

where the rotors are localized on a lattice (e.g., a translationally invariant Bravais lattice, a quasi-crystal, a hierarchical network). If the lattice is a d-dimensional hypercubic one (with periodic boundary conditions) we have $r_{ij} = 1, 2, 3, \ldots$ if $d = 1$, $r_{ij} = 1, \sqrt{2}, 2, \ldots$ if $d = 2$, and $r_{ij} = 1, \sqrt{2}, \sqrt{3}, 2, \ldots$ if $d = 3$. The potential energy has been written in this particular manner so that its value for the ground state (i.e., $\theta_i = \theta_j \; \forall(i, j)$) vanishes in all cases. We have considered unit momenta of inertia (which rescales time) and unit first-neighbor coupling constant (which rescales temperature) without loss of generality, and (p_i, θ_i) are conjugate canonical pairs. Due to the periodic boundary conditions, the model is defined on a d-dimensional torus (i.e., a ring for $d = 1$, a torus for $d = 2$). Consequently, between any (i, j) pair of spins, there are more than one distances; in every case we consider as r_{ij} in the Hamiltonian the minimal of those distances. The model basically is a classsical inertial XY ferro-

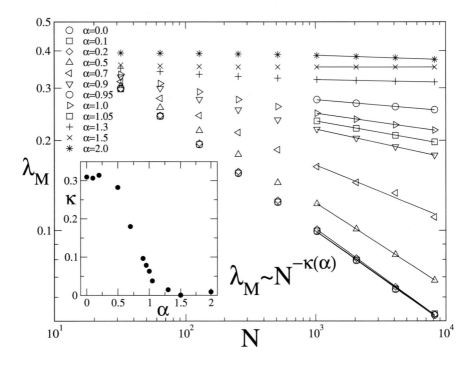

Fig. 5.33 Maximal Lyapunov exponent λ_M dependence on system size N (log-log plot), showing that $\lambda_M \sim N^{-\kappa(\alpha)}$. Initial conditions correspond to $\theta_0 = 0.5$, $\delta\theta = 0.5$, $p_0 = 0.5$ and $\delta p = 0.5$. Fixed parameters are $a = 0.005$ and $b = 2$. We averaged over 100 realisations. *Inset:* κ versus α, exhibiting weak chaos in the limit $N \to \infty$ when $0 \leq \alpha \lesssim 1$. These numerical results appear to be consistent with the expected ones for $d = 1$, namely, $\kappa > 0$ for $0 \leq \alpha < 1$ and $\kappa = 0$ for $\alpha \geq 1$. From [551]

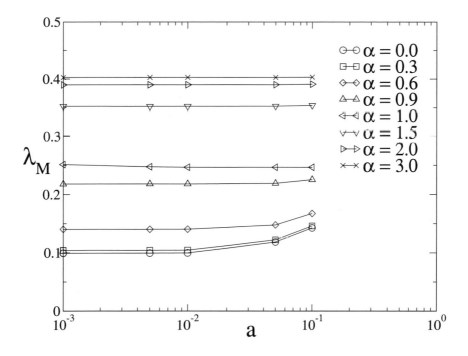

Fig. 5.34 Maximal Lyapunov exponent dependence on a for different values of α. Fixed constants are $N = 1024$ and $b = 2$. Initial conditions correspond to $\theta_0 = 0.5$, $\delta\theta = 0.5$, $p_0 = 0.5$ and $\delta p = 0.5$. We averaged over 100 realisations. From [551]

magnet (coupled rotators), and the limiting cases $\alpha \to \infty$ and $\alpha = 0$ correspond to the first-neighbor and mean-field-like models respectively. Clearly, the $\alpha = 0$ case does not depend on the particular lattice on which the spins are localized. This Hamiltonian is extensive (in the thermodynamical sense) if $\alpha/d > 1$, and nonextensive if $0 \le \alpha/d \le 1$. Indeed, in contrast with its kinetic energy, which scales like N, the potential energy scales like $NN^\star$, where

$$N^\star \equiv \sum_{j=1}^{N} \frac{1}{r_{ij}^\alpha}. \tag{5.49}$$

See also Eq. (3.69). For instance, for $\alpha = 0$, $N^\star = N$, and for $\alpha/d \ge 1$ and $N \to \infty$, $N^\star \to constant$. Since the variables $\{p_i\}$ involve a first derivative with respect to time, if we define $t' = \sqrt{N^\star} t$, Hamiltonian $\mathcal{H}$ in (5.48) is transformed (see details in [219]) into $\mathcal{H}' = \mathcal{H}/N^\star$, where

$$\mathcal{H}' = \frac{1}{2} \sum_{i=1}^{N} (p_i')^2 + \frac{1}{2N^\star} \sum_{i \ne j} \frac{1 - \cos(\theta_i - \theta_j)}{r_{ij}^\alpha} \qquad (\alpha \ge 0). \tag{5.50}$$

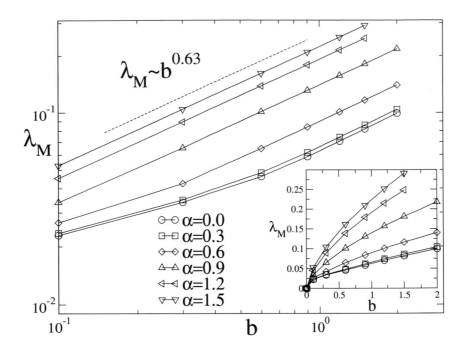

Fig. 5.35 Maximal Lyapunov exponent dependence on b in log-log plot. Fixed constants are $N = 1024$ and $a = 0.005$. Initial conditions correspond to $\theta_0 = 0.5$, $\delta\theta = 0.5$, $p_0 = 0.5$ and $\delta p = 0.5$. We averaged over 100 realisations. *Inset*: Same data in linear-linear plot. From [551]

It is in this form, *and omitting the 'primes'*, that this system is usually presented in the literature. Although physically meaningless (since it involves microscopic coupling constants which, through $N^\star$, depend on N), it has the advantage of being (artificially) extensive, like the familiar short-range-interacting ones. Unless explicitly declared otherwise, we shall from now on conform to this frequent use. For the $\alpha = 0$ instance, it will present the widespread mean-field-like form, frequently referred to in the literature as the *HMF* model [218] (see also [553–555]),

$$\mathcal{H} = \frac{1}{2}\sum_{i=1}^{N}p_i^2 + \frac{1}{2N}\sum_{i\neq j}[1 - \cos(\theta_i - \theta_j)]. \tag{5.51}$$

This model, as well as its generalizations and extensions, are being intensively studied (see [556–559] and references therein) in the literature through various procedures. A particularly interesting one is the molecular dynamical approach of an isolated N-sized system. Its interest comes from the fact that this is a *first-principle* calculation, since it is exclusively based on Newton's law of motion, and therefore constitutes a privileged viewpoint to try to understand in depth the microscopic dynamical founda-

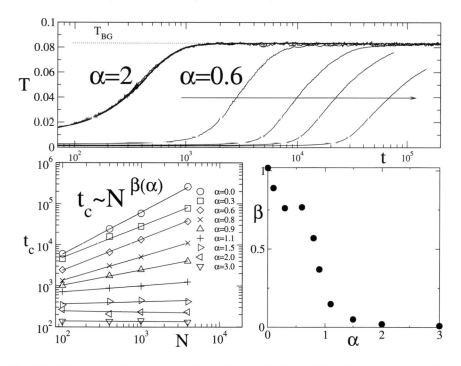

Fig. 5.36 *Upper panel*: Temperature evolution for $\alpha = 2$ and $\alpha = 0.6$ and four system sizes $N =$ 100, 400, 1000, 4000. Initial conditions correspond to $\theta_0 = 0.5, \delta\theta = 0.5, p_0 = 0.3$ and $\delta p = 0.05$. Fixed constants are $a = 0.05$ and $b = 2$. For $\alpha = 2$ the four curves coincide almost completely, all having a relatively fast relaxation to $T_{BG} = 1/12$. For $\alpha = 0.6$ the same sizes are shown, growing in the direction of the arrow. *Bottom left panel*: crossover time t_c (at mid-height of T) versus N, showing a power-law dependence $t_c \sim N^{\beta(\alpha)}$ with $\beta(\alpha) \geq 0$. *Bottom right panel*: β vs α exhibits that for long-range interactions the quasi-stationary-state life-time diverges in the thermodynamic limit. Note that when $\alpha = 0$, $\beta = 1$, and hence $t_c \propto N$. Given the nonglectable error bars due to finite size effects, the relation $\beta = 1 - \alpha$ is not excluded as possibly being the exact one; more precisely, it is non unplausible that $t_c \propto \frac{N^{1-\alpha}-1}{1-\alpha}$. From [551]

tions of statistical mechanics[6] (both the BG and the nonextensive theories). The time evolution of the system depends on the class of initial conditions that are being used. Two distinct such classes are frequently used, namely *thermal-equilibrium*-like ones (characterized by a initial Gaussian distribution of velocities) and the *water-bag*-ones (characterized by a initially uniform distribution of velocities within an interval compatible with the assumed total energy $U(N)$ of the system). The initial angle distribution ranges usually from all spins being aligned (say to the $\theta_i = 0$ axis), which corresponds to maximal average *magnetization* (i.e., $m = 1$), to angularly completely disordered spins, which corresponds to minimal average *magnetization* (i.e., $m = 0$). The simplest model (HMF) presents, in its BG microcanonical ver-

---

[6] This is sometimes referred to as the *Boltzmann program*. Boltzmann himself died without having accomplished it, and rigorously speaking it so remains until today!

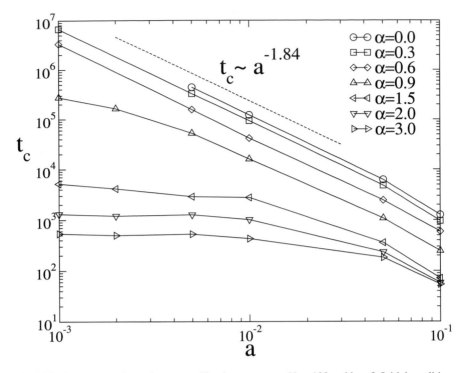

Fig. 5.37 Temperature dependence on a. Fixed constants are $N = 100$ and $b = 2$. Initial conditions correspond to $\theta_0 = 0.5$, $\delta\theta = 0.5$, $p_0 = 0.3$ and $\delta p = 0.05$. We averaged over 100 realisations. From [551]

sion, a second order phase transition at the scaled total energy $u_c = 0.75$, where $u \equiv U(N)/NN^\star$, or $u \equiv U(N)/N$, depending on whether we are adopting Hamiltonian (5.48) or (5.50) respectively. For $0 \leq u \leq u_c$, the system tends to be ordered in a *ferromagnetic* phase, whereas for $u > u_c$ it is in a disordered *paramagnetic* phase.

Metastability, nonergodicity and influence of the initial conditions

The model is analytically solvable in the BG canonical ensemble (equilibrium with a thermostat at temperature T). At first sight, the molecular dynamics approach apparently coincides with it *if* the initial conditions for the velocities are described by a Gaussian. But, *if* we use instead a water-bag, a longstanding *metastable* or *quasistationary state* (QSS), appears at values of u below 0.75 and not too small (typically for $u \in (0.5, 0.75)$). A value at which the effect is numerically very noticeable is $u = 0.69$, hence many studies are done precisely at this value, as shown in Fig. 5.38. In Fig. 5.39 we can see the influence of the initial value of m on T_{QSS}.

The above mentioned coincidence however is only apparent, as we shall see later on. Indeed, the plateau coming after the QSS state occurs, for all $\alpha/d < 1$,

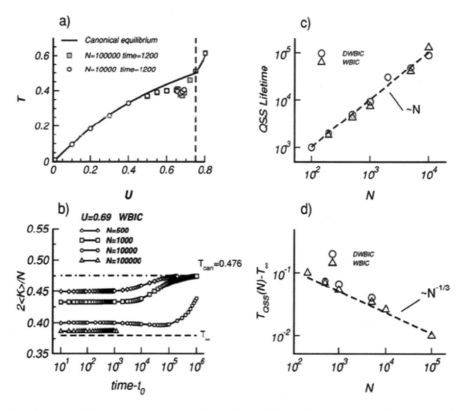

Fig. 5.38 **a** Caloric curve: microcanonical ensemble results for $N = 10000,\ 100000$ are compared with equilibrium theory in the BG canonical ensemble. The dashed vertical line indicates the critical energy: Water bag initial conditions (WBIC) and initial $m = 1$ are used in the numerical simulations. Temperature is computed from $2\langle K(N)\rangle/N$, where $\langle \ldots \rangle$ denotes time averages after a short transient time $t_0 = 100$ (not reported here). The time step used was 0.2 [560–563]. **b** Microcanonical time evolution of T, for the energy density $u = 0.69$ and different sizes. Each curve is an average over typically 100-1000 events (*ensemble average*). The dot-dashed line represents the BG canonical temperature $T_{BG} = 0.476$. The quantity T, which starts from 1.38 ($V = 0$ and $K = UN$ for WBIC), does not relax immediately to the temperature T_{BG}. The system lives in a QSS with a *plateau temperature* $T_{QSS}(N)$ smaller than the canonically expected value 0.476. The lifetime of the QSS increases with N, and the value of their temperature converges, as N increases, to the temperature 0.38, reported as a dashed line. Log-log plots for the QSS lifetime **c** and the difference $T_{QSS}(N) - T_\infty$ (with $T_\infty \equiv T_{QSS}(\infty)$) **d** are reported as functions of N. The QSS lifetime diverges roughly as N, and $T_{QSS} - 0.38$ vanishes roughly as $1/N^{1/3}$ (see fit shown as a dashed line). Note that from the caloric curve one gets $m^2 = T + 1 - 2u = T - 0.38$. Therefore, from the behavior reported in panel (**d**), being $T_\infty = 0.38$, one gets $M_{QSS} \sim 1/N^{1/6}$. Results are similar when we consider double water bag initial conditions (DWBIC), more precisely initial $m = 1$ and velocities uniformly distributed within $(-p_2, -p_1)$ and (p_2, p_1). In the figure, we report the case $p_1 = 0.8$ and $p_2 = 1.51$. From [557]

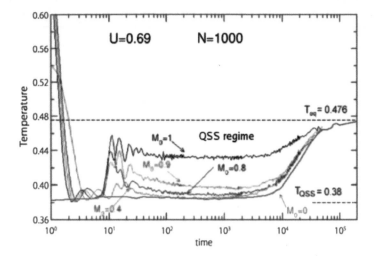

Fig. 5.39 Time evolution of the HMF temperature for the energy density $u = 0.69$, $N = 1000$ and several initial conditions with different magnetizations. After a very quick cooling, the system remains trapped into metastable long-living Quasi-Stationary States (QSS) at a temperature smaller than the equilibrium one. Then, after a lifetime that diverges with the size, the noise induced by the finite number of spins drives the system towards a complete relaxation to the equilibrium value. Although from a macroscopic point of view the various metastable states seem similar, they actually have different microscopic features and correlations which depend in a sensitive way on the initial magnetization. From [59]

at one and the same value T_{BG}, but various other properties sensibly differ from the BG expectation, very particularly the velocity distribution which is in fact non-Maxwellian. Also at the QSS the velocity distribution is non-Maxwellian. We learn here an important lesson: the fact that exact analytical calculations might be doable with BG statistical mechanics does not imply that the system indeed follows the BG theory in all of its aspects.[7]

The behavior is quite complex on the QSS plateau[8] emerging for long-range interactions (i.e., $0 \leq \alpha/d \leq 1$). Indeed, the ensemble- and time-averages do *not*

---

[7] Analogously, the fact that, within the mean-field approximation of the BG theory for critical phenomena, we can analytically calculate the usual critical exponents, by no means implies that they are correct!

[8] The lifetime τ_{QSS} of this QSS plateau has been conjectured (see Fig. 4 in [71], where $\lim_{N\to\infty} \lim_{t\to\infty}$ is expected to yield the standard BG canonical thermal equilibrium and $\lim_{t\to\infty} \lim_{N\to\infty}$ is expected to yield the nonextensive statistical mechanics results) to diverge, for fixed $\alpha \leq d$ if $N \to \infty$. Also, for the $d = 1$ model, it has been suggested [564] that, for fixed N, τ_{QSS} decreases exponentially with α increasing above zero. All these results are consistent with $\tau_{QSS} \propto (N^\star)^a$ with $N^\star$ given in Eq. (3.69) and $a > 0$. Indeed, such scaling yields, for $0 \leq \alpha/d < 1$, $\tau_{QSS}(\alpha/d, N) \propto N^{a[1-(\alpha/d)]}$ $(N \to \infty)$, which implies $\tau_{QSS}(0, N) \propto N^a$, and exponentially decreasing with α/d for fixed N. All authors do not always use the same definition for τ_{QSS}. The definition used in [557] implies $a = 1$; the definitions used by other authors imply $a > 1$.

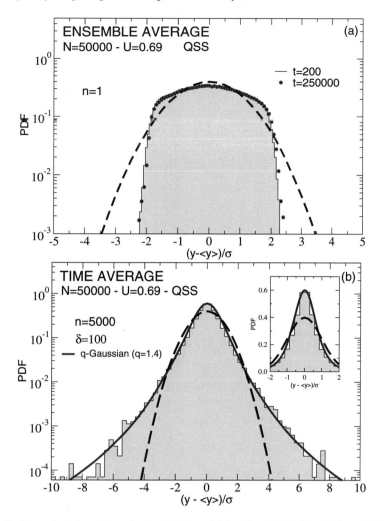

Fig. 5.40 Numerical simulations for the HMF model for N = 50000, U = 0.69 and M1 initial conditions *in the QSS regime*. **a** We plot the Pdfs of single rotor velocities at the times t = 200 and t = 250000 (ensemble average over 100 realizations). **b** We plot the time average Pdf for the variable *y* calculated over *only one single realization* in the QSS regime and after a transient time of 200 units. In this case we used $\delta = 100$ and $n = 5000$, in order to cover a very large portion of the QSS. Again, a q-Gaussian reproduces very well the calculated Pdf both in the tails and in the central part (see inset). See text for further details. From [61]

coincide [61, 565], thus exhibiting *nonergodicity* (which, as we shall see, is consistent with the fact that, along this longstanding metastable state, the entire Lyapunov spectrum collapses onto zero when $N \rightarrow \infty$). The situation is further illustrated in Figs. 5.40, 5.41, 5.42, 5.43, 5.44 and 5.45. In Fig. 5.46 we illustrate the influence of α on the QSS1 and QSS2 plateaux for the $d = 1$ model.

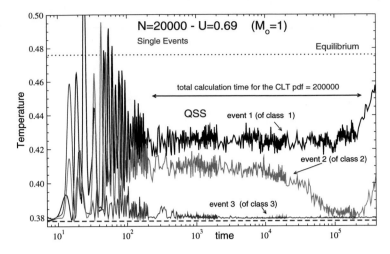

Fig. 5.41 Time evolution of the temperature (calculated as twice the average kinetic energy per particle) for three single events representative of the three different classes observed at $U = 0.69$ for initial magnetization $M_0 = 1$. The size of the system is $N = 20000$. From [565]

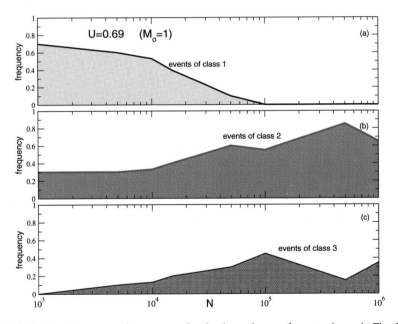

Fig. 5.42 Relative frequency of occurrence for the three classes of events shown in Fig. 1 as a function of N. A total of 20 realizations for each N was considered. The three curves add up to unity. From [565]

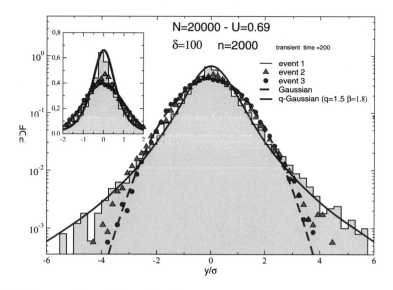

Fig. 5.43 We present for each class of the QSS found, the different central limit theorem behavior observed. A Gaussian (dashed curve) with unitary variance and a q-Gaussian $p(x) = Ae_q(-\beta x^2)$ with $A = 0.66$ $q = 1.5$ and $\beta = 1.8$ (full curve) are also reported for comparison. In the inset a magnification of the central part in linear scale is plotted. From [565]

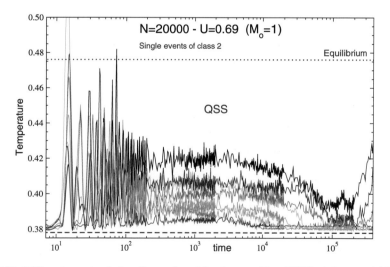

Fig. 5.44 Different events of class 2 are plotted for the case $N = 20000$. A large variability is observed for this class, at variance with the other two. From [565]

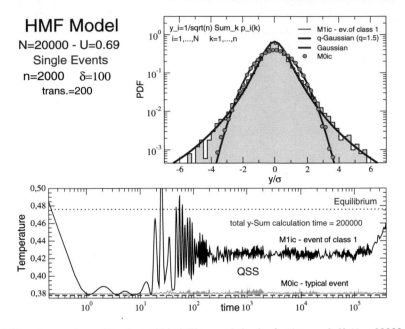

Fig. 5.45 a Comparison of the Central Limit Theorem behavior for the $u = 0.69$ $N = 20000$ case with initial magnetization $m = 1$ and $m = 0$. A Gaussian (dashed curve) with unit variance and a q-Gaussian with $A = 0.66$, $q = 1.5$ and $\beta = 1.8$ (full curve) are also reported for comparison. **b** Temperature time evolutions of the same events shown in panel (**a**). From [565]

Time-averaged distributions of velocities and energies

In the spirit of the q-CLT [477] we focus now on the time-averaged single-particle distributions of velocities and of energies. We do that on the quasi-stationary state (hereafter referred to as QQS2) that emerges with $T = T_{BG}$, *after* the traditional QSS (hereafter referred to as QSS1) that we discussed up to now: see Figs. 5.48, 5.49 and 5.50. We also exhibit one example in the QSS1, i.e., *before* the QSS2: see Fig. 5.51 [567]. In both QSS1 and QSS2 we have strong numerical indications of the applicability of q-statistics. It is obvious that, even if the QSS2 occurs at a BG temperature T_{BG}, its basic thermostatistical properties are non-BG. The thermodynamical limit of the exact stationary state requires both $t \to \infty$ and $N \to \infty$. The corresponding detailed study remains to be done. However, it might well be similar to what it happens for the α-Fermi-Pasta-Ulam model[9] (or α-Fermi-Pasta-Ulam-Tsingou problem [587, 588]), as shown in Fig. 5.81. More precisely, a critical line might exist of the form $N \sim Ct^\gamma$ (with $C > 0$ and $\gamma > 0$) such that, when the double N and t infini-

---

[9] We use this notation to conform to the notation used for the α-XY and α-Heisenberg models. Do not confuse with the notations α-FPU and β-FPU models which make respectively reference to introducing cubic and quartic anharmonic terms in addition to the harmonic two-body interaction. Therefore, the $\alpha \to \infty$ limit of the present α-FPU model coincides with the so called β-FPU model in current literature.

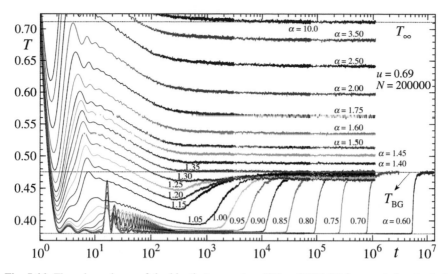

Fig. 5.46 Time dependence of the kinetic temperature $T(t) \equiv 2K(t)/N$ for a waterbag typical single initial condition for $(u, N) = (0.69, 200, 000)$ and various values of α.The upper horizontal line, at $T_\infty = 0.7114\ldots$,corresponds to the BG thermal equilibrium temperature of the $\alpha \rightarrow \infty$ model at $u = 0.69$. The middle (lower) horizontal line, at $T = 0.475$ ($T = 0.380$), indicates the BG thermal equilibrium temperature (the QSS base temperature, corresponding to zero magnetization), at $u = 0.69$ and $0 \leq \alpha < 1$. In the range $1 \leq \alpha < \infty$, no generic analytical solution is available, as far as we know. From [566]

ties are approached, it might occur as $N/t^\gamma < C$ or as $N/t^\gamma > C$. These two cases may respectively correspond to the validity of BG statistics or of q-statistics for the final stationary state. However, in both cases the relaxation appears to be extremely slow. Consistently, the system exhibits long-standing q-statistical distributions before deciding to eventually approach an ultimate $q = 1$ or $q \neq 1$ behavior. In Fig. 5.47 we illustrate the distribution of velocities at the QSS2 plateau for the $d = 1$ model, for both long- and short-range interactions.

Lyapunov spectrum

A set of $2dN$ Lyapunov exponents is associated with the d-dimensional Hamiltonian (5.48), half of them positive and half of them negative (coinciding in absolute value two by two) since the system is symplectic. We focus on the maximal value $\tilde{\lambda}_N^{max}$; if this value vanishes, the entire spectrum vanishes. This property is extremely relevant for the foundations of statistical mechanics. Indeed, if $\tilde{\lambda}_N^{max} > 0$, the system will be *mixing* and *ergodic*, which is the basis of BG statistical mechanics. If $\tilde{\lambda}_N^{max}$ vanishes, there is no such guarantee. This is the realm of nonextensive statistical mechanics, as we have already verified for paradigmatic dissipative and conservative low-dimensional maps. The scenario for the $d = 1$ α- XY model is described in

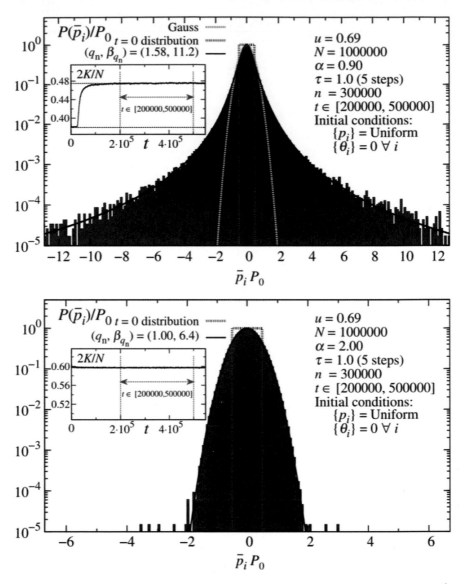

Fig. 5.47 A typical single-initial-condition one-momentum distribution $P(p)$ for $N = 10^6$, $u = 0.69$, $\tau = 1$ (corresponding to 5 molecular-dynamical algorithmic steps), calculated in the region $[t_{min}; t_{max}] = [200,000; 500,000]$ for $\alpha = 0.9$ (top plot), and $\alpha = 2.0$ (bottom plot). The upper temperature indicated in the $\alpha = 0.9$ inset coincides with that analytically calculated within BG statistical mechanics, namely $T_{kin} \equiv 2K(t)/N \simeq 0.475$. The horizontal line of the $\alpha = 2.0$ inset corresponds to the time average calculated numerically; indeed, analytical solutions are only available for $\alpha < 1$ and in the $\alpha \to \infty$ limit. The continuous curves correspond to $P(\tilde{p})/P_0 = e_{q_n}^{-\beta_{q_n}^{(P_0)}[\tilde{p}P_0)]^2/2}$ with $(q_n, \beta_{q_n}^{(P_0)}) = (1.58, 11.2)$ for $\alpha = 0.9$ and $(1.0, 6.4)$ for $\alpha = 2.0$. Notice that, for $\alpha = 0.9$, $1/\beta_{q_n}^{(P_0)} \neq T$. Each distribution has been rescaled with its own P_0. From [566]

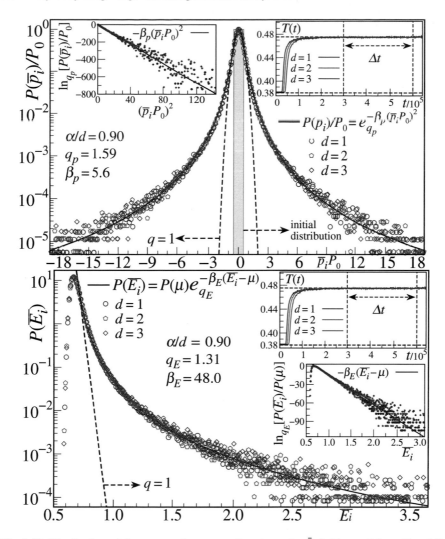

Fig. 5.48 Distributions of time-averaged momenta $\bar{p}_i$ and energies $\bar{E}_i$ (with $\tau = 1$) for $\alpha/d = 0.9$, in $d = 1, 2$ and 3 dimensions. The simulations were carried for the energy per particle $u = 0.69$ and total number of rotators $N = 1000000$. **a** Distribution $P(\bar{p}_i)$ is shown $[P_0 \equiv P(\bar{p}_i = 0)]$; the full line is a q-Gaussian with $q_p = 1.59$ and $\beta_p = 5.6$; the dashed line is a Gaussian ($q = 1$). The left inset shows the same data in a q-logarithm *versus* squared-momentum representation; a straight line is obtained as expected (since $\ln_q (e_q^x) = x$). **b** The full line represents the q-exponential $P(\bar{E}_i) = P(\mu) \exp_{q_E}[-\beta_E(\bar{E}_i - \mu)]$, with $q_E = 1.31$ ($\beta_E = 48.0$, $\mu = 0.69$, and $P(\mu) = 12$); the corresponding exponential (dashed line) is also shown for comparison. Since the density of states is necessary to reproduce the entire range of data, the parameter μ was introduced in the fitting. The bottom inset shows a straight line by using the q-logarithm in the ordinate. The kinetic temperature $T(t) \equiv 2K(t)/N$, and time window $(1 - q)t$ along which the time averages were calculated, coincide in both cases (shown as insets). In all plots one notices the collapse of all dimensions with nearly the same value of q. From [567]

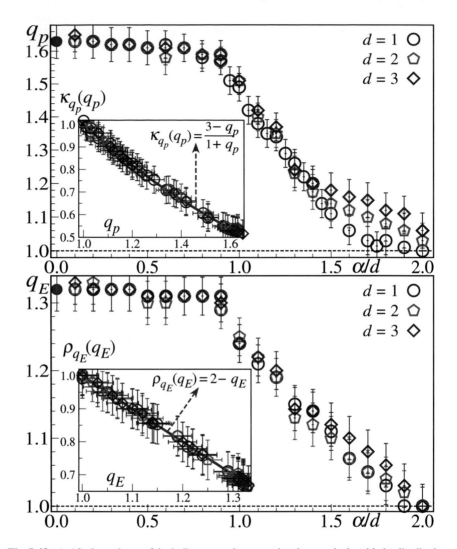

Fig. 5.49 (α/d)-dependence of the indices q_p and q_E associated respectively with the distributions of time-averaged momenta $\bar{p}_i$ and energies $\bar{E}_i$ for $d = 1, 2, 3$ and $u = 0.69$. The insets show the corresponding q-kurtosis (**a**) and q-ratio (**b**), compared to the analytical results (solid curves). The full bullets correspond to the values for $\alpha = 0$. Notice that, within the error bars, the indices q remain constant for $0 \leq \alpha/d \leq 1$, and approach unit only around $\alpha/d = 2$ (see text). From [567]

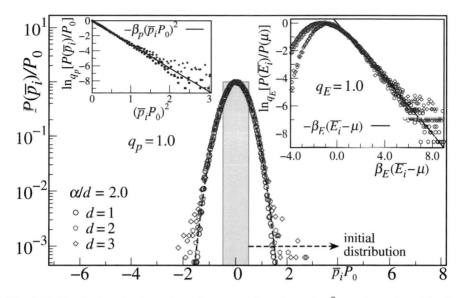

Fig. 5.50 Distributions for time-averaged momenta $\bar{p}_i$ and energies $\bar{E}_i$ are shown for $\alpha/d = 2$ ($d = 1, 2, 3$). For the momenta we have used conveniently scaled variables [like in Fig 5.48a], and the full line is the Maxwellian ($q = 1$); the left inset shows the same data in a logarithm versus squared-momentum representation. The right inset exhibits $\ln[P(\bar{E}_i)/P(\mu)]$ versus $\beta_E(\bar{E}_i - \mu)$. Similarly to Fig 5.48b, we verify the appearance of d-dependent densities of states; the full line is an exponential in the variable $\beta_E(\bar{E}_i - \mu)$. From [567]

Figs. 5.52 and 5.53. The corresponding scenarios for $d = 2, 3$ have been discussed in [568], and are illustrated in Figs. 5.54, 5.55 and 5.56. They are completely analogous to that of the $d = 1$ case, and strongly suggest that the relevant exponent κ does not depend separately on α and d, but, like $N^\star$ (see Eq. (3.69)), only on the ratio α/d.

The above molecular dynamical results concerned the disordered (paramagnetic) phase. Results are also available [558, 570] for the ordered (ferromagnetic) phase of the $d = 1$ model, more precisely for its QSS. For reasons that are not totally transparent, the value for κ obtained on the QSS (below u_c), turns out to numerically be 1/3 of its value above u_c: See Fig. 5.57.

Ageing and anomalous diffusion

The very fact that, for $u < u_c$ and fixed N, a QSS exists which, after a time of the order of τ_{QSS}, eventually goes to a q-statistical or BG stationary state implies that the system has some sort of *internal clock*. This immediately suggests that *ageing* should be expected. More precisely, if we consider a two-time autocorrelation function $C(t + t_W, t_W)$ of some dynamical variables of the system, we expect this quantity to depend not only on time t, *but also on the waiting time t_W*. This is precisely what is verified in [59, 60, 573, 574]: See Figs. 5.58 and 5.59 (see also 5.66). It is quite

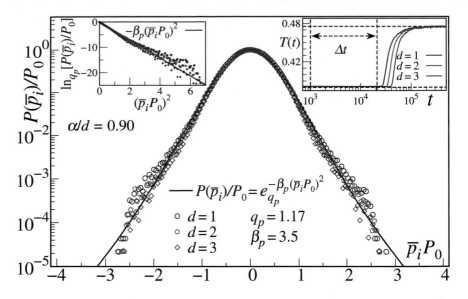

Fig. 5.51 Distributions of time-averaged momenta $\bar{p}_i$ for the same parameters used in Fig. 5.48a, but for a different time window. The left inset presents the same data in a q-logarithm versus squared-momentum representation; the full straight line is a q-Gaussian with $q = 1.17$. One should notice the collapse of all dimensions with nearly the same value of q. The right inset shows $T(t) \equiv 2K(t)/N$ and the time-window δt along which the time averages were calculated ($t \in [1, 10] \times 10^3$). From [567]

remarkable that q-exponential decays are observed in these (and other) cases, and that data collapse, in the form

$$C(t + t_W, t_W) = e_q^{-B\,t/t_W^{\beta}} \quad (B > 0; \ \beta \geq 0) \tag{5.52}$$

is possible (like in usual spin-glasses). The value $q \simeq 2.35$ (corresponding to the (p, θ)-space [60]) is essentially what elsewhere (namely, in the context of the q-triplet to be soon discussed) is noted q_{rel}. Another remarkable fact (see Fig. 5.60) is that, for $u > u_c$, Eq. (5.52) is still satisfied *with the same value of* $q \simeq 2.35$, but with $\beta = 0$, i.e., without ageing. Let us stress that, for a standard BG system (e.g., if $\alpha/d > 1$), one typically observes, both above and below a possible critical point, the form (5.52) with $q = 1$ and $\beta = 0$.

Angular motion

Let us now focus on the diffusion of the angles $\{\theta_i\}$ by allowing them to freely move within $-\infty$ to $+\infty$. In Fig. 5.61 we illustrate the distribution of angles at the QSS2 plateau for the $d = 1$ model, for both long- and short-range interactions. The probability distributions, and corresponding anomalous diffusion exponent γ, can be

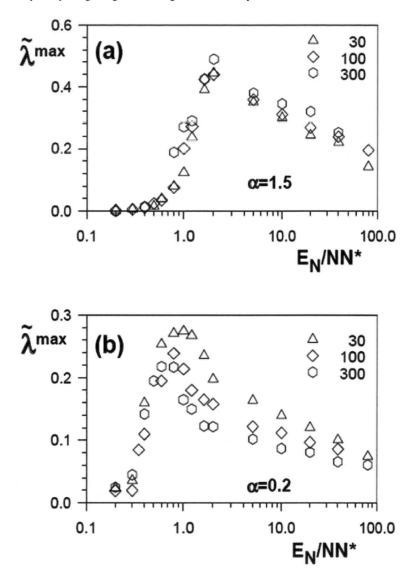

Fig. 5.52 The $u \equiv E_N/NN^*$- dependence of the properly scaled maximal Lyapunov exponent $\tilde{\lambda}_N^{max}$, for the $d = 1$ $\alpha - XY$ model and typical values of N, for $\alpha = 1.5$ (**a**), and $\alpha = 0.2$ (**b**). As illustrated in Fig. 5.53, the $N \to \infty$ limit yields, for high enough values of u (in fact for $u > u_c = 0.75$, $\forall\alpha$), a nonvanishing (vanishing) value for $\tilde{\lambda}_N^{max}$ for $\alpha \geq 1$ ($0 \leq \alpha \leq 1$). From [219]

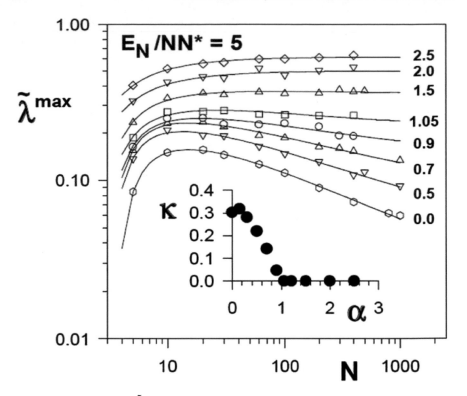

Fig. 5.53 Log-log plots of $\tilde{\lambda}_N^{max}$ versus N for typical values of α and $u = 5$. The full lines are the best fittings with the heuristic forms $(a - b/N)/(N^\star)^c$. Consequently, $\tilde{\lambda}_N^{max} \sim 1/N^{\kappa(\alpha)}$, where κ is positive for $0 \leq \alpha < 1$ and vanishes for $\alpha > 1$. For $\alpha = 1$, $\tilde{\lambda}_N^{max}$ is expected to vanish like some power of $1/\ln N$. From [219]

seen in Figs. 5.62, 5.63, 5.64, 5.65 and 5.66. From the data in Fig. 5.66 we can verify (see Fig. 5.67) the agreement, within a 10% error, with the scaling predicted in Eq. (4.14).

For phenomena occurring at the edge of chaos of simple maps and related to those described above, see [580, 581].

Connection with glassy systems

We have seen above that there is ageing at the QSS below the critical point u_c, whereas no such phenomenon survives above u_c. We expect then to have some sort of glassy behavior during the QSS, and no such behavior above u_c. This is precisely what we see in Fig. 5.68 (see also [580, 581]).

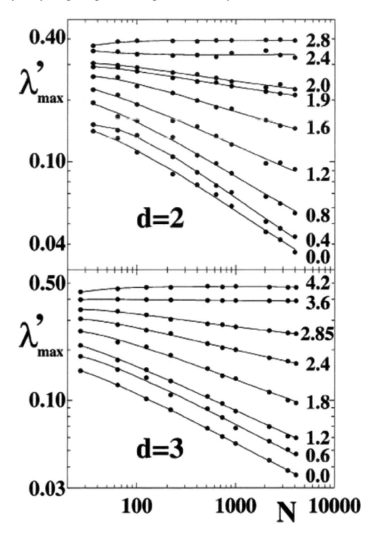

Fig. 5.54 Log-log plots of $\bar{\lambda}_N^{max}$ versus N for typical values of α and $u = 5$: $d = 2$ (top panel) and $d = 3$ (bottom panel). The full lines are the best fittings with the heuristic forms $(a - b/N)/(N^\star)^c$. Consequently, $\bar{\lambda}_N^{max} \sim 1/N^{\kappa(\alpha,d)}$, where κ is positive for $0 \leq \alpha/d < 1$ and vanishes for $\alpha/d > 1$. For $\alpha/d = 1$, $\bar{\lambda}_N^{max}$ is expected to vanish like some power of $1/\ln N$. From [568]

Fig. 5.55 The exponent κ as a function of α/d for the $d = 1, 2, 3$ α- XY model. The points for the $d = 1$ case are those of the inset of Fig. 5.52. The solid line is a guide to the eye consistent with universality. For $\alpha = 0$ we have $\kappa(0) = 1/3$ [569]. From [568]

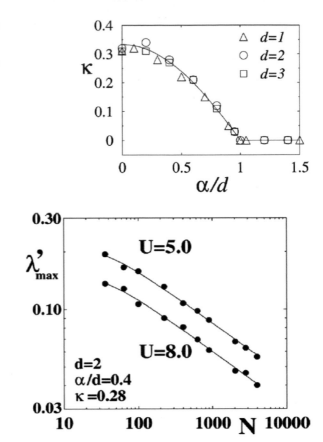

Fig. 5.56 $\tilde{\lambda}_N^{max}(N)$ for the $(\alpha, d) = (0.8, 2)$ model for two different values of energy density u. The asymptotic N behavior, for all values of $u > u_c$ and $0 \leq \alpha/d < 1$, appears to be $\tilde{\lambda}_N^{max}(N) \sim A/N^{\kappa(\alpha/d)}$, where A decreases from a finite, (α, d)-dependent, value to zero when u increases from u_c to infinity. From [568]

Out-of-equilibrium steady state

We focus now on an interesting numerical result [575, 576] concerning thermal conductivity. The $d = 1$ first-neighbor system (i.e., $\alpha \rightarrow \infty$) is placed at two different temperatures at its two ends, and consequently heat flows from the hot end to the cold end. The thermal conductivity is measured at its steady-state, as depicted in Figs. 5.69 and 5.70. We notice a very interesting phenomenon. This system has short-range interactions only (first-neighbors) and has therefore a thermal equilibrium state which definitively is very well described within BG statistical mechanics, this is to say $q = 1$. We see however that its out-of-equilibrium steady-state associated with heat flow is well described with $q \simeq 1.6$. In other words, the appropriate value of q generically depends not only on the system but *also on the specific circumstances under which the system is placed!*.[10] This is a general scenario that we should always bear in mind.

---

[10] This important feature stands as a metaphor for the philosopher José Ortega y Gasset's celebrated statement *Yo soy yo y mi circunstancia* [I am myself and my circumstance].

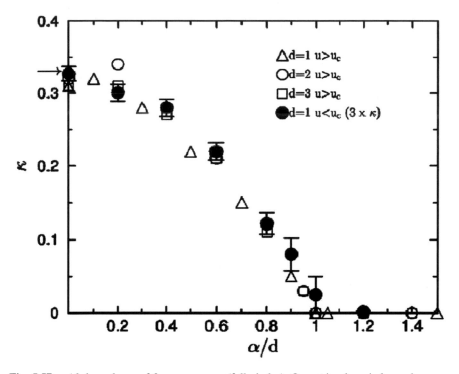

Fig. 5.57 α/d-dependance of $3 \times \kappa_{metastable}$ (full circles). Open triangles, circles and squares respectively correspond to κ_d of the $d = 1, 2, 3$ models [219, 568]. The arrow points to 1/3, value analytically expected [569, 571, 572] to be exact for $\alpha = 0$ and $u > u_c$. From [558]

5.3.2 Three-Dimensional Rotators (Inertial Heisenberg-Like Ferromagnetic Model)

The model

The model that we focus on here is identical to the d-dimensional α-XY ferromagnetic model excepting for the fact that the rotators are now three-dimensional ($n = 3$) instead of planar ($n = 2$) ones. It was introduced and studied in [559, 582–584] and its Hamiltonian is given by

$$\mathcal{H} = \frac{1}{2}\sum_{i=1}^{N}\mathbf{L}_i^2 + \frac{1}{2\tilde{N}}\sum_{i=1}^{N}\sum_{\substack{j=1 \\ j \neq i}}^{N}\frac{1 - \mathbf{S}_i \cdot \mathbf{S}_j}{r_{ij}^{\alpha}} \quad (\alpha \geq 0), \tag{5.53}$$

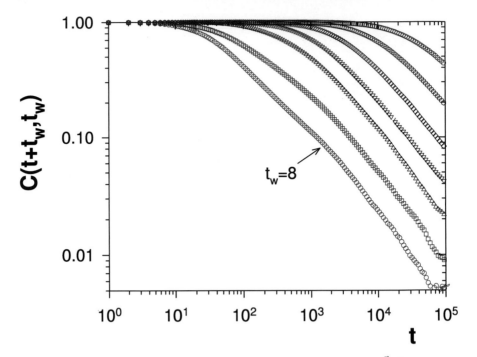

Fig. 5.58 Normalized two-time auto-correlation function C of the state variable $(\vec{\theta}, \vec{p})$ versus time, for $u = 0.69$ (subcritical) and for initial conditions that guarantee that the system will get trapped into a quasi stationary trajectory. Data correspond to averages over 200 of such trajectories. The waiting times are $t_w = 8 \times 4^n$, with $n = 0, \ldots, 6$. The dependence of C on both times is evident. From [573]

with $(\mathbf{S}_i)^2 = (S_i^x)^2 + (S_i^y)^2 + (S_i^z)^2 = 1$, $(\forall i)$, and $\tilde{N} \equiv N^\star$ defined in Eq. (5.49); $0 \le \alpha/d \le 1$ and $\alpha/d > 1$ characterize respectively the long- and short-range regimes.

Distribution of one-particle velocities and energies

The momenta and energy distributions are illustrated in Figs. 5.71 and 5.72 for $\alpha/d = 0.9$.

Lyapunov spectrum

While N gradually increases to infinity, the maximal Lyapunov exponent λ_{max} is expected to remain positive for $\alpha/d > 1$, and vanish like $\lambda_{max} \propto 1/N^\kappa$ for $0 \le \alpha/d < 1$. Consistently we expect $\kappa(\alpha/d) = 0$ for $\alpha/d \ge 1$. We see in Fig. 5.73 that this expectation is fulfilled expecting for the intriguing fact that κ neatly approaches

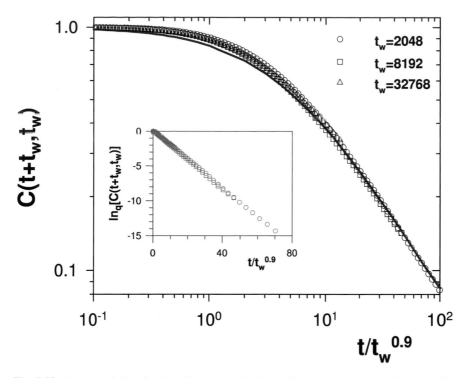

Fig. 5.59 Auto-correlation function C versus scaled time. The data are the same shown in Fig. 5.58 for the three largest t_w, but suitably scaling the time coordinate makes the data collapse into a single curve. The red solid line corresponds to $e_{2.35}^{-0.2t/t_w^{0.9}}$. *Inset*: $\ln_q$-linear representation of the same data, with $q = 2.35$. Linearity indicates q-exponential behavior. From [573]

zero not near $\alpha/d = 1$ but rather at a sensibly higher value. The numerical causes for this anomaly could in principle be the finiteness of N, of t, of the algorithmic precision, of the number of realizations, the class of initial conditions that is used, among others. At the present stage, this remains as an open question. By the way, we may mention here another open question related to this kind of problem: how does $\kappa(0)$ depend on the number n of components of the n vector model? We know that (i) for $n = 2$, $\kappa(0) = 1/3$ [569], (ii) for $n = 3$, $\kappa(0) \simeq 0.3$ (from Fig. 5.73), and (iii) for the spherical model (i.e., $n \to \infty$) it seems reasonable that $\kappa(0) = 0$. For generic n, could it then be something like $\kappa(0) \simeq 3/(7 + n)$?

Duration of the quasi-stationary-state

We illustrate in Fig. 5.74 [585] the behavior of $T(t)$ for fixed values of α/d and u, $d = 1, 2, 3$, and different values of N. The existence of two plateaux (referred to as QSS1 and QSS2) is visible; we note t_{QSS} the characteristic time separating the first

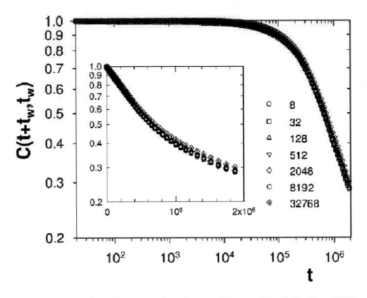

Fig. 5.60 Auto-correlation function versus time for $u = 5$ (supercritical) for $N = 1000$ and various values of t_W. The data correspond to an average over 10 trajectories initialized with water-bag configuration. Notice that (i) like in standard BG systems at thermal equilibrium, there is no ageing; (ii) in spite of the fact that there is no ageing, the function itself appears to be a q-exponential with $q > 1$. *Inset:* Semilog representation of the same data. From [573]

plateau (QSS1, or just QSS) from the second one (QSS2). The time t_{QSS} is defined in the literature through various manners, namely, (i) the value at which $T(t)$ departs from QSS1, (ii) the value at which an inflection point occurs, (iii) the value at which $T(t)$ attains its mid-height between QSS1 and QSS2. For simplicity, we adopt here the third one: t_{QSS} depends essentially on (d, α, N, u).

Figure 5.77 shows, in a single plot, the approximate collapse of 24 different $T(t)$ curves (see also Figs. 5.75 and 5.76): two values of N ($N = 15625$ and 46656, following $N = L^d$, with $d = 1, 2$ and 3) and four values of α/d ($\alpha/d = 0.5, 0.6,$ 0.7, and 0.8). The results in Fig. 5.77 suggest the following scaling:

$$t_{QSS}(N, \alpha, d) = [\lambda(\alpha/d)]^{d-1} \, \nu(N) \, \mu(\alpha/d) N^{A(\alpha/d)} e^{-B(N)(\alpha/d)^2}, \qquad (5.54)$$

with $A(\alpha/d) \simeq 1.8 - 1.5\,(\alpha/d)^2$ and $B(N) \simeq 1.33\,N^{0.19}$; $\mu(\alpha/d)$ and $\nu(N)$ are defined in Figs. 5.75 and 5.76 respectively; $\lambda(\alpha/d)$ remains equal to unit for $0 \le \alpha/d < 0.8$ and close to 1.16 for $\alpha/d = 0.8, 0.9$.

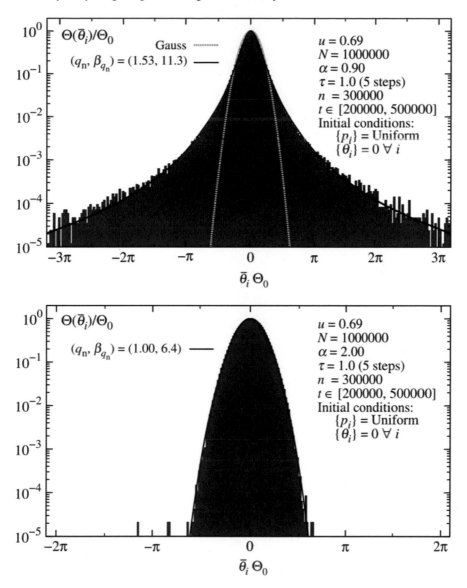

Fig. 5.61 A typical single-initial-condition one-angle distribution $\Theta(\tilde{\theta})$ for exactly the same conditions of Fig. 5.47 ($\alpha = 0.9$ for the top plot; $\alpha = 2.0$ for the bottom plot). Notice that remarkably enough, for both values of α, the present (q_n, β_{q_n}) within $\Theta(\tilde{\theta})/\Theta_0 = e_{q_n}^{-\beta_{q_n}^{(\Theta_0)}[\tilde{\theta}\Theta_0)]^2/2}$ coincide, within small error bars, with those fitting the distribution of velocities in Fig. 5.47. From [566]

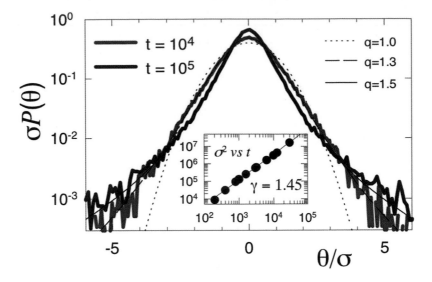

Fig. 5.62 Histogram of normalized angles at different times of the HMF dynamics. Parameters and initial conditions are the same used in previous figures. Notice that at long times, the histogram is of the q-Gaussian form. Inset: squared deviation as a function of time. It follows the super-diffusive law $\sigma^2 \sim t^\gamma$ with $\gamma \simeq 1.45$. From [60]

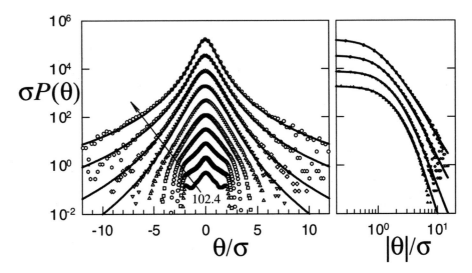

Fig. 5.63 Histograms of rotor phases at different instants of the dynamics (symbols). Simulations for $N = 1000$ were performed starting from fully magnetized initial conditions at $u = 0.69$ (conditions leading to QSSs). Countings were accumulated over 1000 realizations, at times $t_k = 102.4 \times 2^k$, with $k = 1, 2, \ldots, 9$, growing in the direction of the arrow up to $t = 52{,}428.8$. Solid lines correspond to q-Gaussian fittings. Histograms were shifted for visualization. Right-side panel: log-log representation of the fitted data. From [579]

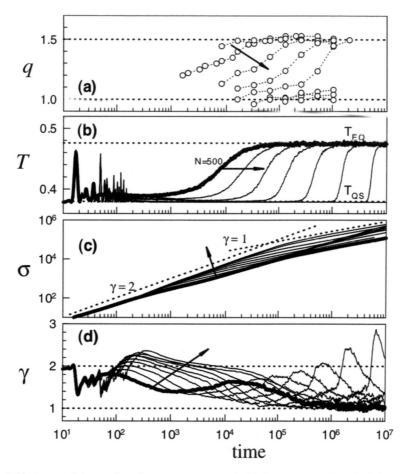

Fig. 5.64 Averaged time series of **a** parameter q (symbols), **b** temperature T, **c** deviation σ, and **d** diffusion exponent γ, $u = 0.69$ and different values of N $N = 500 \times 2^k$, with $k = 0, \ldots, 9$. Bold lines correspond to $N = 500$, as reference, and N increases in the direction of the arrows up to $N = 256,000$. Averages were taken over $2.56 \times 10^5/N$ realizations, starting from a fully magnetized configuration at $t = 0$. Although small numbers of realizations are used for large sizes, curves are reproducible. In panel (**a**), the fitting error is approximately 0.03; dotted lines are drawn as reference. In (**b**), they correspond to temperatures at equilibrium $T_{EQ} \equiv T_{QSS2} = 0.476$ and at QSSs in the TL ($T_{QS} \equiv T_{QSS1} = 0.38$). In (**c**) they correspond to ballistic motion $\gamma = 2$ and to normal diffusion $\gamma = 1$. From [579]

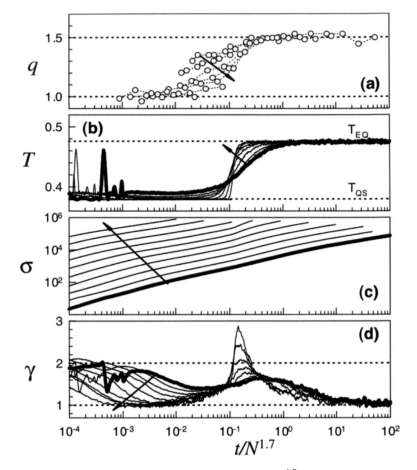

Fig. 5.65 The same data as in Fig. 5.64, but as a function of $t/N^{1.7}$. From [579]

5.3.3 Fermi-Pasta-Ulam-Like Model for Coupled Oscillators

The model

The celebrated Fermi-Pasta-Ulam (FPU) model for coupled anharmonic oscillators was introduced in [586].[11] It was defined in two different versions, namely the α-model which includes both third- and fourth-order anharmonicities, and the β-model which only includes fourth-order anharmonicity. Due to its simpler form, it is the

---

[11] The physical model was introduced by E. Fermi, J. Pasta and S. Ulam. However, the indispensable computational work was done by Mary Tsingou. Further details are available at [587, 588] I had myself the privilege of talking with her, and the importance of her contribution became transparent during our unique conversation.

Fig. 5.66 a Time evolution of the HMF velocity autocorrelation functions for $U = 0.69$, $N = 1000$ and different initial conditions are nicely reproduced by q-exponential curves. The entropic index q used is also reported. **b** Time evolution of the variance of the angular displacement for $U = 0.69$, $N = 2000$ and different initial conditions. After an initial ballistic motion, the slope indicates a superdiffusive behaviour with an exponent γ greater than 1. This exponent is also reported and indicated by dashed straight lines. Anomalous diffusion does not depend in a sensitive way on the size of the system. For both the plots shown, the numerical simulations are averaged over many realizations. From [59]

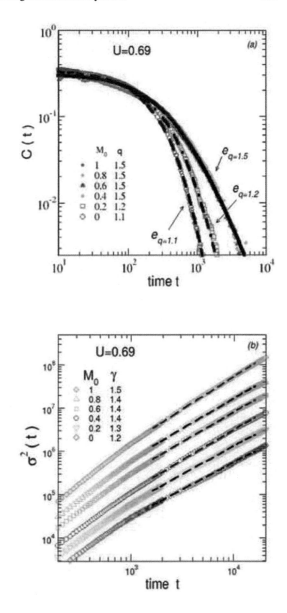

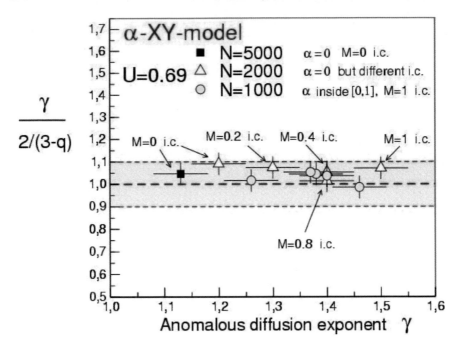

Fig. 5.67 For different system sizes and initial conditions, and for several values of the parameter α which fixes the range of the interaction of a generalized version of the HMF model [12], the figure illustrates the ratio of the anomalous diffusion exponent γ divided by $2/(3-q)$ vs γ. The entropic index q is extracted from the relaxation of the correlation function (see previous figure). This ratio in all cases equals unity within the 10% error bars, thus verifying prediction in [404]. From [59]

latter that we focus on here. The corresponding $d = 1$ classical Hamiltonian with periodic boundary conditions is as follows:

$$\mathcal{H} = \sum_{i=1}^{N} \frac{p_i^2}{2m} + \frac{a}{2} \sum_{i=1}^{N} (x_{i+1} - x_i)^2 + \frac{b}{4} \sum_{i=1}^{N} (x_{i+1} - x_i)^4 \quad (a \geq 0; \ b \geq 0). \quad (5.55)$$

Several studies followed [482, 588–600] among many others which deepened the understanding of this paradigmatic model. In what follows we focus on the Hamiltonian of the d-dimensional system with periodic boundary conditions and long-range interactions between all the $N = L^d$ oscillators:

$$\mathcal{H} = \sum_{i} \frac{\mathbf{p}_i^2}{2m_i} + \frac{a}{2} \sum_{i} (\mathbf{r}_{i+1} - \mathbf{r}_i)^2 + \frac{b}{4\tilde{N}} \sum_{i} \sum_{j \neq i} \frac{(\mathbf{r}_i - \mathbf{r}_j)^4}{d_{ij}^{\alpha}} \quad (5.56)$$

Fig. 5.68 a The magnetization M and the polarization $p \equiv \frac{1}{N} \sum_{i=1}^{N} |\langle s_i \rangle|$ are plotted vs the energy density for $N = 10000$ at equilibrium: the two order parameters are identical. **b** The same quantities plotted in (a) are here reported vs the size of the system, but in the metastable QSS regime. In this case, increasing the size of the system, the polarization remains constant around a value $p \sim 0.24$ while the magnetization M goes to zero as $N^{-1/6}$. From [59]

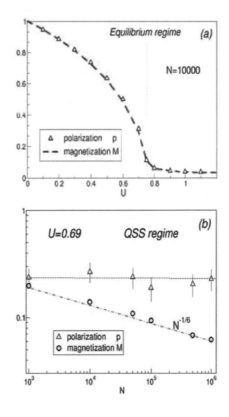

where $\mathbf{r}_i$ and $\mathbf{p}_i$ are the displacement and momentum of the i-th particle with mass $m_i \equiv m$; $a \geq 0$, $b \geq 0$, and $\alpha \geq 0$. Here d_{ij} is the shortest Euclidean distance between the i-th and j-th lattice sites ($1 \leq i, j \leq N$); this distance depends on the geometry of the hypercubic lattice (ring, periodic square or periodic cubic lattices). Thus for $d = 1$, $d_{ij} = 1, 2, 3, ...$; for $d = 2$, $d_{ij} = 1, \sqrt{2}, 2, ...$, and, for $d = 3$, $d_{ij} = 1, \sqrt{2}, \sqrt{3}, 2, ...$ If $\alpha/d > 1$ ($0 \leq \alpha/d \leq 1$) we have short-range (long-range) interactions in the sense that the potential energy per particle converges (diverges) in the thermodynamic limit $N \to \infty$; in particular, the $\alpha \to \infty$ limit corresponds to only first-neighbor interactions, and the $\alpha = 0$ value corresponds to typical mean field approaches, when the coupling constant is assumed to be independent from distance. The instance $(d, \alpha) = (1, \infty)$ recovers the original FPU β-Hamiltonian, namely Eq. (5.55), that has been profusely studied in the literature.

Although not necessary (see [219]), we have followed the current use and have made the Hamiltonian extensive for all values of α/d by adopting the scaling factor $\tilde{N}$ in the quartic coupling, where

$$\tilde{N} \equiv \sum_{i=1}^{N} \frac{1}{d_{ij}^{\alpha}}. \tag{5.57}$$

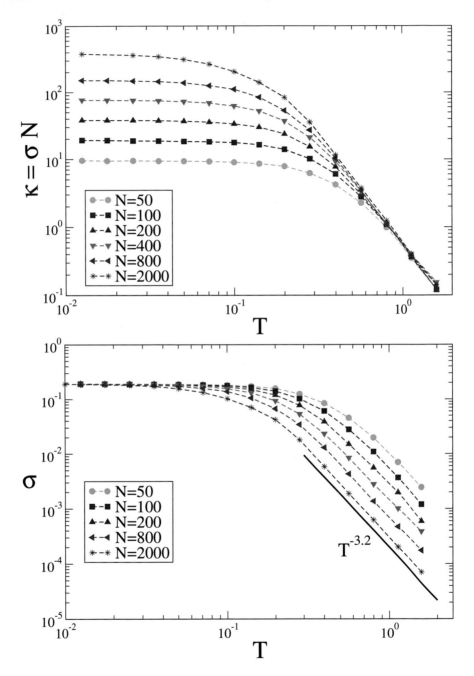

Fig. 5.69 Heat flow at stationary state for typical lattice sizes. *Top*: Thermal conductivity $\kappa \equiv \sigma N$ (notice that data collapse occurs for the high-temperature region); *Bottom*: Thermal conductance σ (notice that data collapse occurs for the low-temperature region). Different colors correspond to the lattice lengths $N = 50, 100, 200, 400, 800$ and 2000. The slope -3.2 indicated in [575] is shown here for comparison. The dashed curves are guides to the eye. From [575]

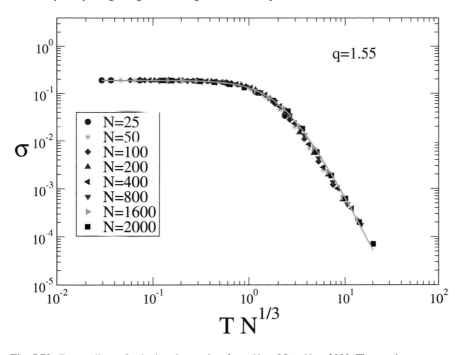

Fig. 5.70 Data collapse for lattice sizes going from $N = 25$ to $N = 2000$. The continuous curve (green line) corresponds to $\sigma = A_q e_q^{-B_q (TN^{1/3})^2}$ with $(q, B_q, A_q) = (1.55, 0.40, 0.189)$. Notice that the maximum value 0.189 has naturally nothing to do with the normalizing factor of a q-Gaussian distribution, which for the present values $(q, B_q) = (1.55, 0.40)$ would approximately be 0.28. Indeed, q-exponential and q-Gaussian functions frequently appear in diverse physical quantities (e.g., the time evolution of the nonlinear dynamical sensitivity to the initial conditions [187, 188]), and not only as probability distributions. The asymptotic slope is given by $2/(1 - q) \simeq -3.64$, in contrast with the intermediate slope -3.2 indicated in [575] (see Fig. 5.69). Notice that the asymptotic slope only becomes visible several decades further down. From [576]. Recent calculations for $d = 1, 2, 3$ ($N = L^d$), with higher precision and also attaining higher values of T [577, 578], appear to be better fitted with the q-stretched exponential $\sigma(T, L) = L^{1-d} A_q(d) e_{q(d)}^{-B_q(d) [TN^{\gamma(d)}]^{\eta(d)}}$, being $(q(1), \eta(1)) \simeq (1.7, 2.36)$, hence the asymptotic slope $\eta(1)/[1 - q(1)] \simeq -3.37$

Consequently, $\tilde{N}$ depends on (N, α, d), and the geometry of the lattice. Note that for $\alpha = 0$ we have $\tilde{N} = N$, which recovers the rescaling usually introduced in mean field approaches. In the thermodynamic limit $N \to \infty$, $\tilde{N}$ remains constant for $\alpha/d > 1$, whereas $\tilde{N} \sim \frac{N^{1-\alpha/d}}{1-\alpha/d}$ for $0 \leq \alpha/d < 1$ ($\tilde{N} \sim \ln N$ for $\alpha/d = 1$); see further details in [219] and references therein.

Let us mention that the analytical thermostatistical approach of the present model is in some sense even harder than that of coupled XY or Heisenberg rotators already addressed in [218, 219, 566, 568, 583]. Indeed, the standard BG approach of these models is analytically tractable, whereas not even that appears to be possible for the original FPU, not to say anything for the present generalization. Therefore, for this

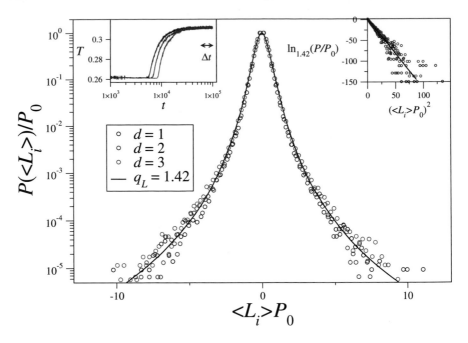

Fig. 5.71 Probability distribution functions of time-averaged angular momenta $P(\langle L_i \rangle)$ are represented in conveniently rescaled variables for one-, two-, and three-dimensional systems with $N = 262144 = 512^2 = 64^3$ rotators and $\alpha/d = 0.9$; for simplicity we use values of N that simultaneously are perfect squares and perfect cubes. The three data sets appear to collapse into a single q-Gaussian (full line), $P(\langle L_i \rangle) = P_0 \exp_{q_L}(-\beta_L \langle L_i \rangle^2)$, with $q_L(\alpha/d) = 1.42$ and $\beta_L = 4.44$; the maximum values $P_0 \equiv P(\langle L_i \rangle = 0)$ varied slightly with the lattice dimensionality (see text). *Left inset*: Evolution in time of the kinetic temperature. A first plateau (occasionally referred to as QSS1) at a temperature close to $T_{QSS}(\alpha/d) = U - \frac{1}{2}$, whose duration depends on d, is observed. After the transition to the second plateau (occasionally referred to as QSS2), the temperature reaches the equilibrium value $T_{BG}(\alpha/d) = 0.3118$ predicted by the BG caloric curve; the time window [60000, 100000] used for the time averages is indicated. *Right inset:* The same data in q-logarithm ($q_L = 1.42$) *versus* $(\langle L_i \rangle P_0)^2$, where the slope of the full straight line yields the value of β_L. From [584]

kind of many-body Hamiltonians, the first-principle numerical approach (i.e., based on Newton's law) appears to be the only generically tractable one.

Lyapunov spectrum

Numerical results for $N \gg 1$ strongly indicate that the maximal Lyapunov exponent λ_{max} remains, for increasing N, constant and positive for $\alpha/d > 1$ (implying strong chaos, mixing and ergodicity), and that it vanishes like $N^{-\kappa}$ for $0 \le \alpha/d < 1$ (thus approaching weak chaos and opening the possibility of breakdown of mixing and/or ergodicity). The suitably rescaled exponent κ exhibits universal scaling, namely

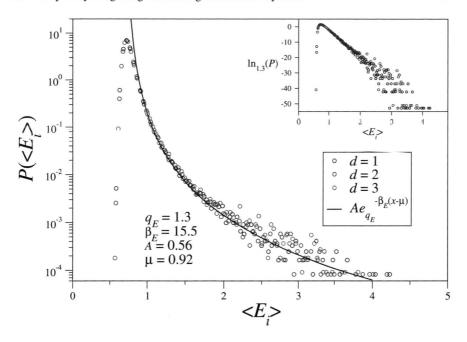

Fig. 5.72 Probability distribution functions of time-averaged individual energies $P(\langle E_i \rangle)$ are represented versus $\langle E_i \rangle$ for one-, two-, and three-dimensional systems. The parameters are the same as in Fig. 5.71, i.e., $\mathcal{N} = 262144$ rotators, $\alpha/d = 0.9$, and the same time window. All data are well fitted by a shifted q−exponential of the form $Ae_{q_E}^{-\beta_E(\langle E_i \rangle)-\mu)}$, with the values of A, q_E, β_E, and μ specified. In the inset we exhibit the same data in the q-logarithm ($q_E = 1.3$) versus $\langle E_i \rangle$, which, as expected, approaches a straight line. From [584]

that $(d + 2)\kappa$ depends only on α/d and, when α/d increases from zero to unity, it monotonically decreases from unity to zero, so remaining for all $\alpha/d > 1$: see Fig. 5.78. The value $\alpha/d = 1$ can therefore be seen as a critical point separating the mixing/ergodic regime from the anomalous one, κ playing a role analogous to that of an order parameter. This scaling law is consistent with Boltzmann-Gibbs statistics for $\alpha/d > 1$, and apparently with q-statistics for $0 \le \alpha/d < 1$, as we shall see here below.

Distribution of velocities and energies

Let us show here first-principle numerical results concerning the one-particle distributions of velocities and of energies. See Figs. 5.79 and 5.80.

A crossover might exist between the BG regime and the q-statistical one. One such example is exhibited in Fig. 5.81 from [482]. It verifies the 1999 conjecture [71] shown in Fig. 5.82.

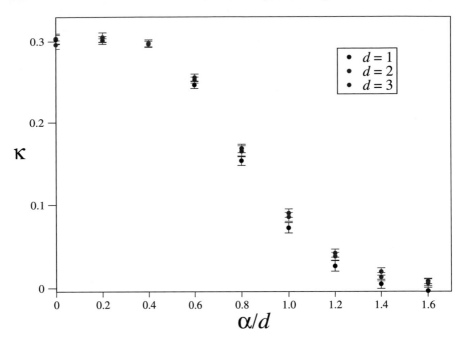

Fig. 5.73 Exponent $\kappa(\alpha/d)$ in the scaling of the maximum Lyapunov exponent $\lambda \sim N^{-\kappa}$. A collapse of the results of the $d = 1, 2, 3$ data is observed. It is an open challenge why they do not neatly approach zero near to $\alpha/d = 1$, but rather near $\alpha/d = 1.6$. From [584]

Realistic ionic-crystal model

By following [598], let us focus now on a realistic Hamiltonian whose first terms in an appropriate expansion resemble the FPU model. We refer precisely to the following $d = 3$ Hamiltonian:

$$\mathcal{H} = \sum_{j,s} \frac{\mathbf{p}_{j,s}^2}{2m_s} + \sum_{j,j',s,s'} V_{s,s'}(|\mathbf{x}_{j,s} - \mathbf{x}_{j',s'}|) \qquad (5.58)$$

with the phenomenological Buckingham potential given by

$$V_{s,s'}(r) = a_{s,s'}e^{-b_{s,s'}r} - \frac{c_{s,s'}}{r^6} + \frac{e_s^{eff}e_{s'}^{eff}}{r}. \qquad (5.59)$$

The parameters $\{a_{s,s'}\}$, $\{b_{s,s'}\}$, $\{c_{s,s'}\}$, and $\{e_s^{eff}\}$ that have been used are those adequate for LiF (Lithium Fluoride). The results for the energy distributions are exhibited in Figs. 5.83 and 5.84. It is certainly interesting to emphasize that q-statistics is obtained for relatively short times whereas BG statistics is obtained for long times. Such a

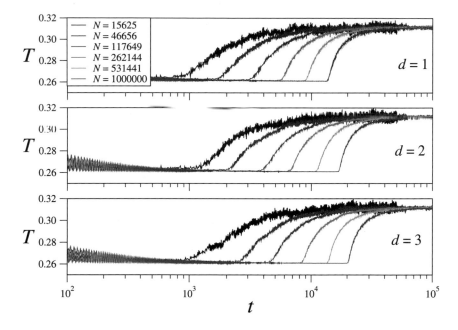

Fig. 5.74 Kinetic temperature, $T(t) = \langle K(t) \rangle / N$ versus time t for $\alpha/d = 0.9$, for widely different values of N on d-dimensional lattices ($d = 1, 2$ and 3) of linear sizes L, with $N = L^d$. Notice that the duration of the QSS (occasionally referred to as QSS1) increases with N and d. From [585]

crossover qualitatively is totally consistent with Fig. 5.81 and with the appealing scenario suggested in Fig. 1 of [336].

5.4   The q-Triplet and More

Let us further consider the ordinary differential equations that we addressed in Sect. 3.1.

The solution of the differential equation

$$\frac{dy}{dx} = a\,y \quad (y(0) = 1) \tag{5.60}$$

is given by

$$y = e^{a x}. \tag{5.61}$$

We may think of it in three different physical manners, related respectively to the sensitivity to the initial conditions, to the relaxation in phase space, and, if the system is Hamiltonian, to the distribution of energies at thermal equilibrium. In the first

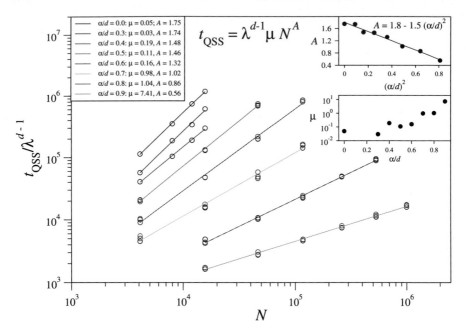

Fig. 5.75 The durations of the QSSs of Fig. 5.74, as well as of others corresponding to different values of α/d, can be rescaled conveniently with $t_{QSS} = [\lambda(\alpha/d)]^{d-1}\mu(\alpha/d)N^{A(\alpha/d)}$. The straight lines fit well for the data. To collapse all the $d = 1, 2, 3$ data we have used $\lambda = 1$, excepting for $\alpha/d = 0.8, 0.9$; indeed, their proximity to the short-range threshold, where no such collapse can exist, has demanded $\lambda = 1.16$. It is expected that, in the thermodynamic limit, λ approaches unity for all $0 \leq \alpha/d \leq 1$. *Inset:* We verify that the exponent A appears to depend linearly with $(\alpha/d)^2$. From [585]

interpretation we reproduce Eq. (2.31). In the second interpretation, we focus some relaxing relevant quantity

$$\Omega(t) \equiv \frac{O(t) - O(\infty)}{O(0) - O(\infty)}, \tag{5.62}$$

where O is some dynamical observable essentially related to the evolution of the system in phase space (e.g., the time evolution of entropy while the system approaches equilibrium). We typically expect

$$\Omega(t) = e^{-t/\tau_1}, \tag{5.63}$$

where τ_1 is the relaxation time. Finally, in the third interpretation, we have Eq. (2.67) (with Eq. (2.68), i.e.,

$$Z_1 p_i = e^{-\beta E_i}, \tag{5.64}$$

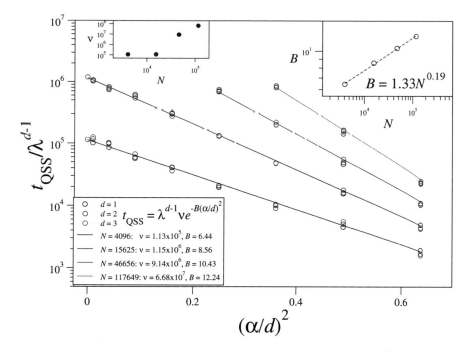

Fig. 5.76 The durations, for $d = 1, 2, 3$, of the QSSs are represented versus $(\alpha/d)^2$ in a log-linear plot. The full straight lines represent best fits for the data. *Inset*: N-dependence of B. From [585]

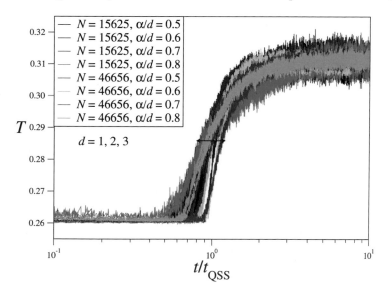

Fig. 5.77 Kinetic temperature $T(t)$ versus the scaled variable t/t_{QSS} for typical values of N ($N = 15625$, and 46656), α/d ($\alpha/d = 0.5, 0.6, 0.7, 0.8$), and $d = 1, 2$ and 3, through Eq. (5.54). The double arrow at mid-height (0.2859) characterizes the numerical abscissa spread of the collapse (from 0.8 to 1.2). From [585]

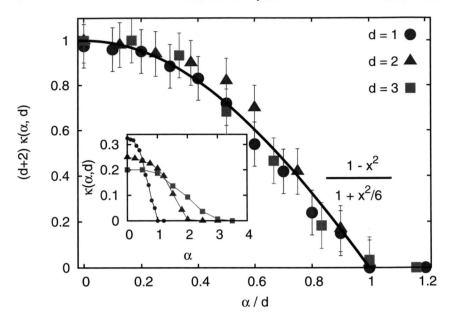

Fig. 5.78 The inset shows the exponent $\kappa(\alpha, d)$ as a function of α for $d = 1, 2, 3$. Note that $\kappa > 0$ for $0 \le \alpha < d$ and $\kappa = 0$ for $\alpha \ge d$. The main figure exhibits the universal law obtained by appropriately rescaling the abscissas and ordinates as indicated on the axes, i.e., $(d + 2)\kappa(\alpha, d) = f(\alpha/d)$. The thick continuous curve is the heuristic scaling function $f(x) = (1 - x^2)/(1 + x^2/6)$ [568], which, within the present precision, is a remarkably close fit to the collapsed data. The present collapse obviously implies $\kappa(0, d) = 1/(d + 2)$, hence $\lim_{d \to \infty} \kappa(0, d) = 0$, thus recovering mixing/ergodicity, as intuitively expected. From [594]

Table 5.1 Three possible physical interpretations of Eq. (5.61) within BG statistical mechanics

	x	a	$y(x)$
Equilibrium distribution	E_i	$-\beta$	$Z_1 p(E_i) = e^{-\beta E_i}$
Sensitivity to the initial conditions	t	λ_1	$\xi(t) = e^{\lambda_1 t}$
Typical relaxation of observable O	t	$-1/\tau_1$	$\Omega(t) = e^{-t/\tau_1}$

where $Z_1 \equiv \sum_{j=1}^{W} e^{-\beta E_j}$ is the partition function. The various interpretations are summarized in Table 5.1.

Let us now generalize these statements. The solution of the differential equation

$$\frac{dy}{dx} = a\, y^q \quad (y(0) = 1) \tag{5.65}$$

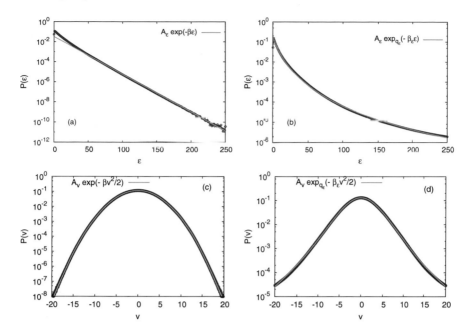

Fig. 5.79 $d = 1$ FPU model. **a** Boltzmann exponential continuous fitting of the one-particle energy ϵ distribution in the presence of short-range anharmonic interactions ($\alpha = 2$): $P(\epsilon) = 0.033\,e^{-0.086\,\epsilon}$. **b** The q-exponential continuous fitting of the one-particle energy ϵ distribution in the presence of long-range anharmonic interactions ($\alpha = 0.9$): $P(\epsilon) = 0.15\,e_{1.22}^{-0.21\,\epsilon}$. For small values of ϵ, a slight departure from the purely exponential (or q-exponential) behavior is observed, as expected due to a regular density of states. **c** Gaussian continuous fitting of the one-particle velocity v distribution in the presence of short-range anharmonic interactions ($\alpha = 2$): $P(\epsilon) = 0.116\,e^{-0.084\,v^2/2}$. **d** The q-Gaussian continuous fitting of the one-particle energy ϵ distribution in the presence of long range anharmonic interactions ($\alpha = 0.9$): $P(\epsilon) = 0.131\,e_{1.23}^{-0.132\,v^2/2}$. The other parameters are $(a, b, N) = (1, 10, 800)$. From [588]

is given by

$$y = e_q^{a\,x}. \tag{5.66}$$

These expressions respectively generalize expressions (5.60) and (5.61).[12] As before, we may think of them in three different physical manners, related respectively to the sensitivity to the initial conditions, to the relaxation in phase space, and, if the

---

[12] For generic $y(0) > 0$, the solution of (5.65) is given by $y(x) = y(0)\,e_q^{a\,[y(0)]^{q-1}\,x}$. We have a monotonically increasing function $y(x)$ for $a > 0$ (with infinite support if $q \leq 1$ and finite support if $q > 1$), and a monotonically decreasing function for $a < 0$ (with infinite support if $q \geq 1$ and finite support if $q < 1$). Notice an important point that will permeate through discussions on related issues: if $q = 1$, *and only then*, the coefficient a (which characterizes the scale of the evolution of $y(x)/y(0)$ with x) is *not* renormalized by the initial condition $y(0)$. If $q \neq 1$, the effective constant $\{a\,[y(0)]^{q-1}\}$ differs from a; the difference can be very important depending on the values of q and $y(0)$ (see, for instance, [601]).

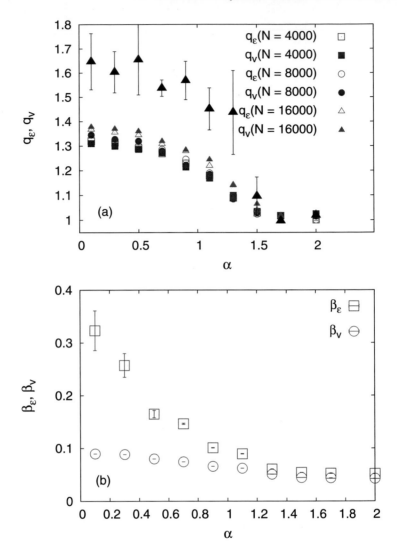

Fig. 5.80 **a** We exhibit here that, within error bars, $q_\epsilon = q_v$ monotonically decreases for increasing α and achieves the Boltzmann value $q_\epsilon = q_v = 1$ for short-range interactions. The $N \to \infty$ values (black triangles) have been extrapolated from the finite N values by following the procedure indicated in [593], namely performing $q(N)$ versus $1/\ln N$ extrapolations for increasing N. **b** we exhibit here that, in contrast with the corresponding q-indices, the inverse temperatures β_ϵ and β_v *do not* coincide unless we are in the Boltzmannian regime $q_v = q_\epsilon = 1$ (short-range interactions); the emergence of temperatures which differ from the usual kinetic one are frequent in such complex systems [314, 566]. From [588]

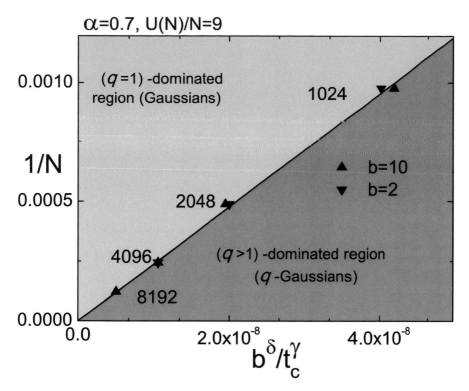

Fig. 5.81 A unified overview of the crossover frontier between BG- and q-statistics for the $d = 1$ FPU model, including diverse values of b. The fitting straight line is $1/N = Db^\delta/t_c^\gamma$, with $D = 2.3818 \times 10^4$, $\delta = 0.27048$, and $\gamma = 1.365$. From [482]

system is Hamiltonian, to the distribution of energies at a stationary state. In the first interpretation we reproduce Eq. (5.8). In the second interpretation, we typically expect

$$\Omega(t) = e_{q_{rel}}^{-t/\tau_{q_{rel}}}, \tag{5.67}$$

where $\tau_{q_{rel}}$ is the relaxation time. Finally, in the third interpretation, we have Eq. (3.241) (with Eq. (3.242)), i.e.,

$$Z_{q_{stat}} p_i = e^{-\beta_{q_{stat}} E_i}, \tag{5.68}$$

where $Z_{q_{stat}} \equiv \sum_{j=1}^{W} e^{-\beta_{q_{stat}} E_j}$ is the partition function. The various interpretations are summarized in Table 5.2.

The set $(q_{sen}, q_{rel}, q_{stat})$ constitutes what we shall refer to as the *q-triplet* (occasionally referred also to as the *q-triangle*). In the *BG* particular case (characterized by full mixing and ergodicity), we should recover $q_{sen} = q_{rel} = q_{stat} = 1$. The existence of these three q-exponentials characterized by the q-triplet was conjectured

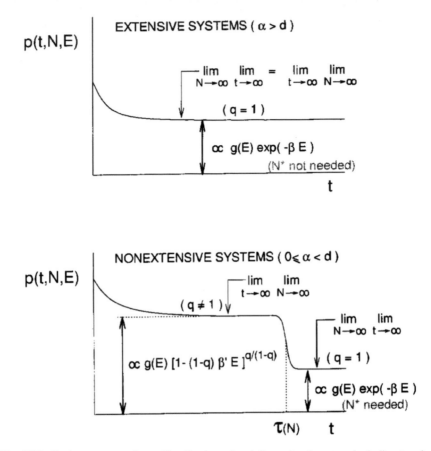

Fig. 5.82 Conjecture assuming a Hamiltonian of a d-dimensional system including two-body (attractive) interactions which decay at long distance as $r^{-\alpha}$. The crossover at $t \sim \tau(N)$ is expected to be slower than indicated in the figure for space reasons; $\lim_{N \to \infty} \tau(N) = \infty$; $N^* \equiv \frac{N^{1-\alpha/d}-1}{1-\alpha/d}$; $g(E)$ is the density of states; $q \geq 1$. From [71]

for complex systems (of the nonextensive type) in 2004 [602] (see also [255]) and confirmed in 2005 by the NASA researchers Burlaga and Vinas [480]. Indeed, the observations accomplished through processing data sent to Earth by the spacecraft Voyager 1 for the solar wind at the distant heliosphere, and also, more recently, in the heliosheath [603–606] (see also [607–610]) are depicted in Fig. 5.85. The Voyager 1 spacecraft was launched in 1977, nearly 45 years ago. It is therefore unreasonable to expect high precision results. This said, the values advanced by Burlaga and Vinas in 2005 [480] were $(q_{sen}, q_{rel}, q_{stat}) = (-0.6 \pm 0.2,\ 3.8 \pm 0.3,\ 1.75 \pm 0.06)$. Since only one of them might be independent, one expects a priori two relations to exist between these three indices. Such relations were heuristically advanced in [274].

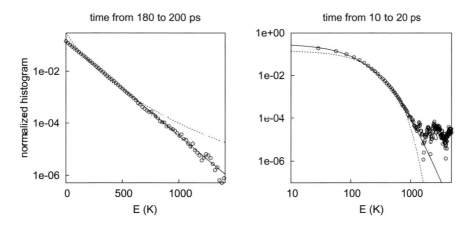

Fig. 5.83 Maxwell-Boltzmann distribution and q-distribution. Left panel. Histogram (in semiloga-rithmic scale) of the energies E (circles) of the modes not initially excited, from time $t = 180$ (after their equipartition is attained) up to $t = 200\ ps$. Only the 15 modes of lowest frequency were initially excited, among the total number 1536 of modes. Specific energy $\epsilon = 120\ K$, $N = 512$. The solid line is the graph of the Maxwell-Boltzmann distribution function $Ce^{-E/\epsilon}$ with $C = 0.15\ K^{-1}$. Right panel. Same as left panel, in logarithmic scale, with data collected for time from 10 to 20 ps. Solid line is the graph of the q-distribution function $C'[1 + (q - 1)\beta E]^{1/(1-q)}$ for $(q, \beta^{-1}, C') = (1.14, 67.8\ K, 0.308\ K^{-1})$. From [598]. The left (right) red dashed curve rep-resents the right (left) black solid curve. This figure has been kindly constructed by the authors of [598]

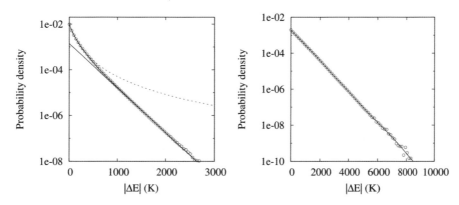

Fig. 5.84 Histograms of the random variable $|\Delta E|$ (circles) for $\epsilon = 500K$, $N = 512$, standard initial conditions. *Left panel*: $\tau = 0.02\ ps$ (non exponential stage); dashed line, fit with the dis-tribution $C[1 + (q - 1)\beta E]^{1/(1-q)}$ with $(q, 1/\beta) = (1.33, 71\ K)$; red line, fit with an exponen-tial distribution with $1/\beta = 223\ K$, for data restricted above 1500 K; blue line, fit with the dis-tribution $C[1 - (\lambda/\mu) + (\lambda/\mu)\ e^{(q-1)\mu E}]^{1/(1-q)}$ with $(q, 1/\lambda, 1/\mu) = (1.33, 71\ K, 223\ K$. *Right panel*: $\tau = 20\ ps$ (full relaxation); red line, best fit with exponential distribution over all data $(1/\beta = 506.6\ K)$. From [600]

Table 5.2 Three possible physical interpretations of Eq. (5.66) within nonextensive statistical mechanics

	x	a	$y(x)$
Stationary state distribution	E_i	$-\beta$	$Z_{q_{stat}} p(E_i) = e_{q_{stat}}^{-\beta E_i}$
Sensitivity to the initial conditions	t	$\lambda_{q_{sen}}$	$\xi(t) = e_{q_{sen}}^{\lambda_{q_{sen}} t}$
Typical relaxation of observable O	t	$-1/\tau_{q_{rel}}$	$\Omega(t) = e_{q_{rel}}^{-t/\tau_{q_{rel}}}$

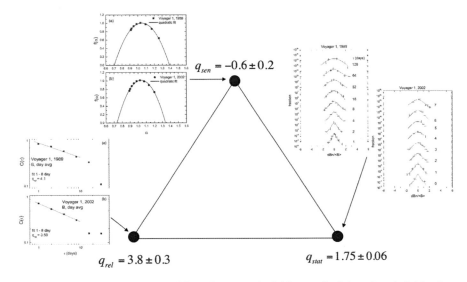

$q_{sen} = -0.6 \pm 0.2$

$q_{rel} = 3.8 \pm 0.3$ $q_{stat} = 1.75 \pm 0.06$

Fig. 5.85 The q-triplet as measured from the magnetic field strength of the solar wind. The three sets of curves correspond to daily averages of the data sent in 1989 (40 AU) and in 2002 (85 AU) by the NASA Voyager 1 spacecraft. See details in [480]. Within the errors bars, these three values have been heuristically approached [274] by the values $(q_{sen}, q_{rel}, q_{stat}) = (-0.5, 4, 7/4)$

The outcome that was found is $(q_{sen}, q_{rel}, q_{stat}) = (-1/2, 4, 7/4)$, which, within the error bars, is consistent with the NASA results.[13]

---

[13] If we have a triplet (x, y, z) of real numbers such that one of them, say x, is the arithmetic average of the other two (i.e., $x = \frac{y+z}{2}$), and one of the other two, say y, is the harmonic average of the other two (i.e., $y^{-1} = \frac{x^{-1}+z^{-1}}{2}$), then, remarkably enough, the third number necessarily is the geometric average of the other two (i.e., $z = \sqrt{xy}$). If we define now $\epsilon \equiv 1 - q$, we have, from [274], that $(\epsilon_{sen}, \epsilon_{rel}, \epsilon_{stat}) = (3/2, -3, -3/4)$. By identifying $(x, y, z) \equiv (\epsilon_{stat}, \epsilon_{rel}, \epsilon_{sen})$, it can be checked that they satisfy the just mentioned remarkable relationships! [616]. In fact, these relations admit only one degree of freedom. In other words, we can freely choose only one number, say x; the other two (y and z) are automatically determined. If $x \geq 0$, the solution is $x = y = z$; if $x < 0$, the solution is $x = y/4 = -z/2$. The set $(\epsilon_{stat}, \epsilon_{rel}, \epsilon_{sen}) = (-3/4, -3, 3/2)$ belongs to this latter case.

This outcome was obtained by introducing the so-called *additive duality*

$$q \leftrightarrow 2 - q \tag{5.69}$$

and *multiplicative duality*

$$q \leftrightarrow 1/q. \tag{5.70}$$

Combining these two transformations, a full algebra can be constructed. These two dualities have in fact emerged more than once in the realm of nonextensive statistical mechanics by several authors and for different motivations (see for instance [69, 389, 611–613]). Both dualities satisfy two crucial properties, namely that $q = 1$ is the unique fixed point (precisely corresponding to BG statistical mechanics), and that its inverse coincides with itself (hence the use of the word *duality*).

Another q-triplet was exhibited some years later, namely at the edge of chaos (frequently referred to as the *Feigenbaum point*) of the logistic map: $(q_{sen}, q_{rel}, q_{stat}) = (0.24448..., 2.24978..., 1.65 \pm 0.05)$ (see [530, 531] and references therein). Although far from transparent, we assume here that the value of q corresponding to the q-Gaussian attractor is to be identified with q_{stat}.

Since then, a plethora of q-triplets have been observed in a great variety of systems. Some illustrative cases are presented in Table 5.3 [479].

These and the solar wind NASA results together seem to indicate that the frequent scheme for the q-triplet is $q_{sen} \leq 1 \leq q_{stat} \leq q_{rel}$. Nevertheless, some examples do exist that appear to be $q_{sen} \leq 1 = q_{rel} \leq q_{stat}$. Further understanding is certainly needed. The basis for this understanding appears to be based on a generalization of the concept of duality. More precisely, transformations (5.69) and (5.70) can be unified through [614, 615]

$$q \leftrightarrow \frac{aq + 1 - a}{(1 + a)q - a} \quad (a \in \mathbb{R}), \tag{5.71}$$

or equivalently

$$\frac{1}{1 - q} \leftrightarrow \frac{q}{q - 1} + a \quad (a \in \mathbb{R}). \tag{5.72}$$

We verify that $a = -1$ and $a = 0$ respectively recover (5.69) and (5.70). Transformation (5.71) is the most general form $\frac{A+Bq}{C+Dq}$ (see Eq. (4.108)) which has an unique fixed point $q = 1$ and satisfies self-duality. We see in [614] the alternative form $q \leftrightarrow \frac{a+2-aq}{a-(a-2)q}$; however, if we make therein the notation change $a \to -2a$, we precisely recover transformation (5.71). Analogously, we see in [615], $q \leftrightarrow \frac{a_1 - a_2 q}{a_2 - (2a_2 - a_1)q}$, i.e., $q \leftrightarrow \frac{1-(a_2/a_1)q}{(a_2/a_1)-[2(a_2/a_1)-1]q}$; however, if we make therein the notation change $(a_2/a_1) \to a/(a - 1)$, once again we precisely recover transformation (5.71). Therefore, transformation (5.71) can indeed be considered as the most general ratio of polynomials that are linear in q.

Table 5.3 Numerical data from a non exhaustive series of observations displaying sensitivity, stationarity, and relaxation q-indices, and their respective auxiliary indices issued from three-term cycle or two-term cycle fixed points q_1. For the three-term cycles, One can observe an interesting closeness between "Solar wind" [274, 480], "Feigenbaum point" [186, 532], and "Brazos river" [617] in the sense that, for all of them, $q_1 \simeq 0$, whilst "Bitcoin" [618], "Standard map" [513, 515–517], and "Ozone layer" [619], are neatly apart from them. With those, as well as with the present two-term cycles [620], one cannot conclude. From [479]

	q_{sens}	q_{stat}	q_{rel}	q_{aux}	q_1 (fixed point)
Solar wind (conjectural)	-1/2	7/4	4	0.5	0
Solar wind (observations)	-0.6 ± 0.2	1.75 ± 0.06	3.8 ± 0.3	0.5158	0.0316
Feigenbaum point (calculations)	0.2444877...	1.65 ± 0.05	2.2497841	0.50375	0.0075
Brazos river (observations)	0.244	1.65	2.25	0.5203	0.0406
Bitcoin (observations)	0.14	1.54	2.25	0.6088	0.2176
Standard map (calculations)	0	1.935	1.4	0.71985	0.4397
Ozone layer (observations)	-8.1	1.32	1.89	0.805	0.61
Solar activity/SN (observations)	-0.71 ± 0.10	1.31 ± 0.07	1	0.725	0.725
Solar activity/MF (observations)	-0.44 ± 0.07	1.21 ± 0.06	1	0.803	0.803
Solar activity/TSI (observations)	-0.52 ± 0.10	1.54 ± 0.03	1	0.544	0.544

Let us generalize now the duality definitions (4.100) and (4.101) respectively as follows [614]:

$$\mu(q) \equiv \frac{aq + 1 - a}{(1 + a)q - a} \quad (a \in \mathbb{R}), \qquad (5.73)$$

and

$$\nu(q) \equiv \frac{bq + 1 - b}{(1 + b)q - b} \quad (b \in \mathbb{R}; \ b \neq a), \qquad (5.74)$$

where we have used transformation (5.71). These relations can, equivalently, be expressed as $\frac{1}{1-\mu(q)} = a + \frac{q}{q-1}$ and $\frac{1}{1-\nu(q)} = b + \frac{q}{q-1}$ respectively. Also, we immediately verify that, for all pairs (a, b) with $b \neq a$, properties (4.102), (4.103) and

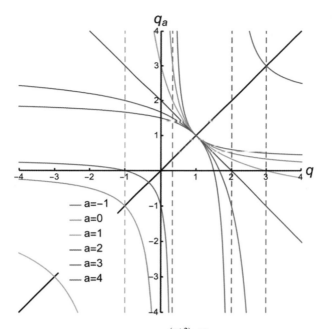

Fig. 5.86 The self-dual transformation $q_a(q) = \frac{(a+2)-aq}{a-(a-2)q}$ for typical values of a; $q_2(q) = 2 - q$ recovers the additive duality; $q_0(q) = 1/q$ recovers the multiplicative duality. For $a > 2$, when q varies within $(-\infty, 1]$, q_a varies biunivocally within $[\frac{a}{a-2}, 1]$, when q varies within $[1, \frac{a}{a-2}]$, q_a varies biunivocally within $[1, -\infty)$, and when q varies within $[\frac{a}{a-2}, \infty)$, q_a varies biunivocally within $(\infty, \frac{a}{a-2}]$. For $a < 2$, when q varies within $(-\infty, \frac{a}{-(2-a)}]$, q_a varies biunivocally within $[\frac{a}{-(2-a)}, -\infty]$, when q varies within $[\frac{a}{-(2-a)}, 1]$, q_a varies biunivocally within $(\infty, 1]$, and when q varies within $[1, \infty]$, q_a varies biunivocally within $[1, \frac{a}{-(2-a)}]$. From [614]

(4.104), as well as definitions $(\mu\nu)^{-n} \equiv (\nu\mu)^n$ and $(\nu\mu)^{-n} \equiv (\mu\nu)^n$, remain valid. It straightforwardly follows that $\frac{1}{1-\nu(\mu(q))} = b - a + \frac{1}{1-q}$ and $\frac{1}{1-\mu(\nu(q))} = a - b + \frac{1}{1-q}$. Consistently, the set of equations (4.105), (4.106) and (4.107) is generalized, for $z = 0, \pm1, \pm2, \pm3, \ldots$, as follows:

$$\frac{1}{1-(\mu\nu)^z(q)} = z(a-b) + \frac{1}{1-q}, \tag{5.75}$$

$$\frac{1}{1-\nu(\mu\nu)^z(q)} = z(b-a) + b + \frac{q}{q-1}, \tag{5.76}$$

$$\frac{1}{1-(\mu\nu)^z\mu(q)} = z(a-b) + a + \frac{q}{q-1}. \tag{5.77}$$

Typical connections are depicted in Figs. 5.86 and 5.87.

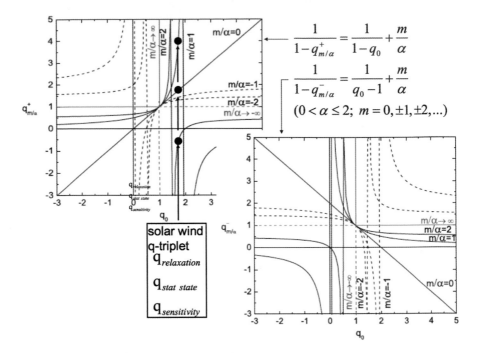

Fig. 5.87 *Left plot:* $q_{m/\alpha}^{+}$ as a function of $q_0^{+} \equiv q_0$ for typical values of m/α. For clarity, the vertical asymptotes are indicated as well. For each value $m > 0$, there are two branches: the left branch is such that $q_{m/\alpha}^{+} > 1$ ($q_{m/\alpha}^{+} < 1$) if $q_0 > 1$ ($q_0 < 1$); the right branch does not contain the point $(q_0, q_{m/\alpha}^{+}) = (1, 1)$. For each value $m/\alpha < 0$, there are two branches: the right branch is such that $q_{m/\alpha}^{+} > 1$ ($q_{m/\alpha}^{+} < 1$) if $q_0 > 1$ ($q_0 < 1$); the left branch does not contain the point $(q_0, q_{m/\alpha}^{+}) = (1, 1)$. *Right plot:* $q_{m/\alpha}^{-}$ as a function of $q_0^{+} \equiv q_0$ for typical values of m/α. For clarity, the vertical asymptotes are indicated as well. For each value $m/\alpha > 0$, there are two branches: the right branch is such that $q_{m/\alpha}^{+} > 1$ ($q_{m/\alpha}^{+} < 1$) if $q_0 < 1$ ($q_0 > 1$); the left branch does not contain the point $(q_0, q_{m/\alpha}^{+}) = (1, 1)$. For each value $m/\alpha < 0$, there are two branches: the left branch is such that $q_{m/\alpha}^{+} > 1$ ($q_{m/\alpha}^{+} < 1$) if $q_0 < 1$ ($q_0 > 1$); the right branch does not contain the point $(q_0, q_{m/\alpha}^{+}) = (1, 1)$. From [621]. The solar wind q-triplet is exhibited: $(q_{sensitivity}, q_{stat\ state}, q_{relaxation}) = (-1/2, 7/4, 4)$ [274, 480]. This figure has been produced in co-authorship with G. Ruiz

   Summarizing at this stage, we realize that, for any fixed (q, a, b) we obtain a countable infinity of q's by running all the integer values of z. This set would in principle characterize the various physical properties of a given complex system, and would be the seed of its universality class. The standard BG behavior, where all q's collapse onto the unique value $q = 1$, corresponds to (i) the BG entropic functional (which yields an extensive entropy, as required by Clausius), (ii) the celebrated BG factor for the distribution of energies, and (iii) an exponential relaxation towards thermal equilibrium, among others. From this perspective, the emergence of at least one q different from unity would reflect the complexity of the system. Within this scenario, the most typical situation occurs when the system exhibits a q-triplet (q_{sens}, q_{stat}, and q_{rel}, as illustrated in Table 5.3). The first q-triplet ever detected in nature has been that in the solar wind, by NASA researchers [480]. It was interpreted [274] within the above structure (with $(a, b) = (0, -1)$), which is that of the Moebius group of transformations [615]. This mathematical structure enabled to establish [479] a three-term-cycle conservation rule, namely

$$\frac{1}{q_{rel} - 1} + \frac{1}{q_{stat} - 1} + \frac{1}{q_{sens} - 1} = \frac{1 + q_1}{1 - q_1}, \tag{5.78}$$

which, for the fixed point $q_1 = 0$ [479], becomes

$$\frac{1}{q_{rel} - 1} + \frac{1}{q_{stat} - 1} + \frac{1}{q_{sens} - 1} = 1. \tag{5.79}$$

As we can check in the Table 5.3), this sum rule is numerically satisfied within less than 10% error for the solar wind, the Feigenbaum point, and the Brazos river. In contrast, the error is larger than 50% for the bitcoin, standard map, and ozone layer data. Clearly, further understanding is necessary along this path.

Concerning transformations of the indices q, it is worthy to mention that an interesting mathematical structure has been found for the family of q-Gaussians [622].

5.5 Connection with Critical Phenomena

It is since long known that systems at criticality (in the sense of standard second-order critical phenomena) exhibit a fractal geometry. Consistently, it is kind of natural to expect that connections would exist between q and critical phenomena: see [623–625]. In particular, an interesting analytical connection has already been established for the Ising ferromagnet, namely [624]

$$q = \frac{1 + \delta}{2}, \tag{5.80}$$

where δ is the critical exponent characterizing the dependance, at *precisely* the critical point, of the order parameter with its thermodynamically conjugate field (e.g., $M \sim H^{1/\delta}$, where M and H are respectively the magnetization and the external magnetic field). Further connections between critical exponents, as well as the associated scaling laws, and q are discussed in [626–628].

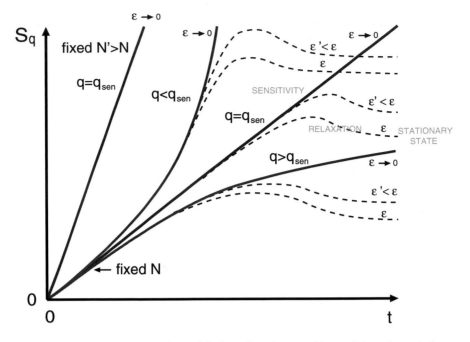

Fig. 5.88 Schematic time-dependence of S_q for various degrees of fine-graining ϵ. Instead of q_{sen}, a better notation would be q_{ent} (we know that for one-dimensional nonlinear dynamical systems, we typically have $q_{ent} = q_{sen}$). We are disregarding in this scenario the influence of possible averaging over initial conditions that might be necessary or convenient. From [275]

5.6 A Conjecture on the Time and Size Dependences of Entropy

We have seen that, for a (not yet fully qualified) large class of systems, there is a special value of q, q_N, such that $S_{q_N}(N, t) \propto N$ $(N \to \infty)$. This is so for all values of time t, including $t \to \infty$, if we are describing the system within some *finite* resolution (or some *finite* degree of fine-graining, i.e., $\epsilon > 0$). We have also seen that, for a (once again not yet fully qualified) large class of systems, there is a special value of q, q_t, such that $S_{q_t}(N, t) \propto t$ $(t \to \infty)$. This is so for an *infinite* resolution (or *ideally precise* degree of fine-graining, i.e., $\epsilon = 0$). The scenario is schematically indicated in Fig. 5.88. If this scenario is correct, then many systems plausibly exist for which $q_N = q_t \equiv q_{ent}$, hence, in the $\epsilon \to 0$ limit, we would have the following form:

$$S_{q_{ent}} \sim s\,N\,t \quad (N \to \infty; \ t \to \infty; \ s \geq 0). \tag{5.81}$$

Chapter 6
Generalizing Nonextensive Statistical Mechanics

Aqui... onde a terra se acaba e o mar começa...
Luís Vaz de Camões
Canto Oitavo – LUSÍADAS

We have schematically represented in Fig. 6.1[1] various possible thermostatistical theories. The present chapter is dedicated to a brief exploration of the *non-q-describable region*.

6.1 Crossover Statistics

Equation (5.60) (paradigmatic for BG statistics), and (5.65) (paradigmatic for nonextensive statistics) can be unified in the following one [459][2]:

$$\frac{dy}{dx} = -a_1\, y - (a_q - a_1)\, y^q \qquad (6.1)$$

We recover the BG equation for $q = 1$ ($\forall\, a_q$), or for $a_q = a_1$ ($\forall q$). We recover the nonextensive Eq. (5.65) for $a_1 = 0$. The instances of Eq. (6.1) for which q is a natural

---

[1] The case of standard critical phenomena deserves a comment. The BG theory explains, as is well known, a variety of properties as close to the critical point as we want. If we want, however, to describe with *finite* quantities certain discontinuities that occur *precisely at the critical point* (e.g., some fractal dimensions connected with the space dimension $d_s = 3$, for say the Ising and Heisenberg ferromagnets at T_c), we need a different theoretical approach.

[2] We could equivalently introduce this unification through $\frac{dy}{dx} = -\bar{a}_1\, y - \bar{a}_q\, y^q$ with $\bar{a}_1 \equiv a_1$ and $\bar{a}_q \equiv a_q - a_1$. Both this form and that of Eq. (6.1) recover Eq. (5.65) for $a_1 = 0$, and both recover Eq. (5.60) for $\bar{a}_q = 0$. The form (6.1) has however a small additional convenience, namely that it recovers Eq. (5.60) *also* for $q = 1$. This advantage is lost in the form $\frac{dy}{dx} = -a_1\, y - \bar{a}_q\, y^q$, which requires an extra precaution whenever writing computer codes.

© Springer Nature Switzerland AG 2023
C. Tsallis, *Introduction to Nonextensive Statistical Mechanics*,
https://doi.org/10.1007/978-3-030-79569-6_6

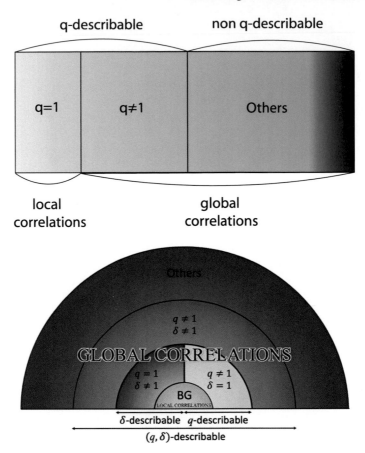

Fig. 6.1 *Top*: Scenario within which nonextensive statistical mechanics is located. At the extreme "left" of the $q = 1$ (BG) region we essentially find the noninteracting systems, such as the ideal gas, and the ideal paramagnet; in the "middle" $q = 1$ region we have systems such as the first-neighbor Ising model and the Lennard-Jones fluid; at its extreme "right" region, we find the critical phenomena associated with standard phase transitions. These systems exhibit, at precisely the critical point, global correlations which bridge with the $q \neq 1$ systems (e.g., [285–287, 624, 625]). In the "middle" of the $q \neq 1$ region we find systems such as Hamiltonian ones with long-range interactions (e.g., [482, 566, 567, 584]), overdamped many-body systems (e.g., [314, 443, 444]), granular matter (e.g., [629, 630]), cold atoms (e.g., [631, 632]), scale-free networks (e.g., [634–637]), logistic and standard maps (e.g., [188, 513, 531]), high-energy particle collisions (e.g., [638]), quantum nonlinearity (e.g., [639, 640]); at its extreme "right" we find very complex systems, for which a statistical mechanics even more general than the nonextensive one, or even neatly different, might be necessary. From [275] (see [274] for more details). *Bottom*: Extension of the above scenario in order to include the entropic functional $S_{(1-q)}$, which differs from S_q, as well as the unifying entropy $S_{q,(1-q)}$ (with $S_{q,1} = S_q$ and $S_{1,(1-q)} = S_{(1-q)}$). In the $S_{(1-q)}$ region we find black holes and holographic dark energy models (e.g., [222, 258, 259, 263, 266, 267]). By *local* correlations we mean both in space and time; by *global* correlations we mean in space and/or time. Nevertheless, it is not unthinkable that, for extremely complex systems, a statistical mechanics compatible with classical thermodynamics could be just impossible to exist. The bottom figure has been produced in co-authorship with A. Pluchino

number are particular cases of the Bernoulli differential equations [641]. The solution of Eq. (6.1) is given by

$$y = \frac{1}{\left[1 - \frac{a_q}{a_1} + \frac{a_q}{a_1} e^{(q-1)a_1 x}\right]^{\frac{1}{q-1}}} \qquad (x \geq 0). \qquad (6.2)$$

It can be straightforwardly verified that it contains, as particular instances, the solutions of Eq. (5.60) and of Eq. (5.65). We can also verify that

$$y \sim \begin{cases} 1 - a_q x & \text{if } 0 \leq x << x_q^* \equiv \frac{1}{(q-1)a_q}, \\[3mm] \frac{1}{[(q-1) a_q x]^{\frac{1}{q-1}}} & \text{if } x_q^* << x << x_1^{**} \equiv \frac{1}{(q-1)a_1}, \\[3mm] \left(\frac{a_1}{a_q}\right)^{\frac{1}{q-1}} e^{-a_1 x} & \text{if } x >> x_1^{**}. \end{cases} \qquad (6.3)$$

As we see, this solution makes a crossover from a q-exponential behavior at low values of x, to an exponential one for high values of x (see Fig. 6.2). If x is to be interpreted as an energy (see Tables 5.1 and 5.2), this constitutes a generalization of the q-statistical weight. It is from this property that this statistics is sometimes referred to as *crossover statistics*.

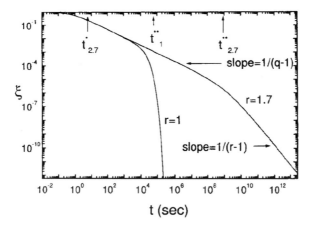

Fig. 6.2 Log–log plot of $\xi \equiv y$ *versus* $t \equiv x$ for $q = 2.7$, $a_q = 1$, $a_1 = 10^{-5}$, and both $r = 1$ and $r = 1.7$. The characteristic values $t_q^* \equiv x_q^*$ and $t_r^{**} \equiv x_r^{**}$ are indicated by arrows (the regions corresponding to short-, intermediate-, and long- abscissa values are clearly exhibited). The slope of the intermediate region is $-1/(q - 1)$. From [459]

Equation (6.1) can be further generalized as follows:

$$\frac{dy}{dx} = -a_r\, y^r - (a_q - a_r)\, y^q \quad (1 \le r \le q) \tag{6.4}$$

The solution of this equation has no *explicit* expression $y(x)$, but only $x(y)$. This expression appears in terms of two hypergeometric functions [459], and also corresponds to a crossover, namely

$$y \sim \begin{cases} 1 - a_q\, x & \text{if } 0 \le x << x_q^* \equiv \frac{1}{(q-1)\,a_q}\,, \\[2ex] \dfrac{1}{[(q-1)\,a_q\, x]^{\frac{1}{q-1}}} & \text{if } x_q^* << x << x_r^{**} \equiv \dfrac{[(q-1)\,a_q]^{\frac{r-1}{q-r}}}{[(r-1)\,a_r]^{\frac{q-1}{q-r}}}\,, \\[2ex] \dfrac{1}{[(r-1)\,a_r\, x]^{\frac{1}{r-1}}} & \text{if } x >> x_r^{**}. \end{cases} \tag{6.5}$$

Because it exhibits a crossover from a q-exponential behavior at low x, to an r-exponential one at high x (see Fig. 6.2), it reinforces its reference as *crossover statistics* in the literature. This type of function has been extremely efficient in fitting a variety of experimental data (see, e.g., [459, 642]).

Let us mention at this point that Eq. (6.4) can be generalized as follows:

$$\frac{dy}{dx} = -\sum_{k=1}^{M} \bar{a}_k\, y^{q_k} \quad (q_1 \le q_2 \le \cdots \le q_M), \tag{6.6}$$

hence

$$x = \int_y^1 \frac{dz}{\sum_{k=1}^{M} \bar{a}_k\, z^{q_k}}. \tag{6.7}$$

As mentioned above, this integral has a closed analytical expression $x(y)$ in terms of two hypergeometric functions for $M = 2$ [459], but needs to be numerically solved for $M \ge 3$.

It should be clear that the generalization of a statistical weight is necessary but not sufficient for having a generalized statistical mechanics. Indeed, the generalization of the entropy is also needed so that the generalized statistical weight can be *deduced* from the entropy through a variational procedure. It is through this path that we can expect to have a smooth matching with thermodynamics itself. In the next section, we further generalize the present approach. We briefly present *spectral statistics* (a straightforward generalization of the crossover statistics), and *Beck-Cohen superstatistics*, which focuses on a possible distribution of parameters such as the direct (or inverse) temperature, assumed to be spatio-temporally fluctuating.

6.2 Further Generalizing

The q-statistical distribution and its generalization, crossover statistics, have been further generalized into *spectral statistics* [383] and Beck-Cohen superstatistics [29, 337, 643, 644] . The exact mathematical connection between spectral and Beck-Cohen statistics is not yet fully clarified. However, as we shall argue later on, indications exist that spectral statistics contains Beck-Cohen statistics as a particular case. Therefore the logical structure appears to be

$$BG\, statistics \subset q - statistics \subset crossover\, statistics$$
$$\subset Beck - Cohen\, superstatistics \subset spectral\, statistics \quad (6.8)$$

It is worthy to emphasize at this point that we are here focusing only on the (stationary state) probability distributions. As previously mentioned, this ingredient is necessary but not sufficient for implementing a full statistical-mechanical theory. It is also necessary to consistently define an entropy functional which, under appropriate constraints, is optimized precisely by that particular distribution. More than that, one of the constraints must have an admissible connection with the concept of energy (another, trivial, constraint is of course normalization). These various steps are going to be illustrated in the next subsections.

6.2.1 Spectral Statistics

Equation (6.4) can be naturally generalized into

$$\frac{dy}{dx} = - \int d\kappa\, F(\kappa)\, y^\kappa \quad (y(0) = 1), \quad (6.9)$$

hence,

$$x = \int_y^1 \frac{dz}{\int d\kappa\, F(\kappa)\, z^\kappa}, \quad (6.10)$$

The nonnegative *q-spectral function* $F(\kappa)$ (QSF) must be integrable, i.e., $\int d\kappa\, F(\kappa)$ must be finite. This (positive) integral does not need to be unity, i.e., $F(\kappa)$ is generically unnormalized. The particular case

$$F(\kappa) = a_r\, \delta(\kappa - r) + (a_q - a_r)\, \delta(\kappa - q), \quad (6.11)$$

$\delta(x)$ being Dirac's delta distribution, recovers Eq. (6.4). Naturally, $F(\kappa) = \sum_{k=1}^{M} \bar{a}_k\, \delta(\kappa - q_k)$ leads to Eqs. (6.6) and (6.7). Unless specified otherwise, for simplicity we shall from now on assume that $F(\kappa)$ is normalized (see in [383] details about how an unnormalized $F(\kappa)$ can be transformed into a normalized one).

The possible solution of Eq. (6.9) will be noted $\exp_{\{F\}}(x)$. In other words,

$$\frac{d\exp_{\{F\}}(x)}{dx} = -\int_{-\infty}^{\infty} d\kappa\, F(\kappa)\,[\exp_{\{F\}}(x)]^{\kappa}. \tag{6.12}$$

By setting $x = \ln_{\{F\}} y$, we have

$$\frac{dy}{d\left[\ln_{\{F\}} y\right]} = \int_{-\infty}^{+\infty} F(\kappa)\, y^{\kappa} d\kappa, \tag{6.13}$$

hence

$$\ln_{\{F\}} x = \int_{1}^{x} \left\{\int_{-\infty}^{+\infty} F(\kappa)\, u^{\kappa} d\kappa\right\}^{-1} du \quad (\forall x \in (0,\infty)), \tag{6.14}$$

which is the generic expression of the *inverse function* of $\exp_{\{F\}}(x)$.

With this definition we can generalize the entropy S_q as follows:

$$S_{\{F\}} = \sum_{i=1}^{W} p_i \ln_{\{F\}} \frac{1}{p_i} \equiv \sum_{i=1}^{W} s_{\{F\}}. \tag{6.15}$$

At equiprobability (i.e., $p_i = 1/W$) we have

$$S_{\{F\}} = \int_{1}^{W} \left\{\int_{-\infty}^{+\infty} F(\kappa)\, u^{\kappa} d\kappa\right\}^{-1} du. \tag{6.16}$$

The generalized logarithm of Eq. (6.14) appears to be isomorphic to the generalized logarithm introduced recently by Naudts [392–395] who started from a different perspective.

We can straightforwardly prove that, assuming that $F(\kappa)$ is normalized, the following properties hold:

$$\ln_{\{F\}} 1 = 0, \tag{6.17}$$

hence

$$\exp_{\{F\}}(0) = 1. \tag{6.18}$$

Also

$$\frac{d}{dx} \ln_{\{F\}} x\Big|_{x=1} = \frac{d}{dx} \exp_{\{F\}}(x)\Big|_{x=0} = 1, \tag{6.19}$$

as well as *monotonicity*, more precisely

$$\frac{d}{dx} \ln_{\{F\}} x > 0, \ \forall x \in (0, +\infty), \tag{6.20}$$

and

$$\frac{d}{dx}\exp_{\{F\}}(x) > 0, \ \forall x \in \mathcal{A}_{exp_{\{F\}}}, \tag{6.21}$$

where $\in \mathcal{A}_{exp_{\{F\}}}$ is the set of admissible values of x for the nonnegative $\exp_{\{F\}}(x)$ function. When only positive values of q contribute (i.e., if $F(\kappa) = 0, \ \forall \kappa \le 0$), then the following properties hold also:

$$\frac{d^2}{dx^2}\ln_{\{F\}} x < 0 \ \ (concavity), \tag{6.22}$$

and

$$\frac{d^2}{dx^2}\exp_{\{F\}}(x) < 0 \ \ (convexity). \tag{6.23}$$

Analogously, when only negative values of q contribute (i.e., if $F(\kappa) = 0, \ \forall \kappa \ge 0$), then

$$\frac{d^2}{dx^2}\ln_{\{F\}} x > 0 \ \ (convexity), \tag{6.24}$$

and

$$\frac{d^2}{dx^2}\exp_{\{F\}}(x) > 0 \ \ (concavity). \tag{6.25}$$

Also, if only values of q below unity contribute (i.e., if $f(\kappa) = 0, \ \forall \kappa \ge 1$), then

$$\lim_{x \to +\infty}\ln_{\{F\}} x = +\infty, \tag{6.26}$$

and, if only values of q above unity contribute (i.e., if $f(\kappa) = 0, \ \forall \kappa \le 1$), then

$$\lim_{x \to 0^+}\ln_{\{F\}} x = -\infty. \tag{6.27}$$

Illustrations of $S_{\{F\}}$ for Gaussian and binary QSF's can be found in [383].

We have seen so far how from a given QSF we can produce the corresponding entropic functional. We will now work in the reverse way: given a specific entropic functional, we will find (if possible) the QSF that produces it. Consider a general entropic functional of the form:

$$S = \sum_{i=1}^{W} s(p_i) \tag{6.28a}$$

$$s(x) = x \ln_{\{F\}} \frac{1}{x} \tag{6.28b}$$

We have:

$$\ln_{\{F\}} x = \int_1^x \frac{du}{\int_{-\infty}^{+\infty} F(\kappa)u^\kappa d\kappa} \quad \Leftrightarrow$$

$$\frac{d}{dx}\left(\ln_{\{F\}} x\right) = \frac{1}{\int_{-\infty}^{+\infty} F(\kappa)x^\kappa d\kappa} \quad \Leftrightarrow$$

$$\int_{-\infty}^{+\infty} F(\kappa)x^\kappa d\kappa = \frac{1}{\frac{d}{dx}\left(\ln_{\{F\}} x\right)} \quad \Leftrightarrow$$

$$\int_{-\infty}^{+\infty} F(\kappa)e^{\kappa \ln x} d\kappa = \frac{1}{\frac{d}{dx}\left(\ln_{\{F\}} x\right)} \tag{6.29}$$

We set:

$$\omega = -i \ln x \tag{6.30}$$

Then, from equation (6.29), we obtain:

$$\frac{1}{\sqrt{2\pi}} \int_{-\infty}^{+\infty} F(\kappa)e^{i\omega\kappa} d\kappa = \frac{1}{\sqrt{2\pi}} \cdot \frac{e^{i\omega}}{\frac{d}{d\omega}\left(\ln_{\{F\}} e^{i\omega}\right)} \tag{6.31}$$

However, the LHS of equation (6.31) is nothing but the Fourier transform of F. Thus, inverting the transform we have:

$$F(\kappa) = \frac{i}{2\pi} \int_{-\infty}^{+\infty} \frac{e^{i\omega(1-\kappa)}}{\frac{d}{d\omega}\left(\ln_{\{F\}} e^{i\omega}\right)} d\omega \tag{6.32}$$

Inserting the entropy functional of Eq. (6.15) into equation (6.32) we finally get:

$$F(\kappa) = \frac{1}{2\pi} \int_{-\infty}^{+\infty} \frac{e^{-i\omega\kappa}}{s\left(e^{-i\omega}\right) - i\frac{d}{d\omega}\left[s\left(e^{-i\omega}\right)\right]} d\omega \tag{6.33}$$

Equation (6.33) is quite important. Indeed, it shows that the QSF corresponding to a large class of entropy functionals can be explicitly calculated. It is straightforward to check that for s given by BG or by nonextensive statistics we get $F(\kappa) = \delta(\kappa - q)$ as anticipated.

In order to derive the normalized QSF associated with a given entropy, it is sufficient that the entropy functional fulfills the following requirements.

1. It must be possible to write the total entropy S as a sum of the entropic function s for each state (Eq. (6.28a)).

2. The function s must satisfy $s(1) = 0$, which is in fact a quite reasonable requirement for an entropy.
3. Furthermore, we must have:

$$\left.\frac{ds(x)}{dx}\right|_{x=1} = -1 \qquad (6.34)$$

This condition is equivalent to having the QSF normalized to unity. If we abandon the normalization of the QSF then we can consistently drop this last requirement.
4. The function $s(x)$ must be defined (or analytically continued) on the unitary circle and it must also be differentiable in the same domain.
5. The integral of Eq. (6.33) must converge.

As a nontrivial illustration, we will now use the present method to find the QSF associated with an exponential entropic form. Let us assume

$$S = \sum_{i=1}^{W} p_i \left(1 - e^{\frac{p_i - 1}{p_i}} \right), \qquad (6.35)$$

hence

$$s(x) = x \left(1 - e^{\frac{x-1}{x}} \right). \qquad (6.36)$$

It is trivial to see that Eq. (6.36) fulfills all the criteria set above, and we can thus find a normalized QSF for it. Using Eq.(6.33) we get:

$$F(\kappa) = \frac{1}{e} \sum_{n=0}^{\infty} \frac{\delta(\kappa - n)}{n!} \qquad (6.37)$$

Although different, entropy (6.35) has some resemblance with that introduced by Curado [178]. We claim, at this stage, no particular physical justification for the form (6.35). In the present context, it has been chosen uniquely with the purpose of illustrating the mathematical procedure involved in the inverse QSF problem.

6.2.2 Beck-Cohen Superstatistics

We may say that Beck-Cohen superstatistics originated essentially from a mathematical remark and its physical interpretation [334, 335]. The basic remark is that there is a simple link, described hereafter, between the q-exponential function (with $q \geq 1$) and the so-called *chi-square (or χ^2) distribution* with n degrees of freedom, which is a particular case of the *Gamma distribution*. Beck and Cohen [337, 643] start from the standard Boltzmann factor but with β being itself a random variable (whence the name "superstatistics") due to possible spatial and/or temporal fluctuations. They define

$$P(E) = \int_0^\infty d\beta' f(\beta') e^{-\beta' E}, \tag{6.38}$$

where $f(\beta')$ is a normalized distribution, such that $P(E)$ also is normalizable under the same conditions as the Boltzmann factor $e^{-\beta' E}$ itself is. They also define

$$q_{BC} \equiv \frac{\langle (\beta')^2 \rangle}{\langle \beta' \rangle^2} = \frac{\int_0^\infty d\beta' \, (\beta')^2 f(\beta')}{\left[\int_0^\infty d\beta' \, \beta' f(\beta') \right]^2}, \tag{6.39}$$

where we have introduced BC standing for *Beck-Cohen*.

If $f(\beta') = \delta(\beta' - \beta)$ we obtain Boltzmann weight

$$P(E) = e^{-\beta E}, \tag{6.40}$$

and $q_{BC} = 1$.

If $f(\beta')$ is the χ^2-distribution, i.e.,

$$f(\beta') = \frac{n}{2\beta \Gamma(\frac{n}{2})} \left(\frac{n\beta'}{2\beta} \right)^{n/2-1} \exp\left\{ -\frac{n\beta'}{2\beta} \right\} \quad (n = 1, 2, 3, \dots), \tag{6.41}$$

we obtain

$$P(E) = e_q^{-\beta E}, \tag{6.42}$$

and $q_{BC} = q$, with

$$q = \frac{n+2}{n} \geq 1. \tag{6.43}$$

Several other examples of $f(\beta')$ are discussed in [337], and it eventually established the following important result: all narrowly peaked distributions $f(\beta')$ behave, in the first nontrivial leading order, as q-statistics with $q = q_{BC}$. Further details and various applications to real systems are now available [29, 53, 643–653] of this theory (which, unless $f(\beta')$ is deduced from first principles, remains phenomenological).

As we mentioned previously, the above discussion concerns the statistics. More than that is needed to have a statistical-mechanical theory, namely it is necessary to introduce an associated generalized entropic functional, as well as an appropriate constraint related to the energy. This program has in fact been carried further out for superstatistics, and details can be seen in [359–361]. An interesting point is worthy mentioning: of all admissible $f(\beta')$, only (6.41) yields a stationary-state distribution optimizing the associated entropy within which the Lagrange parameter (usually noted α) corresponding to the normalization constraint factorizes from the term containing the β Lagrange parameter. In other words, *of all superstatistics, only q-statistics admits a partition function on the usual grounds, i.e., depending on β but not on α.*

In addition to the so-called χ^2-*superstatistics* described above, we have the so-called *inverse* χ^2-*superstatistics*, where it is $1/\beta'$, instead of β', that follows the χ^2 distribution. Finally, a third class is sometimes focused on in the literature. It is referred to as the *log-normal superstatistics*, and corresponds to the case where β' is distributed along a log-normal distribution. These three classes are sometimes referred to as *universality* ones because they are all based on Gaussians, which are attractors in the Central Limit Theorem sense. Being β' (hence $1/\beta'$) *positive* random variables, they may be constructed through sums of squared Gaussians, and also through log-normal distributions.

Let us also mention that, instead of transforming the *unnormalized* BG exponential function as in Eq. (6.38) (referred to as type-*A* superstatistics), we may transform the *normalized* BG weight (referred to as type-*B* superstatistics) [654]. Although both are mathematically well defined, it is plausible that type-B is more naturally found in physical systems. Also, they yield different expressions of the index *q* in terms of *n*, the number of degrees of the χ^2 distribution of the fluctuations of the temperature.

Let us finally conclude this Subsection by focusing on the connection of spectral statistics with the Beck-Cohen superstatistics. An entropic functional has been derived that corresponds to superstatistics. This functional is of the form $S = \sum_i s(p_i)$, with

$$s(y) = \int_0^x \frac{a + K^{-1}(y)}{1 - \frac{K^{-1}(y)}{E^*}},$$ (6.44)

where

$$K(y) = \frac{P(y)}{\int_0^{+\infty} P(u)du}.$$ (6.45)

In Eq. (6.44), E^* stands for the lowest admissible energy for the system and *a* is a Lagrange multiplier. Using Eq. (6.44) and Eq. (6.33) we can in principle get the QSF $F(\kappa)$ for a given temperature distribution function $f(\beta')$. Thus, in principle at least, the various superstatistics can be accommodated into spectral statistics. However, there are certain cases where spectral statistics can go further than superstatistics. For example, it appears that through superstatistics we can, up to now, only produce the nonadditive entropies S_q for $q \geq 1$, while in the spectral formalism we can have them for arbitrary values of *q*. This is the reason for which we have written the last logical inclusion in (6.8) as it stands there.

Finally, let us mention that further connections and generalizations are discussed in [256], which are related, for instance, to the Lambert function.

Part III
Applications or What for the Theory Works

Chapter 7
Thermodynamical
and Nonthermodynamical Applications

Nothing is more practical than a good theory.

The nature of the present chapter is quite different from all the others of the book. In all its other chapters we have privileged the presentation and understanding of nonextensive statistical mechanics itself, and of some of its delicate and unusual concepts. In the present chapter we focus on the concrete and typical applications that are available in the literature, as well as on some connections that have emerged over time with other areas such as quantum chaos, quantum entanglement, random matrices, and theory of networks.[1]

The present list is not an exhaustive one. It is aimed mainly to illustrate the specific types of systems that have been handled in one way or another within the nonextensive framework. Some of them are genuine applications of the theory, others are just possible explanations and connections. Whenever the microscopic, or at least the mesoscopic, dynamics of the system is unknown, it is of course impossible to determine q otherwise than through fitting (like astronomers determine the elliptic eccentricities of the orbits of the planets). This extra difficulty does not exist in BG thermostatistics, since the corresponding value trivially is just $q = 1$. In the more complex systems addressed in this chapter, all types of situations occur. Sometimes the experimental measurements, or observations, or computational results exist through very many numerical decades with satisfactory precision. In these cases, the correctness of the fitness constitutes already a strong argument favoring the applicability of the nonextensive theory, with its predictions and concepts. Sometimes, we have at our disposal only a few numerical decades and/or not very high precision. It might then be disputable whether the system under focus really belongs to the present frame, or to a somewhat different one. Sometimes, it becomes possible to

---

[1] "Nothing is more practical than a good theory" is currently attributed to Lenin.

© Springer Nature Switzerland AG 2023
C. Tsallis, *Introduction to Nonextensive Statistical Mechanics*,
https://doi.org/10.1007/978-3-030-79569-6_7

make precise falsifiable predictions, sometimes not. Sometimes the applications just consist in improved algorithms for optimization, signal analysis, image processing, and similar techniques. In these cases, the quality of the improvement speaks by itself. In all cases, we do achieve a better understanding of the phenomenon, or at least develop some intuition on it.

Before starting with the description of typical applications, let us remind that the knowledge of the microscopic dynamics is *necessary but not sufficient* for the implementation of the entire theory from first principles. Indeed, it is only *in principle* that the microscopic dynamics contains all the ingredients enabling the calculation of the index (or indices) q. It is still necessary to be able to calculate, in the full phase space, quantities such as the sensitivity to the initial conditions or the entropy production. This calculation can be extremely hard. But, whenever tractable, then it provides the value(s) of q. Once q is known, it becomes possible to implement the thermodynamical steps of the theory. This is to say, we can in principle proceed and calculate the partition function of a Hamiltonian system, and, from this, calculate various important thermodynamical quantities such as specific heat, susceptibility, and others. Naturally, the difficulty of this last step of the calculation should not be underestimated. It suffices to remember the formidable mathematical difficulties involved in Onsager's celebrated solution of the square-lattice spin $1/2$ Ising ferromagnet. And he only had to deal with *first-neighbor interactions* and *exponential thermal weights*. In a full q-statistical calculation we have to deal typically with *interactions at all distances* (or similar nontrivial conditions) and *asymptotically power-law weights*! This difficulty might explain why we have, up to now, only partial results for say the many-body long-range-interacting inertial XY ferromagnet addressed in Sect. 5.3.1. This task would be hopeless had we not access to approximate solutions based on variational principles, Green-functions, numerical approaches, and others. As a mathematical exercise, the q-statistics of simple systems such as the ideal gas and the ideal paramagnet are available in the literature. However, these calculations only provide some mathematical hints with modest—phenomenological at best—physical content. Indeed, thermal equilibrium in the absence of interactions mandates $q = 1$.

Other, extremely powerful, hints are also available from the full discussion of simple maps, as shown in Sects. 5.1.1 and 5.2. However, these systems, no matter how useful they might be for various applications, are *non-thermodynamical*. In other words, they do *not* have a natural energy associated, and are therefore useless in order to illustrate all the thermostatistical steps of the full calculation, and their connection to thermodynamics itself.

We present next various applications in various areas of knowledge.

7.1 Physics

7.1.1 Cold Atoms in Optical Lattices

On the basis of nonextensive statistical mechanical concepts, Lutz predicted in 2003 [655] that cold atoms in dissipative optical lattices would have a q-Gaussian distribution of velocities, with

$$q = 1 + \frac{44\, E_R}{U_0}, \tag{7.1}$$

where E_R and U_0 are respectively the *recoil energy* and the *potential depth*. The prediction was impressively verified three years later [631, 632], as shown in Fig. 7.1.

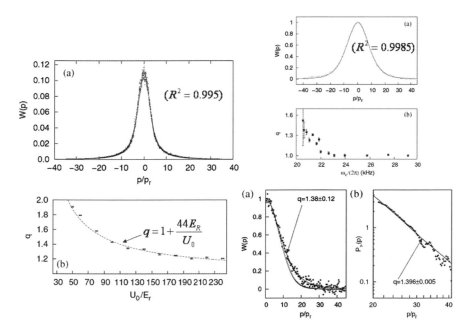

Fig. 7.1 Computational verification with quantum Monte Carlo (left panels), and laboratory verification with C_s atoms (right panels) of Lutz' theory. From [631]

7.1.2 Trapped Ions

In the laser cooling technique, the temperature of a single ion (with mass m_I) in a radio frequency ion trap is determined by a buffer gas of atoms with mass m_B: see Fig. 7.2 [656]. Related results are available in [657].

7.1.3 Granular Matter

Granular matter and similar systems yield many interesting applications (see, for instance [629, 630, 658–661]). They involve inelastic collisions between the particles.

Inelastic Maxwell models

In some simple models, such as the so-called inelastic Maxwell models, analytic calculations can be performed (e.g., in [661]). The velocity distribution that is obtained, from the microscopic dynamics of the system of cooling experiments, for a spatially uniform gas whose temperature is monotonically decreasing with time is given by (see [658] and references therein) following asymptotic (i.e., $t \to \infty$) distribution

$$P(v, t) = \frac{2}{\pi v_0(t)} \frac{1}{\left[1 + \frac{v^2}{(v_0(t))^2}\right]^2} , \tag{7.2}$$

with $v_0(t) = v_0(0)\, e^{-\lambda(r)\,t}$, $\lambda(r)$ being a function of the *restitution coefficient* r of the inelastic collisions (with $\lambda(1) = 0$). Equation (7.2) precisely corresponds to a q-Gaussian with $q = 3/2$.

Silo drainage

Computational simulations have been recently done [659, 660] for the discharge of granular matter out of the bottom of a vertical silo: see Fig. 7.3. Although the outcome precision of the simulations is not very high, our interest in these experiments lies in the fact that they seem to provide one more verification of the predicted scaling (4.14): see Figs. 7.4 and 7.5.

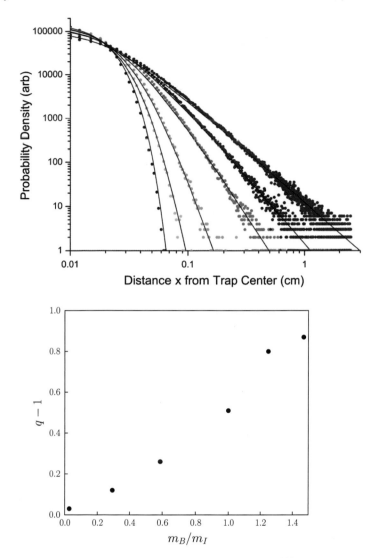

Fig. 7.2 *Top*: Monte Carlo density distributions for a single $^{136}Ba^+$ ion cooled by six different buffer gases at $300\,K$ ranging from $m_B = 4$ (left) to $m_B = 200$ (right). Note the evolution from nearly Gaussian to q-Gaussians as the mass m_B increases: $q = 1.03\,(He)$, $1.12\,(Ar)$, $1.26\,(Kr)$, $1.52\,(Xe)$, $1.80\,(170)$, $1.87\,(200)$. From [656]. *Bottom*: Data taken from Table 1 of [656]. For m_B/m_I approaching zero, q approaches the BG value $q = 1$

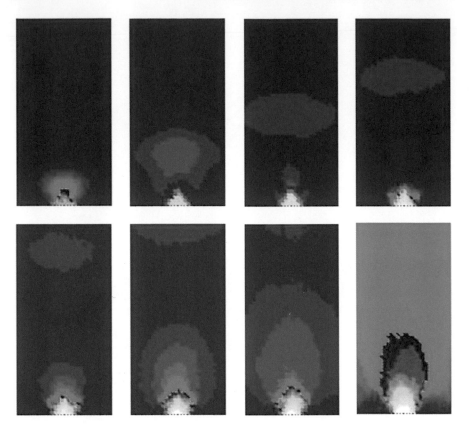

Fig. 7.3 Average vertical velocity profile inside the silo for an aperture $11d$ (d is the diameter of the grains) in eight different stages of its evolution. Time increases from left to right and from top to bottom. In the last one—corresponding to the fully developed flow—a grain on top has fallen a distance equivalent to twice its diameter d. Note the existence of a bounded region in the velocity profile that travel in the vertical direction. From [659] (see also [660])

Bubbling fluidized beds

Fluidized beds are a common form of a chemical reactor, in which a stream of gas is blowing upward through a deep layer of fine solid particles, setting it in motion. From [630]. In Fig. 7.6 we present results obtained through pressure measurements performed at a sampling rate of 200 Hz at different positions inside a pilot-size fluidized bed, 80 cm in diameter, filled with sand particles of size 0.3–0.5 mm up to a settled bed height of 93 cm. The pressure fluctuations are defined through $\Delta P(t, \Delta t) = P(t + \Delta t) - P(t)$.

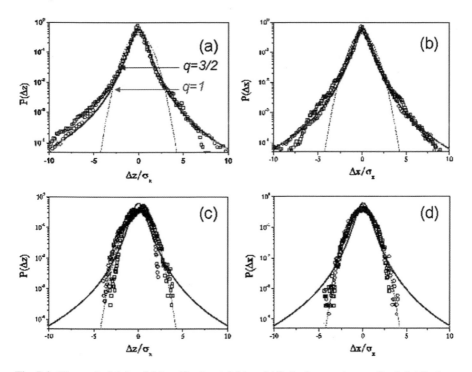

Fig. 7.4 The vertical (**a**) and (**c**) and horizontal (**b**) and (**d**) displacement normalized distributions approach a q-Gaussian with $q \simeq 1.5$ in the *intermediate regime* (**a**) and (**b**), and a Gaussian in the *fully developed regime* (**c**) and (**d**). The symbols indicate the silo aperture: circles for $3.8d$ and squares for $11d$. The blue (red) solid line is a Gaussian (q-Gaussian with $q = 3/2$). From [659] (see also [660])

Fig. 7.5 Time evolution of the second moments at the intermediate regime for a silo aperture $3.8d$. The straight line indicates a slope $\gamma = 4/3$. From [659] (see also [660])

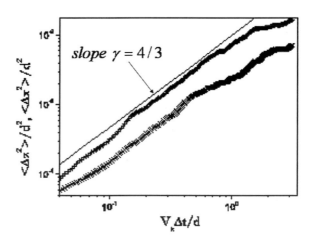

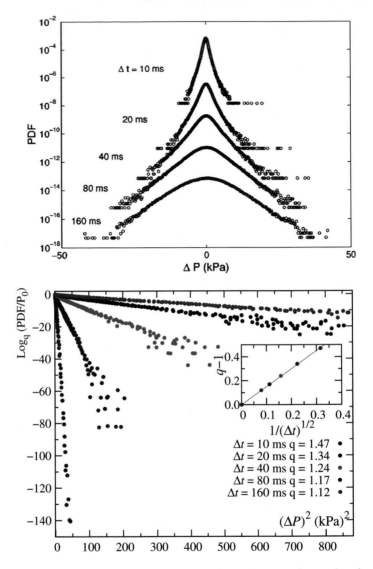

Fig. 7.6 *Top*: Pressure fluctuations measured at intervals Δt. The top set is to scale, other sets are shifted down for clarity. From [630]. *Bottom*: Same data in q-log representation (straight lines correspond to q-Gaussians). The inset shows q as a function of Δt, more precisely $(q-1) \simeq 1.49/\sqrt{\Delta t}$ with $[\Delta t] = ms$ (the red dot corresponds to BG statistics, when all memory of correlations has been lost). This (unpublished) result was obtained in co-authorship with M. Jauregui and L. J. L. Cirto

Fig. 7.7 Probability density functions of the horizontal components of the fluctuating displacements tracked during two different increments of shear strain ($\Delta\gamma = 7.3 \times 10^{-4}$ and $\Delta\gamma = 10^{-1}$). The scatters correspond to experimental data, and the solid lines correspond to q-Gaussian fittings. From [629]

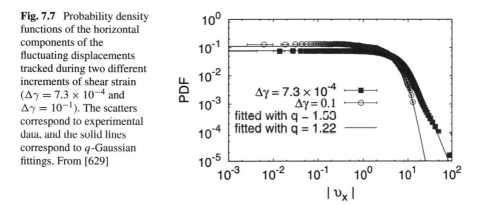

Confined grains under shear

The following scaling law was predicted in 1996 [404]:

$$\mu = \frac{2}{3 - q},$$ (7.3)

where q is the index of the q-Gaussian solution of a relevant nonlinear Fokker–Planck equation initially discussed in [403], and μ is the exponent associated with anomalous diffusion (i.e., x^2 scales like t^μ; $q = 1$ recovers the well known scaling for normal diffusion, i.e., $\mu = 1$).[2] See details in Sect. 4.2.2. This scaling relation has already been verified over the years in restricted experimental conditions, but it was only in 2015 [629] when a neat validation was established within a wide experimental interval in a single experiment. This was accomplished through precise measurement of the fluctuations of $d = 2$ confined grains under shear. See Figs. 7.7 and 7.8 for the determination of the index q as a function of the experimental parameter $\Delta\gamma$. See finally Fig. 7.9 for the determination of the exponent $\alpha \equiv \mu$ as a function of the same experimental parameter $\Delta\gamma$, leading to a neat validation of the prediction (7.3).

7.1.4 High-Energy Physics

Electron–positron annihilation

Electrons and positrons in frontal collisions at high energy typically annihilate and produce two or three hadronic jets. The analysis of the transverse momenta of those jets provides interesting physical information related, among others, to the produc-

---

[2] The statement concerning $x^2 \sim t^\mu$ is stronger than the usual one $\langle x^2 \rangle \sim t^\mu$. It implies $\langle x^2 \rangle \sim t^\mu$ whenever $\langle x^2 \rangle$ is finite, which indeed occurs for q-Gaussians with $q < 5/3$. For all values of q up to $q = 3$, it implies that the q-variance satisfies $\langle x^2 \rangle_q \sim t^\mu$.

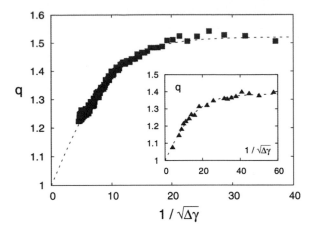

Fig. 7.8 Evolution of the measured q value as a function of the squared inverse of the strain increment for the experiments and simulations. The dashed line corresponds to a regression using the function $q(1/\sqrt{\Delta\gamma}) = 1 + a\tanh(b/\sqrt{\Delta\gamma})$, with $(a, b) = (0.521, 0.096)$. *Inset*: the same plot for data from a simulation that highlights the limit $q = 1$ when $\Delta\gamma \to \infty$. The fitted parameters for simulations were $(a, b) = (0.387, 0.057)$. From [629]

tion of mesons. The process can in principle be described in thermostatistical terms, without entering into microscopic details in the realm of Quantum Chromodynamics. Fermi was the pioneer of this type of approach [662], followed by Hagedorn [663]. According to Hagedorn, such high-energy collisions produce excited hadron *fireballs* that reach some sort of *thermal equilibrium*. An important consequence of this approach would be that increasing the collisional energy would *not* change the involved basic masses (that of mesons that are being produced) but it would only increase their number, like an increase of heat delivery when one boils water does *not* modify the phase-transition temperature, but only increases the amount of liquid that becomes gas. A similar statement was made, a few years later, by Field and Feynman [664]. The use of the Boltzmann weight in the relativistic limit yields [663] a distribution of hadronic transverse momenta which exhibits a reasonably satisfactory agreement with experimental data at relatively low collisional energy, say at 14 Gev (TASSO experiments). But increasing that energy, the temperature (a fitting parameter) did *not* remain constant, as predicted by the theory. This approach was somewhat discredited and abandoned. The idea was revisited in 2000 by Bediaga, Curado, and Miranda [665], but this time assuming a q-statistical weight. The results were very satisfactory this time, even for collisional energies up to 161 Gev (DELPHI experiments), as can be seen in Figs. 7.10 and 7.11. The temperature remained virtually constant all the way long, and the agreement of the theoretical curves (which, though conceptually simple, involve nevertheless eight hypergeometric functions) with the experimental data is quite impressive for the entire range of transverse hadronic momenta. The phenomenological value of q slightly increases from $q = 1$ to $q \simeq 1.2$ (the asymptotic value 11/9 has been suggested [666] for very large energies) when the center-of-mass collisional energy increases from 14 to 161 Gev. See [665] for a possible physical origin of this effect. See Figs. 7.10 and 7.11.

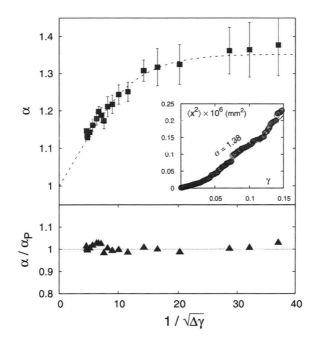

Fig. 7.9 Verification of the scaling law 7.3 in [404] for different regimes of diffusion. *Top*: Evolution of the measured diffusion exponent $\alpha \equiv \mu$ as a function of $1/\sqrt{\Delta\gamma}$; the dashed line is a direct application of the scaling law from the fit of the values shown in Fig. 7.8, $\alpha(1/\sqrt{\Delta\gamma}) = 2/[3 - q(1/\sqrt{\Delta\gamma})]$. *Inset*: a typical diffusion curve showing the mean square displacement fluctuations $\langle x^2 \rangle$ as a function of the shear strain γ; it allows the assessment of the diffusion exponent $\alpha \equiv \mu$ for each strain window tested. In the case shown, it corresponds to the smallest strain window, the rightmost point in the curve at the main panel. Note that for a constant strain rate, γ is proportional to time. *Bottom*: Measure of the deviation of the data relative to the scaling law prediction $\alpha_P = 2/(3 - q)$, as a function of $1/\sqrt{\Delta\gamma}$, showing a remarkable agreement of the order of $\pm 2\%$. From [629]

Electron–positron, proton–proton, and heavy ion collisions

A great variety of high-energy multiparticle production processes (from collisions such as pp, $p\bar{p}$, $Au + Au$, $Cu + Cu$, $Pb + Pb$, etc.) have been analyzed along related lines [667–675, 677–686]. The values of q that emerge (from the BRAHMS, STAR, PHENIX data, for instance) are systematically close to the case discussed above, typically in the range $1 < q < 1.2$. The nonzero values of $(q - 1)$ are frequently interpreted in terms of sizeable temperature fluctuations that exist during the hadronization process (see [334, 335, 337]). See Fig. 7.12 for typical results for the distribution of transverse momenta p_T in pp collisions at the LHC/CERN. These impressive fittings have been discussed in [685] on first-principle grounds within Yang-Mills fields and quantum chromodynamics (QCD), and it was established

$$\frac{1}{q - 1} = \frac{11}{3}N_c - \frac{4}{3}\frac{N_f}{2}, \tag{7.4}$$

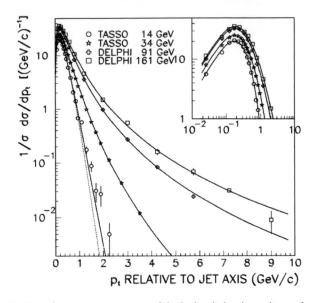

Fig. 7.10 Distribution of transverse momenta of the hadronic jets in positron–electron collisions. From [665]

where N_c and N_f are respectively the numbers of colors and of flavors. By using $N_c = N_f/2 = 3$, it is found $q = 8/7 \simeq 1.14$ [685], which is amazingly close to the LHC experimental values (see [672] and references therein). Also, if we use $N_c = N_f = 3$ we obtain $q = 10/9 \simeq 1.11$, which interestingly enough coincides with the value phenomenologically obtained in 2000 by Walton and Rafelski [676].

Nearly log-periodic small oscillations are well described by complex values of q: see [673–675] and Figs. 7.13 and 7.14, which illustrate that $q \equiv q' + i\, q''$ fits satisfactorily the data.

See also Fig. 7.15 for neat distributions of p_T obtained in PHENIX experiments of hadron production in pp collisions [677].

Many works have focused on outcome distributions of transverse momenta. But longitudinal momenta, directly related to (pseudo)rapidities, have been discussed as well [687, 688]. More specifically, the inelasticity coefficient, directly related to the distribution of rapidities, is defined as $K \equiv \frac{W}{\sqrt{s}} \in [0, 1]$, where $\sqrt{s}$ is the center-of-mass energy of colliding particles; it defines the amount of energy W used effectively for the production of secondaries, and plays an important role in understanding multiparticle production processes. It tells us that in the collisions only a fraction $W = K\sqrt{s}$ of the whole invariant energy $\sqrt{s}$ is spent for production of new particles while the rest is taken by leading particles to the forward and backward phase-space regions. Pseudorapidity distributions $dN/d\eta$ and $K(\sqrt{s})$ have been analyzed in [688]: see Fig. 7.16. If we used $q_{rapid} = 1$ instead of the values of q_{rapid} obtained through the fittings of $dN/d\eta$ (which provide $q_{rapid} \simeq 1.4$ for $\sqrt{s} = 8000\,\mathrm{GeV}$, for

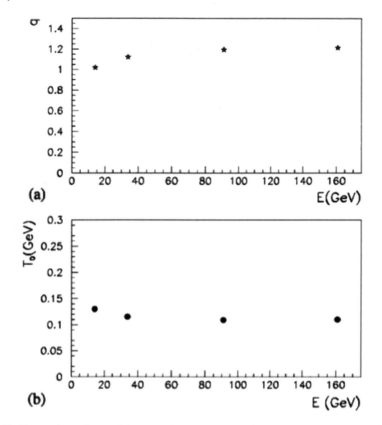

Fig. 7.11 Energy dependence of the entropic parameter q and the temperature T. From [665]

instance), we would obtain a smaller value for K (close to 0.4 instead of close to 0.5), which differs from the experimental value.

Finally, the mass spectrum corresponding to high-energy collisions has been focused on as well: see Fig. 7.17. We see that a wide experimental region (data from the Particle Data Group [689]) of the spectrum is correctly described on q-statistical grounds [680, 683, 690], improving on the Hagedorn approach [691].

Cosmic rays

Cosmic rays arrive to Earth within a vast range of energies, up to values close to 10^{20} ev. Their associated fluxes vary within impressive 33 orders of magnitude: see Fig. 7.18. This curve includes the so-called "knee" and "ankle", at intermediate and very high energies, respectively. It turns out that it is possible, without entering into any specific mechanism, to provide [642] an excellent phenomenological description of these data by assuming a crossover between two q-exponential distribution

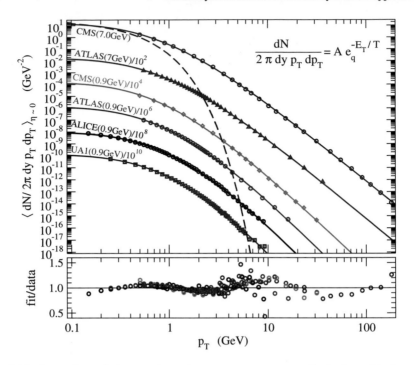

Fig. 7.12 Comparison of the experimental transverse momentum distribution of hadrons in pp collisions at central rapidity y with theoretical q-exponentials with $q \simeq 1.13 \pm 0.02$ and $T \simeq (0.14 \pm 0.01)$ GeV. The corresponding Boltzmann–Gibbs (purely exponential) fit is illustrated as the dashed curve. For a better visualization both the data and the analytical curves have been divided by a constant factor as indicated. The ratios data/fit are shown at the bottom, where a nearly log-periodic behavior is observed on top of the q-exponential one. From [672]

functions. The two corresponding values of q are quite close among them, and also close to 11/9 ([692]).

In recent years there have also been various interesting measurements of cosmic rays from laboratories outside the Earth, namely by the Alpha Magnetic Spectrometer (AMS) on the International Space Station [693, 694]. See Figs. 7.19, 7.20 and 7.21.

Quantum scattering of particles

Entropic bounds for scattering of spinless particles (e.g., pions) by a nucleus have been established and tested [472–475] with available experimental results for phase shifts. Typical results involving ^{4}He, ^{12}C, ^{16}O and ^{40}Ca nuclei are exhibited in Fig. 7.22. Along this line, a conjugation relation naturally emerges for two relevant entropic indices, noted q and $\bar{q}$ (see details in [474, 475]). This relation is given by

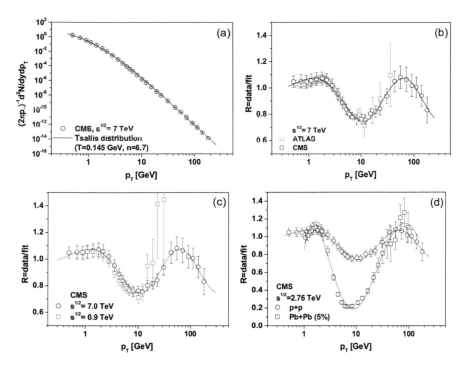

Fig. 7.13 Examples of log-periodic oscillations. **a** dN/dp_T for the highest energy $7\,TeV$; the q-exponential behavior $F(p_T) \propto [1 + p_T/nT]^{-n}$ is evident. Only data from CMS experiment are shown; the others behave essentially in an identical manner. **b** Log-periodic oscillations showing up in different experimental data, like CMS or ATLAS, taken at $7\,TeV$. **c** Results from CMS for different energies. **d** Results for different systems ($p + p$ collisions compared with $Pb + Pb$ taken for 5% centrality. Results from ALICE are very similar. Fits for $p + p$ collision at 7, 2.76 and $0.9\,TeV$ are performed with $q = 1.139 + i\,0.0385$, $1.134 + i\,0.0269$ and $1.117 + i\,0.0307$, respectively. The fit for central $Pb + Pb$ collisions at $2.76\,TeV$ is done with $q = 1.135 + i\,0.0321$. From [674]

$$\frac{1}{q} + \frac{1}{\bar{q}} = 2 \,, \tag{7.5}$$

which can equivalently be written as

$$\bar{q} = \mu\nu\mu(q) \,, \tag{7.6}$$

where the multiplicative and additive dualities μ and ν are those defined in Eqs. (4.100) and (4.101), respectively. A deeper understanding of this intriguing connection would be welcome.

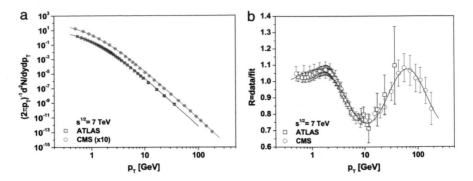

Fig. 7.14 **a** Fit to p_T data for pp collisions at $7\,TeV$ from CMS and ATLAS experiments using distribution $f(x) = \frac{C}{[1+x/nT]^n}$ with parameters $T = 0.145\,\text{GeV}$ and $n = 6.7$. Data points for the CMS experiment are scaled by a factor of 10 for better readability. **b** Fit to p_T dependence of the data/fit ratio for results presented in the panel (**a**) using function $R(E) = a + b\,\cos[c\,\ln(E + d) + f]$ with $a = 0.909$, $b = 0.166$, $c = 1.86$, $d = 0.948$ and $f = -1.462$. From [673]

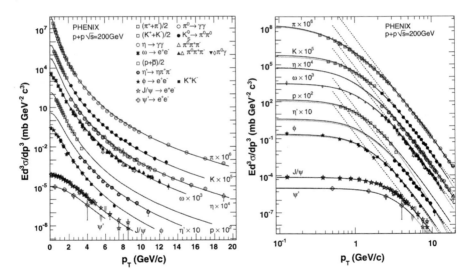

Fig. 7.15 Invariant differential cross sections of different particles measured in $p + p$ collisions at $\sqrt{s} = 200\,\text{GeV}$ in various decay modes. The continuous curves are q-exponential fits. *Left*: Log-linear representation. *Right*: Log–log representation; the dashed lines indicate the power-law asymptotic behavior. The high-quality fitting not only in the power-law asymptotic region but also in the non-asymptotic region numerically excludes other asymptotically power-law distributions such as the Lévy ones. See details in [677]

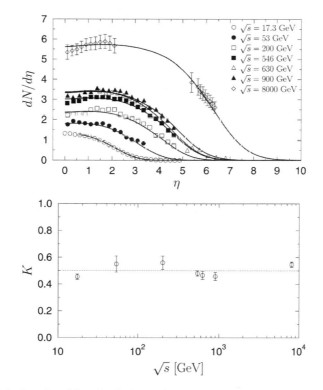

Fig. 7.16 *Left*: Pseudorapidity distributions of charged hadrons produced in proton + proton (proton + antiproton) interactions and registered by the NA49, UA5, UA7, CMS and TOTEM experiments. Continuous lines are (q_{rapid}, T_{rapid})-dependent fits obtained using $dN/d\eta = N_0 \, e_{q_{rapid}}^{-E(\eta)/T_{rapid}} \, \beta(\eta)$, with quantities $(N_0, E(\eta), \beta(\eta))$ that are defined in [688]. *Right*: Energy dependence of the inelasticity coefficient K calculated using the fits presented in the left plot, which led to $q_{rapid} \simeq 1 - 0.104 + 0.058 \ln(\sqrt{s}/\text{GeV})$ and $T_{rapid} \simeq 0.16 \, (\sqrt{s}/\text{GeV})^{0.77}$. From [688]

Diffusion of charm quark

A preliminary analysis of the diffusion of a charm quark in a thermal quark-gluon plasma is available [676]. A direct quantum mechanical calculation and a phenomenological theory based on q-statistics are compared in Fig. 7.23. The two calculations roughly coincide for $q = 1.114$, whereas the discrepancy is considerable if $q = 1$ is adopted instead. Further microscopic dynamical (and surely nontrivial!) studies are certainly necessary in order to understand why the specific value $q = 1.114$ provides a good first approximation, and why, even for this value, a small but visible systematic discrepancy is observed.

As one more admissible application in the area of high-energy physics, let us mention that a detailed literature exists advancing a possible connection between the solar neutrino problem and nonextensive statistics. Indeed, the by now well established neutrino oscillations do not totally explain, in some cases, the discrepancy

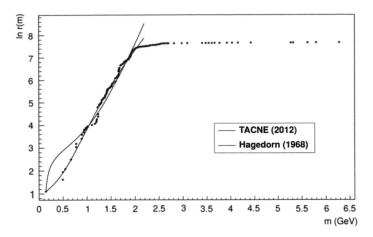

Fig. 7.17 Cumulative hadron spectrum, as a function of the hadron mass. The dots indicate the experimental data [689]. We display also the theoretical results corresponding respectively to the Hagedorn 1968 approach [691] (red curve) and to the q-statistical approach with $q = 1.103 \pm 0.007$ [680, 683, 690] (blue curve) [TACNE is the Portuguese acronym for *Non-Extensive Self-Consistent Thermodynamics*]. From [690]

existing between the theoretical predictions based on the *Standard Solar Model* and the neutrino fluxes measured on Earth. Therefore, some other contributions might be there. It is proposed [695–697] that they come from the fact that most probably the solar plasma is not in thermal equilibrium, but in a sort of stationary state instead, where nonextensive phenomena (basically due to strong spatial-temporal correlations) could be present.

7.1.5 Turbulence

Quite an effort has been dedicated to understanding the ubiquitous connections of nonextensive statistics with turbulence. This includes lattice-Boltzmann models [698, 699], defect turbulence [700], rotations in oceanic flows [701], air turbulence in airports [647, 702, 703], turbulence at the level of the trees of the Amazon forest, [704, 705], turbulent Couette–Taylor flow and related situations [706, 708–712], Lagrangian turbulence [713], two-dimensional turbulence in pure electron plasma [714, 715], the so-called one-dimensional "turbulence" [716, 717], among others. Criticism has also been advanced [718]. For several of the experimental situations that have been studied, q-statistics appears to be a quite good approximation. However, for some experiments, further improvement becomes possible (see, for instance, [647]) whenever many experimental decades are accessible to measurement.

Naturally, we do not intend here to exhaustively review the subject, and the reader is referred to the above literature for details. In what follows we have selected instead

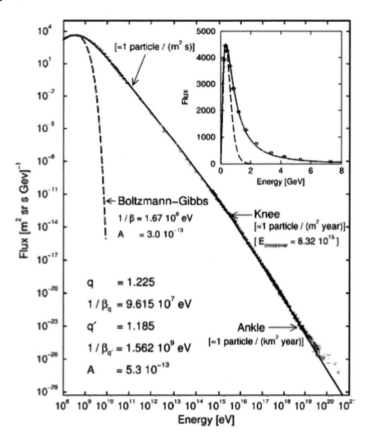

Fig. 7.18 Fluxes of cosmic rays. From [642]

only a few of those studies, with the aim of characterizing the types of approaches that have been developed.

(i) *Lattice-Boltzmann models for fluids*

The incompressible Navier–Stokes equation has been considered [698]on a discretized D-dimensional Bravais lattice of coordination number b. It is further assumed that there is a single value for the particle mass, and also for speed. The basic requirement for the lattice-Boltzmann model is to be *Galilean-invariant* (i.e., invariant under change of inertial reference frame), like the Navier–Stokes equation itself. It has been proved [698] that an H-theorem is satisfied for a trace-form entropy (i.e., of the form $S(\{p_i\}) = \sum_i^W f(p_i)$) only if it has the form of S_q with

$$q = 1 - \frac{2}{D}. \tag{7.7}$$

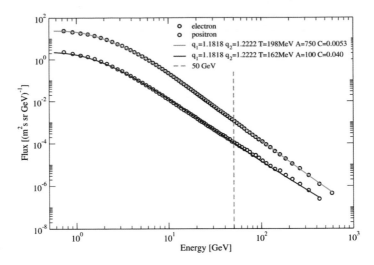

Fig. 7.19 The measured AMS-02 data are very well fitted by linear combination of escort and non-escort distributions (solid lines); $q_1 = 13/11$ and $q_2 = 1/(2 - q_1) = 11/9$. From [693]

Therefore $q < 1$ in all cases ($q > 0$ if $D > 2$, and $q < 0$ if $D < 2$), and approaches unity from below in the $D \to \infty$ limit. This study has been generalized by allowing multiple masses and multiple speeds. Galilean invariance once again mandates [699] an entropy of the form of S_q, with a unique value of q determined by a transcendental equation involving the dimension and symmetry properties of the Bravais lattice as well as the multiple values of the masses and of the speeds. Of course, Eq. (7.7) is recovered for the particular case of single mass and single speed.

(ii) *Defect turbulence*

Experiments have been done [700] in a convection cell which is heated from below and cooled from above, and which is tilted a certain angle with respect to gravity. In such circumstances, defects spontaneously appear in the undulations of the fluid: see Fig. 7.24. The distribution of velocities of these defects as well as their diffusion has been measured: see Figs. 7.25 and 7.26, respectively. The experimental condition is characterized by the *dimensionless driving parameter* $\epsilon \equiv \frac{\Delta T}{\Delta T_c} - 1 \geq 0$, where ΔT is the temperature difference maintained between bottom and top of the cell, and ΔT_c is a characteristic temperature difference of the system. Under many different experimental conditions (in particular, many values of ϵ) it was found that the distribution of velocities is, along six decades, a q-Gaussian with $q \simeq 1.5$, as illustrated in Fig. 7.25. Furthermore, superdiffusion was observed with a diffusion exponent $\alpha \simeq 4/3$, as illustrated in Fig. 7.26. These values satisfy the prediction (4.14) (with the notation change $\mu \equiv \alpha$). Similar results are to be expected [700] for phenomena such as electroconvection in liquid crystals, nonlinear optics, and auto-catalytic chemical reactions.

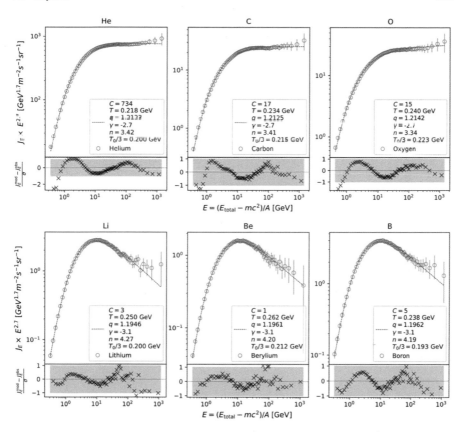

Fig. 7.20 The particle flux of each cosmic rays species was fitted with $J_E^{mod}(E) = CE(E + 2u)e_q^{-E/T}$ using the three parameters (C, T, q); the rest mass m satisfies $m = Au$, where A is the mass number and $u = 0.931\,\text{GeV}$ is the mass unit. The vertical axis in this log–log plot was multiplied with $E^{2.7}$ for better visibility. The accuracy of the fitting can be quantified by the deviation from modeled flux J_E^{mod} to observed flux J_E^{obs} weighted by the respective measurement error σ. Evidently, almost all data points fall within the uncertainty range of $\pm\sigma$ illustrated as grey shaded area. The mean temperature T_0 is defined by $T_0 = T/(4 - 3q)$. The amplitude C has dimensions $[C] = [m^{-2}sr^{-1}s^{-1}\text{GeV}^{-3}]$. From [694]

(iii) *Couette–Taylor flow*

Lagrangian and Eulerian experiments have been done of fluid motion within two rotating concentric cylinders, and results for the velocity distributions have been compared with q-Gaussians. A large literature exists on the subject, but here we only provide a few typical illustrations [335]: see Fig. 7.27. Further details on Eulerian experiments, are indicated in Figs. 7.28 and 7.29. Data collapse for the values of q is possible: see Fig. 7.30. Recent developments suggest that the theory must be somewhat improved in order to match higher precision data: for more details see [647].

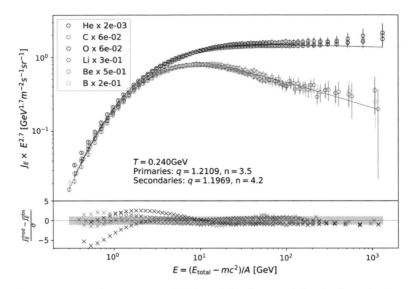

Fig. 7.21 Each particle flux was rescaled with a suitable factor such that the data points (roughly) collapse to a single line at the low energy end and the universal properties of primary and secondary cosmic rays nuclei spectra become visible. For larger energies the spectrum splits into primaries and secondaries which can be distinguished by a single parameter, the entropic index q which can be interpreted by the underlying effective degrees of freedom. From [694]

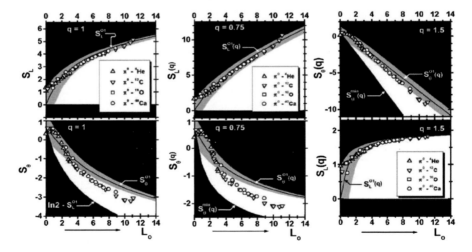

Fig. 7.22 Experimental tests, with data of scattering of pions by various nuclei (^{4}He, ^{12}C, ^{16}O and ^{40}Ca), of the theoretically allowed bands (grey regions) of the angular entropy (S_θ) and the angle-momentum entropy (S_L) for $q = 1$, $q = 0.75$ and $q = 1.5$. From [473]

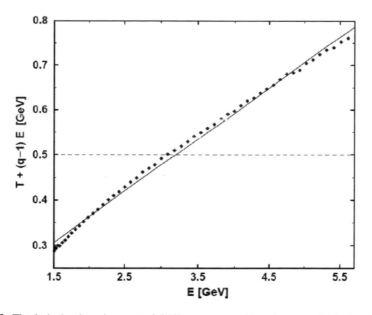

Fig. 7.23 The dashed and continuous straight lines correspond to a phenomenological q-statistical approach with $q = 1$ and $q = 1.114$, respectively. The dots have been obtained from a quantum calculation. From [676]

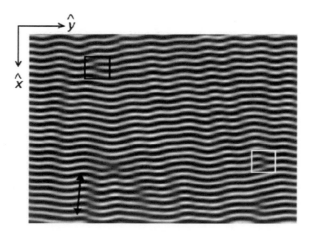

Fig. 7.24 Example shadowgraph image of undulation chaos in fluid (compressed CO_2 with Prandtl number $Pr = \nu/\kappa \simeq 1$, where ν is the *kinematic viscosity* and κ is the *coefficient of thermal expansion*) heated from below and cooled from above, inclined by an angle of $30°$. The dimensionless driving parameter is $\epsilon = 0.08$. The black (white) box encloses a positive (negative) defect. The convection cell has a thickness $d = (388 \pm 2)\,\mu m$ and dimensions $100d \times 203d$, of which only a central $51d \times 63d$ region was used for analysis. From [700]

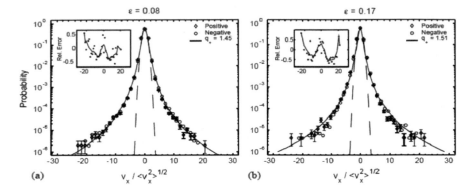

Fig. 7.25 Transverse velocity (v_x) distributions for $\epsilon = 0.08$ (**a**) and $\epsilon = 0.17$ (**b**) for positive and negative defects, rescaled to unit variance. Solid lines are q-Gaussian fittings (q being the only fitting parameter) for positive defects. Dashed lines represent a unit variance Gaussian. *Insets:* Relative errors [$p_{experiment} - p_{theory}$]/$p_{theory}$ for positive defects. From [700]

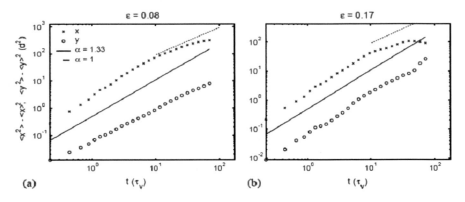

Fig. 7.26 Time evolution of the second moments of position trajectories in x and y. Solid line is the diffusive behavior predicted by Eq. (4.14) with $q = 3/2$, i.e., $\alpha = 4/3$; dotted line corresponds to normal diffusion ($q = 1$). Fits to the data give values of α in the range 1.16–1.5, depending on the region being fit. From [700]

7.1.6 Fingering

When two miscible liquids are pushed one into the other one, it is frequently observed *fingering* (e.g., viscous fingering) [720–722]. By computationally solving an appropriate generalized diffusion equation, this phenomenon has been put into evidence computationally: see Fig. 7.31.

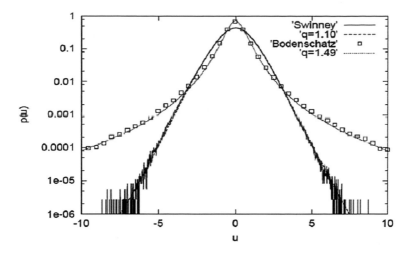

Fig. 7.27 Histogram of horizontal velocity differences as measured and analyzed by Swinney et al. [706], and by Bodenschatz et al. [719] in turbulent Couette–Taylor flow experiments. From [335]

7.1.7 Condensed Matter Physics

Manganites are a family of magnetic materials having "exotic" magnetic and electric properties (such as giant magnetoresistance), as well as ferro-paramagnetic first- and second-order phase transitions. Their theoretical approach has considerable difficulties. Several papers by the same group, [723–728] among others, have adopted a phenomenological approach based on q-statistics, using the index q as a tunable fitting parameter to reflect the consequences of the well known fractal nature (at the level of microstructures) of the family. The attempt has been successful in substances such as $La_{0.60}Y_{0.07}Ca_{0.33}MnO_3$ as can be judged from say Figs. 7.32, 7.33, 7.34, 7.35 and 7.36. The deep understanding of this fact on microscopic or mesoscopic grounds remains, nevertheless, an open question. Especially if one takes into account that many other attempts exist in the literature which exhibit only partial success in spite of the fact that they frequently involve several fitting parameters [725].

Figure 7.32 shows a typical temperature-magnetic field diagram as obtained for $q = 0.1$. In Fig. 7.33 a comparison is done between theory and experimental equations of states. The temperature-dependent parameters of the theory (q and μ) are indicated in Fig. 7.34. Using these phenomenological curves, Figs. 7.35 and 7.36 are obtained, *with no further fitting parameters at all.*

Another interesting phenomenon in condensed matter was observed by [729] in archetypal spin-glasses such as $Cu_{1-x}Mn_x$ ($x = 0.1, 0.16, 0.35$) and $Au_{1-x}Fe_x$

Fig. 7.28 Experimentally
measured probability
distributions of the velocity
differences for the
Couette–Taylor experiment
at *Reynolds number*
Re = 540000 for typical
values of the distance *r* are
compared with theoretical
q-Gaussians: **a** logarithmic
plot; **b** linear plot. The
rescaled distances r/η (η is
the *Kolmogorov length scale*)
are, from top to bottom, 11.6,
23.1, 46.2, 92.5, 208, 399,
830 and 14400. For better
visibility, each distribution in
(**a**) is shifted by –1 unit
along the *y* axis, and each
distribution in (**b**) is shifted
by –0.1 unit along the *y* axis.
From [706]

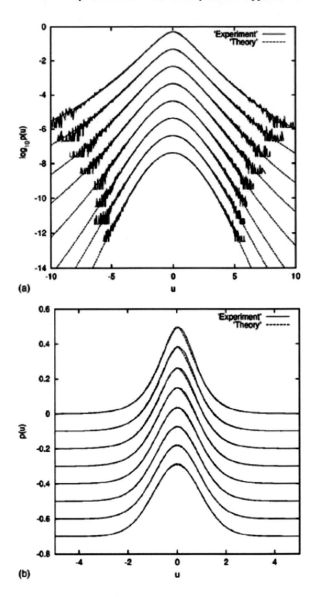

($x = 0.14$). Indeed, the analysis of their neutron spin echo spectra revealed a relaxation function with a stretched *q*-exponential time dependence. The temperature dependence of the corresponding values of *q* are illustrated in Fig. 7.37. This connection may be considered as not really surprising given the well known nonergodic behavior of spin-glasses.

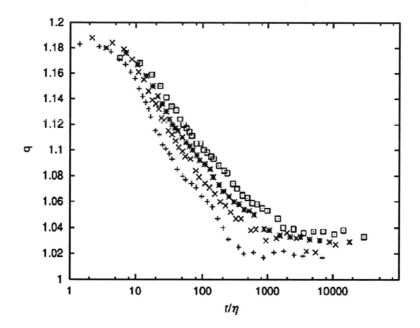

Fig. 7.29 Values of q used in Fig. 7.28. The Reynolds numbers are, from bottom to top, 69000, 133000, 266000, and 540000. From [706]

Fig. 7.30 Same data as in Fig. 7.29. The variable in the abscissa was heuristically found. The variable $[\ln(r/\eta)]/[(\ln Re)^{7/4}]$ leads to data collapse of the central region; the exponent 0.37 makes the data-collapsed region to become roughly a straight line. These features remain unexplained until now. From [707]

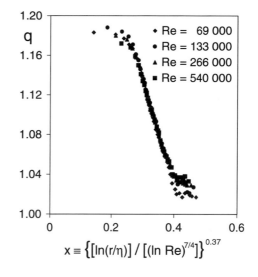

Fig. 7.31 *Top*:
Concentration fluctuations
field in the onset of fingering
between two miscible liquids
showing landscape of
q-Gaussian "hills and wells"
(highest positive values are
red; highest negative values
are magenta). From [720].
These structures identify the
existence of precursors to the
fingering phenomenon as
they develop before any
fingering pattern can be seen.
Bottom: Section plane cut
through the hills and wells.
The dashed line is made
from junctions (at the
successive inflection points)
of q-Gaussian branches

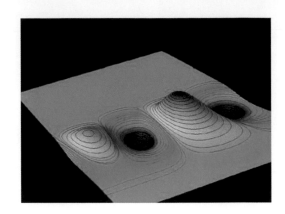

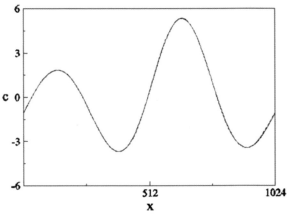

Fig. 7.32 Results from the
theoretical model. Projection
of the phase diagram in the
$h - t$ plane, for $q = 0.1$.
Above certain values of field
$h \geq h_{0_q}$, and temperature
$t \geq t_{0_q}$, the transition
becomes continuous. From
[725]

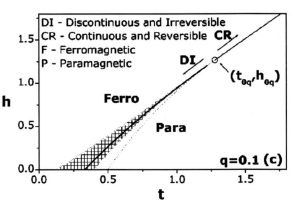

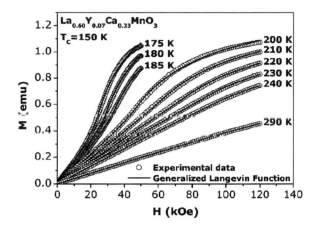

Fig. 7.33 Measured (open circles) and theoretical (solid lines) magnetic moment as a function of magnetic field, for several values of temperature above $T_c = 150\ K$. From [725]

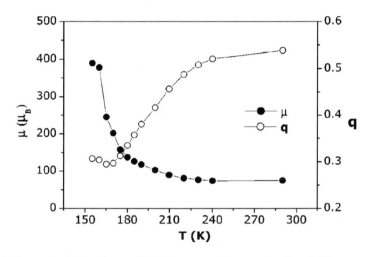

Fig. 7.34 Temperature dependence of the fitting parameters q and μ. From [725]

7.1.8 Plasma

Anomalous diffusion and distribution of displacements have been measured in dusty two-dimensional Ar plasma [730]. The results are respectively exhibited in Figs. 7.38 and 7.39. The numerical values for the anomalous diffusion exponent α and for q are indicated in Fig. 7.40. From these, by plainly averaging α, we obtain an intriguingly precise verification of prediction (4.14) (with the notation change $\mu \equiv \alpha$): see Fig. 7.41.

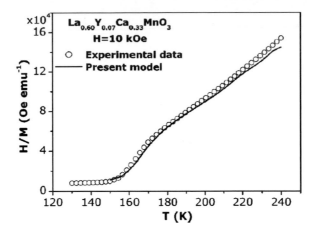

Fig. 7.35 Measured (open circles) and theoretical (solid line) values of the quantity H/M versus T. The solid line in this plot does not include any fitting parameters, and was calculated using only the fitting parameters of Fig. 7.33. From [725]

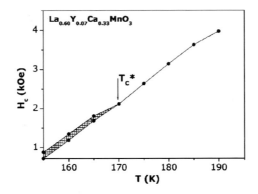

Fig. 7.36 The linear temperature dependence, for $T > T_c^*$, of the characteristic field H_c, which corresponds to the inflection point of the experimental M versus H curves, measured in $La_{0.60}Y_{0.07}Ca_{0.33}MnO_3$. For $T < T_c < T_c^*$ the hysteresis is indicated by the shaded area. The similarity between this experimental plot and the theoretical one, shown in Fig. 7.32, is striking. From [725]

As previously mentioned, several other applications of q-statistics are available in the literature concerning plasmas, e.g., turbulent pure electron plasma [714, 715]. See also [731–751].

7.1.9 Astrophysics and Astronomy

Self-gravitating systems
A vast literature explores the possible connections of q-statistics with self-gravitating systems and related astrophysical phenomena. The first such connection was

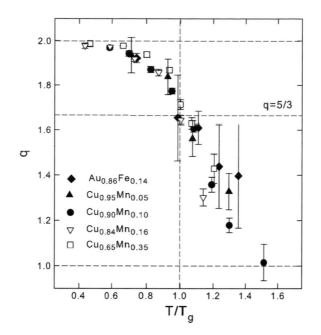

Fig. 7.37 Temperature dependence of the index q obtained from the fits of the neutron spin echo spectra relaxation of AuFe and CuMn samples, T_g being the corresponding glass temperature. From [729]

established in 1993 by Plastino and Plastino [752].[3] It provided a possible way out for an old gravitational difficulty, namely the impossibility of existence of a self-gravitating system maximizing entropy, and having finite total mass and finite total energy. Plastino and Plastino showed that polytropic distributions optimize the S_q entropy (with $q \neq 1$) under the constraints imposed by the total mass and the total energy of a self-gravitating system, paving the way for the application of maximum entropy techniques to the study of this kind of astrophysical systems. Within a Vlasov–Poisson polytropic description of a Newtonian self-gravitating system (i.e., $D = 3$), a connection was put forward between the polytropic index n and the entropic index q, namely (see [753–755] and references therein)

$$\frac{1}{1-q} = n - \frac{1}{2}. \tag{7.8}$$

The limit $n \to \infty$ (hence $q = 1$) recovers the isothermal sphere case (responsible for the paradox mentioned above); $n = 5$ (hence $q_c = 7/9$, where the subindex c stands for *critical*) corresponds to the so-called Schuster sphere; for $n < 5$ (hence $q < q_c =$

---

[3] This contribution constitutes in fact a historical landmark in nonextensive statistical mechanics. Indeed, it was the very first connection of the present theory with *any* concrete physical system. I had the privilege to analyze it with Roger Maynard during a conversation that lasted many hours.

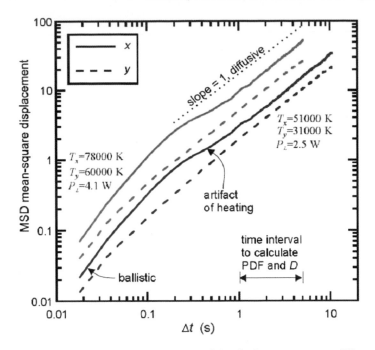

Fig. 7.38 Time evolution of the second moment of the displacements at two different sets of temperatures, namely at $(T_x, T_y) = (78000\,\text{K}, 60000\,\text{K})$, and $(T_x, T_y) = (51000\,\text{K}, 31000\,\text{K})$. The data in Fig. 7.40 have been obtained from such measurements. From [730]

7/9), simultaneous finiteness of mass, energy, and entropy naturally emerges. Eq. (7.8) can be generalized to the D-dimensional Vlasov–Poisson problem, and the following result is obtained [753]

$$\frac{1}{1-q} = n - \frac{D-2}{2}. \tag{7.9}$$

The critical case corresponds to the Schuster D-dimensional sphere, for which

$$n = \frac{D+2}{D-2}. \tag{7.10}$$

Replacing this expression into Eq. (7.9), we obtain

$$q_c(D) = \frac{8 - (D-2)^2}{8 - (D-2)^2 + 2(D-2)}. \tag{7.11}$$

We see that q_c decreases below unity when D increases above $D = 2$. The fact that the limiting case $q_c = 1$ occurs at $D = 2$ is quite natural. Indeed, the D-dimensional gravitational potential energy decays, for $D > 2$, like $-1/r^{D-2}$ with distance r. Con-

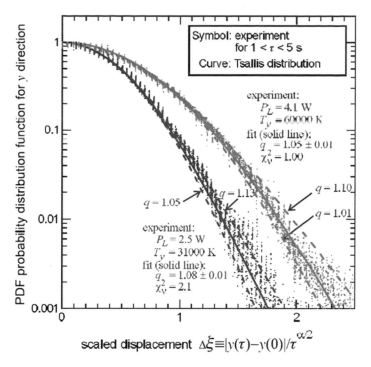

Fig. 7.39 Probability distributions associated with y-displacements. The present best fittings correspond to q-Gaussians with $q = 1.05$ ($q = 1.08$) for $T_y = 60000$ ($T_y = 31000$). From [730]

sequently, the dimension below which BG statistical mechanics can be legitimately used is precisely $D = 2$.

As a different application, let us mention that the rotational velocities of several classes of stars decrease with time due to the so-called magnetic braking. The angular velocity q-exponentially decreases with time as observed, for instance, in the open cluster αPer. The value of q depends on the mass and age of the star, but its mean value is $q \simeq 1.33$ [756].

Many more contributions along these and other lines concerning star, galaxies, black holes, cosmology can be found in the literature [342, 343, 732, 733, 757–810]. We remind, in particular, that black holes constitute a long-standing cosmological theme. [222, 296, 1133–1135].

Temperature fluctuations of the cosmic microwave background radiation
The q-Gaussians are used since many decades in astrophysics with no deep theoretical justification [302]. They are called κ-distributions, and are written as follows:

$$f(v) \propto \frac{1}{\left(1 + \frac{v^2}{\kappa\, v_0^2}\right)^{\kappa+1}} . \tag{7.12}$$

TABLE I: The measure $\bar{q}$ of non-extensivity indicates non-Gaussian statistics. Mean diffusion exponent $\bar{\alpha}_y$ (and p-value for testing the null hypothesis that there is no superdiffusion) indicates superdiffusion.

| | time delay (s) | temperature range T_y (10^3 K) | | |
		$10 - 20$	$20 - 40$	$40 - 60$
$\bar{q}$	$1 < \tau < 5$	1.20 ± 0.09	1.11 ± 0.06	1.05 ± 0.02
$\bar{\alpha}_y$	$1 < \Delta t < 5$	1.052 ± 0.019	1.059 ± 0.011	1.009 ± 0.011
p		0.007	2×10^{-5}	0.210
$\bar{\alpha}_y$	$5 < \Delta t < 9$	1.088 ± 0.048	1.082 ± 0.040	
p		0.042	0.038	
$\bar{\alpha}_y$	$9 < \Delta t < 13$	1.146 ± 0.062		
p		0.028		
$\bar{\alpha}_y$	$13 < \Delta t < 17$	1.183 ± 0.064		
p		0.006		

$\alpha_{average}$	1.117	1.070	1.009
$\dfrac{\alpha_{average}(3-q)}{2}$	1.005	1.011	0.984

Fig. 7.40 The Table is from [730]. The lower box has been calculated from the data of the Table

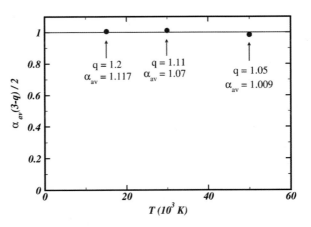

Fig. 7.41 Constructed from the data in the lower box of Fig. 7.40. As wee see, q approaches unity when the temperature increases (we have plotted the mid point of the temperature intervals indicated in the Table of Fig. 7.40). For these average values of $\alpha_{average}$, the prediction (4.14) ($\mu \equiv \alpha$) is satisfied within an overall error bar of 1.6% (intriguingly small, in fact, if we take into account that we are using averages values for α; compare with Fig. 5.67, where the error bar is 10%)

With the notation changes $\kappa = (2 - q)/(q - 1)$ and $1/(\kappa v_0^2) = (q - 1)\beta$, we immediately identify $e_q^{-\beta v^2}$. This appears to be the case of the temperature fluctuations, around the value $T \simeq 2.7$K, of the cosmic microwave background radiation of the universe: See Figs. 7.42 and 7.43 [811]. Many cosmological theories assume (or imply) this distribution to be Gaussian. As we see in Fig. 7.43, this is not correct, the overall value of the index being $q = 1.04 \pm 0.01$. The Gaussian assumption is excluded at a 99% confidence level, and does not constitute more than a first approach to the problem. Moreover, anisotropy is found between the four universe quadrants, the strongest non-Gaussian contribution coming from the South-East universe quadrant, where $q_{SE} \simeq 1.05$.

Dark matter profiles
Nonextensive statistics provide a way to describe Dark Matter cored haloes from basic principles. For example, a set of polytropic, non-Gaussian, Lane–Emden spheres with the central value $q = 0.85$ yield a satisfactory fitting for all the observed rotational curves of some nearby spiral galaxies (see [812] and references therein).

Asteroids, meteorites, and lunar flashes
Let us mention two new astronomical laws, recently established by Betzler and Borges [813], for asteroids, namely that of periods (see Fig. 7.44) and that of sizes (see Fig. 7.45).

It is interesting to notice that the distribution of sizes is a q-exponential whereas that of periods is a q-Gaussian. The reason might be related to the fact that sizes vary from zero to infinity. The frequencies vary instead from minus to plus infinity in we take into account the sense of rotation, therefore their distribution ought to be even functions of the frequency, and the periods are inverses of the *absolute value* of the frequencies.

Also meteor showers and lunar flashes seemingly follow q-statistics [814].

If we focus on the masses of meteorites, the distributions are, for still unknown reasons, quite diverse: exponentials or q-exponentials or stretched q-exponentials [815]. The Whitecourt meteorites constitute, however, a good example of a simple q-exponential with $q \simeq 1.5$. See Figs. 7.46 and 7.47.

Interplanetary coronal mass ejections
Coronal mass ejections (CMEs) are a transient type of solar wind which originate from closed magnetic field regions in the solar corona and are ejected from the solar atmosphere, transporting large quantities of plasma and magnetic flux into interplanetary space. They were identified by images observed in the 1970s when events were detected in images from coronagraphs like Skylab [816]. See Fig. 7.48.

Other astrophysical phenomena might be related with nonextensive concepts as well. Such is the case of solar flares, whose probability distributions of characteristic times appear to be of the q-exponential form [817].

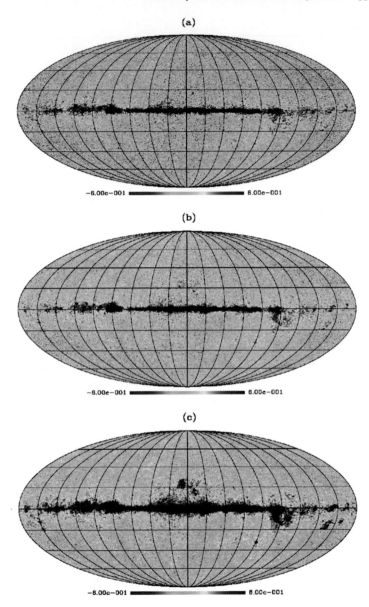

Fig. 7.42 The WMAP1 (WMAP with one-year data) CMBR temperature fluctuations maps, denoted by **a** W (93.5 GHz), **b** V (60.8 GHz), and **c** Q (40.7 GHz). To enhance the effect of the cosmic temperature fluctuations over the galaxy foregrounds, in these plots we consider only those pixels with $(1 - q)T \in [-0.6, 0.6]$ mK. Data from the area contaminated by the Galaxy emissions are usually excluded from the statistical analysis, and, for the regions investigated, the whole range of temperatures measured by WMAP is considered

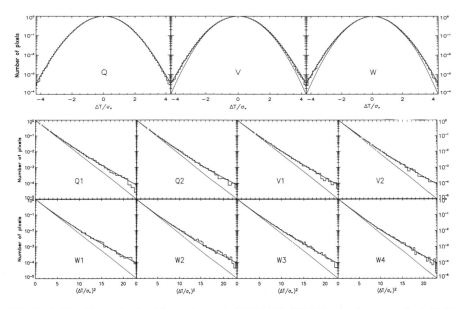

Fig. 7.43 *Top*: Fits to the (positive and negative) WMAP3 (WMAP with three-year data) CMB temperature fluctuations data, corresponding to the Q, V, and W coadded maps (after the Kp0 cut-sky), in the NUMBER OF PIXELS *versus* $(1 - q)T/\sigma_\nu$ plots. We show the χ^2 best-fits: Gaussian distribution (blue curve) with $\sigma_Q = 104\,\mu$K, $\sigma_V = 118\,\mu$K, and $\sigma_W = 131\,\mu$K, respectively, and each nonextensive distribution (red curve) P_q with $q = 1.04$. *Bottom*: Similar analysis, but now with the eight DA WMAP3 maps (Q1,...,W4) after applying the Kp0 mask. We plotted the NUMBER OF PIXELS *versus* $((1 - q)T/\sigma_\nu)^2$ to enhance the non-Gaussian behavior. To avoid possible unremoved Galactic foregrounds, we consider only the negative temperature fluctuations. Again, we show the χ^2 best-fits: Gaussian distribution (blue curve) and nonextensive distribution (red curve) P_q, now with $q = 1.04 \pm 0.01$. From [811]

7.1.10 Geophysics

Earthquakes

The q-statistical theory and functional forms have been successfully applied in many occasions to earthquakes and related phenomena [574, 818–831].

Successive earthquakes in a given geographic area occur with epicenters distributed in a region just below the Earth's surface. We note r the successive distances (measured in three dimensions). It has been verified [819] that, in California and Japan, this distribution, $p(r)$, happens to be well represented (see Fig. 7.49) by a q-exponential form. More precisely, the corresponding accumulated probability is given by the *Abe-Suzuki distance law*.

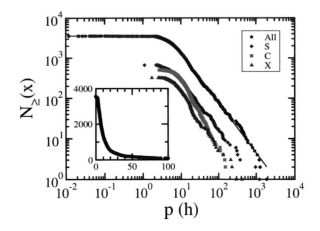

Fig. 7.44 Log–log plot of the decreasing cumulative distribution of periods of 3567 asteroids (dots) with $Rel \geq 2$ taken from the PDS (NASA) and a q-Gaussian distribution $[N_{\geq}(p) = M \exp_q(-\beta_q p^2)]$ (solid line), with $q = 2.6$, $\beta_q = 0.025h^{-2}$, $M = 3567$. The other curves are 663 S-complex asteroids (diamonds, blue online), 503 C-complex asteroids (squares, green online), 321 X-complex asteroids (triangles, magenta online). Inset shows the 3567 asteroids and the q-Gaussian in a linear–linear plot. From [813]

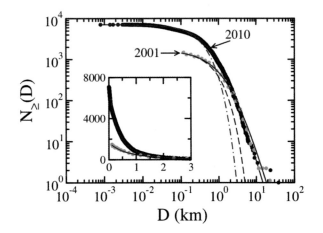

Fig. 7.45 Decreasing cumulative distribution of diameters of known NEAs in 2001 (1649 objects, green dots) and in 2010 (7078 objects, black dots). Solid lines are best fits of q-exponentials $[N_{\geq}(D) = M \exp_q(-\beta_q D)]$. Blue line (2001): $q = 1.3$, $\beta_q = 1.5\text{km}^{-1}$, $M = 1649$, red line (2010): $q = 1.3$, $\beta_q = 3\text{km}^{-1}$, $M = 7078$. Normal exponentials ($q = 1$) are displayed in the main panel for comparison (dashed violet, with $\beta_1 = 1.5\text{km}^{-1}$, $M = 1649$, and dot-dashed magenta, with $\beta_1 = 3\text{km}^{-1}$, $M = 7078$). From [813]

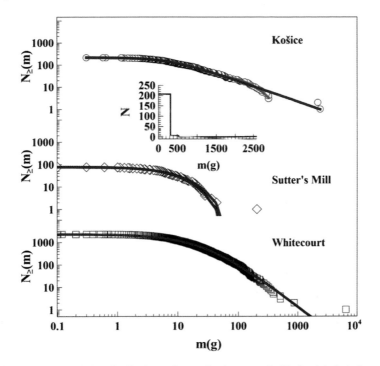

Fig. 7.46 Inverse cumulative distributions of mass for the meteorite Kosice (circles), Sutter's Mill (diamonds), and Whitecourt (squares). The fit of a q-stretched exponential for the Kosice (solid red line) has $N_t = 221$, $\beta_q = 3.79 \times 10^{-2} \, g^{-1}$, $\gamma = 1.45$, and $q = 2.55$. The fit of a bimodal exponential (solid green line) has $N_t = 213$, $\gamma_1 = \gamma_2 = 1.4$, $\mu_1 = 13.1 \, g$, and $\mu_2 = 140 \, g$. The adjustment of a q-exponential for Sutter's Mill (solid red line) has $N_t = 80$, $\beta_q = 1.01 \times 10^{-1} g^{-1}$, and $q = 1.02$. The fit of an exponential (solid blue line) has $N_t = 77$, $\gamma = 1.0$, and $\mu = 11 \, g$. The fit of a q-exponential for the Whitecourt (solid red line) has $N_t = 2308$, $\beta_q = 8.00 \times 10^{-2} \, g^{-1}$, and $q = 1.50$. Inset: histogram of the observed mass distribution of the Kosice meteorite. From [815]

$$P(> r) = e_q^{-r/r_0} \quad (0 < q < 1; \; r_0 > 0) . \tag{7.13}$$

Consequently

$$p(r) = -\frac{dP(> r)}{dr} = \frac{1}{r_0} e_Q^{-r/[r_0(2-Q)]} , \tag{7.14}$$

with

$$Q \equiv 2 - \frac{1}{q} < q . \tag{7.15}$$

Let us address now a different phenomenon, namely the fact that, between successive earthquakes in a given area of the globe, there are calm-times, noted τ, and defined through a fixed threshold m_{th} for the magnitude. It has been verified [823] that, in California and Japan, the calm-time distribution, $p(\tau)$, happens to be well rep-

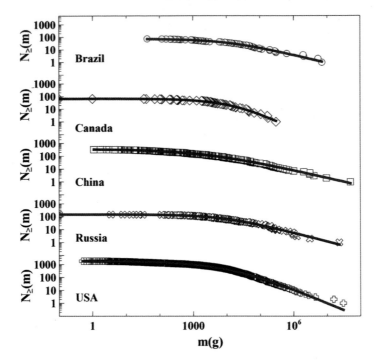

Fig. 7.47 Inverse cumulative distributions of mass for some types of meteorites found in Brazil (circles), Canada (diamonds), China (squares), Russia (crosses), and the USA (plus). The red solid lines represent the fitting of q-stretched exponentials for the meteorites found in Brazil ($q = 1.87$, $\beta - q = 5.43 \times 10^{-3}\,\mathrm{g}^{-1}$, $N_t = 74$, and $\gamma = 0.57$), Canada ($q = 1.47$, $\beta_q = 5.08 \times 10^{-3}\,\mathrm{g}^{-1}$, $N_t = 65$, and $\gamma = 0.61$), China ($q = 1.64$, $\beta_q = 7.48 \times 10^{-2}\,\mathrm{g}^{-1}$, $N_t = 359$, and $\gamma = 0.39$), Russia ($q = 1.91$, $\beta_q = 4.20 \times 10^{-3}\,\mathrm{g}^{-1}$, $N_t = 144$ and $\gamma = 0.61$), and the USA ($q = 1.43$, $\beta_q = 2.60 \times 10^{-2}\,\mathrm{g}^{-1}$, $N_t = 1816$, and $\gamma = 0.48$). From [815]

resented (see Fig. 7.50) by a q-exponential form. More precisely, the corresponding accumulated probability is given by the *Abe-Suzuki time law*

$$P(> \tau) = e_q^{-\tau/\tau_0} \quad (q > 1;\ \tau_0 > 0). \tag{7.16}$$

Moreover, the pair (q, τ_0) depends in a defined form on the threshold m_{th}, as indicated in Fig. 7.51.

As another important application to earthquakes let us focus on the *ageing* which occurs within the nonstationary regime called *Omori regime*, the set of aftershocks that follow a big event. We introduce the following correlation function:

$$C(n + n_W, n_W) \equiv \frac{\langle t_{n+n_W} t_{n_W}\rangle - \langle t_{n+n_W}\rangle \langle t_{n_W}\rangle}{\sigma_{n+n_W} \sigma_{n_W}}, \tag{7.17}$$

where

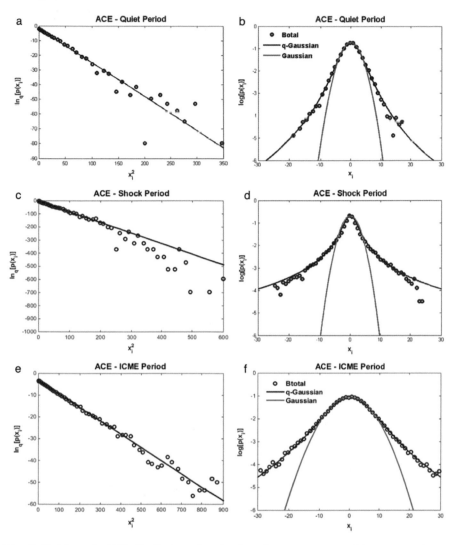

Fig. 7.48 Magnetic field distributions for the interplanetary space. **a** Linear approximant $ln_q\, p(x_i)\ vs\ x_i^2$, where $q_{stat} = 1.31 \pm 0.02$ for the quiet period of magnetic field magnitude, **b** $p(x_i)\ vs\ x_i^2$, with q_{stat}-Gaussian function that fits $p(x_i)$ for the quiet period, **c** Linear approximant $ln_q\, p(x_i)\ vs\ x_i^2$, where $q_{stat} = 1.69 \pm 0.03$ for the shock period of magnetic field magnitude, **d** $p(x_i)\ vs\ x_i^2$, with q_{stat}-Gaussian function that fits $p(x_i)$ for the shock period, **e** Linear approximant $ln_q\, p(x_i)\ vs\ x_i^2$, where $q_{stat} = 1.27 \pm 0.03$ for the ICME period of magnetic field magnitude, **f** $p(x_i)\ vs\ x_i^2$, with q_{stat}-Gaussian function that fits $p(x_i)$ for the ICME period. From [816]

$$\langle t_m \rangle = \frac{1}{N} \sum_{k=0}^{N-1} t_{m+k} \,, \tag{7.18}$$

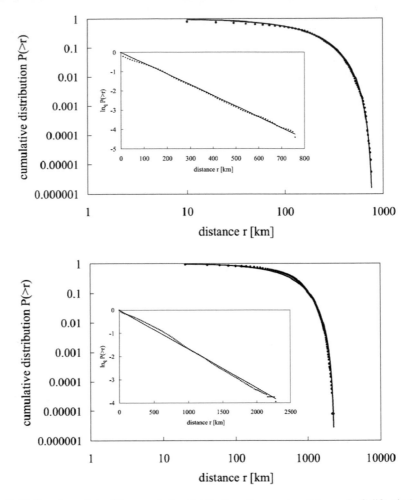

Fig. 7.49 Log–Log plots of the cumulative distribution of successive distances in California (*top*) and Japan (*bottom*). Dots from the California and Japan catalogs respectively; continuous curves from Eq. (7.13). *Insets:* The same in q-log *versus* linear representation (California: $q = 0.773$, $r_0 = 179$ Km, and the linear regression coefficient $R = -0.9993$; Japan: $q = 0.747$, $r_0 = 595$ Km, and $R = -0.9990$). For details see [819]

$$\langle t_m t_{m'} \rangle = \frac{1}{N} \sum_{k=0}^{N-1} t_{m+k} t_{m'+k} \,, \tag{7.19}$$

and

$$\sigma_m^2 = \langle t_m^2 \rangle - \langle t_m \rangle^2 \,, \tag{7.20}$$

N being the number of events that are being considered within the Omori regime, and t_m being the time at which the $m - th$ event occurs; m is sometimes referred to as

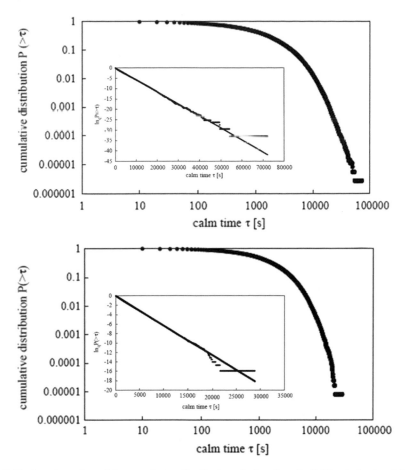

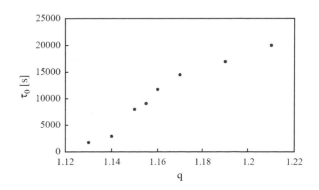

Fig. 7.50 Log–Log plots of the cumulative distribution of calm-times in California (*top*) and Japan (*bottom*). Dots from the California and Japan catalogs respectively; continuous curves from Eq. (7.16). *Insets:* The same in q-log *versus* linear representation (California: $q = 1.13$, $\tau_0 = 1724\,s$, and the linear regression coefficient $R = -0.988$; Japan: $q = 1.05$, $\tau_0 = 1587\,s$, and $R = -0.990$). For details see [823]

Fig. 7.51 Dependence of (q, τ_0) on m_{th}. Data from the California catalog. From the bottom to the top, m_{th}=0.0, 1.4, 2.0, 2.1, 2.2, 2.3, 2.4, 2.5. For further details see [823]

natural time. By definition, $C(n_W, n_W) = 1$. For a stationary state, $C(n + n_W, n_W)$ depends on the natural time n, but not on the *waiting natural time* n_W; if it also depends on n_W, the state is necessarily a nonstationary one, and exhibits *ageing*, one of the most characteristic features of glassy systems. The correlation function of typical earthquakes in Southern California has been discussed in [820]: see Figs. 7.52 and 7.53 for catalog data, respectively inside and outside the Omori regime. Whenever aging is observed, data collapse can be obtained (see Fig. 7.54) through rescaling, more specifically by using as abscissa $n/f(n_W)$ instead of n, where $f(n_W) = an_W^\gamma + 1$, a and γ being fitting parameters. The connection with q-statistics comes from the fact that the type of dependence that we observe in Fig. 7.54 appears to be of the q-exponential form. Let us address this point now. This correlation function has been calculated [574] for a simple mean-field model called the *coherent noise model* (Newman model): the results can be seen in Figs. 7.55, 7.56, and 7.57. Another model for earthquakes has been discussed as well [818], the Olami–Feder–Christensen models. The results for the Newman and the OFC models are respectively $C(n + n_W, n_W) = e_{2.98}^{-0.7 n/n_W^{1.05}}$ and $C(n + n_W, n_W) = e_{2.9}^{-0.6 n/n_W^{1.05}}$: see Fig. 7.57. Summarizing, both earthquake models that have been considered yield virtually the same result, namely that the rescaled correlation function is of the q-exponential form with $q \simeq 2.98$.

Let us address now the most classical quantity for earthquakes, namely the probability of having earthquakes of magnitude m (*Gutenberg-Richter law*). A nontrivial result (generalizing in fact the classical Gutenberg-Richter law) has been analytically obtained [821] along this line for the cumulative probability $G(> m)$ involving two parameters, q and a (a is the constant of proportionality between the released relative

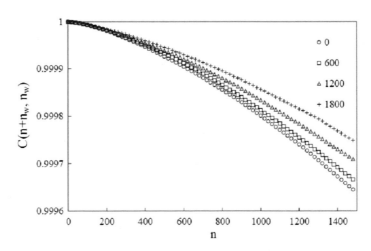

Fig. 7.52 Dependence of the correlation function on the natural time n *inside* the Omori regime following a specific event taken from the California catalog. Aging is visible. For further details see [820]

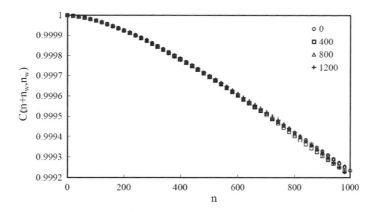

Fig. 7.53 Dependence of the correlation function on the natural time *n outside* the Omori regime following a specific event taken from the California catalog. No aging is visible. For further details see [820]

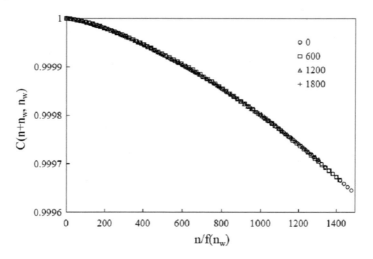

Fig. 7.54 Dependence of the correlation function on a rescaled natural time $n/f(n_W)$ *inside* for the same data of Fig. 7.52 with $f(n_W) = an_W^\gamma + 1$ ($a = 1.37 \times 10^{-6}$, and $\gamma = 1.62$). For further details see [820]

energy ϵ and the linear dimension r of the fragments of the fault plates). These results are much in line with those presented in Figs. 7.58 and 7.59.

Finally, let us focus on the histograms of the avalanche size differences (*returns*, as such quantities are called in finance). These have been focused in [460, 833]. In particular, such probability distributions have been calculated in a dissipative Olami–Feder–Christensen model (Fig. 7.60), and also for real earthquakes (Fig. 7.61). The results for the OFC model have been calculated in both a small-world lattice (referred to as the *critical* case)) and a regular lattice (referred to as the *noncritical* case). The conclusion is highly interesting: at criticality q-Gaussian-like distributions are

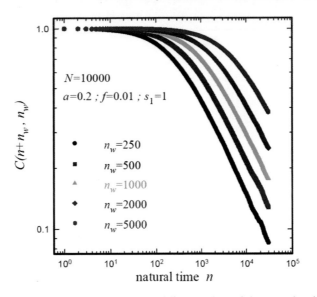

Fig. 7.55 Event-event correlation functions for different values of the natural waiting time n_W: aging is apparent. The ensemble average (which differs, in fact, from the time average, thus exhibiting the breakdown of ergodicity) is performed over 120000 numerical runs with different initial conditions. From [822]

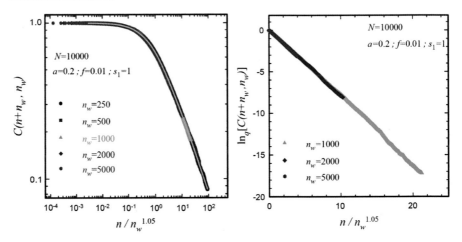

Fig. 7.56 Data collapse of the same numerical data of Fig. 7.55 *Left:* In *log–log* representation. *Right:* In *q-log versus* linear representation; the straight line implies that the scaling function is a *q*-exponential with $q \simeq 2.98$. From [822]

obtained, whereas something close to a Gaussian on top of another (larger) Gaussian is obtained out of criticality.[4] The (analytic) connection between the various *q*'s

---

[4] This fact is quite suggestive on quite different experimental grounds. Indeed, the velocity distribution of cold atoms in dissipative optical lattices has been measured by at least two different groups, namely in [834] and in [631]. The latter obtained a *q*-Gaussian velocity distribution (see

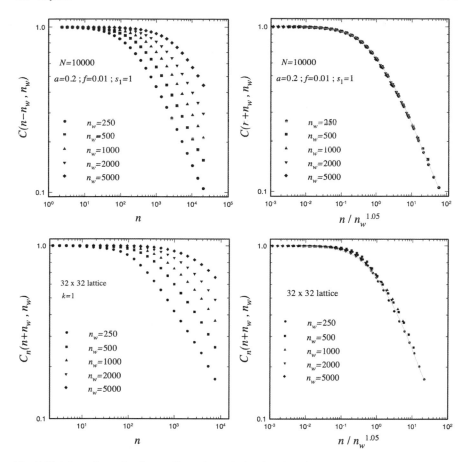

Fig. 7.57 The collapsed and noncollapsed correlation functions for the Newman model (top; averaged over 100, 000 realizations) and the Olami-Feder-Christensen model (bottom; averaged over 20, 000 realizations). From [818]

that have been presented here for earthquakes remains an open worthwhile question (Fig. 7.58).[5]

_____

Fig. 2a in [631]). The former, however, obtained a double-Gaussian distribution (see Fig. 11a in [834]). The reason for such a discrepancy is, to the best of our knowledge, not yet understood. A possibility could be that in the latter experiment, the apparatus is at "criticality", whereas in the former experiment it might be slightly out of it. The point surely is worthy of further clarification.

[5] Along this line, some hint might be obtained from the following observation. Series corresponding to thirteen earthquakes have been analyzed in [828]. It is claimed that the cumulative distribution of the distances between the epicenters of successive events is well fitted by a q_s-exponential (where s stands for *spatial*); analogously, the cumulative distribution of the time intervals between successive events was also well fitted by a q_t-exponential (where t stands for *temporal*). From the data corresponding to the set of 13 earthquakes (see Table 3 of [828]), we can calculate $q_s = 0.73 \pm 0.09$, $q_t = 1.32 \pm 0.08$, and $q_s + q_t = 2.05 \pm 0.07$. If the distances and times between successive events were independent, we should obtain, for the standard deviation of $q_s + q_t$, roughly

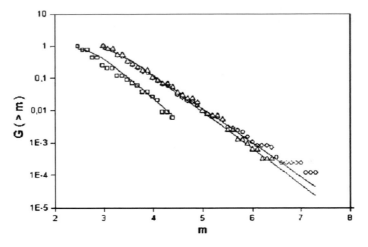

Fig. 7.58 Cumulative probability for having earthquakes with magnitude above m (exceedence). California (*circles*, over 10000 earthquakes, $q = 1.65$, $a = 5.73 \times 10^{-6}$), Iberian Peninsula (*triangles*, 3000 earthquakes, $q = 1.64$, $a = 3.37 \times 10^{-6}$), and Andalusian region (*squares*, 300 earthquakes, $q = 1.60$, $a = 3 \times 10^{-5}$). For further details see [821]

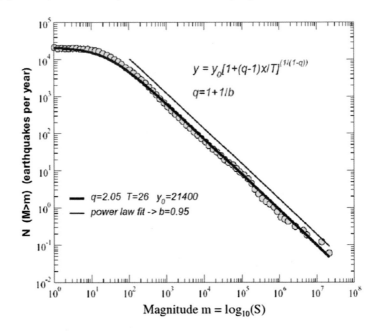

Fig. 7.59 Cumulative number of earthquakes with magnitude above m per year. The dots are from the California catalog and correspond to 335076 earthquakes. The blue curve is a q-exponential with $q = 2.05$. From [832]

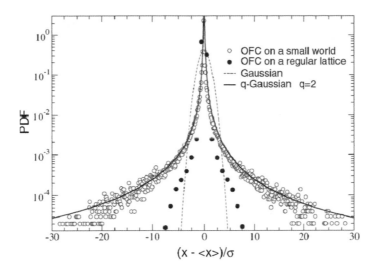

Fig. 7.60 Probability distribution of the avalanche size differences (returns) $x(t) = S(t + 1) - S(t)$ for the OFC model on a small-world topology (critical state, open circles) and on a regular lattice (noncritical state, full circles). For comparison, a Gaussian and a q-Gaussian (with $q = 2$) are indicated as well. All the curves have been normalized so as to have unit area. For further details see [460]

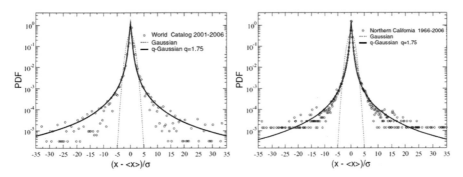

Fig. 7.61 The same as in Fig. 7.60 but for real earthquakes. *Left:* From the World catalog. *Right:* From the Northern California catalog. For comparison, a Gaussian and a q-Gaussian have indicated as well. Both fittings provided $q = 1.75 \pm o.15$. For further details see [460])

El Niño

The *Southern Oscillation Index* (SOI) corresponds to the daily registration of the oceanic temperature (including appropriate pressure corrections) at a fixed point of the Earth. Histograms can be constructed by using values separated by a fixed time

---

$0.08 + 0.07 \simeq 0.17$. Since the data yield 0.07 instead of 0.17, correlation is present, which suggests $q_s + q_t \simeq 2$ for each earthquake.

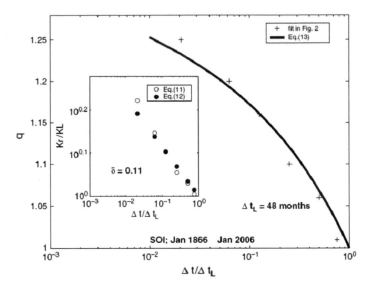

Fig. 7.62 Dependence of q on the (conveniently rescaled) time lag for the SOI. The data correspond to the Jan 1866–Jan 2006 period. From [835]

lag. They are well fitted by q-Gaussians with q depending on the time lag [835, 836]. See Figs. 7.62 and 7.63. Like for financial returns, the index q gradually approaches unity (i.e., Gaussian distribution) when the time lag increases, which corresponds of course to an increasing loss of time correlation of the successive values of the signal. A micro- or meso-scopic theory interpreting the results exhibited in Figs. 7.62 and 7.63 would of course be welcome.

Further geophysical applications, e.g., to clouds [837], the Stromboli volcano [838], geological faults [838], are available in the literature as well.

Geomagnetic field reversals

The paleomagnetic Earth's magnetic field has reversed its polarity, several times, suddenly and disorderly in the last 160 million years. The statistical distribution of time intervals between successive geomagnetic reversals has been analyzed [839] and appears to follow, for yet unknown reasons, q-statistics: see Fig. 7.64.

Sea ice fracturing

Time series of stresses induced by ice motion in the Arctic sea ice have been analyzed [840]. The quantities that have been considered are the increments of the time series $X(t)$ of the two principal stress values σ_1 and σ_2 and of the principal stress direction θ_s. These increments are defined as $X(t) = S(t + 1) - S(t)$, where $S(t)$ is

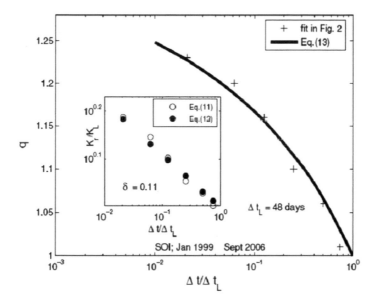

Fig. 7.63 Dependence of q on the (conveniently rescaled) time lag for the SOI. The data correspond to the Jan 1999–Sept 2006 period. From [836]

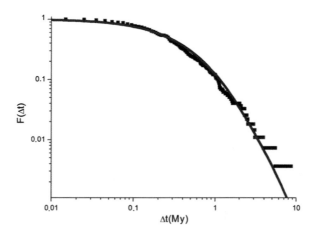

Fig. 7.64 Accumulated frequency distribution function F of time intervals Δt between consecutive reversals in log–log representation (squares). The continuous line is $F = e_q^{-B\Delta t}$ with $q = 1.3$ and $B = 2.94$. From [839]

each one of the parameters $(\sigma_1, \sigma_2, \theta_s)$, respectively. The corresponding probability density function $p(x)$ are shown in Fig. 7.65, where $x \equiv (X - \langle X \rangle)/\sigma_X$, σ_X being the standard deviation of the variable $X(t)$. The complex mechanisms underlying this phenomenology are yet to be elucidated.

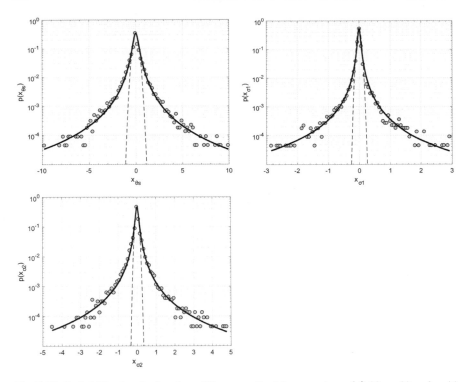

Fig. 7.65 Probability density function of the normalized increments x of $\theta_s(t)$, $\sigma_1(t)$ and $\sigma_2(t)$ (filled circles); the Gaussian fit (dashed black line) and the q-Gaussian fit $p(x) \propto e_q^{-x^2/B}$ (solid purple line) for the parameter values $(q_{\theta_s}, B_{\theta_s}) = (1.74 \pm 0.03, 0.078 \pm 0.006)$, $(q_{x_1}, B_{x_1}) = (1.85 \pm 0.04, 0.0017 \pm 0.0002)$ and $(q_{x_2}, B_{x_2}) = (1.82 \pm 0.12, 0.0068 \pm 0.0006)$, respectively. From [840]

7.1.11 Quantum Chaos

The quantum kicked top (QKT) is a paradigmatic system showing quantum chaos. In its regular regime, the overlap function O behaves roughly constant with time, and in the strongly chaotic regime, it decreases exponentially with time before the emergence of quantum interference effects. It is therefore possible that, precisely in the frontier between both regions, the exponential time dependence of the overlap function be replaced by a q-exponential form. This conjecture has indeed been verified numerically [841, 842] (see also [843]), as can be seen in Figs. 7.66, 7.67 and 7.68.

7.1.12 Quantum Information

A considerable effort has been dedicated to the connections between generalized entropic forms and the location of the critical frontier which has separable states on one side and quantum entangled states on the other one. A remarkably simple, and

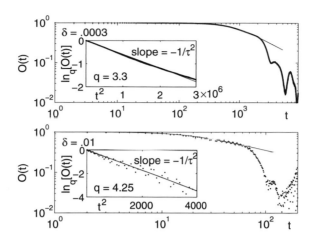

Fig. 7.66 Overlap versus time for an initial angular momentum coherent state located at the border between regular and chaotic zones of the QKT of spin 240 and $\alpha = 3$. This region, the edge of quantum chaos, shows the expected power law decrease in overlap. The top figure is for a perturbation strength in the weak perturbation regime, $\delta = 0.0003$ and the bottom figure is for a perturbation strength of $\delta = 0.01$, within the FGR regime. On the log–log plot the power law decay region, from about 600–2500 in the weak perturbation regime and 20–70 in the FGR regime is linear. We can fit the decrease in overlap with the expression $[1 + (q_{rel} - 1)(t/\tau_{q_{rel}})^2]^{1/(1-q_{rel})}$ where, in the weak perturbation regime, the entropic index $q_{rel} = 3.3$ and $\tau_{q_{rel}} = 1300$ and in the FGR regime $q_{rel} = 4.25$ and $\tau_{q_{rel}} = 34$. The insets of both figures show $\ln_{q_{rel}} O \equiv (O^{1-q_{rel}} - 1)/(1 - q_{rel})$ versus t^2; since $\ln_q x$ is the inverse function of $e_q^x \equiv [1 + (1 - q) x]^{\frac{1}{1-q}}$, this produces a straight line with a slope $-1/\tau^2$ (also plotted). From [842]

sometimes quite performant, criterium based on the conditional form of the entropy S_q was advanced by Abe and Rajagopal in [844]. For some systems, this procedure enabled the exact calculation of the separable-entangled frontier. Such is the case illustrated in Fig. 7.69 (from [845]). An entire literature exists in fact exploring this and related questions [162, 846–868].

Another interesting application consists in the experimental demonstration of robust quantum steering [869, 870].

7.1.13 Random Matrices

The classical Gaussian ensembles of random matrices can be alternatively obtained by maximizing the Boltzmann–Gibbs–von Neumann entropy under appropriate constraints. By optimizing instead the entropy S_q it is possible to q-generalize such ensembles. This has been done in [871] and elsewhere [872–876], and interesting generalizations of the *semi-circle law* for the eigenvalue density, and of *Wigner's surmise* for the level-spacing distribution are obtained. The index q determines the degree of confinement, in such a way that $q \leq 1$ corresponds to strong localization and $q > 1$ corresponds to weak localization. It is possible that this generalization has

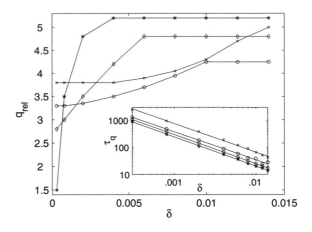

Fig. 7.67 Values of q_{rel} and τ_q (inset) for $J = 120$ (x), 240 (circles), 360 (diamonds) and 480 (stars). q_{rel} remains constant for perturbation strengths below the critical perturbation and above the saturation perturbation. In between q_{rel} increases with a rate dependent on J. The values of q_{rel}^c, q_{rel}^s, δ_c and δ_s can be seen in the figure. In addition, the rate of growth of q_{rel} with increased perturbation strength can be seen. The inset shows a log–log plot of the value of τ_q versus δ for the above values of J. The data can be fit with lines of slope -1.06, -1.03, -1.07, and -1.08 for $J = 120, 240, 360,$ and 480 (top to bottom). From [842]

Fig. 7.68 Values of q_{rel}^c versus $1/J$. These are determined by exploring a number of perturbations much less than δ_c. We note that q_{rel}^c of the $J = 480$ QKT is larger than q_{rel} reported in Fig. 7.67. It is unclear why in this instance the value of q_{rel} decreases with increased perturbation strength. From [842]

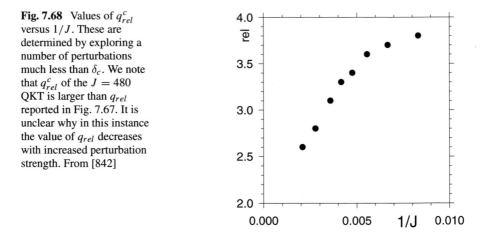

some connection with quantum weak chaos [841, 842], but this remains to be studied. In any case, an application for classifying proteins has already emerged [877].

7.1.14 Nonlinear Quantum Mechanics

Let us consider here possible nonlinear and/or inhomogeneous generalizations of quantum mechanics, either relativistic or nonrelativistic [639, 640, 878–893].

The d-dimensional q-plane wave is defined as follows:

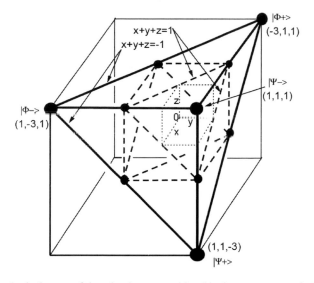

Fig. 7.69 The physical space of the mixed state considered in the present paper is the tetrahedron determined by the four big circles. Every big circle and its three neighboring small circles determine a region (small tetrahedron) where no separability is possible. The four small tetrahedra delimit a central octahedron where the system is separable. The $x + y + z = 1$ plane (dashed) generalizes the $x_c = 1/3$ Peres criterion, and plays the role of a critical surface. The entanglement "order parameter" $\eta \equiv 1/q_I$ is zero inside the central octahedron, and continuously increases when we approach the four vertices of the big tetrahedron, where $\eta = 1$. From [845]

$$\Phi(\vec{x}, t) = \Phi_0 \, \exp_q \left[i (\vec{k} \cdot \vec{x} - \omega t) \right] , \qquad (7.21)$$

which, due to the well known properties of the scalar product $\vec{k} \cdot \vec{x} = \sum_{n=1}^{d} k_n x_n$, exhibits invariance under rotations. If we take into account that $d \exp_q(z)/dz = [\exp_q(z)]^q$ and $d^2 \exp_q(z)/dz^2 = q[\exp_q(z)]^{2q-1}$ we obtain, for the d-dimensional Laplacian,

$$\nabla^2 \Phi(\vec{x}, t) = -q \left(\sum_{n=1}^{d} k_n^2 \right) \Phi_0 \{ \exp_q[i(\vec{k} \cdot \vec{x} - \omega t)] \}^{2q-1} . \qquad (7.22)$$

From the results above one sees that the d-dimensional q-plane wave satisfies the linear wave equation,

$$\nabla^2 \Phi(\vec{x}, t) = \frac{1}{c^2} \frac{\partial^2 \Phi(\vec{x}, t)}{\partial t^2} \quad \Rightarrow \quad \omega = ck , \qquad (7.23)$$

with $k = \sqrt{\sum_{n=1}^{d} k_n^2}$, showing that the resulting dispersion relation is q-invariant.

We will now focus on nonlinear equations for which the above d-dimensional q-plane wave is an exact solution. It should be emphasized at this point that many nonlinear equations in the literature are usually formulated in one dimension and that their extension to d dimensions may not be always an easy task. Accordingly, the following d-dimensional nonlinear generalization of the Schrödinger equation for a free particle of mass m was introduced in [639]

$$i\hbar \frac{\partial}{\partial t}\left[\frac{\Phi(\vec{x},t)}{\Phi_0}\right] = -\frac{1}{2-q}\frac{\hbar^2}{2m}\nabla^2\left[\frac{\Phi(\vec{x},t)}{\Phi_0}\right]^{2-q}. \qquad (7.24)$$

We notice that the scaling of the wave function by Φ_0 guarantees the correct physical dimensionalities for all terms. This scaling becomes irrelevant only for linear equations [e.g., in the particular case $q = 1$ of (7.24)].

Consistently, the energy and momentum operators are generalized as $\hat{E} = i\hbar D_t$ and $\hat{p}_n = -i\hbar D_{x_n}$ respectively, where $D_u f(u) \equiv [f(u)]^{1-q} df(u)/du$. These operators, when acting on the q-exponential $\exp_q\left[i(\vec{k}\cdot\vec{x} - \omega t)\right]$, yield the energy $E = \hbar\omega$ (Planck relation) and momentum $\vec{p} = \hbar\vec{k}$ (de Broglie relation). Now, if one considers the q-plane wave solution of (7.21) by using $\vec{k} \to \vec{p}/\hbar$ and $\omega \to E/\hbar$, one verifies that this new form is a solution of the equation above, with $E = p^2/2m$, for all values of q, and not only for the standard case ($q = 1$).

It was also proposed in [639] a new nonlinear Klein–Gordon relativistic equation in d dimensions, namely

$$\nabla^2\Phi(\vec{x},t) = \frac{1}{c^2}\frac{\partial^2\Phi(\vec{x},t)}{\partial t^2} + q\frac{m^2c^2}{\hbar^2}\Phi(\vec{x},t)\left[\frac{\Phi(\vec{x},t)}{\Phi_0}\right]^{2(q-1)}. \qquad (7.25)$$

One may verify easily that the same q-plane wave used for the NL Schrödinger equation is a solution of (7.25), preserving *for all q* the Einstein relation

$$E^2 = p^2c^2 + m^2c^4. \qquad (7.26)$$

It should be mentioned that, to our knowledge, the nonlinear term in (7.25) is new and different from those of previous formulations, even in the particular cases $q = 2$ and $q = 5/2$, where one gets cubic and quartic terms in the field, respectively. In previous works, the nonlinear term is constructed by multiplying the wave function by a power of its modulus, leading to different types of solutions and *different energy spectra*. Another important aspect of (7.25) concerns its Lorentz invariance: since this property is directly related to the first two terms (containing derivatives) [894], which have not been changed herein, then (7.25) remains invariant under Lorentz transformation.

Finally, along the same lines it was introduced in [639] a new nonlinear form for the free-particle Dirac equation. In this case, we will restrict ourselves to $d = 3$ spatial dimensions; let us then introduce the following generalized Dirac equation,

$$i\hbar\frac{\partial\Phi(\vec{x},t)}{\partial t} + i\hbar c(\vec{\alpha}\cdot\vec{\nabla})\Phi(\vec{x},t) = \beta mc^2 A^{(q)}(\vec{x},t)\,\Phi(\vec{x},t), \qquad (7.27)$$

where $\alpha_x, \alpha_y, \alpha_z$ (written in terms of the Pauli spin matrices) and β (written in terms of the 2×2 identity matrix I) are the standard 4×4 matrices [894]. The new, q-dependent, term is given by the 4×4 diagonal matrix $A_{ij}^{(q)}(\vec{x}, t) = (1 - q)_{ij}[\Phi_j(\vec{x}, t)/a_j]^{q-1}$, where $\{a_j\}$ are complex constants $(A_{ij}^{(1)}(\vec{x}, t) = (1 - q)_{ij})$. The solution of (7.27) we focus on is the following four-component column matrix

$$\Phi(\vec{x}, t) \equiv \begin{pmatrix} \Phi_1(\vec{x}, t) \\ \Phi_2(\vec{x}, t) \\ \Phi_3(\vec{x}, t) \\ \Phi_4(\vec{x}, t) \end{pmatrix} = \begin{pmatrix} u_1 \\ a_2 \\ a_3 \\ a_4 \end{pmatrix} \exp_q \left[\frac{i}{\hbar} (\vec{p} \cdot \vec{x} \quad Et) \right]. \tag{7.28}$$

Substituting this four-component vector into (7.27), we get, for the coefficients $\{a_j\}$, *precisely the same set of four algebraic equations corresponding to the linear case* (see page 803, Eq. (15.45b) of [894]). These equations have, for all q, a nontrivial solution only if the Einstein energy-momentum relation (7.26) is satisfied. The proposal in (7.27) differs from previous ones [895] in the same sense of the above nonlinear Schrödinger and Klein–Gordon equations, i.e., the nonlinearity is introduced by generalizing an existing term, rather than adding an extra nonlinear term. As a consequence, our solutions are given by the above-mentioned q-exponential, thus preserving relation (7.26), in contrast to those in Ref. [895], which are typically written in terms of the standard exponential. The positive and negative parts of the present energy spectrum are naturally expected to be respectively associated to a particle and its corresponding antiparticle. Contrary to the standard plane waves (case $q = 1$), one has q-plane wave solutions, which are square-integrable, i.e., with a finite norm for $1 < q < 3$, representing localized particles.

7.2 Chemistry

7.2.1 Generalized Arrhenius Law and Anomalous Diffusion

The *Arrhenius law* plays a fundamental role in chemistry as it characterizes the speed at which a chemical reaction occurs. It is directly related to BG statistics, and can be written as follows:

$$k = Ae^{-E_0/RT} \quad (A > 0), \tag{7.29}$$

k and E_0 being respectively the *rate constant* and the *activation energy* (R denotes the universal gas constant).

This law has been interestingly q-generalized in [896–898] as follows:

$$k_d = A(1 - dE_0/RT)^{1/d} \equiv Ae_q^{-E_0/RT} \quad (A > 0), \tag{7.30}$$

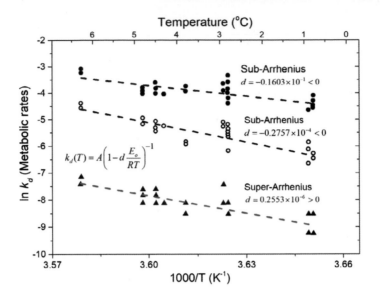

Fig. 7.70 Effect of temperature on the metabolism of Arctic zooplankton: respiration (fitting in red), ammonia excretion rate (fitting in blue), and phosphate excretion rate (fitting in green). The fittings have been done by using Eq. (7.30). From [898]

k_d being the q-generalized rate constant (with $k_0 = k$), and $d \equiv 1 - q$; the cases $d < 0$ and $d > 0$ are respectively referred to as *sub-Arrhenius* and *super-Arrhenius*; the $d \to 0$ limit recovers the usual Arrhenius law (7.29). See Fig. 7.70 for three illustrations of such anomalies, concerning the metabolism of Arctic zooplankton: two of them (respiration and ammonia excretion rate) corresponding to $q > 1$, and the third one (phosphate excretion rate) corresponding to $q < 1$.

Let us focus now on a directly related problem, namely the escape rates associated with the following nonlinear Fokker–Planck equation (currently referred to as the Plastino and Plastino equation [403]):

$$
\begin{aligned}
\frac{\partial \rho(x,t)}{\partial t} &= \frac{\partial}{\partial x}\left[\frac{\partial U(x)}{\partial x}\rho(x,t)\right] + D\frac{\partial^2 [\rho(x,t)]^{2-q}}{\partial x^2} \\
&= \frac{\partial}{\partial x}\left[\frac{\partial U(x)}{\partial x}\rho(x,t)\right] + (2-q)D\frac{\partial}{\partial x}\left\{[\rho(x,t)]^{1-q}\frac{\partial \rho(x,t)}{\partial x}\right\} \qquad (7.31)
\end{aligned}
$$
$$
[q \in \mathbb{R};\ (2-q)D > 0;\ t \geq 0],
$$

with

$$
\int dx\,\rho(x,t) = 1,\ \forall t, \qquad (7.32)
$$

$U(x)$ being a potential whose global minimal value is U_0. The stationary solution is given by

Fig. 7.71 a Double well potential $V(x)$. The stationary-state distributions are depicted in (**b**) for $q < 1$ and **c** for $q > 1$. Observe in (**b**) that, as D decreases, the motion becomes more confined until only the neighborhood of the deepest valley is allowed. The horizontal lines in (**a**) represent the cutoff condition $V(x) = 1/\beta$, which defines the allowed regions for $q = 0$ and the same values of D as in (**b**). All quantities are dimensionless. From [896]

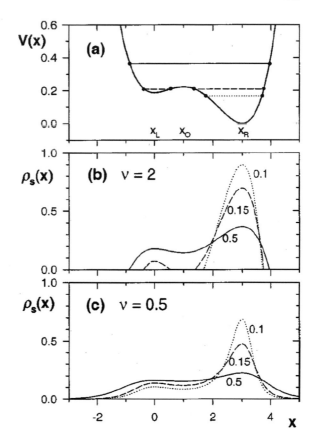

$$\rho_s(x) \equiv \lim_{t \to \infty} \rho(x, t) = \frac{e_q^{-\beta V(x)}}{Z}, \qquad (7.33)$$

with $V(x) \equiv U(x) - U_0$, $\beta = \frac{Z^{1-q}}{(2-q)D}$, Z being a positive normalization constant. The calculations have been illustrated by using the (dimensionless) double-well potential indicated in Fig. 7.71a, namely $V(x) = ax^4 + bx^3 + cx^2 + d$, with $(a, b, c, d) = (1/48, -1/9, 1/8, 3/16)$. Consequently, the left (local) minimum occurs at $x = x_L = 0$, the right (global) minimum occurs at $x = x_R = 3$, and the central maximum occurs at $x = x_0 = 1$. For $q \geq 1$ the full phase space is covered (by power-law tails if $q > 1$). For $q < 1$ a cutoff restricts the attainable space. The stationary distribution is shown, for typical values of D (indicated in the figure), in Fig. 7.71b for $\nu = 2 - q = 2$ and in Fig. 7.71c for $\nu = 0.5$. By using the associated Ito–Langevin equation one can consider a large amount of stochastic trajectories each of them starting at x_L. We note $T(x) \equiv T(x_L \to x)$ the average time for the *first passage* to a value larger than x, with $x \geq x_L$: see Fig. 7.72. In particular, we can focus on the *escape time* $T \equiv T(x_R)$: see Fig. 7.73, where T and $1/D$ play, respectively, the roles of $1/k$ and of the inverse temperature $1/T$ in Eq. (7.29). We verify that, for

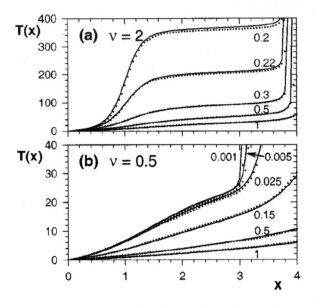

Fig. 7.72 $T(x)$ for typical values of D (indicated in the figure) for $\nu = 2 - q = 2$ (**a**), and $\nu = 2 - q = 0.5$ (**b**). Circles correspond to numerical experiments (mean values over 1000 realizations), and full lines to the analytical predictions of the theory. From [896]

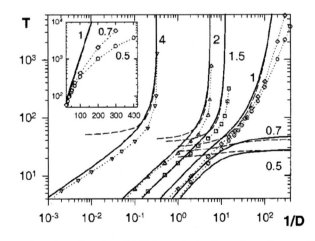

Fig. 7.73 Escape time T as a function of $1/D$ for typical values of $\nu = 2 - q > 0$ (as indicated in the figure). Full lines are analytical results of the theory. Dashed lines correspond to the analytical low-D approximation. Symbols correspond to the initial condition where all the particles (at least 1000) are injected at the same time at x_L. Dotted lines are guides for symbols. *Inset*: Detail (semi-log) of the low-D region for $\nu = 2 - q \leq 1$. The particular case $\nu = q = 1$ recovers Arrhenius law and normal diffusion. From [896]

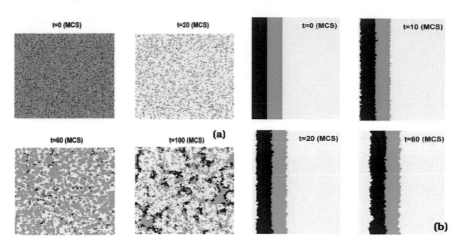

Fig. 7.74 $d = 2$ (**a**) and $d = 1$ **b** time evolution of typical initial conditions. In (**a**) we observe the spontaneous tendency toward clusterization. MCS stands for *Monte Carlo steps*. From [899]

all values of ν, the escape time T asymptotically behaves as a power-law for large values of D.

7.2.2 Lattice Lotka–Volterra Model for Chemical Reactions and Growth

The *lattice Lotka–Volterra* (LLV) model is a paradigmatic one for two-constituent chemical reactions, growth, prey–predator kinetics, political polarization in election periods, and other phenomena. Its mean-field approximation (classical Lotka–Volterra model) is conservative, but its exact microscopic dynamics is not. A large literature is devoted to its study. Here we focus on the time dependence of its configurational entropy by following [899, 900] for a L^2 square lattice with isotropic and strip-like initial conditions: see Figs. 7.74, 7.75 and 7.76, where the red and green colors indicate the two constituents, and the white color indicates that that cell is empty. In Fig. 7.75 we focus on very localized isotropic initial conditions, and verify that only S_q with $q = 1/2$ increases linearly with time before achieving saturation (which occurs later and later for increasingly large values of L); it can be also verified that all sizes can be collapsed by representing $S_{0.5}/L$ versus t/L. In Fig. 7.76 we can see the same phenomenon for strip-like initial conditions, now having $q = 0$; it can be also verified that all sizes can be collapsed by representing S_0/L versus t/L.

Similar results are shown in Figs. 7.77, 7.78 and 7.79 for $D = 2, 3, 4$. The entropy S_q of the D-dimensional model (simple hypercubic lattice with L^D sites) with isotropic very localized initial conditions asymptotically increases *linearly with time* only for

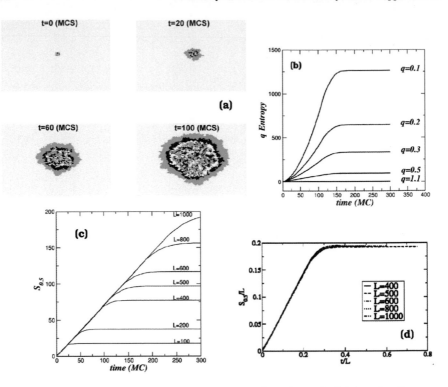

Fig. 7.75 Time evolution for initial conditions very localized (isotropically) in a $L \times L$ square lattice. From [899]

$$q = 1 - \frac{1}{D}, \tag{7.34}$$

which recovers the BG result $q = 1$ in the $D \to \infty$ limit. This expression for q is, essentially, a sort of trivial consequence of the fact that the number W of possibilities increases like the available D-dimensional hypervolume, i.e., $W \propto t^D$. Consequently, if we consider the simplest case, namely equal probabilities, $S_q = \ln_q W \sim W^{1-q}/(1-q) \propto t^{(1-q)D}$. Then, in order to have $S_q \propto t$, Eq. (7.34) must be satisfied. This equation is, in fact, but the particular case of Eq. (3.132) with $\rho = D$. Let us also note that the possibly nontrivial entropic effects associated with the roughness of the overall contour of the system and of its internal evolving clusters remain to be studied.

7.2.3   Re-association in Folded Proteins

Re-association of say $C O$ molecules in heme-proteins has been experimentally studied, and the results are discussed in [459]. See Figs. 7.80 and 7.81. The rate ξ of non re-associated molecules was proposed in the literature to be given by [901]

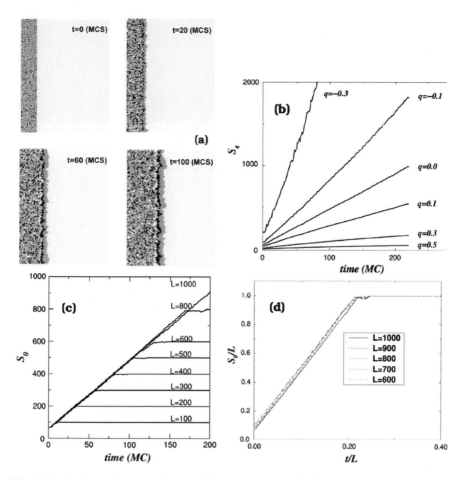

Fig. 7.76 Time evolution for initial conditions very localized within a L-sized strip in a $L \times L$ square lattice. From [899]

$$\xi \equiv \frac{N(t)}{N(0)} = \frac{1}{(1 + t/t_0)^n} . \tag{7.35}$$

But, with the identifications $n \equiv 1/(q - 1)$ and $1/t_0 \equiv (q - 1)/\tau$, this equation can be rewritten as

$$\xi = e_q^{-t/\tau} . \tag{7.36}$$

As verified in Fig. 7.80, this function describes satisfactorily the experimental data for all times that are not too long. In the long-time region, some sort of crossover occurs similar to the type described in Eq. (6.4). The temperature dependences of the corresponding fitting parameters are indicated in Fig. 7.81.

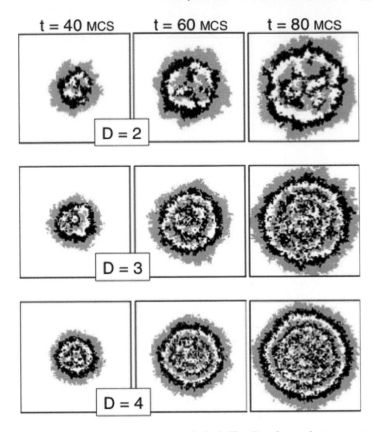

Fig. 7.77 Snapshots of the dynamics for $D = 2, 3, 4$. The $D = 2$ snapshots correspond to the square lattice itself. The $D = 3, 4$ snapshots correspond to a two-dimensional section of the D-dimensional lattice. From [900]

7.2.4 Ground State Energy of the Chemical Elements (Mendeleev's Table) and of Doped Fullerenes

There is nothing more basic in modern chemistry than Mendeleev's Table of elements. However, its standard implementation makes no reference at all to a very basic quantity, namely the energy of the ground state of each specific element. This has been addressed in [902]. The outcome is quite astonishing. The ground state energy of the free atom (as calculated by a performant ab-initio Hartree–Fock method) has been heuristically found to be given, from the hydrogen to the lawrencium, by

$$E = E_H \, e_{0.58145}^{2.4333\,(Z-1)} , \tag{7.37}$$

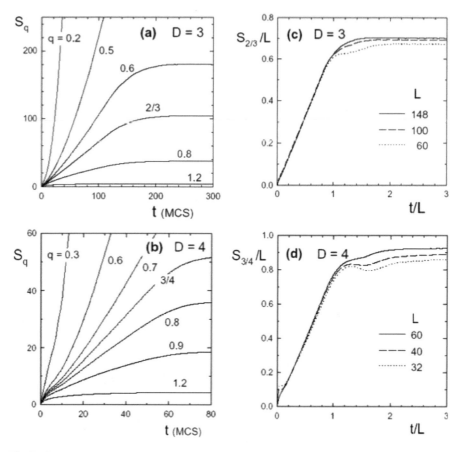

Fig. 7.78 Time evolution of S_q for the $D = 3$, 4 models. The entropy is calculated over the entire phase space of the D-dimensional system, and *not* of its two-dimensional sections, like those shown in Fig. 7.77. From [900]

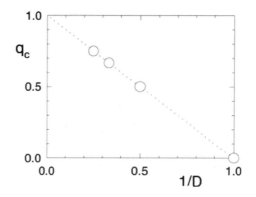

Fig. 7.79 Dependence of q_c on dimensionality. Equation (7.34) is thus numerically verified. From [900]

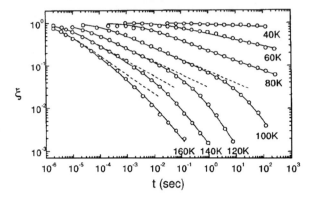

Fig. 7.80 Log–log plot of the time evolution of $\xi \equiv N(t)/N(0)$ associated with $MbCO$ in glycerol-water. The dots are the experimental data (Figs. 2a and 14 of [901]). The dashed lines indicate the fittings with Eq. (7.35) with the same $n(T) \equiv 1/[q(T) - 1]$ used in [901]. The full lines correspond to our best present fittings, with the same $n(T)$ used in [901], and $r(T)$, $a_q(T)$, and $a_r(T)$ as shown in Fig. 7.81. From [459]

where $E_H = -13.60534\,ev$, and Z is the atomic number of the element. See Fig. 7.82. The degree of precision of this Ansatz can be estimated through the procedure indicated in Fig. 7.83.

The ground state energy of doped fullerenes (as calculated now through a density functional theory method) has also been addressed in [902]. The doping atoms that have been studied are the covalent atoms ^6C, ^7N, ^8O, ^9F, ^{14}Si, $615P$, ^{16}S, ^{17}Cl and ^{35}Br, and the transition metals ^{21}Sc, ^{22}Ti, ^{23}V, ^{24}Cr, ^{25}Mn, ^{26}Fe, ^{27}Co, ^{28}Ni and ^{29}Cu. Discounting the energy of pure fullerene C_{60} (i.e., without doping), which is 62.21 Kev, the energies are given by precisely (!) the same Eq. (7.37) by substituting E_H by $E_{FUL} = -14.98$ ev: see Fig. 7.84.

Summarizing, for both Mendeleev Table and doped fullerenes, the ground state total energies (calculated respectively by ab-initio Hartree–Fock and Functional Density Theory methods) are described (with all presently available precision) by one and the same equation, namely the q-exponential form (7.37)! The deep understanding of these two results constitutes, in our opinion, a fantastic challenge in the chemical science. The derivation of the nontrivial index $q = 0.58145$ from quantum first principles would be more than welcome. In fact, such q-statistical effects (coming most probably from the Coulombian interactions between electrons and protons) have the potential of ubiquitously influencing the entire field of atomic physics, as conjectured in [598, 599].

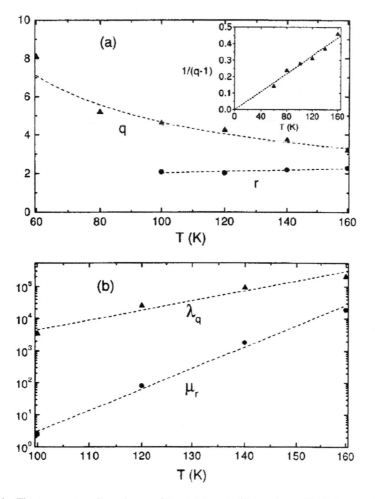

Fig. 7.81 The temperature dependences of (q, r) (a), and of (a_q, a_r), used to fit the experimental data of Fig. 7.80. *Inset of (a): T-dependence of* $n(T) \equiv 1/(q - 1)$ (from Fig. 15 of [901]). From [459]

7.2.5 Compton Profiles

Let us focus now on Compton profiles of alkali metals [903, 904]. As we see in Figs. 7.85 and 7.86, in all cases the profiles oscillate around q-Gaussians. The values of q indicated in Fig. 7.85 monotonically increase from 1.829 to 1.936, and those of β monotonically decrease from 0.2669 to 0.1337 while going from Li to Cs. Similarly, the data indicated in Fig. 7.86 also follow a monotonic trend, exhibiting additionally a slight anisotropy along the different crystal directions. These systematic facts, together with the analytical results in [905], suggest that the cause might well be

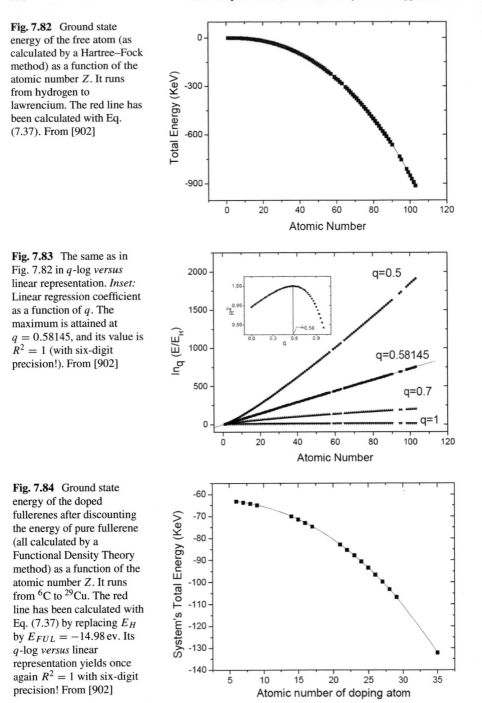

Fig. 7.82 Ground state energy of the free atom (as calculated by a Hartree–Fock method) as a function of the atomic number Z. It runs from hydrogen to lawrencium. The red line has been calculated with Eq. (7.37). From [902]

Fig. 7.83 The same as in Fig. 7.82 in q-log *versus* linear representation. *Inset:* Linear regression coefficient as a function of q. The maximum is attained at $q = 0.58145$, and its value is $R^2 = 1$ (with six-digit precision!). From [902]

Fig. 7.84 Ground state energy of the doped fullerenes after discounting the energy of pure fullerene (all calculated by a Functional Density Theory method) as a function of the atomic number Z. It runs from ^6C to ^{29}Cu. The red line has been calculated with Eq. (7.37) by replacing E_H by $E_{FUL} = -14.98$ ev. Its q-log *versus* linear representation yields once again $R^2 = 1$ with six-digit precision! From [902]

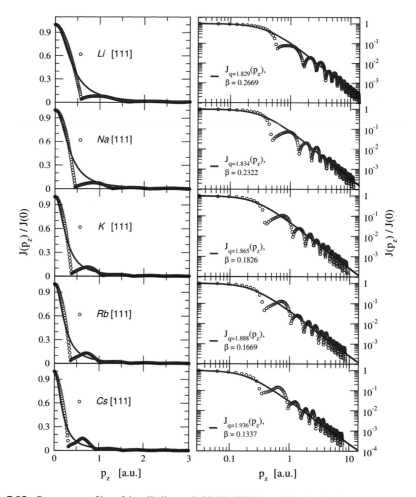

Fig. 7.85 Compton profiles of the alkali metals Li, Na, K, Rb, and Cs calculated along the principal direction [111]. q-Gaussian fit to the lineshape (red solid line) normalized to $J(p_z = 0) = 1.0$; β is given in atomic units ($a.u.$). From [903]

the (long-ranged) Coulombian interactions between the electrons and protons, and between electrons themselves. The precision of the experimental results undoubtedly points toward an open theoretical challenge.

7.3 Economics

A sensible amount of papers have used q-statistical concepts to discuss and/or extend various financial and economical quantities, such as distributions of returns, distributions of stock volumes, Black–Scholes equation, volatility "smile", pricing, risk

Fig. 7.86 Rescaled Compton profiles vs. $\tilde{p}_z^2 = (p_z a_0)^2$ for different scattering directions **a**: [111]; **b**: [110]; **c**: [001]. The element specific $(q, \tilde{\beta})$-parameters are seen to be anisotropic. From [903]

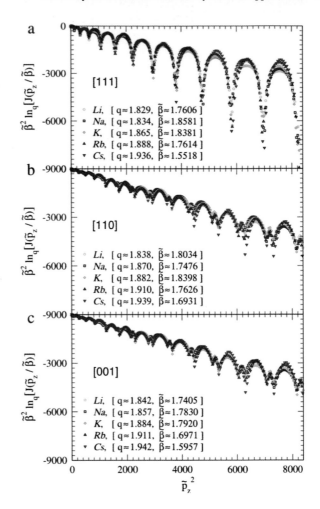

aversion in economic transactions, wealth inequalities, and various others [423, 906–948]. In what follows we present a selection of such results.

In Fig. 7.87 a typical return distribution of stocks is shown to be satisfactorily fitted with a q-Gaussian with $q = 1.43$. The definition of *return* r_t is as follows

$$r_t \equiv \ln \frac{p_t}{p_{t-1}} \sim \frac{\Delta p_t}{p_{t-1}}, \tag{7.38}$$

if $\Delta p_t \equiv p_t - p_{t-1} << p_t$, p_t being the price at (discretized) time t. Returns are convenient variables since, on one hand, they correspond to the popular concept of *relative* price variation and, on the other hand, they erase the effect of inflation.

In Fig. 7.88, it is shown that the same value of q produces "volatility smiles" which consistently fit typical stock empirical data. Similar results are depicted [910]

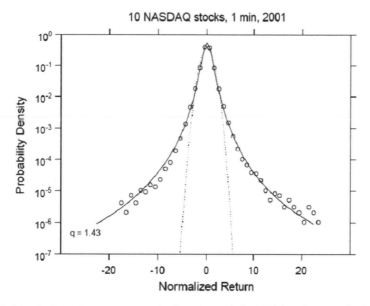

Fig. 7.87 Distributions of log returns over 1 minute intervals for 10 high-volume stocks. Solid line: q-Gaussian with $q = 1.43$. Dashed line: Gaussian. From [907]

in Figs. 7.89 and 7.90, where the expected dependence of q on the time scale Δt is verified. In particular, when Δt attains very large values, q critically approaches unity (i.e., Gaussian limit), more precisely like $q - 1 \propto 1/(\Delta t)^{\tau}$ with $\tau \simeq 0.08$. One more illustration along the same lines is provided by Shanghai Stock Exchange typical return data as discussed in [911]: see Fig. 7.91. Not only the stock prices are well described by q-statistics but also their volumes v (number of stocks that are negotiated), as illustrated in Figs. 7.92, 7.93, and 7.94. The pre-factor v^{λ} that we observe in Figs. 7.92 and 7.93 is inspired by the traditional density of states in condensed matter physics, and has probably no statistical-mechanical origin, in contrast with the q-exponential weight and its associated effective temperature θ.

   Not only the prices of stocks but various others as well seemingly follow q-statistics. Such is the case of the prices of lands in Japan, as shown in Fig.7.95.

   In a different vein, the wealth distributions in typical countries and their time evolution have been studied [922], and the results are shown in Figs. 7.96 and 7.97. The distribution corresponding to a set of 167 countries at a given year is depicted in Fig. 7.98. This type of description of wealth/income inequalities (large values of q corresponding to large inequality) may be seen as an alternative to the traditional *Gini* coefficient.

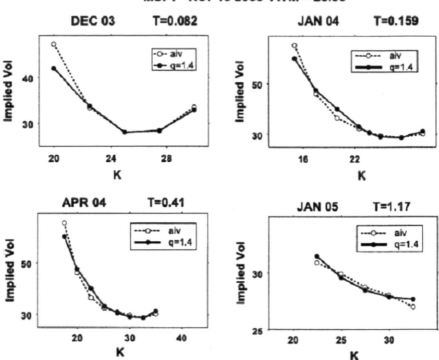

Fig. 7.88 Quantitative comparison between the skewed implied volatilities obtained from a set of Microsoft options traded on November 19, 2003, and the theoretical model with $q = 1.4$, which fits well the returns distribution of the underlying stock. From [920]

7.3.1 Inter-Occurrence Times in Economics, Geophysics, Genome Structure, Civil Engineering, and Turbulence

The distribution of inter-occurrence times constitutes a classical stylizing property in financial theory [949–951]. As we shall see, this property is in fact largely shared in the same form in many other areas such as earthquakes [952], DNA structure [953], civil engineering [954, 955], and turbulence [956].

Let us start now by focusing on financial inter-occurrence times [949, 950]. We choose the IBM price returns as the financial quantity to be analyzed, and show in Fig. 7.99 the inter-occurrence times corresponding to a chosen value Q of the threshold. Typical averaged inter-occurrence times R_Q as functions of Q are shown in Fig. 7.100; q-exponentials fit quite well the empirical data. The q-exponential behavior of the distributions of inter-occurrence times is illustrated in Fig. 7.101, which includes the dependence of (q, β) on R_Q. Fig. 7.102 exhibits the intriguing universality of the distributions of inter-occurrence times through diverse classes

Fig. 7.89 Cumulative
distributions of absolute
normalized returns that
correspond to different time
scales Δt for the 100
American companies with
the highest market
capitalization (points), and
the fitted cumulative
q-Gaussian distributions
(lines). In order to better
visualize the results, each
q-Gaussian CDF and the
respective experimental data
have been multiplied by a
positive factor $c \neq 1$. From
[910]

Δt	q	β
4	1.53	1.78
8	1.526	1.67
16	1.48	1.52
30	1.46	1.42
60	1.45	1.33
120	1.42	1.25
240	1.39	1.14
390	1.365	1.10
780	1.35	1.03

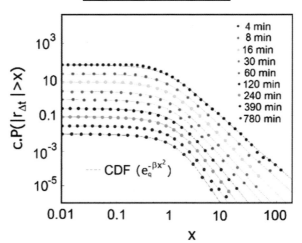

Fig. 7.90 Dependence of
the index q on the time scale
Δt, for the estimated
q-Gaussian pdfs of
normalized absolute returns
in Fig. 7.89. *Inset:* log–log
representation exhibiting a
power-law dependence of the
type $q - 1 \propto (\Delta t)^{-\tau}$, with
$\tau = 0.081 \pm 0.004$. From
[910]

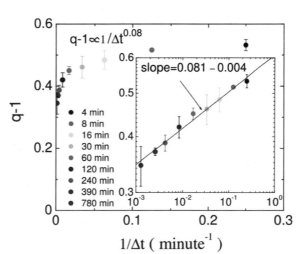

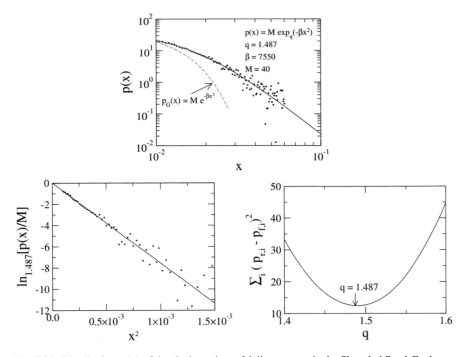

Fig. 7.91 Distribution $p(x)$ of the closing prices of daily returns x in the Shanghai Stock Exchange during the period from 02/09/2015 to 01/10/2020. The dots are referred to as "real" in [912]. The red curves correspond to $p(x) = Me_q^{-\beta x^2}$ with $q = 1.487$. *Top:* log–log representation; a Gaussian distribution is displayed for comparison. *Bottom left:* $\ln_{1.487}$ versus x^2 representation of the same data, the slope being $-\beta$. *Bottom right:* Sum of squares of the differences between the real values $\{p_{r,i}\}$ and the corresponding fitting values $\{p_{f,i}\}$ for the region with low dispersion, namely $x < 0.02$. This procedure yields the value $q = 1.487$. The same procedure was used to determine the present values of M and β. From [911]

of financial assets. Finally, in Fig. 7.103, the associated (universal) *risk function* $W_Q(t; \Delta t) \equiv \frac{\int_t^{t+\Delta t} P_Q(r)dr}{\int_t^{\infty} P_Q(r)dr}$ is illustrated for $R_Q = 100$, its analytical form being the following one (see details in [949]):

$$W_Q(t; \Delta t) = 1 - \left[1 + \frac{\beta(q-1)\Delta t}{1 + \beta(q-1)t}\right]^{\frac{q-2}{q-1}}. \tag{7.39}$$

The fact that this useful function is a simple explicit expression in terms of (t, Δ) clearly constitutes an operational advantage.

Let us focus now on earthquake inter-occurrence times as discussed in [952]. Typical results (earthquakes in Greece spanning the period 1901 to 2009) are shown in Fig. 7.104, where we see a log–log plot of the inter-event time distribution $P_M(T)$ as a function of the inter-event time T (in days) for earthquake magnitudes $M \geq M_c$ for the entire dataset considered, where the threshold magnitude $M_c = 4.1$ has been

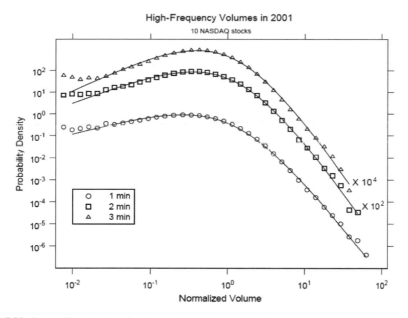

Fig. 7.92 Probability density of volumes of 10 high-capitalization stocks in NASDAQ, compared to the theoretical curves $p(v) = \frac{1}{Z} (\frac{v}{\theta})^\lambda e_q^{-v/\theta}$ with λ, $\theta > 0$, $q > 1$, and the normalization constant $Z > 0$ (full lines). From [918]

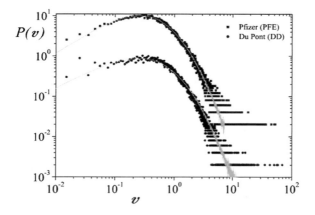

Fig. 7.93 Probability density of volumes of two specific stocks, compared to stochastic results following essentially $p(v) = \frac{1}{Z} (\frac{v}{\theta})^\lambda e_q^{-v/\theta}$ with λ, $\theta > 0$, $q > 1$, and the normalization constant $Z > 0$ (full lines). From [932]

conventionally chosen. The fitting dashed curve is a q-exponential with $q_{T1} = 1.24 \pm 0.054$ (the corresponding declustered data are fitted with $q_{T2} = 1.14 \pm 0.057$).

Let us next focus on DNA structures as discussed in [953]. Typical results are shown in Figs. 7.105 and 7.106. It is undoubtedly a remarkable fact that the values $q =$

Fig. 7.94 Cumulative distribution of the daily net exchange of shares (between all pairs of two institutions), at the London Stock Exchange (Block market). Data from I. I. Zovko; fitting by E. P. Borges [using the analytic form (in red) emerging within *crossover statistics*; see Eq. (6.4)]; unpublished (2005)

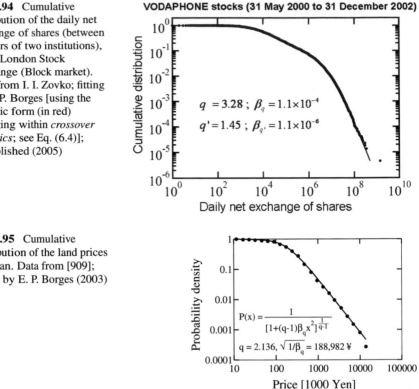

VODAPHONE stocks (31 May 2000 to 31 December 2002)

$q = 3.28$; $\beta_q = 1.1 \times 10^{-4}$

$q' = 1.45$; $\beta_{q'} = 1.1 \times 10^{-6}$

Daily net exchange of shares

Fig. 7.95 Cumulative distribution of the land prices in Japan. Data from [909]; fitting by E. P. Borges (2003)

$$P(x) = \frac{1}{[1+(q-1)\beta_q x^2]^{\frac{1}{q-1}}}$$

$q = 2.136$, $\sqrt{1/\beta_q} = 188{,}982$ ¥

Price [1000 Yen]

1.11 and $\beta = 1.5$ play a universal role in so diverse biological beings. Its explanation certainly remains as an open problem. Moreover, it would be extremely interesting to check whether the double q-exponential behavior that is observed in Homo Sapiens is also present in other mammals. Such a study has the potential of playing a crucial role in the evolution of species.

Let us finally focus on fractures of civil engineering materials as discussed in [955]: see Figs. 7.107, 7.108, 7.109, 7.110, and 7.111. In Fig. 7.107 we see the time evolution of the loading (in red) and of the acoustic emission (AE, in black) of concrete (top panel) and of basalt (bottom panel). The loading apparatus on the specimen as well as the microphone which detects the AE are illustrated in Fig. 7.108. In Fig. 7.109 we see the empirical data corresponding to the last loading of the concrete (left) and basalt (right) specimens depicted in Fig. 7.107. In Fig. 7.110 we show twelve typical cumulative distributions of inter-occurrence times and their q-exponential fittings (we remind that a q-exponential distribution also yields a q-exponential cumulative distribution, though with a different value of q).

Finally, the critical-like curves in the $(q, 1/\beta_q)$ space of Fig. 7.111 are analogous to those indicated in Figs. 7.89, 7.147, 7.148, 7.152 and 7.160f for very different systems. Indeed, such curves are in fact very frequent in q-statistical systems. For all

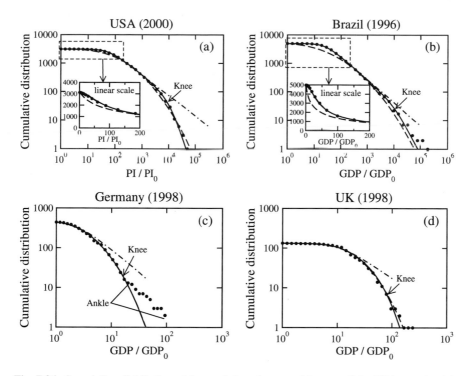

Fig. 7.96 Cumulative distribution of the *scaled total personal income* of the USA *counties* (**a**), and the *scaled gross domestic product* of the Brazilian (**b**), German (**c**), and the United Kingdom (**d**) *counties*. From [922]

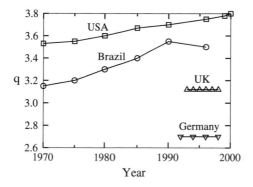

Fig. 7.97 Time evolution of the index *q* for USA (squares), Brazil (circles), the United Kingdom (up triangles), and Germany (down triangles). This means that the economic inequalities are larger in USA, then in Brazil, then the United Kingdom, and finally Germany. We also see that inequalities are increasing in USA and Brazil, whereas they remain at the same level in the United Kingdom and Germany. From [922]

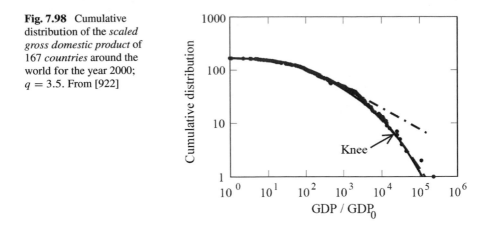

Fig. 7.98 Cumulative distribution of the *scaled gross domestic product* of 167 *countries* around the world for the year 2000; $q = 3.5$. From [922]

the specimens that have been tested in [955], the macroscopic fracture systematically occurs when the $1/\beta_q \to 0$ limit is practically attained in each one of the critical-like curves of Fig. 7.111. This phenomenon opens the door to practical protocols aiming the establishment of the health status of civil engineering structures (buildings, bridges, tunnels, monuments).

7.4 Computer Sciences

7.4.1 Optimization Algorithms

Global optimization consists in numerically finding a global minimum of a given (not necessarily convex) cost/energy function, defined in a continuous D-dimensional space. Such algorithms have a plethora of useful applications. A well known classical procedure, referred to as the *Boltzmann machine*, is the so-called *Simulated Annealing*, introduced in 1983 [957], which visits phase space with a Gaussian distribution. A few years later, a faster procedure, referred to as the *Cauchy machine* because it visits phase space using a Cauchy-Lorentz distribution, was introduced [958]. Finally, inspired by q-statistics, an algorithm was introduced [959, 960], named *Generalized Simulated Annealing* (GSA), which recovers the two just mentioned ones as particular cases.

GSA consists, like the Boltzmann and Cauchy machines, of two algorithms that are to be used with alternation. These are the *Visiting algorithm* and the *Acceptance algorithm*. The visiting algorithm is based on exploring the phase space using a q_V-Gaussian (instead of using either a Gaussian or a Cauchy distribution), and the acceptance algorithm is based on a q_A-exponential weight (instead of the Monte Carlo Boltzmann weight). Therefore a GSA machine is characterized by the pair (q_V, q_A). The choice $(1, 1)$ is the Boltzmann machine, and the choice $(2, 1)$ is the Cauchy

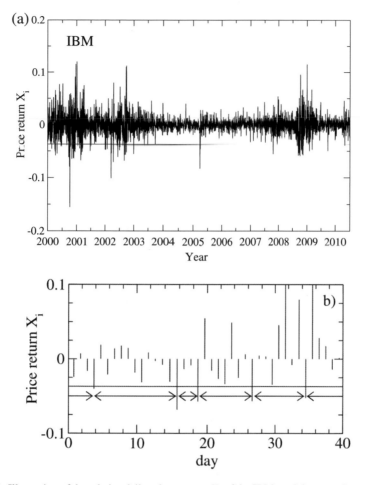

Fig. 7.99 Illustration of the relative daily price returns X_i of the IBM stock between January 2000 and June 2010 (**a**) and between August 27 and October 23, 2002 (**b**) (40 working days). The red line shows the threshold $Q \simeq -0.037$, which corresponds to an average inter-occurrence time of $R_Q = 70$. In (**b**) the inter-occurrence times are indicated by arrows. From [949]

machine. In practice, the most performant values have been shown to be $q_V > 1$, slightly below the maximal admissible value for the D-dimensional problem (for $D = 1$ the maximal admissible value is $q_V = 3$, and a performant value is $q_V \simeq 2.7$; for D dimensions, the maximal admissible value is $q_V = (D + 2)/D$), and $q_A < 1$. A very convenient random number generator following a q-Gaussian distribution has been proposed in [961].

Part of the simulated annealing procedure consists in the *Cooling algorithm*, which determines how the *effective temperature T* is to be decreased with time, so that the global minimum is eventually attained within the desired precision. A quick cooling is of course computationally desirable. But not too quick, otherwise the rate

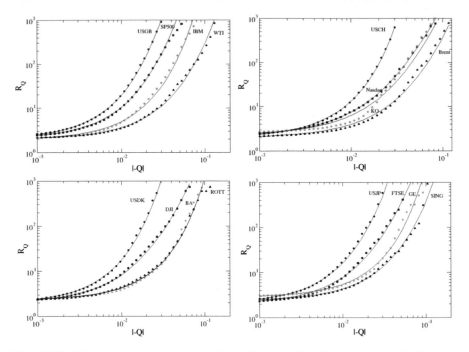

Fig. 7.100 The mean inter-occurrence time R_Q vs. the absolute value of the loss threshold $-Q$. The continuous curves are fittings with $R_Q = A\, e_{q_{inter}}^{B_{inter}|Q|} = A[1 + (1 - q_{inter})B_{inter}|Q|]^{1/(1-q_{inter})}$. *Top left:* For the exchange rate of the U.S. Dollar against the British Pound, the index S&P500, the IBM stock, and crude oil (West Texas Intermediate (WTI)), from left to right in the plot; the corresponding values for q_{inter} are 0.95, 0.92, 0.97, 0.927 (with $A = 2$, 2.04, 1.95, 2.02 and $B_{inter} = 240$, 175, 95, 60). Similarly for the Top right, Bottom left, and Bottom right plots. From [951]

of success of ultimately arriving to the real global minimum decreases sensibly. The optimal cooling procedure appears to be given by [959, 960]

$$\frac{T(t)}{T(1)} = \frac{2^{q_V - 1} - 1}{(1 + t)^{q_V - 1} - 1} = \frac{\ln_{q_V}(1/2)}{\ln_{q_V}[1/(t + 1)]} \quad (t = 1, 2, 3, \ldots),\tag{7.40}$$

where $T(1)$ is the initial high temperature imposed onto the system. We verify that, for $q_V = 1$, we have

$$\frac{T(t)}{T(1)} = \frac{\ln 2}{\ln(1 + t)} \quad (t = 1, 2, 3, \ldots),\tag{7.41}$$

and that, for $q_V = 2$, we have

$$\frac{T(t)}{T(1)} = \frac{2}{t} \quad (t = 1, 2, 3, \ldots).\tag{7.42}$$

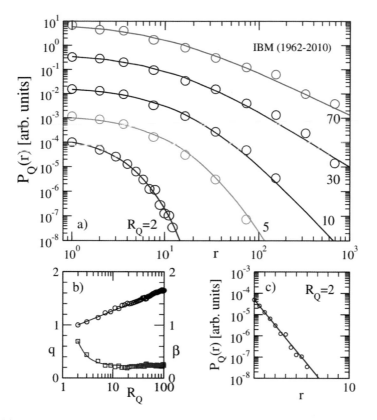

Fig. 7.101 a The distribution function of the inter-occurrence times r for the relative daily price returns Xi of IBM in the period 1962–2010. The data points belong to $R_Q = 2, 5, 10, 30$, and 70 (in units of days), from bottom to top. The full lines show the fitted q-exponentials. The dependence of the parameters β (squares, lower curve) and q (circles, upper curve) on R_Q in the q-exponential is shown in Panel **b**, where $q = 1 + q_0 \ln(R_Q/2)$ with $q_0 \simeq 0.168$. Panel **c** confirms that for $R_Q = 2$ the distribution function is a simple exponential. The straight line corresponds to 2^{-r}. From [949]

For the $D = 1$ upper limit, we have $q_V = 3$, hence

$$\frac{T(t)}{T(1)} = \frac{3}{(1+t)^2 - 1} \quad (t = 1, 2, 3, \ldots). \tag{7.43}$$

We see therefore a strong influence of q_V on the cooling allowed speed, which can ultimately benefit (decrease) quite strongly the necessary computational time. To test the method, a toy model has been studied. The cost function is shown in Fig. 7.112, and typical runs are shown in Fig. 7.113. The influence of (q_V, q_A) is depicted in Fig. 7.114.

Another toy model [960] is to use the $D = 4$ cost function

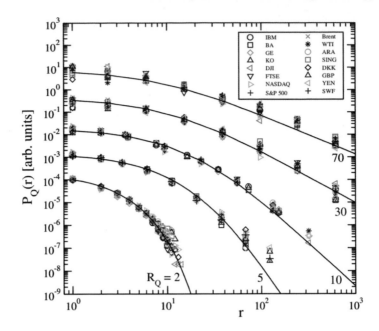

Fig. 7.102 The distribution function of the inter-occurrence times (as in Fig. 7.101) for the relative daily price returns of 16 examples of financial data, taken from different asset classes (stocks, indices, currencies, commodities). The assets are (i) the stocks of IBM, Boeing (BA), General Electric (GE), Coca-Cola (KO), (ii) the indices Dow Jones (DJI), Financial Times Stock Exchange 100 (FTSE), NASDAQ, S&P 500, (iii) the commodities Brent Crude Oil, West Texas Intermediate (WTI), Amsterdam-Rotterdam-Antwerp gasoline (ARA), Singapore gasoline (SING), and iv) the exchange rates of the following currencies vs. the US Dollar: Danish Crone (DKK), British Pound (GBP), Yen, Swiss Francs (SWF). The full lines show the fitted q-exponentials, which are the same as in Fig. 7.101. From [949]

$$E(x_1, x_2, x_3, x_4) = \sum_{i=1}^{4}(x_i^2 - 8)^2 + 5\sum_{i=1}^{4} x_i , \qquad (7.44)$$

which has 15 *local* minima and one *global* minimum. Typical results can be seen in Figs. 7.115 and 7.116.

An extension of these ideas has been advanced which leads to a q-generalization of the so-called *Pivot method* [962, 963]. Typical results are indicated in Figs. 7.117 and 7.118, where we see that, for a given system size, the necessary computer time is sensibly smaller and the success rate (percentage of trials which lead to the correct global minimum) is dramatically larger if we use an adequate value for the index q.

The first use of the GSA in quantum systems was done in [964]. Since those early times a large number of algorithmic methods inspired by q-statistics have been implemented, for chemical, neural network, and other purposes [965–1061].

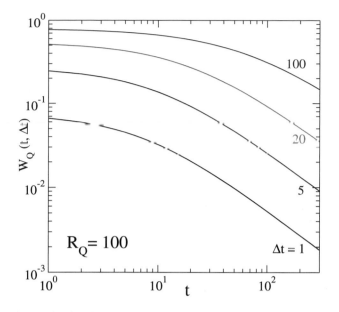

Fig. 7.103 Universal risk function $W_Q(t, \Delta t)$ from Eq. (7.39) for the mean inter-occurrence time $R_Q = 100$, and for the intervals $\Delta t = 1, 5, 20, 100$ days (from bottom to top). From [949]

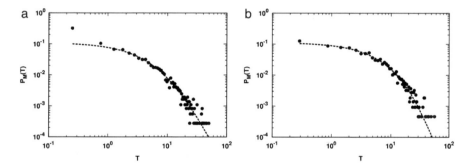

Fig. 7.104 a $P_M(T)$. **b** Same as in (**a**) for the corresponding declustered dataset. Data (filled circles), and q-exponential fittings (dashed curves). From [952]

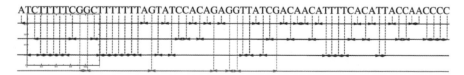

Fig. 7.105 The procedure of the assessment of the four inter-nucleotide interval sequences from the DNA primary sequence. From [953]

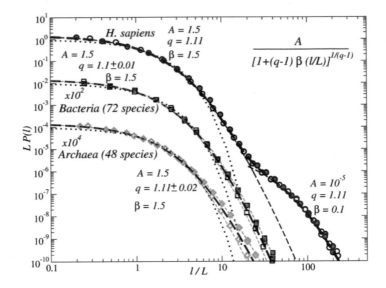

Fig. 7.106 PDFs of the inter-nucleotide intervals A-A, T-T (open symbols); G-G, C-C (full symbols) in the DNA sequences from H. Sapiens and Bacteria full genomes (in scaled form). Dashed lines show the best fits by a q-exponential distribution $A = 1/[1 + (q-1)\beta(l/L)]^{\frac{1}{q-1}}$. While in Bacteria the approximation by a single q-exponential with $q \sim 1.1$ and $\beta \sim 1.5$ is possible, in H. Sapiens a sum of two q-exponentials with $q \sim 1.11$ and $\beta \sim 1.5$ and 0.1 makes the best fit. To avoid overlapping, the PDFs for Bacteria are shifted downward by two decades. For comparison, dotted lines show corresponding exponential PDFs. From [953]

7.4.2 Analysis of Time Series and Signals

Concepts of q statistics have inspired several methods for processing time series and signals, such as electroencephalograms (EEG), electrocardiograms (ECG), machine-learning, and various others. It has been possible to focus on some specific features of epilepsy (in humans and turtles), Alzheimer's disease, and other complex circumstances [1062–1090].

The analysis of the tonic-clonic transition of some types of epilepsy constitutes a typical illustration [1078]. The EEG during a crisis can be seen in Fig. 7.119. At time 125 s, a clinically dramatic transition occurs with the patient. However, nothing special can be seen in the direct EEG at that moment. In contrast, as we verify in Fig. 7.120, after appropriate processing, the tonic-clonic transition becomes absolutely visible. The discrimination becomes even stronger if $q < 1$ is used. If no specialized medical agents are present at the precise moment of the crisis of the patient, the existence of such a neat peak makes possible the automatic start of computer-controlled administration of appropriate drugs during the emergency.

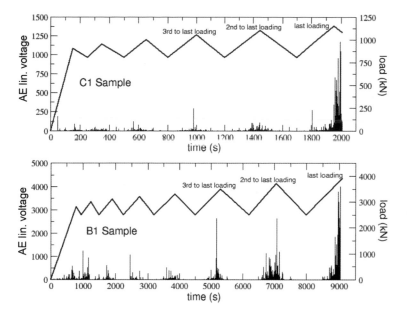

Fig. 7.107 Experimental recording of the acoustic emission (AE) linear voltage as a function of time during the cyclic compression test for two samples of, respectively, concrete (C1, top panel) and basalt (B1, bottom panel). From [955]

Fig. 7.108 A concrete specimen inside the hydraulic press after the compression test and the final rupture. From [955]

7.4.3   Analysis of Images

Various applications exist in the literature concerning image processing, such as *segmentation* or *thresholding* (Fig. 7.121), *edge detection* (Fig. 7.122), *fusion* (Fig. 7.123), *Magnetic Resonance* and *Computed Tomography* (Fig. 7.124), facial expression recognition (Fig. 7.125), among others [1091–1108].

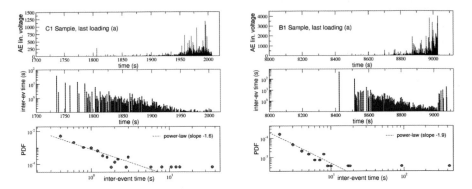

Fig. 7.109 The AE recordings of the last loading of the two compression test shown in Fig. 7.107 for the two specimens C1 (concrete) and B1 (basalt) are reported, together with the inter-event time series and the corresponding probability distribution functions. From [955]

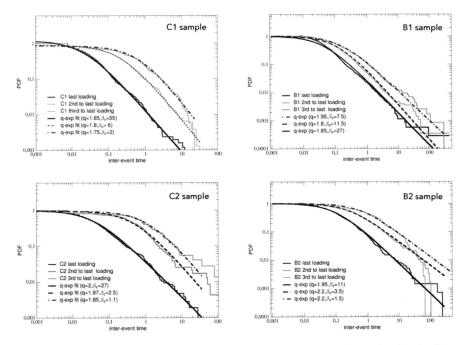

Fig. 7.110 Complementary cumulative pdfs of the inter-event times of AE for the last three loadings of the specimens *left:* C1, C2 (concrete) and *right:* B1, B2 (basalt). The fits with q-exponential curves are also indicated. From [955]

In Fig. 7.121 we see little lateral spikes that are clearly visible for $q = 3$ but not so much for $q = 1$. In Fig. 7.122 we see that the $q = 1.5$ image is the best among the computer-processed ones used in the paper for comparison.

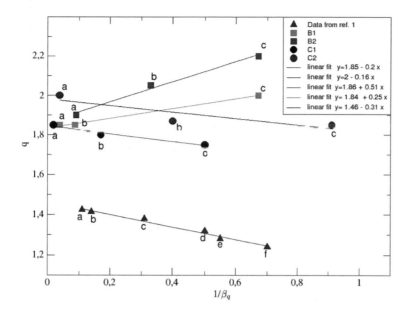

Fig. 7.111 The values of the entropic index of the q-exponential fits, reproducing the complementary cumulative pdfs obtained from experimental data about the AE inter-event time series for both the concrete (C1, C2—full circles) and basalt (B1, B2—full squares) specimens, are reported as function of $1/\beta_q$. Linear fits of the reported values are also shown. We report also (triangles) the values for q and $1/\beta_q$ extracted from analogous failure tests with cement mortar specimens in [954]. From [955]

In Fig. 7.123 we see two panels, the goal of the top one being to focus on the human profile, and the goal of the bottom one being to focus on the background. The top left and top middle images in both panels are taken with infrared and visible light, respectively, the top right images being their simple averages; the three bottom images (left, middle, and right) show the results using different algorithmic processing. For both the figure and the background, the better subjective quality fusion is that indicated in the bottom middle image, which was obtained with $q \simeq 1.85$.

In Fig. 7.124, the processing time performance is illustrated for *Magnetic Resonance* and *Computed Tomography*. The algorithm uses a q-generalized mutual information. The goal is to make a fast and accurate image registration. Using $q = 0.7$, instead of $q = 1$, yields up to *seven* times faster convergence and *four* times more precise registration.

In Fig. 7.125 we see typical facial images. Their recognition and classification can be done by using several types of algorithms (ALBP, q-based, NLDAI, see details in [1099]). Several resolution levels have been used, namely 64×64, 48×48, 32×32, and 16×16. The *classification accuracy* is indicated in the Table for each case: top right for 64×64 resolution and bottom right for 48×48, 32×32, and 16×16 resolutions. In all cases, the best classification accuracy is obtained with

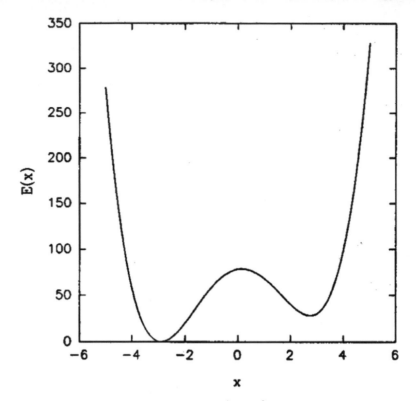

Fig. 7.112 The cost function is given by $E(x) = x^4 - 16x^2 + 5x + 78.3323$. From [959]

the combination "ALBP + q-based + NLDAI" (e.g., 94.59 % accuracy for 64 × 64 resolution).

The detection of possibly pathological microcalcifications as revealed in mammograms can be sensibly improved by using q-entropy with $q \neq 1$ [1109]: see Fig. 7.126. The $q \neq 1$ outcome exhibits 96.55 % true positives (compared to 80.21 % without using $q \neq 1$), and 0.4 % false positives (compared to 8.1 %).

Brain tissue segmentation using q-entropy improves the diagnosis of multiple sclerosis in Magnetic Resonance Images [1110]: see Fig. 7.127.

The image of Computer Tomography scans revealing fibrosis due to COVID-19 can be sensibly improved by incorporating in the algorithm the q-entropy with $q = 0.5$ [1111]: see Fig. 7.128.

7.4.4   Ping Internet Experiment

PING is a quick internet procedure which enables, from a given computer, to check whether any other specific computer is online at that moment. There is naturally

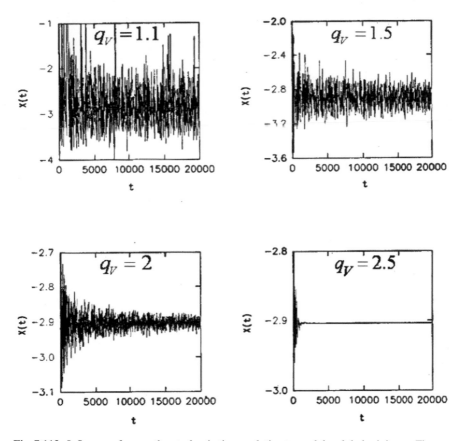

Fig. 7.113 Influence of q_V on the stochastic time evolution toward the global minimum. The runs start with $x(1) = 2$ and $T(1) = 100$. From [959]

a time delay (*sparseness time interval*) before the answer arrives. Abe and Suzuki [1112] devised an interesting experiment which consisted in automatically repeating the ping instruction many times in order to measure the distribution of the sparseness time interval. The results can be seen in Figs. 7.129 and 7.130. They are relatively well fitted by the expression $P(> \tau) = e_q^{-\tau/\tau_0}$. If we plot the four pairs $(q, \ln \tau_0)$, it does not suggest a monotonic *curve*, but rather something which could be closer to a "cloud", would we have many such points. But, of course, with only four points it is hard to advance any behavior with some degree of reliability. Interestingly enough, in these tests both $q > 1$ and $q < 1$ were observed.

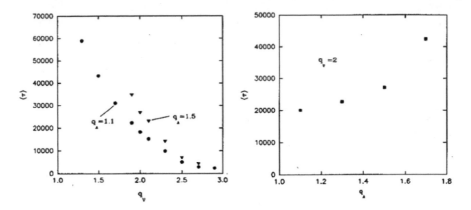

Fig. 7.114 Influence of (q_V, q_A) on the average computer time needed for finding the global minimum with a given precision. For the present problem the optimal choice consists in $q_V \simeq 2.8$ and $q_A < 1$. From [959]

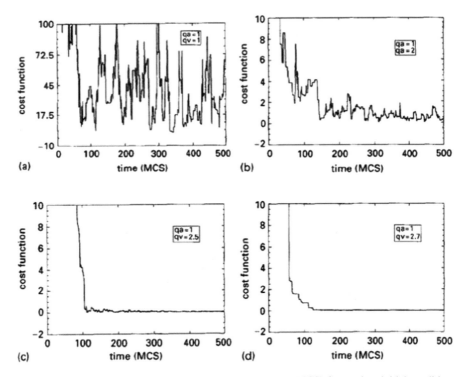

Fig. 7.115 Typical runs of the GSA algorithm. E_t $versus$ t (MCS) for random initial conditions and $T(1) = 100$. Acceptance parameter $q_A = 1$ and (a) $q_V = 1$, (b) $q_V = 2$, (c) $q_V = 2.5$, and (d) $q_V = 2.7$. From [960]

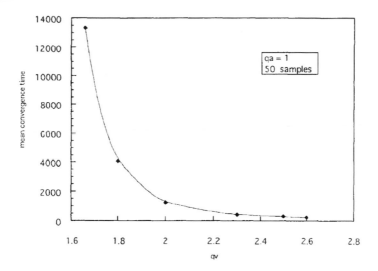

Fig. 7.116 Mean convergence time *versus* q_V. The solid line is a guide to the eye. The mean convergence time for $q_V = 1$ is about 50000. By taking $q_V \simeq 2.6$, there is a gain in computer time of a factor close to 100. From [960]

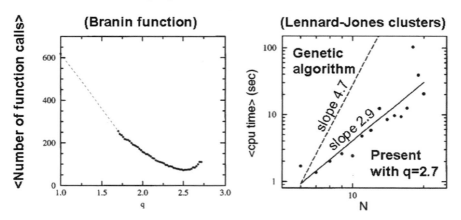

Fig. 7.117 Typical results of a q-generalized pivot method: for the Branin function (left plot) and for Lennard-Jones clusters (right plot). The method is sensibly more performant for $q > 1$ than for $q = 1$, and even more performant than the popular Genetic Algorithm. From [962]

7.5   Biosciences

Biological systems (e.g., fishes, insects, bacteria, viruses, cells) exhibit many kinds of complex motion. We focus here on some of them, namely emerging in cells.

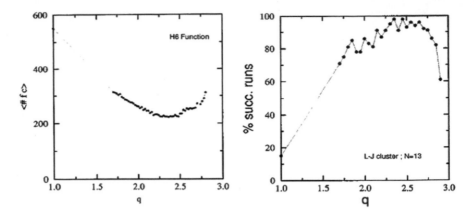

Fig. 7.118 Typical results of a q-generalized pivot method. Using $q > 1$ instead of $q = 1$ provides a double advantage: the computer time decreases (from 600 to 200 in the left panel), and the success rate (dramatically) increases (from 15% to 95% in the right panel). From [963]

Fig. 7.119 Electroencephalogram (including the contribution of muscular activity) during an epileptic crisis which starts at 80 s, and ends at 155 s. By direct inspection of the EEG, it is virtually impossible to detect the (clinically dramatic) transition (at 125 s) between the tonic stage and the clonic stage of the patient. From [1078]

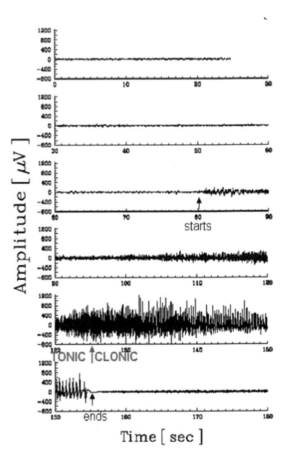

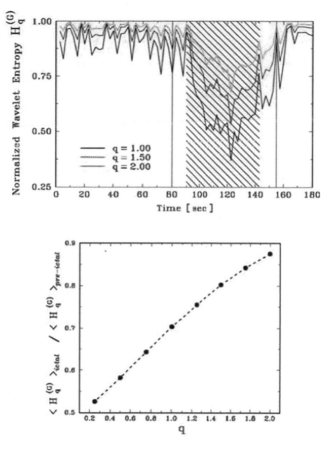

Fig. 7.120 *Top panel*: After processing (of the EEG signal) which includes the use of the entropic functional S_q, the precise location of the tonic-clonic transition becomes very visible. *Bottom panel*: The effect is even more pronounced for values of q going below unity. From [1078]

Motion of cells

Hydra viridissima is a small organism which may live in "dirty" (feeding) water. Experiments are described in [1113] which enabled the study of its motion, particularly the measure of the distribution of the velocities and the (anomalous) diffusion. The results are indicated in Figs. 7.131 and 7.132. It turned out that the distribution of velocities is *not* Maxwellian, but a q-Gaussian with $q \simeq 1.5$ instead Also, the diffusion was shown to occur with an exponent $\gamma \simeq 1.24 \pm 0.1$. Therefore the prediction (4.14) is verified within the experimental error bars. This turned out to be, in fact, the first experimental verification of the scaling relation (4.14) predicted in [404].

Dictyostelium discoideum constitutes another such example. Data from [1114] have been satisfactorily discussed in terms of q-Gaussians [1115], as shown in Fig. 7.133. Interestingly enough, the value of q differs from the vegetative to the starving state.

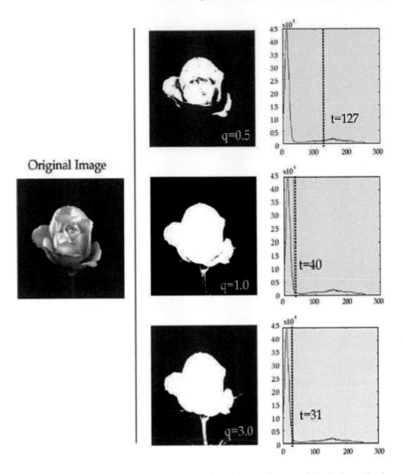

Fig. 7.121 Segmentation using the entropic functional S_q. Influence of the index q in the natural image of a rose. From [1091]

When starving, the cells adopt a higher value for q, thus making longer jumps to find food, like a sort of ultimate strategy.

Finally, an interesting study of collective migrating cells has been undertaken in [1116, 1117]. The velocity distributions obtained for Madin Darby canine kidney (MDCK) multicellular monolayer are depicted in Fig. 7.134. They yield q-Gaussians with $q = 1.183 \pm 0.035$. Very similar results are obtained for human umbilical vein endothelial cells (HUVEC) with $q = 1.208 \pm 0.025$, for C2C12 mouse myoblasts with $q = 1.184 \pm 0.014$, and for NIH-3T3 mouse embryo fibroblast with $q = 1.176 \pm 0.024$.

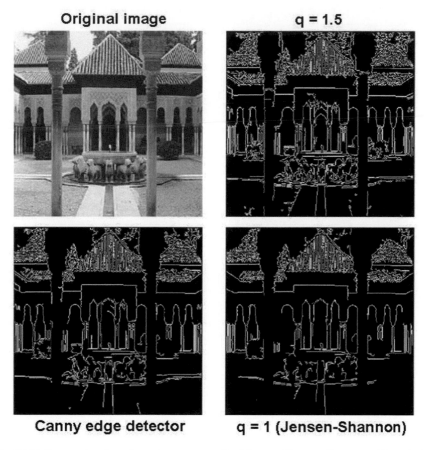

Fig. 7.122 Image edge detection using a q-generalized Jensen–Shannon divergence. The $q = 1.5$ image shows more details than both the $q = 1$ image and the Canny edge detector image. From [1100]

Medical applications

Signal processing of the EEG for direct medical use has been proposed for brain injury following severe situations such as cardiac arrest or asphyxia [1119]. Typical results are indicated in Figs. 7.135 (where the highest sensitivity of the recovery EEG is achieved for $q \simeq 3$) and (7.136) (where artificial low-amplitude spikes become detectable after (entropic) processing with $q \geq 3$). Further biomedical applications can be seen in [1120–1131].

INFRARED   VISIBLE

INFRARED   VISIBLE

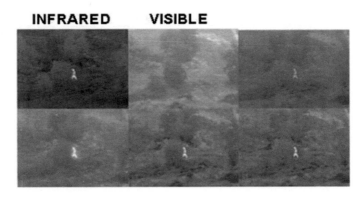

Fig. 7.123 Image fusion metric based on q-generalized mutual information. *Top panel*: The goal is to better distinguish the human profile. *Bottom panel*: The goal is to better distinguish the background. The best correlation with the subjective quality of fused images is obtained for $q \simeq 1.85$. From [1101]

Epidemiology and COVID-19

Satisfactory approaches to the pandemic COVID-19 peaks in various countries has been possible using typical q-statistical functions and relations [601, 1132] having provided successful results in high-frequency stock trading at NYSE and NASDAQ [423]. See Fig. 7.137, where the fittings have been done with the following form:

$$N = C(t - t_0)^{\alpha} e_q^{-\beta (t-t_0)^{\gamma}} = \frac{C(t - t_0)^{\alpha}}{[1 + (q - 1)\beta (t - t_0)^{\gamma}]^{1/(q-1)}}, \qquad (7.45)$$

Moreover, the dynamics of epidemiological multilinear models such as SIR and SEIR have been q-generalized by introducing non-multilinear models [601]. These models exhibit anomalous kinetics in the sense that they present power-law increase

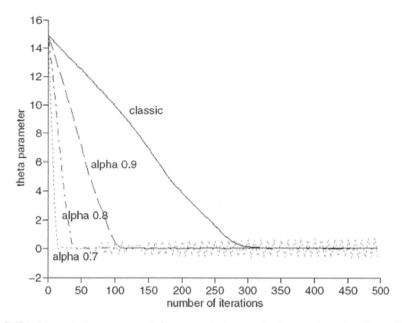

Fig. 7.124 Magnetic Resonance and Computered Tomography images. The algorithm achieves up to *seven* times faster convergence and *four* times more precise registration for $q \equiv \alpha < 1$ when compared to the classic case ($q = 1$). From [1092]

toward a peak (of quantities such as the number of active cases or of deaths) and also power-law decrease out of it. These behaviors are in variance with those of the standard SIR and SEIR models, which yield exponential increase and decrease.

7.6 Cellular Automata

The first connection between cellular automata (CA) and q-concepts has been attempted in [1136], by introducing a long memory in some typical Wolfram Class II CA. We focused on Rules 61, 99, and 111. The weight of the memory decays toward the past as $1/\tau^\alpha$ ($\tau = 1, 2, 3, ...;$ $\alpha \geq 0$), so that $\alpha \to \infty$ has no other memory than that of the previous step, i.e., the model recovers the simple Wolfram CA. If $\alpha = 0$ instead, we have infinitely long memory. Since the memory function is summable for $\alpha > \alpha_c$ and nonsummable for $0 \leq \alpha \leq \alpha_c$ with $\alpha_c \simeq 1$, we expect important changes to occur while crossing $\alpha \simeq 1$. This is indeed observed in the time behavior of the *Hamming distance*. Since this quantity plays a role totally analogous to the sensitivity to the initial conditions, it is natural to expect $H(t) \propto e_{q_{sen}}^{\lambda_q t} \propto t^{1/(1-q_{sen})}$ ($q_{sen} < 1$). Typical results are shown in Figs. 7.138, 7.139, and 7.140. The space-time behavior strongly depends on the initial conditions. We show the plots corresponding to the Rules 61 and 99 for random initial conditions and also for very localized initial con-

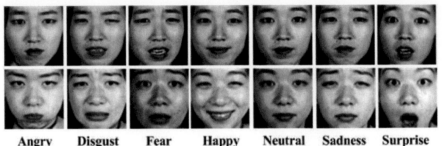

| Angry | Disgust | Fear | Happy | Neutral | Sadness | Surprise |

Features	Classification Accuracy %
AMGFR [15]	82.46
LBP [6]	85.57
ALBP	88.26
Tsallis	85.36
ALBP + Tsallis	91.89
ALBP + Tsallis + NLDAI	94.59

Features	Classification accuracy (%)		
	48×48	32×32	16×16
AMGFR [15]	78.13	67.83	56.35
LBP [6]	81.44	77.28	68.02
ALBP	84.27	82.74	75.39
Tsallis	79.25	71.04	63.81
ALBP + Tsallis	87.31	85.73	80.40
ALBP + Tsallis + NLDAI	90.54	88.82	84.62

Fig. 7.125 Facial expression recognition using *Advanced Local Binary Patters* (ALBP), entropy S_q and *global appearance features*. Sample images from the JAFFE database. At all resolution levels (64×64, 48×48, 32×32, and 16×16), the combination "ALBP + Tsallis + NLDAI" yields the highest accuracy. From [1099]

ditions instead, in Figs. 7.138 and 7.139 respectively, for typical values of α. The α-dependence of q_{sen} for Rules 61 and 111 is depicted in Fig. 7.140. In both cases we observe criticality at $\alpha \simeq 1.5$.

7.7 Self-Organized Criticality

Several studies are available in the literature in connection with self-organized criticality (SOC), in connection with biological evolution [1137–1141], imitation games [1142], atmospheric cascades [1143], earthquakes [460, 833], and others [1144].

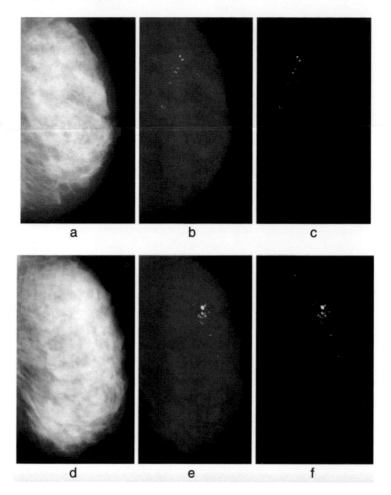

Fig. 7.126 Without q-entropy enhancement with $q \neq 1$, detection of microcalcifications is meager: 80.21% Tps (true positives) with 8.1 Fps (false positives), whereas upon introduction of the q-entropy, the results surge to 96.55 % Tps with 0.4 Fps. Detection results from the experiment: **a** mdb236, **b** output with the Mcs enhanced, **c** output with the Mcs extracted; **d** mdb216, **e** output with the Mcs enhanced, **f** output with the Mcs extracted. From [1109]

For example, the sensitivity to the initial conditions of the Bak–Sneppen model of biological evolution exhibits a time evolution similar to that focused on within the q-generalized Pesin identity. More precisely, the Hamming distance $D(N, t)$ corresponding to a system with N sites scales [1137] like

$$D(N, t) \sim N^\beta F(t/N^\gamma) \tag{7.46}$$

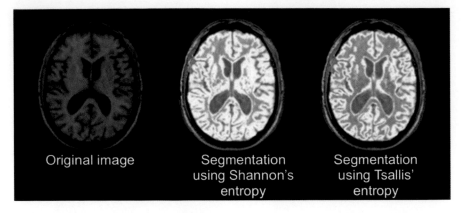

Fig. 7.127   Image segmentation using Shannon ($q = 1$) and q-entropy with $q \neq 1$. From [1110]

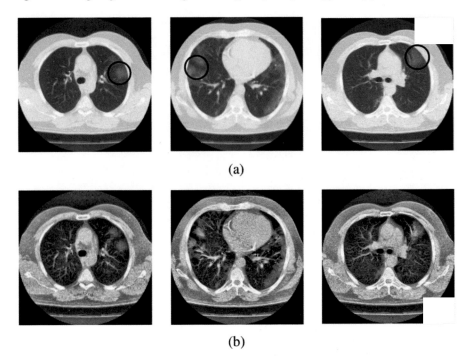

Fig. 7.128   Sample scans from the dataset before and after enhancement showing infected lungs.
a Original Computer Tomography scans, with red circles highlighting some regions where fibrosis
can be seen; **b** enhanced Computer Tomography scans using $q = 0.5$. From [1111]

with $(\beta, \gamma) \simeq (0.54, 1.7)$, $F(z)$ being a function whose form is nearly the same as
the time evolution of the q-entropy for paradigmatic models like the logistic map at
the Feigenbaum point (see Sect. 2.1.2.10), but with $q \simeq -2.1$ in the present case.

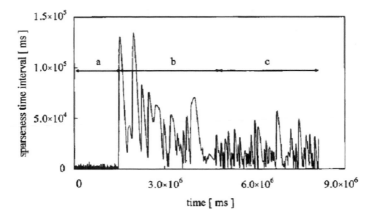

Fig. 7.129 Time series data of the sparseness time interval. Roughly, three different nonequilibrium stationary states (denoted a, b, and c) may be recognized. From [1112]

7.8 Asymptotically Scale-Free Networks

Networks exist of various types [1145–1149]. They are typically characterized by sets of *nodes* (*sites*) and sets of directed or nondirected *links* (*bonds*) joining the nodes. These are the most studied, although it is clear that it is easy to generalize the concept by also including plaquettes and other many-node, many-link, and mixed connections. Networks can be *topological* in nature, in the sense that we may be allowed to arbitrarily deform them as long as we do not modify the connections between nodes and links. But they can also be *metrical*, in the sense that they may have a "geography" with a concept of *distance*, which can sensibly influence a variety or properties. *Bravais lattices* can be thought as networks which are invariant through discrete *translations*. Through the concept of unitary crystalline cell, we can attribute to them a nonzero Lebesgue measure. *Hierarchical networks* typically are scale-invariant, and can be characterized through a *Hausdorff* or *fractal dimension*. More complex networks can exhibit a multifractal structure, and can thus be characterized by a $f(\alpha)$ function [163]. In what follows we focus on the so-called *scale-free networks* (more precisely, *asymptotically scale-free networks*). Indeed, they play an interesting role as systems which can be (at least for some of their properties) addressed by the entropy S_q and nonextensive statistical mechanics.

These networks are of the hierarchical kind, made of hubs, sub-hubs, sub-sub-hubs, and their links, the whole constituting a connected structure which exhibits (statistical) invariance under *dilation*. Their basic characterization is done through the *degree distribution* $p(k)$, defined as the probability of a node having k links ($k = 1, 2, ...$). It happens that many of them exhibit a power-law dependence in k for large values of k. And many among those, precisely have the form

$$p(k) = p(0)\, e_q^{-k/\kappa} \quad (\kappa > 0), \tag{7.47}$$

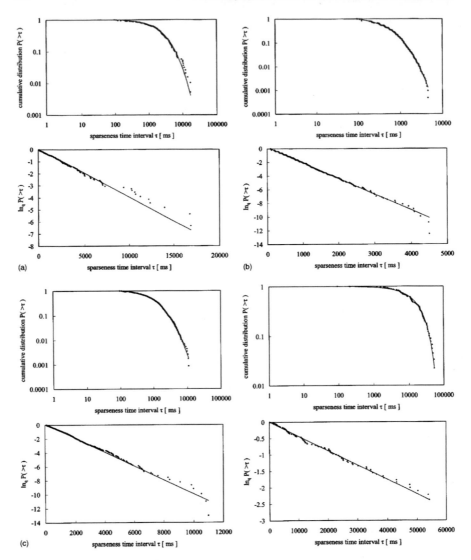

Fig. 7.130 Cumulative probability of the measured sparseness time interval corresponding to four different nonequilibrium stationary states (the first three are precisely the states a, b, and c of Fig. 7.129; the fourth is still a different one. All four upper panels are in log–log representation; all four lower panels are the same data, in q-log *versus* linear representation. The continuous curves are q-exponentials with $q = 1.7$, 1.12, 1.16 and 0.73 respectively. Notice that values of q both above and below unity occur. From [1112]

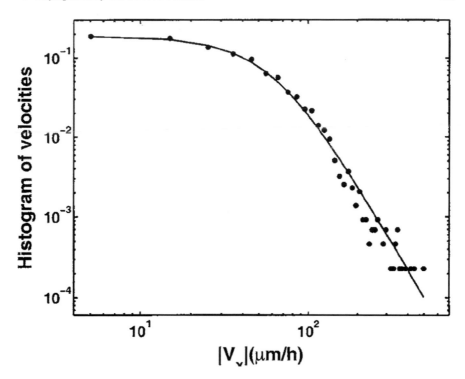

Fig. 7.131 Probability distribution for the horizontal component of velocity for endodermal cells in an ectodermal aggregate. The solid line is a fit with a q-Gaussian using $q = 1.5$. From [1113]

where $p(0)$ is a normalizing factor. This form is known to extremize S_q with simple constraints (see Sect. (3.5)). It appears frequently in the literature as

$$p(k) \propto \frac{1}{(k_0 + k)^{\mu}}, \tag{7.48}$$

which is identical to Eq. (7.47) through the transformation

$$\mu \equiv \frac{1}{q-1}, \qquad k_0 \equiv \frac{\kappa}{q-1}. \tag{7.49}$$

Let us exhibit now a few systems whose degree distribution is precisely of this type, in order to show later what the connection is between this type of networks and nonextensive statistical concepts [1150].

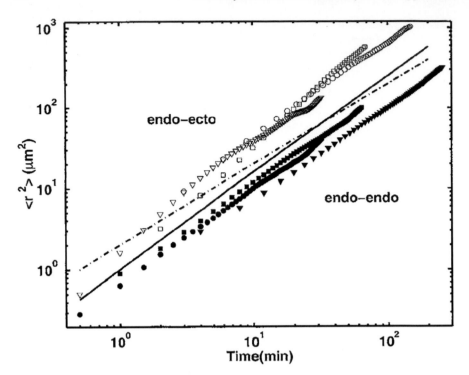

Fig. 7.132 $\langle r^2 \rangle$ *versus* t plot for endodermal cells in an endodermal aggregate (filled symbols), and endodermal cells in an ectodermal aggregate (open symbols). The solid line has a slope of 1.23, while the dashed line has a slope of 1.0 (which would correspond to normal diffusion). From [1113]

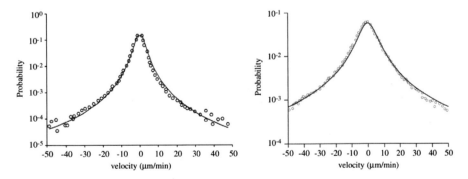

Fig. 7.133 The distribution of one-component of velocity of vegetative (left) and starved (right) *Dictyostelium discoideum* cells in the absence of external chemical stimuli. The data from [1114] are well fitted by q-Gaussians, respectively, with $q = 5/3$ and $q = 2$ (solid line). From [1115]

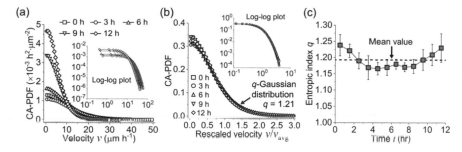

Fig. 7.134 The circumferentially averaged probability distribution of cell velocities in the MDCK (Madin Darby canine kidney) multicellular monolayer is well fitted by a q-Gaussian distribution. **a** Evolution with time. *Inset:* log–log plot of the same data. **b** Collapse obtained with rescaled cell velocities at different times. *Inset:* log–log plot of the CA-PDFs. **c** Temporal evolution of the effective index q. From [1116]

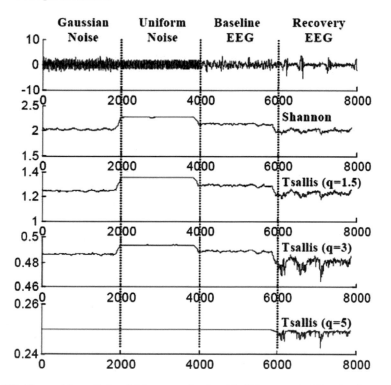

Fig. 7.135 The goal is to distinguish between signals with different probability distributions, and between EEG from different physiological conditions. The optimal is achieved for $q \simeq 3$. From [1119]

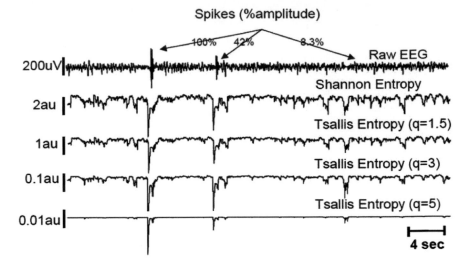

Fig. 7.136 The goal is to detect the existence of three (artificially introduced) spikes which corrupt the raw EEG. Even small spikes become detectable after processing with $q \geq 3$. From [1119]

7.8.1 The Natal Model

For convenience—and also as a homage—we shall refer to this growth model [1152] as the *Natal* one because all four co-authors have deep connections with that sea-shore town of the North-East of Brazil. Let us follow [635, 1153] to describe the d-dimensional Natal model.

Our growing model starts with one site at the origin. We then stochastically locate a second site (and then a third, a fourth, and so on up to N) through the d-dimensional isotropic distribution

$$p(r) \propto \frac{1}{r^{d+\alpha_G}} \quad (\alpha_G > 0; \; d = 1, 2, 3, 4) , \tag{7.50}$$

where $r \geq 1$ is the Euclidean distance from the newly arrived site to the center of mass of the pre-existing system (in one dimension, $r = |x|$; in two dimensions, $r = \sqrt{x^2 + y^2}$; in three dimensions $r = \sqrt{x^2 + y^2 + z^2}$, and so on); we assume angular isotropy; $p(r)$ is zero for $0 \leq r < 1$; the subindex G stands for *growth*. We consider $\alpha_G > 0$ so that the distribution $P(r)$ is normalizable; indeed, $\int_1^\infty dr \, r^{d-1} \, r^{-(d+\alpha_G)} = \int_1^\infty dr \, r^{-1-\alpha_G}$, which is finite for $\alpha_G > 0$, and diverges otherwise.

Every new site which arrives is then attached to one and only one site of the pre-existing cluster. The choice of the site to be linked with is done through the following *preferential attachment* probability:

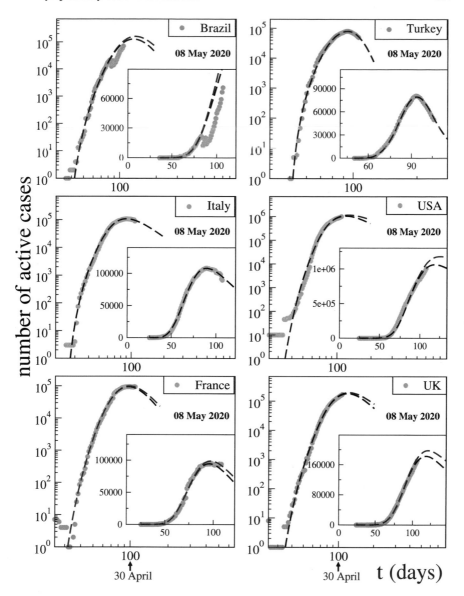

Fig. 7.137 Fittings with Eq. (7.45) of the time evolution of the active cases data available in 28 April 2020 (green dots) for various severely affected countries around the world. From [1132]

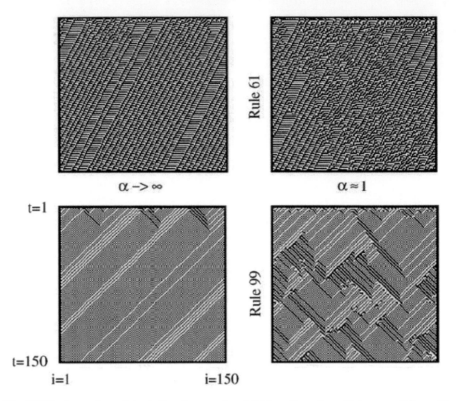

Fig. 7.138 Space-time plots starting from random initial configurations. States $\sigma_i = 0$ ($\sigma_i = 1$) are shown yellow (red). From [1136]

$$\Pi_{ij} = \frac{k_i\, r_{ij}^{-\alpha_A}}{\sum_i k_i\, r_{ij}^{-\alpha_A}} \in [0, 1] \quad (\alpha_A \geq 0)\,, \tag{7.51}$$

where k_i is the connectivity of the i-th pre-existing site (i.e., the number of sites that are already attached to site i), and r_{ij} is the Euclidean distance from site i to the newly arrived site j; subindex A stands for *attachment*. The role of α_A is illustrated in Fig. 7.141.

For α_A approaching zero and arbitrary d, the physical distances gradually loose relevance and, at the limit $\alpha_A = 0$, all distances becomes irrelevant in what concerns the connectivity distribution, and we therefore recover the Barabási–Albert (BA) model [1151], which has topology but no metrics. It will be shown later on that the degree distribution $P(k)$ behaves asymptotically as $1/k^\gamma$. We further show that, for arbitrary dimensionality, $\gamma \equiv 1/(q - 1) > 0$ can be controlled in a simple manner, namely by metric changes through only one control parameter (specifically, the ratio α_A/d) in the structure of the network.

Large-scale simulations have been performed for the ($d = 1, 2, 3, 4$) models for fixed (α_G, α_A), and we have verified in all cases that the degree distribution $P(k)$ is

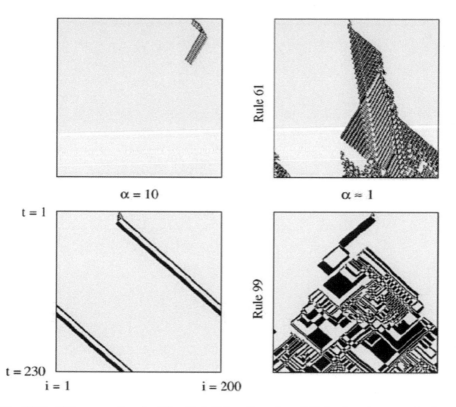

Fig. 7.139 Difference patters for CA with initial configurations differing in only one randomly chosen bit. Cells with difference in both configurations at time t are shown in red. From [1136]

completely independent from α_G: see Fig. 7.142. Using this fact, we have arbitrarily fixed $\alpha_G = 2$, and have numerically studied the influence of (d, α_A) on $P(k)$: see Figs. 7.143 and 7.144. In all cases, the q-exponential fittings $P(k) = P(0)e_q^{-k/\kappa}$ with $q > 1$ and $\kappa > 0$ have proved to be remarkably good. The corresponding values of (q, κ) are shown in Figs. 7.145, 7.146 and 7.147.

Let us next address the analogous generalization [636], also for $d = 1, 2, 3, 4$, of the Bianconi–Barabási model [1154], where $\eta_i \in (0, 1]$ characterizes the ith site *fitness*, randomly chosen once for ever. The results are shown in Figs. 7.148, 7.149 and 7.150.

Finally, we address a quite general growth $d = 1, 2, 3$ model [637], which unifies the models introduced in [635, 1152] and in [636], and contains, consequently, both Barabási–Albert and Bianconi–Barabási models as particular instances. To do this unification we introduce a fitness distribution $P(\eta)$ ($\eta \in [0, 1]$) defined as follows:

$$P(\eta) = (1 + \rho)\eta^{\rho}, \qquad for \ \rho > 0 \qquad (7.52)$$

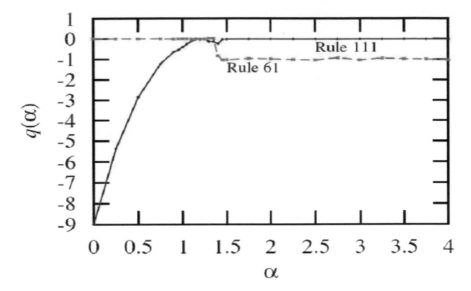

Fig. 7.140 The α-dependence of $q \equiv q_{sen}$ for Rules 61 and 111. From [1136]

$$P(\eta) = 1, \qquad\qquad for \ \rho = 0 \qquad\qquad (7.53)$$

$$P(\eta) = (1 - \rho)(1 - \eta)^{-\rho}, \ \ for \ \rho < 0 \qquad\qquad (7.54)$$

where the pre-factors $(1 + \rho)$ and $(1 - \rho)$ come from normalization, and $\rho \in (\infty, \infty)$. See Fig. 7.151. The degree distributions are once again given by the q-exponential form $e_q^{-k/\kappa}$ with (q, κ) illustrated in Figs. 7.152 and 7.153.

7.8.2 Albert–Barabási Model

Another growth model, also including *preferential* attachment, has been introduced and analytically solved in 2000 by Albert and Barabasi [1155] as a prototype of emergence of the ubiquitous scale-free networks. At each time step, m new links are added with probability p, or m existing links are rewired with probability r, or a new node with m links is added with probability $1 - p - r$; all linkings are done with probability $\Pi(k_i) = (k_i + 1)/\sum_j (k_j + 1)$, where k_i is the number of links of the $i - th$ node. The exact stationary-state distribution of the number k of links at each site is given [1155] by Eq. (7.48) with

$$k_0 = 1 + (p - r)\left[1 + \frac{2m(1 - r)}{1 - p - r}\right] > 0. \qquad\qquad (7.55)$$

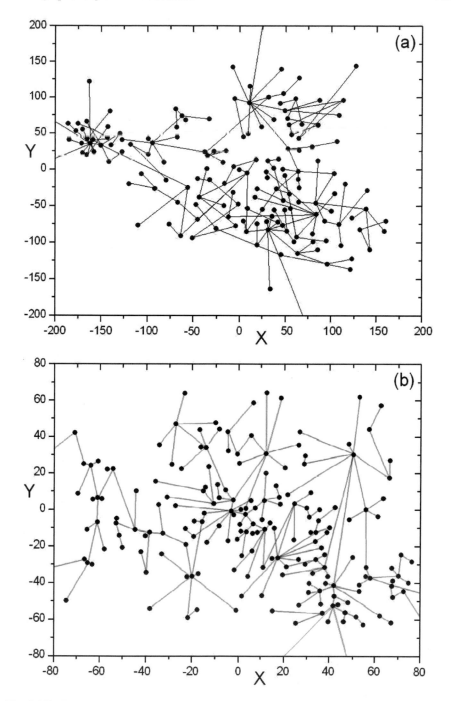

Fig. 7.141 Typical $N = 250$ network for (a) $(\alpha_G, \alpha_A) = (1, 0)$ and (b) $(\alpha_G, \alpha_A) = (1, 4)$. The starting site is at $(X, Y) = (0, 0)$. Notice the spontaneous emergence of hubs. From [1152]

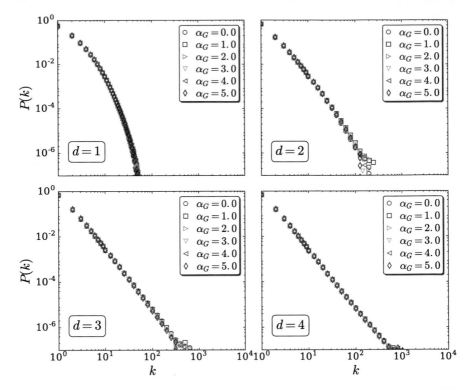

Fig. 7.142 Connectivity distribution for $d = 1, 2, 3, 4$, $\alpha_A = 2.0$ and typical values for α_G. The simulations have been run for 10^3 samples of $N = 10^5$ sites each. We verify that $P(k)$ is independent from α_G ($\forall d$). Logarithmic binning was used whenever convenient

and

$$\mu = \frac{m(3 - 2r) + 1 - p - r}{m} > 0. \qquad (7.56)$$

With the notation change (7.49), this degree distribution can be rewritten in the form of Eq. (7.47) with

$$q = \frac{2m(2 - r) + 1 - p - r}{m(3 - 2r) + 1 - p - r} \geq 1, \qquad (7.57)$$

with $\kappa > 0$ given by Eqs. (7.55) and (7.56) replaced into $\kappa = k_0(q - 1)$.

Along the present lines let us emphasize some simple connections. If we have

$$y \propto [a + bx]^\mu \quad (a > 0; \; x \in \mathbb{R}), \qquad (7.58)$$

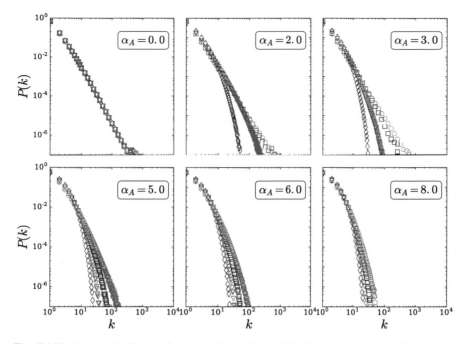

Fig. 7.143 Degree distribution for $d = 1$ (*blue diamonds*), 2 (*green triangles*), 3 (*magenta squares*), 4 (*grey circles*), and typical values of α_A, with $\alpha_G = 2.0$. The simulations have been run for 10^3 samples of $N = 10^5$ sites each. Logarithmic binning was used whenever convenient

it follows $y \propto [1 + (b/a)x]^\mu = e_q^{-\beta x}$, with

$$\mu \equiv 1/(1 - q); \quad b/a \equiv -\beta(1 - q). \tag{7.59}$$

Analogously, if we have

$$y \propto [a + bx + cx^2]^\mu \quad (a - b^2/4c > 0; \; x \in \mathbb{R}), \tag{7.60}$$

it follows $y \propto [1 + \frac{4c^2}{4ac - b^2}(x + b/2c)^2]^\mu = e_q^{-\beta(x - x_0)^2}$, with

$$\mu \equiv 1/(1 - q); \quad \frac{4c^2}{4ac - b^2} \equiv -\beta(1 - q); \quad x_0 \equiv -b/2c. \tag{7.61}$$

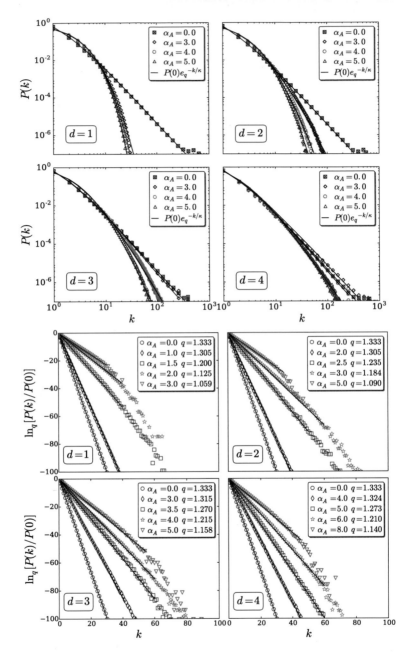

Fig. 7.144 Fittings of the $d = 1, 2, 3, 4$ connectivity distributions with the function $P(k) = P(0)e_q^{-k/\kappa}$, where $e_q^z \equiv [1 + (1 - q)z]^{1/(1-q)}$. *Top:* log–log representation. *Bottom:* $\ln_q[P(k)/P(0)]$ versus k representation.4. Notice that straight lines in a $\ln_q$–linear representation univocally determine the q-exponential function. The fitting parameters are exhibited in Fig. 7.145. The numerical failure, at large enough values of k, with regard to straight lines are finite-size effects that gradually disappear when we approach the thermodynamic limit $N \to \infty$. Logarithmic binning was used whenever convenient. From [635]

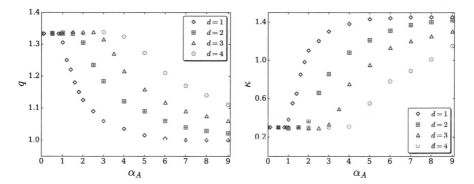

Fig. 7.145 q and κ for $d = 1, 2, 3, 4$. For $\alpha_A = 0$ and $\forall d$, we recover the Barabási–Albert universality class $q = 4/3$ (hence $\gamma = 3$) [1151], which has no metrics. From [635]

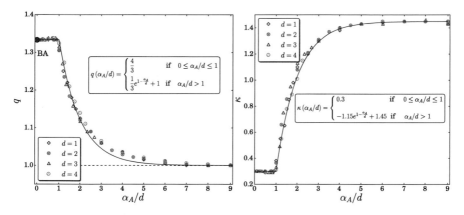

Fig. 7.146 q and κ versus α_A/d (same data as in Fig. 7.145). We see that $q = 4/3$ for $0 \leq \alpha_A/d \leq 1$, and a nearly exponential behavior emerges for $\alpha_A/d > 1$ ($\forall d$); similarly for κ. These results exhibit the universality of both q and κ. The red dot indicates the Barabási–Albert (BA) universality class $q = 4/3$ [1151]. In what concerns the universal $q = 4/3$ cut-off (i.e., the $1/(q - 1) = 3$ cut-off), see [1156] and references therein. From [635]

7.8.3 Non-growing Model

Asymptotically scale-free networks *without growth* are known since long [633]. We focus here on a recent one [634], on which a q-exponential degree distribution has numerically emerged in the limit of infinitely large, though fixed, size. The results are indicated in Figs. 7.154, 7.155 and 7.156.[6]

---

[6] Such examples illustrate inadvertences on page 74 of Review of Modern Physics **74**, 47 (2002) by R. Albert and A.L. Barabasi and elsewhere, where it is stated that growth is *necessary* for having (asymptotically) scale-free networks.

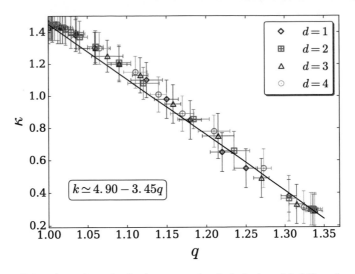

Fig. 7.147 All the values of q and κ for the present $d = 1$, 2, 3, 4 models follow closely a linear relation (continuous straight line). The upmost value of q is 4/3, yielding $\kappa \simeq 0.3$ ($\forall d$). From [635]

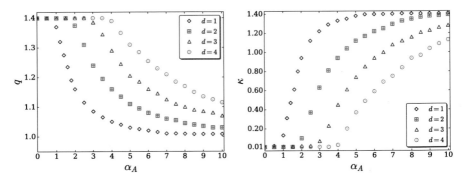

Fig. 7.148 q and κ for $d = 1, 2, 3, 4$. We can see how q and κ vary with α_A and d. q has an upper limit ($q = 7/5$) and κ has a lower limit ($\kappa = 0.01$) regardless of the system dimension. From [636]

7.8.4 Connection Between Asymptotically Scale-Invariant Networks with Weighted Links and q-Statistics

We make here more explicit and general the deep connection between classes of asymptotically scale-free networks and nonextensive statistical mechanics. Let us follow [1159]. As before, our growing d-dimensional network starts with one site at

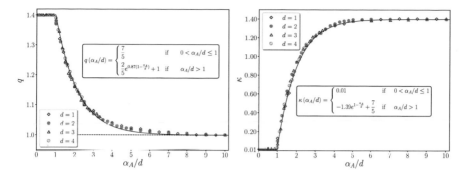

Fig. 7.149 q and κ versus α_A/d. We see that $q = 7/5$ and $\kappa = 0.01$ for $\mathbf{0 < _A/d \leq 1}$. Nearly exponential behavior gradually emerges for $\alpha_A/d > 1$ ($\forall d$) and similarly for κ. These results exhibit the universality of both q and κ. From [636]

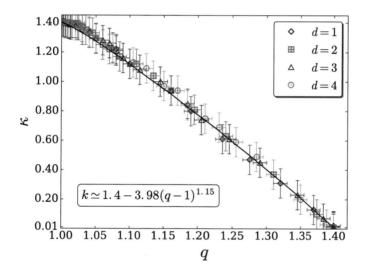

Fig. 7.150 q and κ, for the present models ($d = 1, 2, 3, 4$), follow nearly linear relation given by $\kappa \simeq 1.4 - 3.98(q - 1)^{1.15}$ (continuous straight line). The up most value of q is $7/5$, yielding $\kappa = 0.01$ ($\forall d$). From [636]

the origin. We then stochastically locate a second site (and then a third, a fourth, and so on up to N) through an isotropic probability $p(r) \propto 1/r^{d+\alpha_G}$ ($\alpha_G > 0$), where $r \geq 1$ is the Euclidean distance from the newly arrived site to the center of mass of the pre-existing cluster; α_G is the *growth* parameter and $d = 1, 2, 3$ is the dimensionality of the system (large α_G yields geographically concentrated networks); each site is assumed to have around it an excluded unity radius.

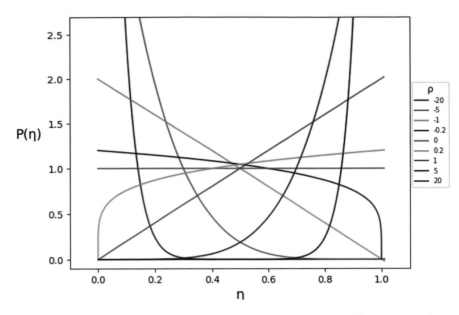

Fig. 7.151 Fitness distributions $P(\eta)$ for typical values of $\rho \in (-\infty, \infty)$. The $\rho \to \infty$ and $\rho \to -\infty$ limits correspond respectively to $P(\eta) = \delta(\eta - 1)$ and $P(\eta) = \delta(\eta)$, $\delta(x)$ being the Dirac delta distribution; $\rho = 0$ corresponds to the uniform distribution $P(\eta) = 1$, $\eta \in [0, 1]$. From [637]

The site $i = 1$ is then linked to the site $j = 2$. We sample a random number w_{ij} from a distribution $P(w)$ that will give us the corresponding link weight. Each site will have a *total energy* ε_i that will depend on how many links it has, noted k_i, and the widths $\{w_{ij}\}$ of those links. At each time step, the site i is associated to its *local energy* ε_i defined as:

$$\varepsilon_i \equiv \sum_{j=1}^{k_i} \frac{w_{ij}}{2} \quad (w_{ij} \geq 0), \tag{7.62}$$

where the factor $1/2$ is introduced to avoid double counting in the total energy of the network. The value of ε_i will directly affect the probability of the site i to acquire new links. Indeed, from this step on, the arriving sites $j = 3, 4, \ldots$ will be linked to one of the previous ones with probability

$$\Pi_{ij} \propto \frac{\varepsilon_i}{d_{ij}^{\alpha_A}} \quad (\alpha_A \geq 0), \tag{7.63}$$

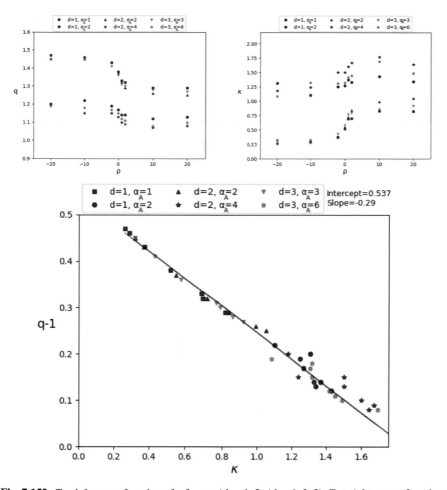

Fig. 7.152 *Top left: q* as a function of ρ for $\alpha_A/d = 1, 2$ $(d = 1, 2, 3)$. *Top right:* κ as a function of ρ for $\alpha_A/d = 1, 2$ $(d = 1, 2, 3)$. *Bottom:* Nearly linear relation between q and κ (for typical values of ρ and $d = 1, 2, 3$), whose monotonicity denotes the criticality of the system. From [637]

where d_{ij} is the Euclidean distance between i and j, where i runs over all pre-existing sites which can be linked to the new arriving site j. The *attachment* parameter α_A controls the importance of the distance in the preferential attachment rule (7.63). When $\alpha_A \gg 1$ the sites tends to connect to close neighbors, whereas $\alpha_A \simeq 0$ tends

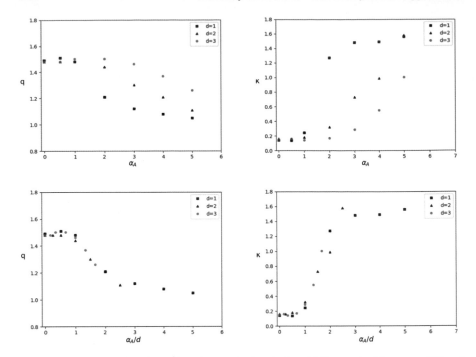

Fig. 7.153 q and κ as functions of α_A (upper plots) and of α_A/d (lower plots) for $\rho = -100$ and $d = 1, 2, 3$. From [637]

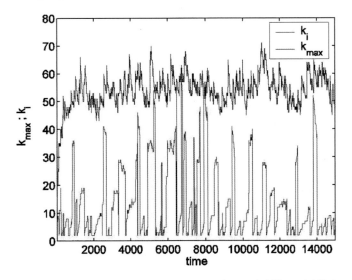

Fig. 7.154 A node collapsing (gas-like) model with a merging probability $\propto 1/d_{ij}^\alpha$ ($\alpha \geq 0$), where d_{ij} is the shortest topological distance between sites i and j on the network. We illustrate here the time evolution of the number of links of both the most important hub (blue) and of a typical node (red) of a network with $N = 2^7 = 128$ nodes and $\alpha = 0$. In the present model the most linked hub maintains its "leadership" for ever. From [634]

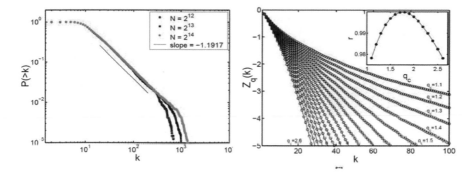

Fig. 7.155 Cumulative degree distribution of the same model as in Fig. 3 but for $\alpha \to \infty$ and typical values of N, where the finite-size effects are visible. *Left:* log–log scale. *Right:* The same data in (q–log)—(linear) scale, for various values of q, the optimal value being $q = 1.84$ [Inset: The q-dependance of the linear correlation r, which achieves its maximal value ($r > 0.9999$) for $q = 1.84$]. From [634]

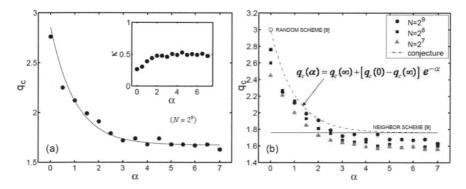

Fig. 7.156 Same model as in Figs. 3 and 4. *Left:* α-dependance of the values of q and κ for the best q-exponential fitting of the numerical results for the $N = 2^9$ network. *Right:* The same for the values of q for increasingly large networks. In the limits $\alpha = 0$ and $\alpha \to \infty$ we recover the *random* and *neighbor* schemes of [1157], respectively. The dashed curve corresponds to a possible heuristic analytical behavior. From [634]

to generate distant connections all over the network. Notice that, while the network size increases up to N nodes, the variables k_i and ε_i (number of links and *total energy* of the ith node; $i = 1, 2, 3 \ldots, N$) also increase in time (see Fig. 7.157 for a sample of the ever growing network).

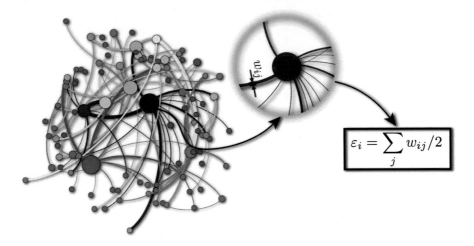

Fig. 7.157 Sample of a $N = 100$ network for $(d, \alpha_A, \alpha_G, \eta, w_0) = (2, 1, 5, 1, 1)$. As can be seen, for this choice of parameters, hubs (highly connected nodes) naturally emerge in the network. Each link has a specific width w_{ij} and the total energy ε_i associated to the site i will be given by half of the sum over all link widths connected to the site i (see zoom of site i). From [1159]

If we consider the particular case $P(w) = \delta(w - 1)$, where $\delta(z)$ denotes the Dirac delta distribution, Eq. (7.63) becomes $\Pi_{ij} \propto k_i/d_{ij}^{\alpha_A}$ $(\alpha_A \geq 0)$, thus recovering the usual preferential attachment rule. Consequently, the present model recovers the one in [635, 636, 1153] as a particular instance (notice however that, usually, the definition of site degree does not exclude the double counting we mentioned in definition (7.62)). Note that, if we additionally consider the particular case $\alpha_A = 0$, we recover the standard Barabási–Albert model with $\Pi_i \propto k_i$ [1147, 1151].

We focus here on the case where $w \geq 0$ is given by the following stretched-exponential distribution:

$$P(w) = \frac{\eta}{w_0 \, \Gamma\left(\frac{1}{\eta}\right)} e^{-(w/w_0)^\eta} \quad (w_0 > 0; \, \eta > 0), \tag{7.64}$$

which satisfies $\int_0^\infty dw \, P(w) = 1$. As particular cases of Eq. (7.64) we have: $\eta = 1$, which corresponds to an exponential distribution, $\eta = 2$, which corresponds to a half-Gaussian distribution, and $\eta \to \infty$, which corresponds to a uniform distribution within $w \in [0, w_0]$. The results are shown in Figs. 7.158, 7.159 and 7.160, where

$$p_q(\varepsilon) = \frac{e_q^{-\beta_q \varepsilon}}{Z_q}. \tag{7.65}$$

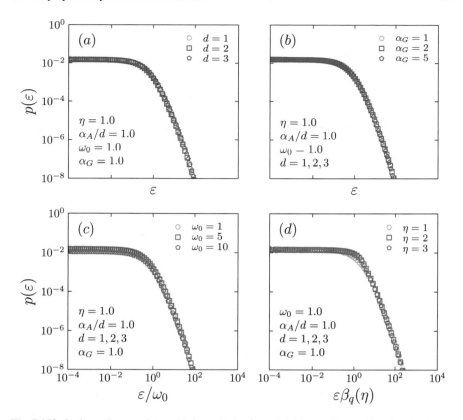

Fig. 7.158 In these plots we show $p(\varepsilon)$ for typical values of d (a), α_G (b), w_0 (c) and η (d). (a) By fixing $(\alpha_G, \eta, w_0, \alpha_A/d) = (1, 1, 1, 1)$ the dimensionality d does not modify $p(\varepsilon)$. (b) By fixing $(\eta, w_0, \alpha_A/d) = (1, 1, 1)$, α_G has no influence on $p(\varepsilon)$. (c) We show that variations of w_0 yield a $p(\varepsilon)$ which remains invariant when expressed in terms of ε/w_0. (d) We show that for variations of η the curves of $p(\varepsilon)$ versus $\varepsilon\beta_q(\eta)$ collapse once again. For simplicity, the values of the fixed variables were set equal to unity, but the results remain independent from this choice. The numerical precision of all the collapses is verified to be quite impressive. Very tinny discrepancies might be due to the fact that both N and the number of realizations are finite, and/or to high-order metric-topological terms. The simulations were averaged over 10^3 realizations for $N = 10^5$. From [1159]

The distribution $p_q(\varepsilon)$ represents the generalization, within nonextensive statistical mechanics, of the BG energy weight with ε, β_q and Z_q playing respectively the roles of energy, inverse temperature, and normalization factor (see Fig. 7.159).

For all $(d, \alpha_A, \alpha_G, w_0, \eta)$, we heuristically found that

$$q = \begin{cases} \frac{4}{3} & \text{if } 0 \leq \frac{\alpha_A}{d} \leq 1 \\ \frac{1}{3} e^{1-\alpha_A/d} + 1 & \text{if } \frac{\alpha_A}{d} > 1 \end{cases}, \tag{7.66}$$

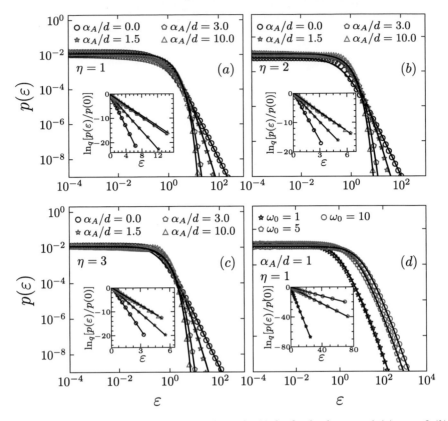

Fig. 7.159 In these plots we show the variations of $p(\varepsilon)$ for fixed values $\eta = 1$ (*a*) , $\eta = 2$ (*b*) and $\eta = 3$ (*c*), for $\alpha_A/d = 0, 1.5, 3, 10$. In (*d*) we show the variations of $p(\varepsilon)$ for fixed values of $(\alpha_A/d, \eta) = (1, 1)$ and $w_0 = 1, 5, 10$. In all figures, the black continuous lines are given by Eq. (7.65) with (q, β_q) given by Eqs. (7.66), (7.67) respectively. *Insets:* $\ln_q$-linear representation of the same data; the slopes of the straight lines precisely yield the corresponding values of $(-\beta_q)$. The simulations were averaged over 10^3 realizations for $N = 10^5$. From [1159]

$$\beta_q = \begin{cases} \beta_{q_0} & \text{if } \leq \frac{\alpha_A}{d} \leq 1 \\ (\beta_{q_0} - \beta_{q_\infty}) \, e^{2(1-\alpha_A/d)} + \beta_{q_\infty} & \text{if } \frac{\alpha_A}{d} > 1 \end{cases}, \tag{7.67}$$

with $\beta_{q_0} \simeq (-10.81e^{-1.36\eta} + 6.04)/w_0$ and $\beta_{q_\infty} \simeq (-4.81e^{-1.22\eta} + 2.56)/w_0$. As can be seen, q does not depend on (η, w_0), but only on the scaled variable α_A/d. In contrast, β_q is less universal and depends on all three parameters $(w_0, \eta, \alpha_A/d)$. Strong numerical evidence exists [1160] which indicates that, for $\eta = 1$, $\forall(\alpha_G, w_0)$,

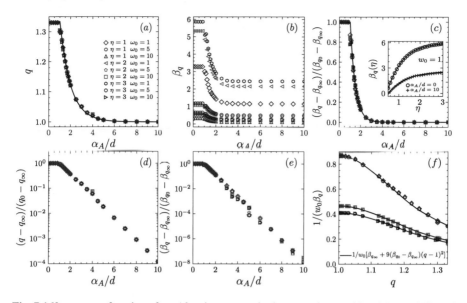

Fig. 7.160 **a** q as a function of α_A/d; q is constant in the range $0 \leq \alpha_A/d \leq 1$ ($q_0 = 4/3$) and decreases exponentially with α_A/d for $\alpha_A/d > 1$, down to $q_\infty = 1$ (black solid line). **b** β_q as a function of α_A/d for $\eta = 1, 2, 3$ and $w_0 = 1, 5, 10$, for typical values of α_A/d; β_q increases with η and decreases with w_0 and α_A/d. **c** By plotting $(\beta_q - \beta_{q_\infty})/(\beta_{q_0} - \beta_{q_\infty})$, all curves collapse and exponentially decrease with $\alpha_A/d > 1$ (black straight line). *Inset:* β_{q_0} and β_{q_∞} exponentially vary with η; β_{q_∞} was estimated by fixing $\alpha_A/d = 10$. In (**d**) and (**e**) we present log-linear representations of the same data as in (**a**) and (**c**) respectively, exhibiting the exponential dependence of both q and β_q on α_A/d, when $\alpha_A/d \geq 1$. **f** By eliminating the variable α_A/d, we show $1/(w_0\beta_q)$ as functions of q for the same set of data shown in the previous plots; q appears to be related with $1/(w_0\beta_q)$ through the equation $1/w_0[\beta_{q_\infty} + 9(\beta_{q_0} - \beta_{q_\infty})(q-1)^2]$ that is valid for all values of (w_0, η). From [1159]

$\alpha_A/d \in [0, 1]$, and $N \to \infty$, the distribution is *exactly* given by $p(\varepsilon) = p(0)e_q^{-\beta_q \varepsilon}$ with $(q, \beta_q) = (4/3, 10/(3w_0))$. This yields a possible isomorphism between a Hamiltonian model (the total energy being equal to $\sum_{i=1}^N \varepsilon_i$) within nonextensive statistical mechanics and a random geometry growth model, similarly to what happens between Hamiltonian models within BG statistical mechanics [for the $\lambda \to 1$ (Kasteleyn–Fortuin theorem) and $\lambda \to 0$ limits of the λ-state Potts ferromagnet, and for the $n \to 0$ limit of the n-vector ferromagnet (de Gennes isomorphism)] and random geometry systems (bond percolation, random resistor network, and self-avoiding walk in linear polymer physics respectively) [1160].

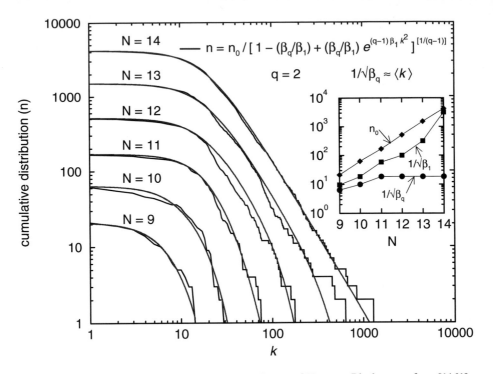

Fig. 7.161 Degree distribution for Lennard-Jones clusters of N atoms. Black curves from [1161] (see also [1162]). Red curves: fittings with the function indicated in the figure. *Inset*: N-dependence of the parameters of the fitting function

7.8.5 Lennard-Jones Cluster

Lennard-Jones small clusters (with N up to 14) have been numerically studied [1161, 1162]. The distributions of the number of local minima of the potential energy with k neighboring saddle-points in the configurational phase space can be quite well fitted with q-exponentials with $q = 2$. No explanation is still available for this suggestive fact. Qualitatively speaking, however, the fact that we are talking of very *small* clusters makes that, despite the fact that the Lennard-Jones interaction is not a long-range one thermodynamically speaking (since $\alpha/d = 6/3 > 1$), all the atoms sensibly "see" each other, therefore fulfilling roughly a nonextensive scenario. See Fig. 7.161.

7.9 Linguistics

In any written or oral text, there are words that appear more frequently than others, i.e., that are characterized by a high *ranking* (or *score*) $s = 1, 2, 3, \ldots$. The large values of s, correspond to small values for their frequency $f(s)$, which depends in principle on the language, author, particular corpus of texts. However, it has been empirically verified that, in very many instances, $f(s) \propto 1/s$ roughly, which is usually referred to as Zipf's law [1163]. Previous empirical studies by J. B. Estoup and by F. Auerbach are available in the literature, but Zipf has the merit of having attempted to derive it from the so-called *Principle of Least Effort*. B. B. Mandelbrot refined this law, which is sometimes referred to as the Zipf–Mandelbrot law. It turns out that Zipf's law has interesting connections with q-statistics as illustrated in Figs. 7.162, 7.163 and 7.164 from [1164]. Let us notice that $f(s)$ can be equivalently expressed as a

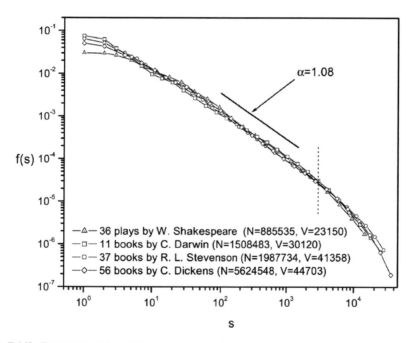

Fig. 7.162 Frequency-rank distribution of words for four large text samples. In order to reveal individual variations these corpora are built with literary works of four different authors respectively. From [1164]

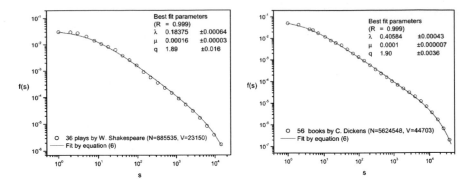

Fig. 7.163 Frequency-rank distributions for corpus of Shakespeare and Dickens. The solid lines are fittings using Eq. (6.2). From [1164]

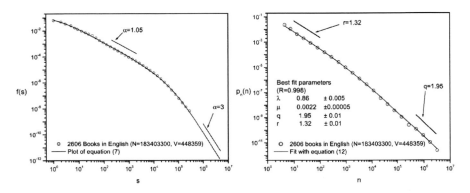

Fig. 7.164 Data from a corpus of 2606 books in English: frequency-rank distribution (*left*), and probability density function (*right*). The solid lines are fittings using crossover statistics: see details in [1164]

probability distribution $p(n)$ as illustrated in Fig. 7.164(right), where large values of n correspond to the most frequent words. Let us finally mention that the nature of Zipf's law appears to be similar to the so-called Benford's law [1165] for the frequency of digits in vast classes of numbers.

7.10 Other Sciences

Citations

The statistical analysis of the citations of scientific papers has become possible thanks to internet research tools like those implemented by ISI-Web of Science and Scopus/Elsevier. Some of these analyses [1166–1168, 1170, 1171] exhibit connections

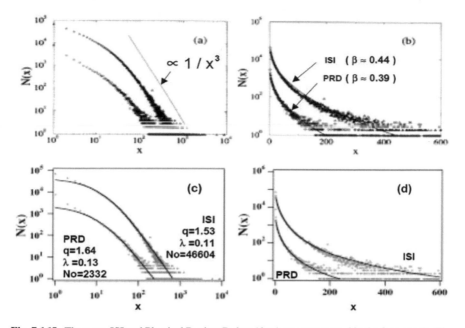

Fig. 7.165 The same ISI and Physical Review D data (dots) are represented in the four panels. The two upper panels have been fitted with stretched exponentials [1169], whereas the two lower ones have been fitted (with improved success) with q-exponentials. From [1166]

with nonextensive statistics. Illustrative results are shown in Figs. 7.165, 7.166, 7.167 and 7.168. A summary of this remarkable connection, including its possible explanation, is given in [1172]: see Fig. 7.169.

Transportation and navigation

The train delays of the British railway network have been relatively well fitted by q-exponential forms [1173]: see Fig. 7.170.

Similarly, flight delays (at arrival) have been studied as well [1174]: see Fig. 7.171. The fittings have been done with

$$
p(t) = \begin{cases} A_e\, e^{-\lambda \sqrt{C+(t-t_{peak})^2}} & \text{if } t < t_{peak}\,, \\[2mm] A_q\, e_q^{-\lambda_q \sqrt{C+(t-t_{peak})^2}} & \text{if } t \geq t_{peak}\,. \end{cases} \tag{7.68}
$$

with $A\,e^{-\lambda\sqrt{C}} = A_q\,e^{-\lambda_q\sqrt{C}}$ and $\int_{t_0}^{\infty} dt\, p(t) = 1$ (t_0 nearly $-\infty$ in practice). This ansatz can be easily generalized by adopting $e_{q_{left}}^{-\lambda_{q_{left}}\sqrt{C+(t-t_{peak})^2}}$ for $t < t_{peak}$, and

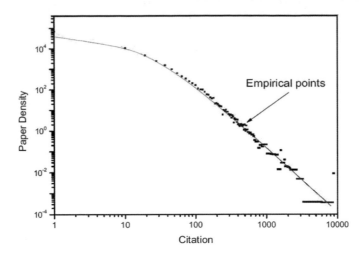

Fig. 7.166 Publication density (publications per citation) versus citation using 783339 papers from the ISI data base. The continuous line is a fitting based on nonextensive-statistical-mechanical analytical expressions. See details in [1167]

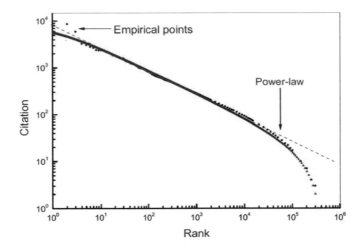

Fig. 7.167 Zipf plot (number of citations of the nth ranked paper) using the ISI data base. The continuous line is a fitting based on nonextensive-statistical-mechanical analytical expressions. The dashed line represents a power-law. See details in [1167]

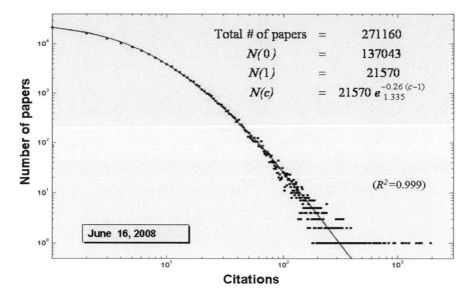

Fig. 7.168 ISI citations of all papers ($N(c)$ is the number of papers that have been cited c times) involving at least one Brazilian institution (more precisely, having the word "Brazil" in the field "Address"), from 1945 on. From [1171]

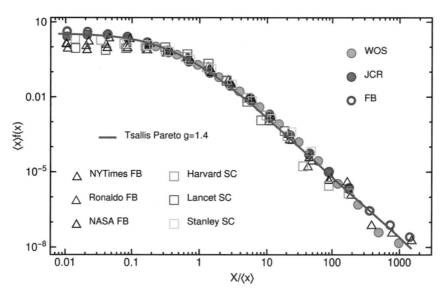

Fig. 7.169 Rescaled distribution $f(x)$ for one paper or Facebook post to have x scientific citations (SC) or shares (FB) for diversified datasets, $\langle x \rangle$ being its first moment. The entire data set can be fitted with the *universal* q-exponential $f(x)\langle x \rangle = \frac{2-q}{3} e_q^{-x/3\langle x \rangle}$ with $q \equiv \frac{2+g}{1+g} = \frac{17}{12} \simeq 1.4$. From [1172]

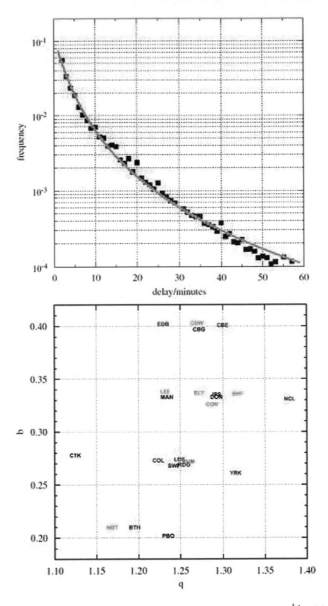

Fig. 7.170 *Top*: All train data and best-fit q-exponential $frequency = c\,e_q^{-bt}$, with $q = 1.355 \pm 8.8 \times 10^{-5}$ and $b = 0.524 \pm 2.5 \times 10^{-8}$ (c is a normalization factor). *Bottom*: The estimated pairs (q, b) for 23 stations. From [1173]. Notice that we have here a *cloud* of points (and *not* a *curve*), kind of similarly to what was obtained in [1112] for the *internet-quakes*

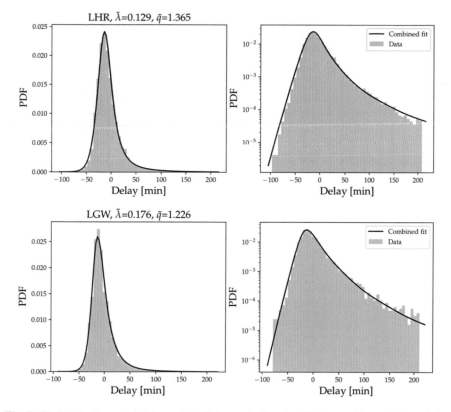

Fig. 7.171 PDF's for arrival delays of air flights at the London Heathrow Airport (top) and the London Gatwick Airport (bottom), in linear scales (left) and logarithmic scales (right). From [1174]

$e_{q_{right}}^{-\lambda_{q_{right}} \sqrt{C+(t-t_{peak})^2}}$ for $t \geq t_{peak}$. Such generalization could conveniently describe epidemiological peaks such as the COVID-19 ones in their first waves.

Autonomous navigation

Drones are used to handle many useful tasks, such as spreading seeds and fungicides, observing the Usu volcano activity and the Fukushima nuclear disaster, to mention but a few. Its safe and precise navigation is a delicate issue. See [1175] and Fig. 7.172 for details.

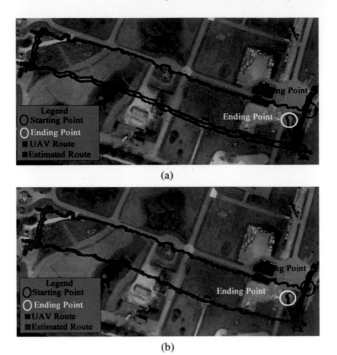

Fig. 7.172 Drone true (red) and estimated (blue) trajectories with: (a) estimation by Visual odometry (VO), (b) estimation by Nonextensive particle-filter (NExt-PF) with $q = 2.57$. The averages errors are 6.7755 % (Visual odometry), 4.4458 % (Computer vision), 3.5313 % (NExt-PF with $q = 1$), and 2.8339 % (NExt-PF with $q = 2.57$). From [1175]

Air pollution

Air contamination is a general and undesirable phenomenon, in particular the urban air pollution. Among the many particles that pollute the air of towns we have NO, NO_2, CO, and CO_2 molecules, and other hazardous chemicals. Typical measurements of NO pollution are presented in Fig. 7.173 from [1176], which are shown to be well fitted with q-exponentials. A possible explanation of this fact remains elusive at the present stage.

At this point we may mention that the entropy S_q has been used in ecology as well, more specifically to measure ecological diversity and species rarity [1118].

Social sciences

Many social phenomena have been addressed on grounds related to q-statistics, such as urban agglomerations [1177], circulation of magazines and newspapers [1178], football dynamics[1179], among others. Some typical results are shown in Figs.

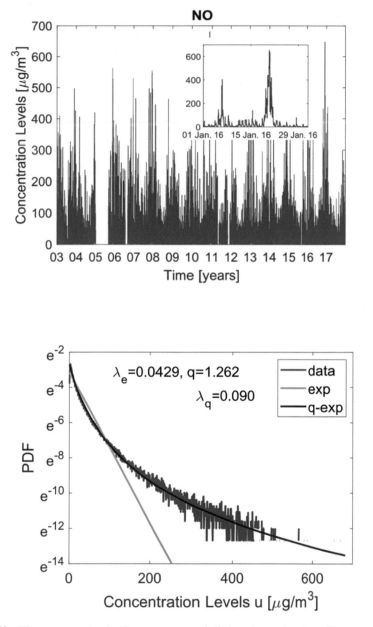

Fig. 7.173 The concentration levels vary on several distinct timescales. *Top*: Concentration of NO, recorded at the Sir John Cass School location (London), measured in 15 min intervals. The inset reveals large variations within 1 month. *Bottom*: Pollution statistics is not exponentially but rather q-exponentially distributed. We display a typical histogram of NO data, together with the best exponential fit (green) and a q-exponential fit with $q = 1.262$ (red). From [1176]

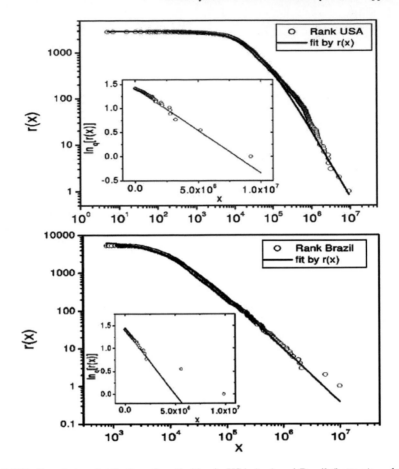

Fig. 7.174 Cumulative distributions for all cities in USA (top) and Brazil (bottom); x denotes the number of inhabitants. The solid lines are q-exponential fittings. Curiously enough, for both countries it has been found the same value for q, namely $q = 1.7$. From [1177]

7.174 and 7.175. In the context of other sciences as well, nonextensive concepts have been evoked, such as musicology [1180] and cognitive sciences [1181–1184].

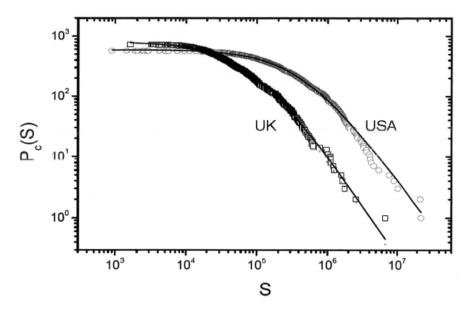

Fig. 7.175 Cumulative distributions for 570 USA magazines and 727 UK magazines in 2004, S denotes the circulation of the magazine. The solid lines are q-exponential fittings with $q = 1.66$ for USA and $q = 1.60$ for UK. From [1178]

Part IV
Last (But Not Least)

Chapter 8
Final Comments and Perspectives

I think it is safe to say that no one understands Quantum Mechanics

Richard Feynman

8.1 Falsifiable Predictions and Conjectures, and Their Verification

According to the deep epistemological observations of Karl Raimund Popper, a scientific theory cannot be considered as such if it is not capable of providing *falsifiable predictions*. This is to say predictions that *can* in principle be checked to be true or false. And a successful theory is of course that one which accumulates predictions that have been *verified to be correct*, and whose basic hypothesis has *not been proved to be violated* within the restricted domain of conditions for which the theory is thought to be applicable.

It is needless to say that nonextensive statistical mechanics, like its Boltzmann–Gibbs particular case, must naturally comply with this basic epistemological requirement. Although several such illustrations have already been presented in the body of this book, let us briefly and systematically list here some of the falsifiable predictions or conjectures of the theory, as well as their verification in recent years. This list is not exhaustive: for simplicity, I restrict in what follows to those examples in which I have been personally involved.

8.1.1 The Scaling Relation $\gamma = \frac{2}{3-q}$

Within the context of the nonlinear Fokker–Planck equation in the absence of external forces, and its exact q-Gaussian solution for all space-time (x, t), it was analytically

© Springer Nature Switzerland AG 2023
C. Tsallis, *Introduction to Nonextensive Statistical Mechanics*,
https://doi.org/10.1007/978-3-030-79569-6_8

proved in 1996 [404] that x^2 scales like t^γ with (Eq. (4.14)) $\gamma = \frac{2}{3-q}$ (hence, for instance, if $\langle x^2 \rangle$ is finite, it must be $\langle x^2 \rangle \propto t^{\frac{2}{3-q}}$). Through the perception of the crucial role that this equation plays in the complex systems addressed by nonextensive statistical mechanics, the wide applicability of this scaling relation was conjectured, and also illustrated, in 2004 [1185]. Many verifications are available at the present date, namely in

- the experiments with *Hydra viridissima* reported in 2001 [1113] (the measured value $q = 1.5 \pm 0.05$ implies, through Eq. (4.14), $\gamma = 1.33 \pm 0.05$, which is consistent with the measured value $\gamma = 1.24 \pm 0.1$; see Figs. 7.131 and 7.132);
- the experiments in defect turbulence reported in 2004 [700] (the measured value $q \simeq 1.5$ implies, through Eq. (4.14), $\gamma \simeq 1.33$, which is consistent with the measured value $\gamma = 1.16 - 1.50$; see Figs. 7.24, 7.25 and 7.26);
- the molecular dynamical simulations for the long-range classical inertial α-XY ferromagnet reported in 2004/2005 ($\gamma (3 - q)/2 = 1.0 \pm 0.1$; see also [562, 1186], and Figs. 5.66 and 5.67);
- the computational simulations for silo drainage reported in 2007 [659, 660]($q \simeq 3/2$ and $\gamma \simeq 4/3$; see Figs. 7.4 and 7.5);
- the experiments with dusty plasma reported in 2008 [730] ($\bar{\gamma} (3 - q)/2 = 1.00 \pm 0.016$, where $\bar{\gamma}$ is an averaged value; see Figs. 7.38, 7.39, 7.40 and 7.41);
- the experiments with granular matter [629], where it was measured, along a wide range of the experimental parameter, $\gamma (3 - q)/2 = 1.00 \pm 0.02$; see Fig. 7.9;

8.1.2   q-Gaussian Distributions of Velocities

- Lutz predicted in 2003 [655] that the distribution of velocities of cold atoms in dissipative optical lattices should be q-Gaussian with $q = 1 + \frac{44 E_R}{U_0}$ (Eq. (7.1)). The prediction was validated in 2006 [631] through quantum Monte Carlo calculations, as well as through experiments with Cs atoms: see Fig. 7.1. The Monte Carlo calculations neatly confirmed both the q-Gaussian shape of the distribution (with a correlation factor $R^2 = 0.995$), and Lutz formula (Eq. (7.1) within the range $50 \leq U_0/E_R \leq 240$. The laboratory experiments provided a laser-frequency dependence of q qualitatively the same as Lutz formula; the quantitative check would have demanded the direct measure of E_R and of U_0, which was out of the scope of the experiment. In what concerns the form of the distribution, the experiments verified the predicted q-Gaussian shape with $R^2 = 0.9985$, and obtained (in the illustration that is presented in [631]) $q = 1.38 \pm 0.12$ from the body of the distribution, and the consistent value $q = 1.396 \pm 0.005$ from the tail of the distribution.
- The circumferentially averaged probability distribution of cell velocities in the Madin Darby canine kidney multicellular monolayer is well fitted by a q-Gaussian distribution with $q = 1.21$ [1116].

- The rotation periods of asteroids follow a q-Gaussian with $q = 2.6$ regardless of taxonomy, diameter, or region of the solar system of the object [813].

8.1.3 Generalized Central Limit Theorem Leading to Stable q-Gaussian Distributions

The possible generalization of the standard and the Levy-Gnedenko Central Limit Theorems (CLT) was suggested in 2000 [424], and was then formally conjectured in 2004 [271]. Its proof, on grounds of *sufficient but not necessary* conditions, started in 2006 and was finally published in 2008 [477] (see also [240, 492, 500, 501]).

- The CLT attractor at the integrable case ($K = 0$) of the standard map has been recently proved to be a Cauchy-Lorentz distribution (q-Gaussian with $q = 2$) [516].
- The CLT attractor of the logistic map at its Feigenbaum point appears to be a q-Gaussian with $q \simeq 1.65 \pm 0.05$ [532].

8.1.4 Existence of q, λ_q and K_q, and the Identity $K_q = \lambda_q$

It was argued in 1997 [186] that, whenever the Lyapunov exponent λ_1 vanishes, (the upper bound of the) the sensitivity is given by $\xi = e_{q_{sen}}^{\lambda_{q_{sen}} t}$, which determines a special value of q, noted q_{sen}. It was further argued that, at the edge of chaos of systems having not more than one Lyapunov exponent, $S_q(t)$ would increase *linearly* with t only for $q = q_{sen}$, and that the *slope* (entropy production per unit time) would satisfy $K_{q_{sen}} = \lambda_{q_{sen}}$, thus q-generalizing the $q = 1$ Pesin-like identity $K_1 = \lambda_1$. This scenario was numerically verified in various systems since 1997 and, in some cases, analytically proved since 2002: see [187–201, 333], among others. An unifying overview of the three traditional routes to chaos (doubling-period, tangent, and quasi-periodic) and their deep connections with nonextensive statistical mechanics and the nonadditive entropy S_q is given in [1190].

8.1.5 Scaling with N^* for Long-Range-Interacting Systems

An important class of two-body potentials $V(r)$ in d dimensions consists in being smooth or integrable at short distances, and satisfying $V(r) \sim -\frac{A}{r^\alpha}$ ($A > 0$; $\alpha \geq 0$) at long distances. If the system is classical, such potentials are said *short-range-interacting* if $\alpha/d > 1$, and *long-range-interacting* if $0 \leq \alpha/d \leq 1$ (see, for instance,

Eq. (3.69)). The usual thermodynamical recipes address the short-range cases. Special scalings must be used in the long-range cases.

Since the successful verification done in 1995 [226] for ferro-fluids, it became natural to conjecture that, in order to have *finite* equations of states in the $N \to \infty$ limit, it was necessary to divide by N^* (defined in Eq. (3.69)) quantities such as temperature, pressure, external magnetic field, chemical potential, etc., by N quantities such as volume, magnetization, entropy, number of particles, etc., and by NN^* quantities such as the internal energy and all the thermodynamical potentials. These prescriptions were verified since 1996 in many kinds of systems, such as Lennard-Jones-like fluids [227, 1188–1190], magnets [219, 221, 228–230, 1191–1193], anomalous diffusion [231], and percolation [232, 233].

8.1.6   *Vanishing Lyapunov Spectrum for Classical Long-Range-Interacting Many-Body Hamiltonian Systems*

It was first realized in 1997 [186] that the q-exponential functions emerge when the maximal Lyapunov exponent vanishes (see Sect. 8.1.4 here above). It then became natural to conjecture that, in any anomalous stationary (or quasi-stationary) state, the Lyapunov spectrum should exhibit a generic tendency to approach zero at the $N \to \infty$ limit for classical long-range-interacting Hamiltonian systems (whereas it is of course expected to be positive for short-range-interacting Hamiltonians). This was indeed verified, first in 1998 [219] for the α-XY ferromagnet (see Fig. 5.52 and 5.53), and since then in many other systems [558, 559, 568, 582] (see Figs. 5.54, 5.55, 5.56 and 5.57). In all these cases, there are numerical indications that, in the $N \gg 1$ limit, the Lyapunov spectrum tends to vanish for $0 \le \alpha/d \le 1$, and tends to be nonzero for $\alpha/d > 1$.

8.1.7   *Nonuniform Convergence for Long-Range Hamiltonians Associated with the $(N, t) \to (\infty, \infty)$ Limits*

It was conjectured in 1999 (see Fig. 4 in [71]) that classical long-range-interacting many-body systems could evolve, before attaining thermal equilibrium, through one (or more) nonequilibrium long-standing states. The departure from the long-standing states toward equilibration would occur (slowly, as indicated in that Fig. 4, along something like a logarithmic scale for time) at $t_{crossover}(N)$. Furthermore, it was conjectured that $\lim_{N \to \infty} t_{crossover}(N) = \infty$. Since 1999, this scenario or parts of it were verified many times: see [61, 557–563, 565, 573, 582, 1186, 1212], among others.

In recent years, the picture has been further clarified [482]. In the long-range regime of the Fermi-Pasta-Ulam-Tsingou problem, for fixed N, increasing t makes the system to crossover from $q \neq 1$ to $q = 1$, whereas, for fixed t, increasing N makes the system to crossover from $q = 1$ to $q \neq 1$. Moreover, if $(N, t) \to (\infty, \infty)$ simultaneously, the final state appears to be $q = 1$ ($q \neq 1$) for $N/t^{\gamma} \to 0$ ($N/t^{\gamma} \to \infty$) with $\gamma > 1$. See also [585, 1194–1196].

8.1.8   q-Duplets, q-Triplets and q-Nplets

In Sect. 5.4, several q-duplets and q-triplets, selected among the many available in the literature, have been listed in Table 5.3.[1] They have been discussed within the Moebius context of [479], which constitutes an important step. However, some crucial issues still remain elusive. For example, to what precise physical property (e.g., generic-order momenta of measurable quantities), each of those q-indices is associated, and how many of them are independent. The full understanding of such issues would clarify relevant aspects of the whole theory of nonadditive entropies and nonextensive statistical mechanics, as well as the deep reasons for the validity and the failure of BG statistical mechanics.

8.1.9   Degree Distributions of the q-Exponential Type for Scale-Free Networks

The so-called scale-free networks (which are in fact only asymptotically scale-free) exhibit very frequently a degree distribution of the form $k^{\delta} e_q^{-k/k_0}$ ($q > 1$; $k_0 > 0$, with an exponent δ than can be either zero, or positive or negative. This was first noticed in 2004 [1201] with $\delta = 0$. The scale-invariance being a basic ingredient of nonextensive statistics (in particular in relation to the q-CLT), it was kind of natural to expect that this q-exponential degree distribution would be rather frequent in networks. Indeed, it has been so verified since 2005 in many d-dimensional models [62, 634, 1152, 1202–1204]. However, it is yet elusive what motivates δ to be zero, or nonzero. Recent years have brought better understanding about the connection between q-statistics and random geometrical probabilistic models, very particularly in the realm of weighted links [1158, 1159].

---

[1] At least one example of q-quadruplet already exists [1198], namely $(Q, q, q_r, q_{entropy})$ with $(Q, q, q_r) \neq (1, 1, 1)$ and $q_{entropy} = 1$, where (Q, q) play roles analogous to $(q_{stationary}, q_{relaxation})$, respectively.

8.2 Frequently Asked Questions

As the history of sciences profusely shows to us, every possible substantial progress
in the foundations of any science is unavoidably accompanied by doubts and controversies (see, for instance, [1200]). This is a common and convenient mechanism for
new ideas to be checked and better understood by the scientific community. Clearly,
objections and critiques have frequently helped the progress of science. There is
absolutely no reason for expecting that statistical mechanics, and more specifically
nonextensive statistical mechanics, would be out of such a process. Quite on the
contrary—remember the words of Nicolis and Daems [4] that were cited in the
Preface!—given the undeniable fact that entropy is one among the most subtle and
rich concepts in physics. Some frequently asked points are addressed here. Indeed,
we believe that some space dedicated here to such issues might well be useful at this
stage (see also [1201]).

   This section only includes frequently asked questions, or critiques, the (basic)
answer of which is believed to be known. Questions, frequent or not, whose answer
is still a matter of research have been considered instead as "open questions", and as
such have been included in Sect. 8.3.

(a) *Finally, the entropy S_q is extensive or nonextensive?*

As it stands, this question is not posed in a completely satisfactory manner. What
can be straightforwardly answered is whether an entropic functional such as S_q is
additive or not, the correct answer being that S_1 is *additive*, whereas S_q for $q \neq 1$
is *nonadditive*. Extensivity is a more complex question. Indeed, the answer depends
not only on the entropic functional but *also* on the system (more precisely, on the
nature of the correlations between the elements of the system) to which it is applied.
If the elements have no correlation at all, or only local correlations, then typically
S_1 is *extensive* in the thermodynamical sense, whereas S_q for $q \neq 1$ is *not*. But if
the correlations are nonlocal, then it can happen (e.g., the quantum magnetic chain
analytically discussed in [284]) that S_q is *nonextensive* for all values of q (including
$q = 1$), *excepting a special value of $q \neq 1$ for which S_q is extensive.*

(b) *If the entropic index q is chosen such that the entropy S_q is extensive, why this
   theory is named "nonextensive statistical mechanics"?*

This mismatch has its historical roots on the fact that, during over one century of
BG statistical mechanics, the entropy S_{BG}, known to be *additive*, was also *extensive*
for all those systems (known today as *extensive systems*) for which the BG theory
is plainly valid. This led imperceptibly to the abusive use of the words *additive* and
extensive as practically synonyms. Later on, starting with the 1988 paper [47], the
distinctive *nonadditivity* property (Eq. (3.21)) was wrongly, but rather frequently,
referred to as the *nonextensivity property*. The expression *nonextensive statistical
mechanics* was coined from there. When, many years later (see, for instance, the end
of the Introduction of Chap. 1 in [18]), this matter became gradually clear, the idea

of course emerged to rather rename this theory as *nonadditive statistical mechanics*. But, on the other hand, the expression *nonextensive statistical mechanics* has already been used in thousands of papers. Moreover, statistical mechanics involves not only *entropy* but also *energy*. And the typical Hamiltonian systems for which the present theory was devised are those involving long-range two-body interactions, for which the total energy is definitively *nonextensive*. The expression *nonextensive statistical mechanics* was therefore maintained. Nowadays, many authors call *nonextensive systems* those whose nonequilibrium stationary-state distribution (or similar properties, such as relaxation functions, and sensitivity to the initial conditions) is of the q-exponential form (or other similarly deformed exponential forms), in contrast with *extensive systems*, which are therefore those whose stationary-state (thermal equilibrium) distribution (or similar properties) is of the usual BG exponential form. Summarizing, in extensive (BG) statistical mechanics, both the total energy and the total entropy are extensive, whereas, in nonextensive statistical mechanics, the total energy typically is nonextensive but the total entropy remains extensive (thanks to the use of a nonadditive entropic functional)! Regretfully, there was an inadvertence when the book [18] was published with the title "Nonextensive Entropy" instead of the correct title "Nonadditive Entropy"!

(c) *How come ordinary differential equations play an important role in nonextensive statistical mechanics?*

Some remarks related to ordinary differential equations might surprise some readers, hence the matter deserves a clarification. Indeed, in virtually all the textbooks of statistical mechanics, functions such as the energy distribution at thermal equilibrium are discussed *using a variational principle*, namely referring to the entropy functional, and *not using ordinary differential equations* and their solutions. In our opinion, it is so *not* because of some basic (and unknown) principle of exclusivity, but rather because the first-principle dynamical origin of the BG factor still remains, mathematically speaking, at the status of a *dogma* [42]. Indeed, as already mentioned, to the best of our knowledge, no theorem yet exists which establishes the necessary and sufficient *first-principles* conditions (at the level of Hamiltonians, for instance) for being legitimate the use of the celebrated BG factor. Nevertheless, one must not forget that it was precisely through a differential equation that Planck heuristically found, as described in his famous October 1900 paper [371–373, 1205, 1206],[2] the black-body radiation law. It was only in his equally famous December 1900 paper that

---

[2] The celebrated equation in Planck's 19 October 1900 paper is $-(\frac{\partial^2 S}{\partial U^2})^{-1} = \alpha U + \beta U^2$ (where α and β are constants), the heuristic interpolation between a term proportional to U and one proportional to U^2. By replacing in this equation the thermodynamic relation $\frac{\partial S}{\partial U} = T^{-1}$, one obtains $\frac{\partial U}{\partial(1/T)} = -\alpha U - \beta U^2$, which is precisely the $q = 2$ particular case of the differential equation (6.1)! From the solution of this equation (see Eq. (6.2)), Planck readily arrived to his famous black-body radiation law $u(\nu, T) = (a\nu^3/c^3)/(e^{b\nu/T} - 1)$. Two months later, in his 14 December 1900 paper, by incorporating a discretized energy within Boltzmann's thermostatistical theory, he obtained the form which is used nowadays, namely $u(\nu, T) = (8\pi\nu^2/c^3)h\nu/(e^{h\nu/kT} - 1)$ (where b was replaced by h/k). The constant k (introduced and named *Boltzmann constant* by Planck himself) was the ratio between the gas constant R and the Avogadro number $\mathcal{N}$; the constant h was

he made the junction with the—at the time, quite controversial— Boltzmann factor by assuming the—at the time, totally bizarre—hypothesis of discretized energies.

A further point which deserves clarification is *why* have we *also* interpreted the linear ordinary differential equation in Sect. I.A as providing the typical time evolution of both the sensitivity to the initial conditions and the relaxation of relevant quantities. Although the bridging was initiated by Krylov [1207], the situation still is far from completely clear on mathematical grounds. However, intuitively speaking, it seems quite natural to think that the sensitivity to the initial conditions is precisely what makes the system to relax to equilibrium, and therefore opens the door for the BG factor to be valid. In any case, although some of the statements in Sect. 5.4.4 are (yet) not proved, this by no means implies that they are generically false. Furthermore, they provide what we believe to be a powerful metaphor for generalizing the whole scheme into the nonlinear ordinary differential equations discussed in Sect. 3.1 (see also [1208, 1209]). Interestingly enough, the q-exponential functions thus obtained have indeed proved, in most of the cases, to be the correct answers for a sensible variety of specific situations reviewed in the present book. This is so for *all three* basic interpretations—the q-triplet—as energy distribution for the stationary state, time evolution of the sensitivity to the initial conditions, and time evolution of basic relaxation functions. Further details, as well as some exceptions, can be found in [479].

(d) *Do we really need to generalize BG statistical mechanics to handle many-body long-range-interacting Hamiltonians? Is it not possible to do that just within the current BG theory?*

This surely is a very interesting and central issue. But we absolutely need to clearly distinguish two quite different questions that are here involved. The first question concerns the mathematical validity (either easily tractable or not) of a BG-like calculation. The second one—*which is the really relevant one since it focuses on the core of the BG theory!*—concerns whether a BG-like calculation is able to correctly predict the appropriately averaged results obtained from first-principle dynamics (typically Newton's law for a classical system, Schroedinger equation for a quantum system).[3] When the interactions are short-ranged (typically integrable two-body potential), these two different questions seem to collapse as if they were only one. For such systems, the $t \to \infty$ and the $N \to \infty$ limits are typically commutable, which eventually determines the validity of the celebrated usual statistical mechanics. The situation is by far more subtle for long-ranged interactions (i.e., basically nonintegrable two-body potential). For such systems, the results typically depend, among others (e.g., initial conditions, boundary conditions), on the relative speed at which t

---

obtained by fitting the black-body experimental data available at the time. Historically, it is still not completely clear why Planck called it h.

[3] As an elementary metaphor, we may think of an iron ball. If the interior of the ball is fully constituted by iron we can correctly calculate the weight of the ball by knowing its volume and the iron density. However, if the interior of the ball happens to be partly empty and we are unaware of that, the same, perfectly tractable, calculation will now provide a wrong answer for the weight.

and N approach infinity (e.g., on whether N/t^γ remains above or below some special value, γ being some model-dependent positive value). This is of course related to the observation by Gibbs himself, that a possible divergence of the partition function, which precisely is what happens for nonintegrable potentials, makes the BG theory inapplicable, *illusory* in his own words (see Sect. 1.3).

Kac-like potentials and similar techniques have been used, along the years, to mathematically approach nonintegrable potentials within the BG recipe. An illustrative such approach can be seen in [1210] for a d-dimensional classical fluid with two-body interactions exhibiting a hard core as well as an attractive potential proportional to $r^{-\alpha}$ with $0 \leq \alpha/d < 1$ (logarithmic dependence for $\alpha/d = 1$). A Kac-like long-distance cutoff R is introduced in the calculation such that no interactions exist for $r > R$, and eventually R is allowed to approach infinity. It is shown that an analytical solution within Boltzmann–Gibbs statistical mechanics is possible and that—no surprise—it exhibits a mean field criticality. Moreover, it is therein argued that very similar considerations hold for lattice gases, $O(n)$ and Potts models.

The technique used in [1210] preserves of course the BG exponential distribution of energies, as well as the Maxwellian distribution of velocities whenever the model admits a microscopic Newtonian dynamics. Truth is, however, quite different, as profusely illustrated in Sect. 5.3 and elsewhere.

Summarizing, we must always keep in mind that the mathematical tractability of some problem within a given theory by no means implies the physical correctness of the calculation. Notorious examples are the mean field approach for critical phenomena in all kinds of systems, and the classical approach of ideal fermionic or bosonic gases.

(e) *Is there any need or advantage to introduce and refer to the q-entropy in nonlinear dynamical systems, where the $q = 1$ entropy (Boltzmann–Gibbs-Shannon) is known to correctly apply to such systems?*

To address this interesting question, it is convenient to have in mind the illustrative case of the z-logistic map in Eq. (5.1). For values of a where the Lyapunov exponent λ_1 is positive, the entropy production K_{BG} is positive and equals λ_1. Consistently, if the system is initially put in some small region of its phase-space, $S_{BG}(t)$ increases, after a short transient, *linearly* with time. But if a is chosen to be at the Feigenbaum-like value $a_c(z)$, $S_{BG}(t)$ briefly increases and then saturates at a finite positive value, hence K_{BG} vanishes. There is nothing wrong about that, but in terms of description it is rather poor. Indeed, for *all values* of $z \in [1, \infty)$ it happens exactly the same, which does not allow for a distinctive characterization of the central phenomenon of loss of information. In other words, there are zeros and zeros, but $S_{BG}(t)$ does not distinguish among them. The introduction of $S_q(t)$ substantially enriches the discussion since, for a fixed value of z, there is a unique value of $q < 1$ such that K_q is positive and finite, which conveniently characterizes the speed at which S_q increases with time. The fact that the BG concepts can remain, in a certain context, well defined and correctly apply does not mean that they conveniently and advantageously apply.

A second illustration is provided by standard second-order d-dimensional critical phenomena. Both above and below T_c, BG statistical mechanics is of course valid, which enables the calculation of quantities such as the diverging specific heat, susceptibility and correlation length, and the vanishing order parameter (e.g., magnetization). The situation is more complex at *precisely* T_c, where ergodicity in the entire phase space is replaced by ergodicity in a part of phase space (say half of it). At this point, for Ising, XY and Heisenberg ferromagnets, specific heat, susceptibility and correlation length are infinite, and the magnetization is zero. But, as before, there are infinities and infinities, and zeros and zeros, and the BG theory does not distinguish among them. It is our expectation that the discussion of these quantities by introducing S_q would enable a richer identification. In fact, one such example does exist, and concerns the critical exponent δ (see Eq. (5.80)), which is defined *precisely* at T_c and not just close to it, like the traditional exponents α, γ, ν and β. The critical exponent δ (hence q) is known to be different for say the $d = 3$ Ising, XY and Heisenberg ferromagnets.

(f) *Is the zeroth principle of thermodynamics valid in nonextensive statistical mechanics?*

This important question was raised up to me for the first time in 1993 by Oscar Nassif de Mesquita [1213], and addresses whether the zeroth principle of thermodynamics and thermometry are consistent with nonextensive statistical mechanics. The answer is specific to the class of systems that we are focusing on. Two particular classes have been up to now studied in this respect, namely overdamped many-body systems, and long-range-interacting Hamiltonian systems at quasi-stationary states. The first class has been repeatedly studied analytically and numerically (see, for instance, [314, 434, 439] and references therein, as well as Sect. 4.5.3), and the validity of all the principles of classical thermodynamics, including the zeroth one, Carnot cycle (with effective temperatures which become the usual kinetic temperature in the $q = 1$ limit), H-theorem, equations of states, has been successfully verified.

In what concerns the second class, namely many-body long-range-interacting Hamiltonian systems, we may say that it is still in its infancy. Given their mathematical intractability, only some related numerical results are available. This issue was critically focused on in 2003 [1214]; the many and diverse critiques are rebutted in [1215] and elsewhere. An issue which possibly plays a crucial role concerns the discussion of "weak coupling" in such many-body conservative systems. Indeed, if we call c the coupling constant associated with long-range interactions (i.e., $0 \le \alpha/d \le 1$), we have $\lim_{N \to \infty} \lim_{c \to 0} c\tilde{N} = 0$, whereas $\lim_{c \to 0} \lim_{N \to \infty} c\tilde{N}$ diverges, with $\tilde{N} \equiv \frac{N^{1-\alpha/d}-1}{1-\alpha/d}$. No such anomaly exists for short-range interactions (i.e., $\alpha/d > 1$). Indeed, in this simpler case, we have that $\lim_{N \to \infty} \lim_{c \to 0} c\tilde{N} = \lim_{c \to 0} \lim_{N \to \infty} c\tilde{N} = 0$. The nonuniform convergence that, for long-range interactions, exists at this level is possibly related to the concomitant nonuniform convergence associated with the $t \to \infty$ and $N \to \infty$ limits previously discussed.

The strict verification of the zeroth principle of thermodynamics in many-body Hamiltonian systems demands checking the transitivity of the concept of temperature

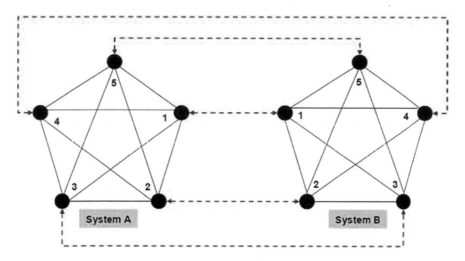

Fig. 8.1 Systems A and B that will be put in thermal contact at a certain moment by allowing the coupling constant l to become different from zero (see Eq. (8.1)). Here $N = 5$. From [1216]

through successive thermal contacts between three, initially disconnected systems, A, B, and C. Some preliminary results along these lines for the paradigmatic HMF model (which corresponds to infinitely long-range interactions) are available [1216]. As a first step, two (equal) systems, A and B, are put into thermal contact. The Hamiltonian is given by (see Fig. 8.1)

$$
\begin{aligned}
\mathcal{H} = &\sum_{i=1}^{N} \frac{(L_i^A)^2}{2} + \frac{1}{N} \sum_{i=1}^{N} \sum_{j=1}^{N} [1 - \cos(\theta_i^A - \theta_j^A)] \\
&+ \sum_{i=1}^{N} \frac{(L_i^B)^2}{2} + \frac{1}{N} \sum_{i=1}^{N} \sum_{j=1}^{N} [1 - \cos(\theta_i^B - \theta_j^B)] \\
&+ l \sum_{k=1}^{N} [1 - \cos(\theta_k^A - \theta_k^B)].
\end{aligned}
\tag{8.1}
$$

As we see, there are long-range interactions within system A and within system B, but only short-range interactions connecting systems A and B through the coupling constant l. See in Fig. 8.2 the time evolution of the temperatures of A and B. We verify that, after the thermal contact being established, the two temperatures merge into an intermediate one, as they would do if they were at thermal equilibrium... *but they are not!* Indeed, those are quasi-stationary states. Only later, the two systems go together toward another quasi-stationary state, which might or might not terminate eventually as thermal equilibrium. A discussion such as the present one, but involving three instead of two systems, remains to be done in order to possibly illustrate the validity of the zeroth principle for the present anomalous systems.

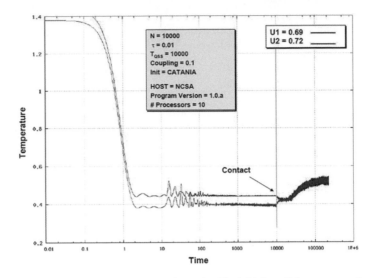

Fig. 8.2 Time evolution of the temperatures of A and B. The initial conditions are water bag for both A and B, at slightly different initial internal energies, hence slightly different initial temperatures. Here $N = 10000$, and l is taken zero until the moment indicated with a green vertical line, and $l = 0.1$ after that moment. From [1216]

The zeroth principle and its connections with nonextensive statistical mechanics has also been addressed in several other papers such as [1217–1221].

(g) *What is the nature of the quasi-stationary state in long-range-interacting Hamiltonians such as the XY one?*

A re-analysis was done by Zanette and Montemurro [1222] of the molecular-dynamics approach and results presented in [557] for the infinitely long-range interacting planar rotators already discussed here. They especially focus on the time-dependence of the temperature $T(t)$ defined as the mean kinetic energy per particle. For total energy slightly below the second-order critical point and a nonzero-measure class of initial conditions, a long-standing nonequilibrium state emerges before the system achieves the terminal BG thermal equilibrium. When T versus $\log t$ is plotted, an inflection point exists. If we call $t_{crossover}$ the value of t at which the inflection point is located, it has been repeatedly verified numerically by various authors, including in [1222], that $\lim_{N \to \infty} t_{crossover}(N)$ appears to diverge. Therefore, if the system is very large (in the limit $N \to \infty$, mathematically speaking) it remains virtually for ever in the anomalous state, currently called *quasi-stationary state* or *metastable state*. It is correctly pointed out in [1222] that, if a linear scale is used for t, the inflection point disappears.[4] For increasingly large N, $T(t)$ remains constant, and *different from the BG value*, within a quite small error bar. This effect appears in an even more

---

[4] From this, these authors conclude that this well known metastable state is but a kind of mathematical artefact, and no physically relevant quasi-stationarity exists. Such an argument is mathematically

pronounced way because of a slight minimum that $T(t)$ presents just before going up to the BG value. This intriguing minimum had already been observed in [557], and has been detected with higher precision in [1222]. Further details are presented in [1186]. It remains nevertheless a fact that the nature of this quasi-stationary state is quite unusual (with aging and other indications of glassy-like dynamics [1223–1226]), and surely deserves more extended studies. Similar remarks might as well be applicable to the inertial long-range-interacting Heisenberg ferromagnet, among others.

(h) *Are the q-exponential distributions compatible with the central limit theorems which only allow, in the thermodynamic limit, for Gaussian and Lévy distributions?*

This interesting issue has been raised in several occasions by several people. For example, soon after their previous critique, Zanette and Montemurro advanced a second one [1227] objecting the validity of nonextensive statistical mechanics for thermodynamical systems. This line of critique addresses the possibility of the ubiquity of the q-exponential form as a stable law in nature. The argument essentially goes that only Gaussians and Lévy distributions would be admissible, because of the respective central limit theorems. Such question has been preliminarily addressed long ago in [424], and once again in [423] as a rebuttal to [1227]. The answer basically reminds that the stability observed in the usual central limit theorems is intimately related to the hypothesis of independence of the random variables that are being composed. If important global correlations are present even in the $N \to \infty$ limit, different central limit theorems are applicable, as proved in [240, 477, 491, 492].

(i) *Is entropy S_q "physical"?*

Another question (or line of critique) that occasionally emerged in the past (very rarely at the present stage of development of the theory) concerns the "physicality" of S_q (see [1228]). Or whether it could exist a "physical" entropy different from S_{BG}. Since such issues appear to be of a rather discursive/philosophical nature, we prefer to put these critiques on slightly different, more objective, grounds. We prefer to ask, for instance, (i) whether S_q is useful in theoretical physics in a sense similar to that in which S_{BG} undoubtedly is useful; (ii) whether q necessarily is a

---

similar to stating that the high-to-low energies crossing occurring, at a given temperature, in Fermi–Dirac statistics would have no physical meaning! Indeed, if instead of using the linear scale for the energies we were to use a faster scale (e.g., an exponential scale), the well known inflection point would disappear. Nevertheless, there is no point to conclude from this that the textbook crossing in Fermi–Dirac statistics is but a mathematical artefact. In fact, *any* inflection point on *any* such curve will disappear by sufficiently "accelerating" the abscissa in an ad hoc manner. The "crossover" will obviously remain.

fitting parameter, or whether it can be determined a priori, as it should if the present theory is to be considered a complete one[5]; (iii) whether there is no other way of addressing the thermal physics of the anomalous systems addressed here, very specifically whether one could not do so by just using S_{BG}; (iv) whether S_q is special in some physical sense, or whether it is to be put on the same grounds as the over forty entropic functionals popular in cybernetics, control theory, engineering, and information theory generally speaking.

Such questions have received answers in [200, 1229–1234] and elsewhere. (i) The usefulness of this theory is neatly being answered by the large amount of applications it has already received, and by the ubiquity of the q-exponential form in nature. (ii) The a priori calculation of q from microscopic dynamics has been specifically illustrated in Chap. 5 (see also point (n) here below). (iii) The optimization of S_q, as well as of almost any other entropic form, with a few constraints has been shown in [1230] to be equivalent to the optimization of S_{BG} with an infinite number of appropriately chosen constraints. Therefore we could in principle restrain to the exclusive use of S_{BG}. If we followed that line, we would be doing like a hypothetical classical astronomer who, instead of using *one*, extremely convenient, Keplerian elliptic form for the planetary orbits, would (equivalently) use an *infinite* number of Ptolemaic (circular) epicycles (the Hoyle 1973 expansion of one elliptic orbit into an infinite number of circular orbits is explored in [1230]). It is obviously appreciably much simpler to characterize, whenever possible, a complex structure of constraints with a single index $q \neq 1$ (in analogy with the fact that the ellipticity of a Keplerian orbit can be simply specified by a single parameter, namely the eccentricity of the ellipse). (iv) The entropy S_q shares with S_{BG} an impressive set of important properties (see also point (k) here below), which includes, among others, concavity, extensivity, Lesche-stability, composability, trace-form, and finiteness of the entropy production per unit time, $\forall q > 0$. The difficulty of simultaneously satisfying all these properties can be illustrated by the fact that the (additive) Renyi entropy (usefully used in the geometric characterization of multifractals) satisfies, under the hypothesis of probabilistic independence or quasi-independence (and only then), extensivity $\forall q$, and only a few of the other properties for all $q > 0$. Such features point S_q as being very special,[6] although probably not unique, for thermostatistical purposes.

(j) *By adjusting the constraints under which the entropy optimization is done, one can obtain virtually any desired distribution. Is that not a serious problem?*

Soon after their second critique, Zanette and Montemurro advanced a third one [1235]. This time the objection addresses nonthermodynamical systems, in contrast to the previous critiques which mainly addressed thermodynamical ones. It is argued

---

[5] We are using here "fitting parameter" in a sense which differs from the empirical one in which the fundamental physical constants (G, c, k, h) are (possibly inevitably) determined through fittings matching experimental results with basic mathematical expressions.

[6] Let us also remind at this point that the Enciso-Tempesta theorem [236] proves that S_q is the unique entropic functional which is simultaneously trace-form, composable and containing S_{BG} as a particular case.

by these authors that nonthermodynamical applications of nonextensive statistics are ill-defined, essentially because of the fact that *any* probability distribution can be obtained from the nonadditive entropy S_q by conveniently adjusting the constraint used in the optimization. We argue here that, since it is well known to be so for *any* entropic form and, in particular, for the (additive) Boltzmann–Gibbs entropy S_{BG} (see [1236]), the critique brings absolutely no novelty to the area. In other words, it has *nothing* special to do with the entropy S_q. In defence of the usual simple constraints, typically averages of the random variable x_i or of x_i^2 (where x_i is to be identified according to the nature of the system), we argue, *and this for all entropic forms*, that they can hardly be considered as arbitrary, as the authors of [1235] seemingly consider. Indeed, once the natural variables of the system have been identified (e.g., constants of motion of the system, such as the energy for Hamiltonian systems), the variable itself and, in some occasions, its square obviously are the most basic quantities to be constrained. Such constraints are used in hundreds (perhaps thousands) of useful applications outside (and also inside) thermodynamical systems, along the information theory lines of Jaynes and Shannon, and more recently of A. Plastino and others. And this is so for S_{BG}, S_q, and any other entropic form. If, however, other quantities are constrained (e.g., an average of x^σ or of $|x|^\sigma$) for specific applications, it is clear that, at the present state of the art of information theory, *and for all entropic forms*, this must be discussed case by case. Rebuttals of this critique can be found in [1201, 1237].

As a final comment let us mention that statistical mechanics is much more than just a stationary-state (e.g., thermal equilibrium) distribution. Indeed, under exactly the same constraints, the optimization of S_{BG} and $(S_{BG})^3$ yields precisely the same distribution. This is obviously not a sufficient reason for using $(S_{BG})^3$, instead of S_{BG}, in a thermostatistical theory which must also satisfy various thermodynamical requirements, including its Legendre transformation structure.

(k) *What properties are common to S_{BG} and S_q?*

The additive entropy S_{BG} and the nonadditive entropy S_q share a huge amount of mathematical properties. These include nonnegativity, expansibility ($\forall q > 0$), optimality for equal probabilities, concavity ($\forall q > 0$), extensivity for wide classes of systems, Lesche-stability (or experimental robustness) ($\forall q > 0$), finiteness of the entropy production per unit time, existence of partition function depending only on temperature, composability, trace-form, the Topsoe factorizability property [1238] ($\forall q > 0$), the mathematical relationship of the Helmholtz free energy with the partition function is the same as the microscopic energies with their probabilities, the function (namely $\ln_q x$) which (through a standard probabilistic mean value) defines the entropy is precisely the inverse of the function (namely e_q^x) which provides the energy distribution at the stationary state, the Amari-Ohara-Matsuzoe conformally invariant geometry [122], the Biro-Barnafoldi-Van universal thermostat independence [1240]. Another property, related to generalized logarithms and generalized duality, also leads to a new uniqueness theorem [1239]. We are unaware of the existence of *any* other entropic functional form having all these properties in common with S_{BG}.

Let us stress, at this point, that a crucial property that S_{BG} and S_q do *not* share is additivity. This difference is extremely welcome. It is precisely this fact which makes it possible for *both* entropies to be *thermodynamically extensive* for a special value of q, more specifically $q = 1$ for extensive systems (i.e., those whose correlations are generically short-ranged), and $q < 1$ for nonextensive ones (i.e., a large class among those whose correlations are generically long-ranged).

(1) *Is nonextensive statistical mechanics necessary, convenient or something else?*

Let us first address a somewhat simpler question, namely: *Is Boltzmann–Gibbs statistical mechanics necessary or just convenient?* The most *microscopic* level at which collective properties of a system can be answered is that of *mechanics* (classical, quantum, or any other that might be appropriate for the case). Let us illustrate this with classical Hamiltonian systems. Let us consider a system constituted of N well defined interacting particles. Its time evolution is fully determined by the initial conditions. So, for every admissible set of initial conditions we have a point evolving along a unique trajectory in the *full phase space* Γ. We can in principle calculate all its mechanical properties, its time averages, its ensemble averages (over well defined sets of initial conditions). For example, its time-dependent "temperature" can be defined as being proportional to the total kinetic energy of N particles divided by N. If we wish to approach a more thermodynamical definition of temperature, we might wish to consider the average of this quantity over an ensemble of initial conditions, even all admissible initial conditions. This ensemble can be uniformly distributed over the entire Γ space, or be as special or particular as we wish. Of course, in practice, this road is almost always analytically intractable; moreover, it quickly becomes computationally intractable as well when N increases above some number... immensely below the Avogadro number!

Another approach, which is not so powerful but surely is more tractable (both analytically and computationally), consists in considering the projection of the Γ into the *single-particle phase space* μ, where the coordinates and momenta of only one particle are taken into account. In other words, we might be interested in discussing only those properties that are well defined in terms of the single-particle marginal probabilities. Such is the case of the *Vlasov equation* (see, for instance, [50]), and analogous approaches such as the *Boltzmann transport equation* itself. These procedures are expected to be very useful whenever mixing and ergodic hypotheses are (strictly or nearly) verified in Γ space. This surely is the case of almost all Hamiltonian systems whose many elements interact through a potential which is nowhere singular, and which decays quickly enough at long distances. In other cases, the situation might be more complex. For example, such an approach is not expected to be very reliable if the microscopic dynamics are such that structures (e.g., hierarchical ones) emerge in Γ space, which might or might not reflect into nontrivial structures in the μ space itself.[7] This could be the case if the interactions decay too slowly with distance, at least for various classes of initial conditions.

---

[7] Nontrivial structures in μ space imply nontrivial ones in Γ space. The other way around is not true: structures could exist in Γ space which would not be seen in μ space (the "shadow" of a fractal sponge on a wall can be a quite smooth surface).

A third possible approach is that of *stochastic equations*. The paradigm of such a level of description is the *Langevin equation*. One particle is selected (and followed) from the entire system, and part of the action of all the others is seen as a noise, typically an additive white Gaussian-like one. Such a description has the advantage of being relatively simple. It has however the considerable disadvantage of being partially phenomenological, in the sense that one has to introduce quite *ad hoc* types of noises (e.g., additive, multiplicative, non-Gaussian). If we are not interested in following the possible trajectories of a single particle, but rather in following the time evolution of probability distributions associated with such particles, we enter into the level of description of the *Fokker–Planck equation*, and the alike. At this *mesoscopic level*, exact analytical calculations, or relatively easy numerical ones, are relatively frequent.

A fourth possible approach is that of *statistical mechanics*. It directly connects—and this is where its beauty and power come from—the relevant microscopic information contained, for instance, in the Hamiltonian (with appropriate boundary conditions), to useful macroscopic quantities such as equations of states, specific heats, susceptibilities, and even various important correlation functions. In some epistemological sense, it superseeds all the previous types of approaches, excepting the fully microscopic one *with which it should always be consistent*. This last point is kind of trivial since statistical mechanics is nothing but a "shortcuted path" from the microscopic world to the macroscopic one. Let us precisely qualify the sense in which statistical mechanics "superseeds" other approaches such as those of Vlasov, Langevin, and Fokker–Planck. We mean that, whenever the collective states (usually at thermal equilibrium) and the quantities that are being calculated are *exactly the same*, no admissible mesoscopic description could be inconsistent with the statistical-mechanical one.

A fifth possible approach is that of *thermodynamics*. It directly connects many types of *macroscopic* quantities with sensible simplicity. However, it is uncapable of calculating from first principles quantities such as specific heats, susceptibilities, among many others. One expects, of course, that the results and connections obtained at the thermodynamical level will be consistent with those obtained at any of the previous levels, whenever comparison is justified and possible.

After this brief overview, it becomes kind of trivial to answer *part* of our initial question. Indeed, statistical mechanics, even the BG statistical mechanics, is *not necessary*, but it can be extremely *convenient*; also, it provides an *unifying description* of a great variety of useful questions. A point which remains to be answered is the following one. Given the fact that we *do* have—since more than one century—*BG* statistical mechanics, *do we need, or is it convenient, a more general one?* We can say that it is *not necessary* in the very same sense that, as we saw above, *BG* statistical mechanics is *not necessary* either. Is it convenient? We may say that, whenever possible, it is so in the same sense that the *BG* theory is convenient. The next point that has to be addressed in order to satisfactorily handle our initial question follows. Assuming that—because of its convenience and unifying power—we indeed want to make, whenever possible, a statistical-mechanical approach of a given problem, *do we need a generalization of the BG theory?* The answer is *yes*. For instance, quasi-stationary and other intermediate states are known to exist for long-range interacting classical Hamiltonian systems whose one-particle velocity distributions (both ensemble-averaged and time-averaged) are *not* Gaussians. This *excludes* the exponential form of the BG distribution law for the stationary state. Indeed, the marginal probability for the one-particle velocities derived from an exponential of the total Hamiltonian *necessarily is Gaussian*. Therefore, we definitely need something more general, if it can be formulated. Nonextensive statistical mechanics (as well as its variations such as the Beck-Cohen superstatistics, and others) appears to be at the present time the strongest operational paradigm. And this is so because of a variety of reasons which include the following interrelated facts: (i) Many of the functions that emerge in long-range interacting systems are known to be precisely of the q-exponential form; (ii) The entropy S_q is consistent with nonergodic (and/or slowly mixing) occupancy of the Γ space; (iii) The entropy S_q is, in many nonlinear dynamical systems, appropriate when the system is weakly chaotic (vanishing maximal Lyapunov exponent); (iv) In the presence of long-range interactions, the elements of the system tend to evolve in a rather synchronized manner, which makes virtually impossible an exponential divergence of nearby trajectories in Γ space: this prevents the system from quick mixing, and, in some cases, violates ergodicity, one of the pillars of the BG theory; (v) The central limit theorem, on which the BG theory is based, has been generalized in the presence of a (apparently large) class of global correlations, and the $N \to \infty$ basic attractors are q-Gaussians (see, for instance, [61, 530, 531, 565]); (vi) The block entropy S_q of paradigmatic Hamiltonian systems in quantum entangled collective states is extensive only for a special value of q which differs from unity (see [284, 285]).

(m) *Why do we need, in some contexts, to use escort distributions and q-expectation values instead of the ordinary ones?*

The essential mathematical reason for this can be seen in the set of Eqs. (4.141) and the following ones, and is based on the analytical results proved in [505]. When we are dealing with distributions that decay quickly at infinity (e.g., an exponential decay), then their characterization can be done with standard averages (first, second, and higher-order moments). This is the typical case within BG statistical mechanics, and low-order such moments precisely are the constraints that are normally imposed for the extremization of the entropy S_{BG}. But if we are dealing with distributions that decay slowly at infinity (e.g., power-law decay), the usual characterization becomes inadmissible since all the moments above a given one (which depends on the asymptotic behavior of the distribution) diverge. The characterization can, however, be done with mathematically well defined quantities by using q-expectation values (i.e., with escort distributions). This is the typical case within nonextensive statistical mechanics.

Let us illustrate with the $q = 2$ q-Gaussian (i.e., the Cauchy-Lorentz distribution) $p_2(x) \propto 1/(1 + \beta x^2)$. Its width is characterized by $1/\sqrt{\beta}$. However, its second moment diverges. At variance, its $q = 2$ q-expectation value is finite and given by

$\langle x^2 \rangle_q \propto 1/\beta$. This is therefore a natural constraint to be used for extremizing the entropy S_q.

Further arguments yielding consistently the escort distributions as the appropriate ones for expressing the constraints under which the entropy S_q is to be extremized can be found in [356, 1201], and in Appendix B.

(n) *Is it q just a fitting parameter? Does it characterize universality classes?*

From a first-principle standpoint, the basic universal constants of contemporary physics, namely c, h, G and k_B, are fitting parameters, but q *is not*. The indices q are in principle determined a priori from the microscopic or mesoscopic dynamics of the system. Very many examples illustrate this fact. However, when the micro- or meso-scopic dynamics are unknown (which is virtually always the case in real, empirical systems), or when, even if known, the problem turns out to be mathematically intractable (also this case is quite frequent), then and only then q is handled, *faute de mieux*, as a fitting parameter.

To make this point clear cut, let us remind here a nonexhaustive list of examples for which indices q are analytically known in terms of microscopic or mesoscopic quantities, or similar indices:

Standard critical phenomena at finite critical temperature: $q = \frac{1+\delta}{2}$ (see Eq. (5.80));

Zero-temperature critical phenomena of quantum entangled systems: $q = \frac{\sqrt{9+c^2}-3}{c}$ (see Eq. (3.163));

Lattice Lotka-Volterra models: $q = 1 - \frac{1}{D}$ (see Eq. (7.34));

Boltzmann lattice models: $q = 1 - \frac{2}{D}$ (see Eq. (7.7));

Probabilistic correlated models with cutoff: $q = 1 - \frac{1}{d}$ (see Eq. (3.155));

Probabilistic correlated models without cutoff: $q = \frac{\nu-2}{\nu-1}$ (see Eq. (4.125));

Unimodal maps: $\frac{1}{1-q} = \frac{1}{\alpha_{min}} - \frac{1}{\alpha_{max}}$ (see Eq. (5.9));

The particular case of the z-logistic family of maps: $\frac{1}{1-q(z)} = (z-1)\frac{\ln \alpha_F(z)}{\ln b}$ (see Eq. (5.11));

The $z = 2$ particular case of the z-logistic maps: $q = 0.244487701341282066198....$ (see Eq. (5.13));

Scale-free networks: $q = \frac{2m(2-r)+1-p-r}{m(3-2r)+1-p-r}$ (see Eq. (7.57));

Nonlinear Fokker–Planck equation: $q = 2 - \nu$ (see Eq. (4.8)), and $q = 3 - \frac{2}{\mu}$ (see Eq. (4.14));

Langevin equation including multiplicative noise: $q = \frac{\tau+3M}{\tau+M}$ (see Eq. (4.80));

Langevin equation including colored symmetric dichotomous noise: $q = \frac{1-2\gamma/\lambda}{1-\gamma/\lambda}$ (see Eq. (4.82));

Ginzburg–Landau discussion of point kinetics for $n = d$ ferromagnets: $q = \frac{d+4}{d+2}$ (see Eq. (4.173));

The q-generalized central limit theorems: $q_{\alpha,n} = \frac{(2+\alpha)q_{\alpha,n+2}-2}{2q_{\alpha,n+2}+\alpha-2}$ (see Eq. (4.151));

The connection with a many-body system in the presence of overdamping: $q = 1 - \frac{\lambda}{d}$ (see Eq. (4.62)).

Further analytical expressions for q in a variety of other physical systems are available
in the literature (see for instance [1241]).

As we readily verify, in some cases q characterizes universality classes (of nonadditivity), in total analogy with those of standard critical phenomena. Relations
(5.80), (3.163), and (5.11) constitute such examples. In other cases, analogously to
the two-dimensional short-range-interacting isotropic XY ferromagnetic model and
to the Baxter line of the square-lattice Ashkin-Teller ferromagnet [1242] (whose
critical exponents depend on the temperature and on the details of the Hamiltonian),
q depends on model details. Relations (7.57), (4.80), and (4.82) constitute such
examples. Cases which are believed to be of the universality class type are those
of classical long-range Hamiltonian systems (e.g., the α-XY, the α-Heisenberg, and
the α-Fermi-Pasta-Ulam models). The relevant indices q appear to depend only on
α (which characterizes the range of the forces) and on d (spatial dimension of the
system), possibly even only through the ratio α/d. However, analytically speaking,
this remains an open problem at the time when this book is being written.

(o)   *Why are there so many different values of q for the same system?*

The basic function ubiquitously emerging in the BG theory is a very universal one,
namely the exponential function. It is present in the sensitivity to the initial conditions,
in the relaxation of many physical quantities, in the distribution of energy states at
thermal equilibrium (in particular, in the distribution of velocities), in the solution
of the linear Fokker–Planck equation in the absence of external forces (and even for
linear external forces), in the attractor in the sense of the Central Limit Theorem
(CLT). In all these cases, the only quantity which is not universal is the *scale* of the
independent variable (e.g., the temperature). Of course, functions different from the
exponential also appear in BG statistical mechanics, but at the crucial and generic
points we find it again and again.

For many complex systems (the realm of nonextensive statistical mechanics), this
function is generalized into a less universal one, namely the q-exponential function
(a power-law, in the asymptotic region). It is this one which ubiquitously emerges
now at the same crucial and generic points. The q-exponential function depends not
only on the *scale*, but also on the *exponent* (i.e., the value of q) of the power-law.
Therefore, for a given system, different physical quantities are associated with different values of q. The indices q are expected to appear in the theory in infinite
number. However, only a few of them are expected to be necessary to characterize
the most important features of the system under given circumstances. And several
of these few are expected to be interrelated in such a way that only few of them
would be independent. A paradigmatic case has been analytically shown to occur
in the context of the q-generalization of the CLT: see Eq. (4.151) and Fig. 4.53.
Once the values of α and $q \equiv q_{\alpha,0}$ are fixed, the entire family of infinite countable
indices q is uniquely determined. Analogously, it is expected that, for wide classes
of classical long-range-interacting many-body d-dimensional Hamiltonians, most
relevant values of q would be fixed once the ratio $\alpha_{interaction}/d$ (where $\alpha_{interaction}$

fixes how quickly the force decays with distance, independently from the intensity of the force as long as it is nonzero) is fixed. Such possibility is reinforced by the Moebius structure put forward recently in [479]. See also point (a) in Sect. 8.3.

(p) *Do we need to microscopically discuss every single new dynamical system in order to know the numerical values associated with say its q-triplet?*

The examples for which analytical and/or numerical results are available today (e.g., Eqs. (3.163) and (5.80), and Fig. 5.57) suggest that the generic answer might be *no*. What we need is to know the relevant values of q for the universality classes of nonextensivity. In this step of the problem being solved, we just use the values associated with the universality class to which our specific system belongs. But, on general grounds, this issue still is in its infancy.

(q) *Are q-Gaussians ubiquitous?*

In the same sense that Gaussians are said to be ubiquitous (meaning by this that they appear very frequently, and in very diverse occasions, *not* in the strict etymological sense that they appear everywhere), the answer is *yes*. The q-Gaussians are well defined distributions which extremize the entropy S_q under quite generic constraints, and which, for positive β (inverse 'temperature'), are normalizable for $q < 3$, with finite (diverging) variance for $q < 5/3$ ($q \geq 5/3$), and with compact (infinite) support for $q < 1$ ($q \geq 1$). They are *analytical extensions* of the Student's t-distributions (r-distributions) for $q \geq 1$ (for $q \leq 1$). The cause of their ubiquity presumably is the fact that, within the q-generalization of the central limit theorem (under sufficient but not necessary conditions), q-Gaussians are *attractors* in probability space [1243] (see also [493, 1244–1248]). Through a related viewpoint, q-Gaussians are *stable* distributions (i.e., independent from the initial conditions) of an ubiquitous nonlinear Fokker–Planck equation. Moreover, these distributions are deeply related to scale-invariance (see, for instance, [470]), an ubiquitous property of many natural, artificial, and social systems. Finally, they have already been detected under a large variety of experimental and computational circumstances (see [61, 480, 530, 531, 565, 604, 631, 659, 660, 700, 730, 907, 908, 1113] among others).

An interesting analysis involving q-Gaussian distributions for $q \leq 1$ deserves to be mentioned here. Two (physically and mathematically interesting) probabilistic models were introduced and numerically analyzed in 2005-2006, namely the MTG [468] and the TMNT [469], which were thought to yield q-Gaussian distributions (with $q \leq 1$) in the $N \to \infty$ limit. However, the *exact* limiting distributions were analytically found in 2007 [471], and, although amazingly close numerically to q-Gaussians, they are *not* q-Gaussians.[8] Further news were to come along this fruitful line. Indeed, three more probabilistic models were introduced in 2008 [470] (see

---

[8] This was immediately commented in [1249] in a quite misleading manner, which generated a vague impression that there was something wrong with the q-Gaussian distributions themselves. This critique was soon rebutted [1250], the confusing point being hopefully clarified.

details in Sect. 4.8.5). Let us refer to them as RST1, RST2, and RST3. The models RST1 and RST2 exactly yield q-Gaussian limiting distributions (RST1 for $q \leq 1$, and RST2 for arbitrary values of q, both above and below unity), the first one on a probabilistic first-principle basis, the second one by construction. The model RST3, like the MTG and TMNT ones, approach limiting distributions which are not q-Gaussians. So, as we see, all types of situations can occur, and the whole picture surely deserves further clarification, especially since all five models are scale-invariant (strictly so for the MTG, TMNT, RST1, and RST3 models, and only asymptotically for the RST2 model).

(r) *Can we have some intuition on what is the physical origin of the nonadditive entropy S_q, hence of q-statistics?*

Yes, we can. Although rarely looked at through this perspective, a very analogous phenomenon occurs at the emergence, for an ideal gas, of Fermi–Dirac and Bose–Einstein quantum statistics. Indeed, their remarkably different mathematical expressions compared to Maxwell-Boltzmann statistics come from a drastic reduction/modification of the admissible physical states. Indeed, let us note $\mathcal{E}_H^{(N)}$ the Hilbert space associated with N particles; the N-particle wavefunctions are of the form $|m_1, m_2, ..., m_N\rangle = \Pi_{i=1}^{N} \phi_{m_i}(\mathbf{r}_i)$, where $\phi_{m_i}(\mathbf{r}_i)$ represents the wavefunction of the i-th particle being in the quantum state characterized by the quantum number (or set of quantum numbers) m_i . If for any reason (e.g., localization of the particles) we are allowed to consider the N particles as *distinguishable*, then Boltzmann–Gibbs equal-probability hypothesis for an isolated system at equilibrium is to be applied to the entire Hilbert space $\mathcal{E}_H^{(N)}$. At thermal equilibrium with a thermostat, we consistently obtain, for the occupancy of the quantum state characterized by the wave-vector $\mathbf{k}$ and energy $E_\mathbf{k}$, $f_\mathbf{k}^{MB} = e^{-\beta(E_\mathbf{k}-\mu)} = Ne^{-\beta E_\mathbf{k}}$, where μ is the chemical potential, and MB stands for *Maxwell-Boltzmann*. If however, the particles are to be considered as *indistinguishable*, then only *symmetrized (anti-symmetrized)* N-particle wavefunctions are physically admissible for bosons (fermions). For example, for $N = 2$, we have $|m_1, m_2\rangle = \frac{1}{\sqrt{2}}[\phi_{m_1}(\mathbf{r}_1)\phi_{m_2}(\mathbf{r}_2) + \phi_{m_1}(\mathbf{r}_2)\phi_{m_2}(\mathbf{r}_1)]$ for bosons, and $|m_1, m_2\rangle = \frac{1}{\sqrt{2}}[\phi_{m_1}(\mathbf{r}_1)\phi_{m_2}(\mathbf{r}_2) - \phi_{m_1}(\mathbf{r}_2)\phi_{m_2}(\mathbf{r}_1)]$ for fermions. For the general case of N particles, let us note, respectively, $\mathcal{E}_H^{(N)}(S)$ and $\mathcal{E}_H^{(N)}(A)$ the Hilbert spaces associated with symmetrized and anti-symmetrized wavefunctions. We have that $\mathcal{E}_H^{(N)}(S) \bigoplus \mathcal{E}_H^{(N)}(A) \subseteq \mathcal{E}_H^{(N)}$, the equality holding only for $N = 2$. For increasing N, the reduction of both $\mathcal{E}_H^{(N)}(S)$ and $\mathcal{E}_H^{(N)}(A)$ becomes more and more relevant. It is precisely for this reason that statistics eventually is profoundly changed. Indeed, the occupancy is now given by $f_\mathbf{k}^{BE} = 1/[e^{\beta(E_\mathbf{k}-\mu)} - 1]$ for bosons (BE standing for *Bose-Einstein*), and by $f_\mathbf{k}^{FD} = 1/[e^{\beta(E_\mathbf{k}-\mu)} + 1]$ for fermions (FD standing for *Fermi-Dirac*). The corresponding entropies are consistently changed from $S^{MB}/k_B = -\sum_\mathbf{k} f_\mathbf{k} \ln f_\mathbf{k}$ to $S^{BE}/k_B = \sum_\mathbf{k}[-f_\mathbf{k} \ln f_\mathbf{k} + (1 + f_\mathbf{k}) \ln(1 + f_\mathbf{k})]$ for bosons, and $S^{FD}/k_B = -\sum_\mathbf{k}[f_\mathbf{k} \ln f_\mathbf{k} + (1 - f_\mathbf{k}) \ln(1 - f_\mathbf{k})]$ for fermions. The need, in nonextensive statistical mechanics, for an entropy more general than the BG one, comes from essentially the same reason, i.e., a restriction of the space of

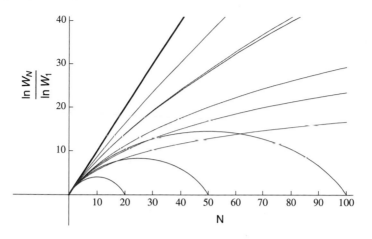

Fig. 8.3 All one-particle W_1 states ($W_1 = 1, 2, 3, ...$) are assumed non-degenerate. We consider the N-particle case assuming no interaction energy between the particles. $W_N^{(MB)} = W_1^N$, $N > 0$ (black curve); $W_N^{(FD)} = \frac{W_1!}{N!(W_1-N)!}, 0 < N \leq W_1$ (red curves); $W_N^{(BE)} = \frac{(N+W_1-1)!}{N!(W_1-1)!}, N > 0$ (blue curves). $N = 20, 50, 100, 1000, 100000$. In the present scale, the FD and BE curves for $N = 100000$ appear superimposed. In the limit $N \to \infty$ and $W_1 \to \infty$ with $N/W_1 \to 0$, $W_N^{(FD)}$ and $W_N^{(BE)}$ collapse onto the $W_N^{(MB)}$ result; they both satisfy $W_N \propto (W_1/N)^N$

the physically admissible states. Indeed, for the classical case for instance, vanishing Lyapunov exponents possibly generate, in regions of Γ-space, sets of orbits which are (multi)fractal-like. Since such sets of orbits are generically expected to have zero Lebesgue measure, an important restriction emerges for the physically admissible space (see also [29]). The basic ideas are illustrated for the microcanonical entropy in Fig. 8.3 for ideal Maxwell-Boltzmann, Fermi–Dirac, and Bose-Einstein N-particle systems (W_1 being the number of states, assumed non-degenerate, of the one-particle system), and in Fig. 8.4 for a highly correlated N-body system. We verify in Fig. 8.3 that, in the limit of large systems ($N \to \infty$ and $W_1 \to \infty$), the MB, FD, and BE systems yield a BG entropy which is extensive, i.e., thermodynamically admissible. This is not the case for the highly correlated N-body system. Indeed, the BG entropy asymptotically becomes independent from N, whereas the nonadditive entropy S_q exhibits extensivity for a special value of q, therefore becoming thermodynamically admissible. In other words, when the reduction of the (physically) admissible number of states is inexistent (MB model), or moderate (FD and BE models), the BG entropy is extensive. But if this reduction is very severe (like in the present highly correlated model), then we are obliged to introduce a different entropy in order to satisfy thermodynamics. Obviously this point is most important, since it basically makes legitimate the use of virtually all general formulas of textbooks of thermodynamics.

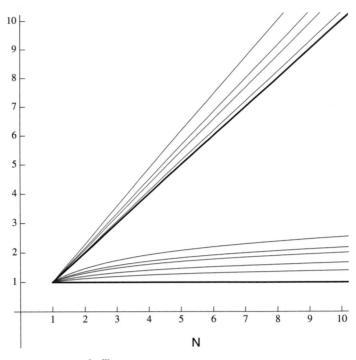

Fig. 8.4 N-dependence of $\frac{\ln_q W_N}{\ln_q W_1}$, where $W_N = W_1 N^\rho$ $(\rho > 0)$ with $q = 1 - \frac{1}{\rho}$ $(\rho = 2$ hence $q = 1/2$; $W_1 > 1$, $N \geq 1)$. $\lim_{N \to \infty} \frac{\ln_{1-1/\rho} W_N}{\ln_{1-1/\rho} W_1} = \frac{W_1^{1/\rho}}{W_1^{1/\rho}-1} N$, which asymptotically approaches N in the limit $W_1 \to \infty$. Under the same conditions $\lim_{N \to \infty} \frac{\ln W_N}{\ln W_1}$ approaches unity, $\forall N$. Blue (red) set of curves for $q = 1/2$ (for $q = 1$), with $W_1 = 20, 50, 100, 1000, 100000$ from top to bottom. *Black curves:* $\frac{\ln_q W_N}{\ln_q W_1} = N$ for $q = 1 - 1/\rho$, and $\frac{\ln_q W_N}{\ln_q W_1} = 1$ for $q = 1$

(s) *Since Renyi entropy is a monotonic function of S_q, can we not use it equally well to generalize BG statistical mechanics?*

The same question applies in fact for any entropic functional which is a monotonic function of S_q, such as, for instance, the Sharma-Mittal entropy $S_{q,r}^{SM}$, or simply S_q^3, assuming that the constraints under which optimization is accomplished are precisely the same.

This is obviously true in what concerns the optimizing distribution (e.g., q-exponential or q-Gaussian depending on the constraints that are imposed). But thermodynamics and statistical mechanics are, as argued earlier in this book, much more than just an optimizing distribution! To illustrate this, let us remind that nobody has ever used say $(S_{BG})^3$, instead of S_{BG}, to successfully connect statistical mechanics to thermodynamics. To be more specific, S_q and Renyi entropy S_q^R strongly differ in many other (interrelated) crucial aspects.

- To start with, S_q is concave for any value $q > 0$, whereas S_q^R is concave only for $0 < q < 1$.
- If $q < 1$, S_q and S_q^R can not be simultaneously extensive. Assuming equal probabilities, for the family $W(N) \sim N^\rho$ with $\rho > 0$ (instead of the traditional $W(N) \sim \mu^N$ with $\mu > 1$), S_q is extensive for an unique value of q, namely $q = 1 - 1/\rho < 1$. The Renyi entropy is nonextensive unless $q = 1$. It is in fact proportional to $\ln N$ for any arbitrary value of $0 < \rho < \infty$, hence $q < 1$. This logarithmic dependence of S_q^R for $q < 1$ definitively violates the Legendre structure of (generalized) thermodynamics, whereas S_q satisfies it.
- If the sensitivity $\xi(t)$ to initial conditions is a power-law (i.e., with maximal Lyapunov exponent $\lambda_1 = 0$), hence if $\xi \sim [(1 - q)\lambda_q t]^{1/(1-q)}$, then the growth of the entropy is given by $S_q \sim K_q t$ where $K_q = \lambda_q$ (q-generalized Pesin identity). Nothing like this occurs for the Renyi entropy, which increases with time as $\ln t$. Therefore S_q has, for a unique value of q which depends on the specific weakly chaotic dynamics of the system, a *finite* entropy production per unit time, whereas this rate indistinctively *vanishes* for S_q^R for any value of $q \neq 1$.
- If we consider the famous Porous Medium Equation $\partial p(x, t)/\partial t = D[\partial^2/\partial x^2] [p(x, t)]^\nu$, whose paradigmatic solution is a q-Gaussian with $q = 2 - \nu$, then S_q satisfies an H-theorem whereas Renyi entropy has not been analyzed on these grounds.
- If you consider the Large Deviation Theory, it has been verified for various examples whose limiting distributions (in the sense of a generalized Central Limit Theorem) are Q-Gaussians with $Q > 1$, that $P(N) \sim e_q^{-r_q N}$, where the generalized rate function r_q is some kind of relative nonadditive entropy per particle. Therefore $(r_q N)$ definitively behaves like an extensive entropy. In this sense, Renyi entropy cannot succeed if $q \neq 1$ since it increases logarithmically with N whenever S_q increases linearly with N.
- If you consider a special nonlinear master equation, generalizing p_i into some form of the type p_i^q in the right-hand term, you obtain precisely the Porous Medium Equation, which, as already said, satisfies an H-theorem. Neither here Renyi entropy has been fully analyzed.

However, all the above said, we cannot definitively exclude the possibility of validity of the Renyi entropy in some specific thermodynamical grounds to be qualified, but the discussion clearly goes far beyond the mere monotonicity of S_q^R as a function of S_q.

(t) *Are there extensions of the q-exponential and the q-Gaussian functions which also appear in natural or other systems?*

Yes, very many of them, and various examples can be found in the present book. Let us mention at this point three important extensions of the $d = 1$ q-Gaussians.

One of them is *asymmetric q-Gaussians*. They can emerge in the form presented in Eq. (4.68), and also in the following form:

$$F(x) = e_q^{-(ax^2+bx^3+cx^4)} \quad (a \in \mathbb{R},\ b \neq 0,\ c > 0). \tag{8.2}$$

The second one concerns the relativistic generalization of the Maxwellian distribution of velocities, following the Lorentz-invariant generalization as discussed in [1251]. The one-dimensional Maxwellian distribution unphysically allows for velocities v surpassing, in absolute value, c, the velocity of light. Its relativistic generalization can only allow for velocities up to c, and it was tentatively generalized by F. Jüttner in 1911, very frequently used since then in high-energy physics (collisions of elementary particles, astrophysics, among others). It was recently argued however that the Jüttner distribution can not be physically correct since it is not Lorentz-invariant. The Lorentz-invariant one is presented in [1251], and it basically consists in the (normalized) Gaussian form $e^{-\beta\sigma^2}$, where the *rapidity* σ is defined as $\sigma \equiv tanh^{-1}(v/c) \in (-\infty, \infty)$, both σ and $\beta > 0$ being Lorentz-invariant. The corresponding Lorentz-invariant q-generalization is naturally given by the (normalized) q-Gaussian form $e_q^{-\beta_q\sigma^2}$,[9] which should hopefully find applications in relativistic generalizations of the distributions of longitudinal momenta in particle collisions, astrophysics, plasmas, and others. Whenever worked out in the literature, the $d = 2, 3$ relativistic generalizations of the $d = 1$ Maxwellian distribution will naturally follow the same path in what concerns their q-generalization.

The third one is the *stretched q-exponential* one, defined as follows:

$$F(x) = e_q^{-a|x|^\eta} \quad (a > 0, \eta > 0), \tag{8.3}$$

which recovers the q-exponential for $\eta = 1$, the q-Gaussian for $\eta = 2$, and the stretched exponential for $q = 1$.

(u) *If one value of q of a q-multiplet (q-duplet, q-triplet, etc) equals unity, must the values of all q's also equal unity?*

Not necessarily. It appears that, for given classes of properties of a given class of systems with a given class of boundary and/or initial conditions, a specific q-multiplet (q-duplet, q-triplet, q-quadruplet, etc.) emerges from first principles. A q-multiplet generically contains a countable infinity of q values, typically only a few of them being independent (all the others being hopefully simply related to the independent ones). If the system is well described by BG statistical mechanics, then typically all values of q equal unity. If the system is well described by q-statistics, then some of those q's equal unity while most of them differ from unity. We may illustrate this through two paradigmatic examples. The first of them is observational and concerns solar activity: we see in Table 5.3 that q_{sens} and q_{stat} are different from unity, whereas $q_{rel} = 1$. The second one concerns a mathematically defined probabilistic model: we see in [1198] that Q (related to the q-generalized Central Limit Theorem), q (related

---

[9] This distribution optimizes $S_q = -k \int_{-\infty}^{\infty} d\sigma\, p(\sigma) \ln_q p(\sigma)$ with the constraints $\int_{-\infty}^{\infty} d\sigma\, p(\sigma) = 1$ and $\int_{-\infty}^{\infty} d\sigma\, p(\sigma)\sigma^2 = constant$.

to a possible q-generalized Large Deviation Theory), and q_r (emerging within a q-generalized rate function) are different from unity, whereas $q_{entropy} = 1$ (due to the fact that the model is constructed upon *exchangeable* random variables). Although essentially related, as shown in [479], to Moebius transforms, the full structure of generic q-multiplets remains elusive until present date.

8.3 Open Questions

Like in any physical theory in intensive development, a large amount of open questions still exists within nonextensive statistical mechanics. Since we do not intend here to make a lengthy description, we will simply mention some of those few points that we find particularly intriguing and fruitful.

(a) *What are the q-indices relevant to the stationary-state associated with a d-dimensional classical many-body Hamiltonian including (say attractive) interactions that are not singular (or are, at least, integrable) at the origin and decay with distance r like $1/r^\alpha$ ($\alpha \geq 0$)?* We know that, for $\alpha/d > 1$ (i.e., short-range interactions), $q = 1$ (hence $q_{sen} = q_{rel} = q_{stat}^{(velocities)} = q_{stat}^{(energies)} = 1$). What happens for $0 \leq \alpha/d \leq 1$ (i.e., long-range interactions) ? What would be the possible (α, d)-dependences (perhaps (α/d)-dependences) of indices such as $(q_{sen}, q_{rel}, q_{stat}^{(velocities)} = q_{stat}^{(energies)})$?

(b) *Compatibility between the (presumably) scale-invariant correlations leading to an extensive S_q and the q-exponential form for the stationary-state distribution of energy for many-body Hamiltonian systems* More precisely, what must be satisfied by the interaction Hamiltonian $\mathcal{H}_{AB}$ within the form $\mathcal{H}_{A+B} = \mathcal{H}_A + \mathcal{H}_B + \mathcal{H}_{AB}$ when A and B are two large systems? Let us be more concrete and discuss the $q = 1$ case. Assume that we are dealing with short-range interactions, and that A and B are two equally sized d-dimensional systems. Let L be the linear size of each of them. Then the energy corresponding to $\mathcal{H}_A$ increases like L^d, and the same happens with system B. Let us also assume that A and B are in contact only through a common $(d - 1)$-dimensional surface. Then the energy corresponding to $\mathcal{H}_{AB}$ increases like L^{d-1}. In the limit $L \to \infty$, we can neglect the interaction energy, i.e., consider $\mathcal{H}_{AB} = 0$. Then $\mathcal{H}_{A+B} = \mathcal{H}_A + \mathcal{H}_B$ is clearly compatible with $p_i^A = e^{-\beta E_i^A}/Z_A$, $p_i^B = e^{-\beta E_i^B}/Z_B$ and $p_{ij}^{A+B} = p_i^A p_j^B$. The question we would like to answer is what exactly happens for $q \neq 1$?

(c) *What is the geometrical-dynamical interpretation of the escort distribution?* This is a question frequently asked whose full answer is still unclear. We have presented in a previous section a variety of mathematical reasons pointing the relevance of the escort distributions within the present theory. However, a clearcut physical interpretation in terms of the dynamics and occupancy geometry within the full phase space Γ is still lacking. Some important hints can be found in [63–65].

(d) *What is the logical connection between the class of systems whose extensivity requires the adoption of the entropy S_q with $q \neq 1$, and the class of systems whose probabilities distributions of occupancy of phase space leads, in the limit $N \to \infty$, to anomalous central limit theorems?* The present scenario is that asymptotic scale-invariance is necessary but not sufficient. Hints can be found in [283], in Sect. 4.8.5, and in the nonlinear Fokker-Planck equation.

(e) *Under what generic conditions nonlinear dynamics such as those emerging at the edge of chaos as well as in long-range-interacting many-body classical Hamiltonians at their quasi-stationary state tend to create, in the full phase space, structures geometrically similar to scale-free networks?* The scenario is that probabilistic correlations of the q-independent class are sufficient (but possibly not necessary) to create a (multi)fractal occupation of phase space. The clarification of this point would most probably also provide an answer to the above point (c).

(f) *What are the precise physical quantities associated with the infinite set of inter-related values of q emerging in relations such as Eq. (4.150)? What is their precise connection to sets such as the q-triplet?* This is a most important open question. The scenario is that somehow the usual q-triplet (q-multiplet in general) essentially corresponds to central elements (such as $n = 0, \pm 1, \pm 2$, etc) of the relation (4.150). The solution, or at least crucial hints pointing along that direction, of these and other similar questions would be more than very welcome!

(g) *Why, in several d-dimensional classical models like the α-XY, α-Heisenberg, α-Fermi-Pasta-Ulam ones, the numerical first-principle approaches that are available in the literature still yield $q > 1$ above $\alpha/d = 1$?* Up to now, the answer does not seem to be neatly related to insufficiently large values of N and/or of t in the simulations. However, we can not exclude such possibility unless extremely large values of (N, T) become computationally accessible in the future. Indeed, the duration of the relevant long-standing states can be amazingly long for increasing values of N (see [585, 1194, 1195]). Another cause could be the class of initial conditions that are usually implemented (such behavior has been observed in other long-range-interacting systems [1197]). This possibility remains to be systematically explored. Last but not least, one more operational ingredient—the *precision* within which classical calculations are performed—makes the puzzle even harder. Indeed, subtle numerical effects such as those depicted in Figs. 5.19 and 5.20 could be playing a nontrivial role. In fact, this key issue has been recently revisited in [1261]

Appendix A
Useful Mathematical Formulae

$$\ln_q x \equiv \frac{x^{1-q} - 1}{1 - q} \quad (x > 0, \, q \in \mathbb{R}) \tag{A.1}$$

$$\ln_q x = x^{1-q} \ln_{2-q} x \quad (x > 0; \, \forall q) \tag{A.2}$$

$$\ln_q(1/x) + \ln_{2-q} x = 0 \quad (x > 0; \, \forall q) \tag{A.3}$$

$$q \ln_q x + \ln_{(1/q)}(1/x^q) = 0 \quad (x > 0; \, \forall q) \tag{A.4}$$

$$e_q^x \equiv [1 + (1-q)\,x]_+^{\frac{1}{1-q}} \equiv \begin{cases} 0 & \text{if } q < 1 \text{ and } x < -1/(1-q)\,, \\[2mm] [1 + (1-q)\,x]^{\frac{1}{1-q}} & \text{if } q < 1 \text{ and } x \geq -1/(1-q)\,, \\[2mm] e^x & \text{if } q = 1 \quad (\forall x)\,, \\[2mm] [1 + (1-q)\,x]^{\frac{1}{1-q}} & \text{if } q > 1 \text{ and } x < 1/(q-1)\,. \end{cases} \tag{A.5}$$

$$e_q^x \, e_{2-q}^{-x} = 1 \quad (\forall q) \tag{A.6}$$

$$(e_q^x)^q \, e_{(1/q)}^{-qx} = 1 \quad (\forall q) \tag{A.7}$$

$$e_q^{x+y+(1-q)xy} = e_q^x \, e_q^y \quad (\forall q) \tag{A.8}$$

© Springer Nature Switzerland AG 2023
C. Tsallis, *Introduction to Nonextensive Statistical Mechanics*,
https://doi.org/10.1007/978-3-030-79569-6

$$e_q^{x+y} = e_q^x \otimes e_q^y \quad (\forall q) \tag{A.9}$$

$$e_q^x = {}_1F_0(1/(q-1), -, (q-1)x) \quad (q \in \mathcal{R}), \tag{A.10}$$

${}_1F_0$ being the hypergeometric function.

$$\int_{x_c}^{\infty} dx \, e_q^{-\beta x^s} = \frac{x_c^{\frac{q-s-1}{q-1}} \Gamma\left(\frac{s+1-q}{s(q-1)}\right) {}_2\tilde{F}_1\left(\frac{1}{q-1}, \frac{s+1-q}{s(q-1)}, \frac{(s-1)q+1}{s(q-1)}, \frac{x_c^{-s}}{\beta(1-q)}\right)}{s[\beta(q-1)]^{\frac{1}{q-1}}}$$
$$(\beta > 0; \ s > 0; \ x_c \geq 0; \ 1 < q < s+1), \tag{A.11}$$

$$\sim \frac{\Gamma\left[\frac{1-q+s}{(q-1)s}\right]}{s \, \beta^{\frac{1}{q-1}} (q-1)^{\frac{1}{q-1}} \Gamma\left[\frac{1+q(s-1)}{(q-1)s}\right]} \frac{1}{x_c^{\frac{s+1-q}{q-1}}} \quad (x_c \to \infty), \tag{A.12}$$

where ${}_2\tilde{F}_1$ is the regularized hypergeometric function. Consequently

$$\int_{x_c}^{\infty} dx \, e_q^{-\beta x} = \frac{\left[1 + (q-1)\beta x_c\right]^{\frac{2-q}{1-q}}}{\beta(2-q)} \quad (1 < q < 2), \tag{A.13}$$

$$\sim \frac{1}{(2-q)(q-1)^{\frac{2-q}{q-1}} \beta^{\frac{1}{q-1}} x_c^{\frac{2-q}{q-1}}} \quad (x_c \to \infty), \tag{A.14}$$

and

$$\int_{x_c}^{\infty} dx \, e_q^{-\beta x^2} = \frac{x_c^{\frac{q-3}{q-1}} {}_2F_1\left(\frac{3-q}{2(q-1)}, \frac{1}{q-1}, \frac{q+1}{2(q-1)}, \frac{-1}{(q-1)\beta x_c^2}\right)}{(q-1)^{\frac{2-q}{q-1}} (3-q)\beta^{\frac{1}{q-1}}} \quad (1 < q < 3), \tag{A.15}$$

$$\sim \frac{(q-1)^{\frac{q-2}{q-1}}}{(3-q)\beta^{\frac{1}{q-1}}} \frac{1}{x_c^{\frac{3-q}{q-1}}} \quad (x_c \to \infty), \tag{A.16}$$

${}_2F_1$ being the hypergeometric function.

$$e_Q^x = e_q\left\{\frac{[e_Q^x]^{1-q} - 1}{1-q}\right\} \quad (\forall q, \forall Q, \forall x) \tag{A.17}$$

$$x \oplus_q y \equiv x + y + (1-q)x y \tag{A.18}$$

For $x \geq 0$ and $y \geq 0$:

$$x \otimes_q y \equiv [x^{1-q} + y^{1-q} - 1]_+^{\frac{1}{1-q}} \equiv \begin{cases} 0 & \text{if } q < 1 \text{ and } x^{1-q} + y^{1-q} < 1, \\ [x^{1-q} + y^{1-q} - 1]^{\frac{1}{1-q}} & \text{if } q < 1 \text{ and } x^{1-q} + y^{1-q} \geq 1, \\ xy & \text{if } q = 1 \quad \forall (x, y), \\ [x^{1-q} + y^{1-q} - 1]^{\frac{1}{1-q}} & \text{if } q > 1 \text{ and } x^{1-q} + y^{1-q} > 1. \end{cases}$$

$$\text{(A.19)}$$

$$x \otimes_q y = [1 + (1 - q)(\ln_q x + \ln_q y)]^{\frac{1}{1-q}} \tag{A.20}$$

$$e_q^{x \oplus_q y} = e_q^x e_q^y \quad (\forall q) \tag{A.21}$$

$$e_q^{x+y} = e_q^x \otimes_q e_q^y \quad (\forall q) \tag{A.22}$$

$$\frac{d \ln_q x}{dx} = \frac{1}{x^q} \quad (x > 0; \forall q) \tag{A.23}$$

$$\frac{d e_q^x}{dx} = (e_q^x)^q \quad (\forall q) \tag{A.24}$$

$$D_{q,\alpha} e_q^x = e_q^x \tag{A.25}$$

with[1]

$$D_{q,\alpha} f(x) \equiv [1 + (1 - q)x]^\alpha [f(x)]^{[(1-q)(1-\alpha)]} \frac{df(x)}{dx} \tag{A.26}$$

$$(e_q^x)^q = e_{2-(1/q)}^{qx} \quad (\forall q) \tag{A.27}$$

$$(e_q^x)^a = e_{1-(1-q)/a}^{ax} \quad (\forall q) \tag{A.28}$$

$$x^a e_q^{-\frac{x}{b}} = \left[\frac{b}{q-1}\right]^{1/(q-1)} x^{a-\frac{1}{q-1}} e_q^{-\frac{b/(q-1)^2}{x}} \quad (b > 0; q > 1) \tag{A.29}$$

---

[1] $D_{q,1}$ recovers Eq. (17) of [239]. $D_{q,0}$ recovers the generalized derivative in [639], whose factor $[f(x)]^{1-q}$ precisely causes the (appropriate) nonlinearity in the q-Fourier Transform as indicated in Eq. (2.5) of [477].

$$e_q^x = e^x \left[1 - \frac{1}{2}(1-q)x^2 + \frac{1}{3}(1-q)^2 x^3 \left(1 + \frac{3}{8}x\right) - \frac{1}{4}(1-q)^3 x^4 \left(1 + \frac{2}{3}x + \frac{1}{12}x^2\right)\right.$$
$$+ \frac{1}{5}(1-q)^4 x^5 \left(1 + \frac{65}{72}x + \frac{5}{24}x^2 + \frac{5}{384}x^3\right)$$
$$\left. - \frac{1}{6}(1-q)^5 x^6 \left(1 + \frac{11}{10}x + \frac{17}{48}x^2 + \frac{1}{24}x^3 + \frac{1}{640}x^4\right) + \dots \right] \quad (q \to 1; \forall x)$$

(A.30)

$$\ln_q x = (\ln x) \left[1 + \frac{1}{2}(1-q)\ln x + \frac{1}{6}(1-q)^2 \ln^2 x + \frac{1}{24}(1-q)^3 \ln^3 x \right.$$
$$\left. + \frac{1}{120}(1-q)^4 \ln^4 x + \frac{1}{720}(1-q)^5 \ln^5 x + \dots \right]$$
$$= \sum_{k=1}^{\infty} \frac{\ln^k x}{k!}(1-q)^{k-1} \quad (q \to 1; x > 0)$$

(A.31)

$$x \otimes_q y = xy\left[1 - (1-q)(\ln x)(\ln y)\right.$$
$$+ \frac{1}{2}(1-q)^2 \left[(\ln^2 x)(\ln y) + (\ln x)(\ln^2 y) + (\ln^2 x)(\ln^2 y)\right]$$
$$- \frac{1}{12}(1-q)^3 \left[2(\ln^3 x)(\ln y) + 9(\ln^2 x)(\ln^2 y) + 2(\ln x)(\ln^3 y)\right.$$
$$\left. + 6(\ln^3 x)(\ln^2 y) + 6(\ln^2 x)(\ln^3 y) + 2(\ln^3 x)(\ln^3 y)\right]$$
$$+ \frac{1}{24}(1-q)^4 \left[(\ln^4 x)(\ln y) + 14(\ln^3 x)(\ln^2 y)\right.$$
$$+ 14(\ln^2 x)(\ln^3 y) + (\ln x)(\ln^4 y)$$
$$+ 7(\ln^4 x)(\ln^2 y) + 24(\ln^3 x)(\ln^3 y) + 7(\ln^2 x)(\ln^4 y)$$
$$\left.\left. + 6(\ln^4 x)(\ln^3 y) + 6(\ln^3 x)(\ln^4 y) + (\ln^4 x)(\ln^4 y)\right] + \dots \right] \quad \text{(A.32)}$$

$$x \,_q \oplus y \equiv e_q^{\ln[e^{\ln_q x} + e^{\ln_q y}]} \quad (x > 0, \ y > 0)$$

(A.33)

$$x \otimes_q (y \,_q \oplus z) = (x \otimes_q y) \,_q \oplus (x \otimes_q z)$$

(A.34)

$$x \,_q\oplus y = (x + y)\left\{1 + \frac{1}{2}(1 - q)\left[\frac{x \ln^2(x) + y \ln^2(y)}{x + y} - \ln^2(x + y)\right]\right.$$

$$+ \frac{1}{24}(1 - q)^2 \frac{1}{x + y}\left[3x \ln^4(x + y) + 3y \ln^4(x + y)\right.$$

$$+ 8x \ln^3(x + y) + 8y \ln^3(x + y)$$

$$- 6x \ln^2 x \log^2(x + y) - 12x \ln^2 x \ln(x + y)$$

$$- 6y \ln^2 y \ln^2(x + y) - 12y \ln^2 y \ln(x + y)$$

$$\left.\left. + 3x \ln^4 x + 4x \ln^3 x + 3y \ln^4 y + 4y \ln^3 y\right] + \ldots\right\} \qquad \text{(A.35)}$$

$$e_q^x = 1 + x + \frac{1}{2}x^2 q + \frac{1}{6}x^3 q(2q - 1) + \frac{1}{24}x^4 q(2q - 1)(3q - 2)$$

$$+ \frac{1}{120}x^5 q(2q - 1)(3q - 2)(4q - 3) + \ldots \quad (x \to 0; \forall q) \qquad \text{(A.36)}$$

$$\ln_q(1 + x) = x - \frac{1}{2}qx^2 + \frac{1}{6}q(1 + q)x^3 - \frac{1}{24}q(1 + q)(2 + q)x^4$$

$$+ \frac{1}{120}q(1 + q)(2 + q)(3 + q)x^5 + \ldots \quad (x \to 0; \forall q) \,\text{(A.37)}$$

$$\ln_{q,q'}(1 + x) = x + \frac{1}{2}(1 - q - q')x^2 + \frac{1}{6}(1 - 2q + q^2 - 2q' + 3qq' + q'^2)x^3$$

$$+ \frac{1}{24}\left(1 - 4q + 4q^2 - q^3 - 3q' + 8qq' - 7q^2q' + 3q'^2 - 6qq'^2 - q'^3\right)x^4$$

$$+ \frac{1}{120}(1 - 4q + 11q^2 - 9q^3 + q^4 - 4q' + 20qq' - 25q^2q' + 15q^3q' + 6q'^2$$

$$- 20qq'^2 + 25q^2q'^2 - 4q'^3 + 10qq'^3 + q'^4)x^5 + \ldots (x \to 0; \forall(q, q')) \quad \text{(A.38)}$$

$$\ln_{q,q'} x = \ln x\left[1 + \frac{1}{2}((1 - q) + (1 - q'))\ln x\right.$$

$$+ \frac{1}{6}((1 - q)^2 + 3(1 - q)(1 - q') + (1 - q')^2)\ln^2 x$$

$$+ \frac{1}{24}((1 - q)^3 + 7(1 - q)^2(1 - q') + 6(1 - q)(1 - q')^2 + (1 - q')^3)\ln^3 x$$

$$+ \frac{1}{120}((1 - q)^4 + 15(1 - q)^3(1 - q') + 25(1 - q)^2(1 - q')^2$$

$$\left. + 10(1 - q)(1 - q')^3 + (1 - q')^4)\ln^4 x + \ldots\right] \quad (x > 0) \qquad \text{(A.39)}$$

$$x^a e_q^{-\frac{x}{b}} = \left[\frac{b}{q-1}\right]^{\frac{1}{q-1}} x^{a-\frac{1}{q-1}} e_q^{-\frac{b/(q-1)^2}{x}} \qquad (q > 1; \, b > 0) \qquad\qquad (\text{A}.40)$$

$$\ln_{q,q'} x \equiv \ln_{q'} e^{\ln_q x} \qquad (x > 0, \, (q,q') \in \mathbb{R}^2) \qquad\qquad (\text{A}.41)$$

$$\ln_{q,q'}(x \otimes_q y) = \ln_{q,q'} x \oplus_{q'} \ln_{q,q'} y \qquad (x > 0, \, (q,q') \in \mathbb{R}^2) \qquad\qquad (\text{A}.42)$$

$$e_q^{-\beta z} = \frac{1}{\Gamma\left(\frac{1}{q-1}\right)} \frac{1}{[\beta(q-1)]^{\frac{1}{q-1}}} \int_0^\infty d\alpha \, \alpha^{\frac{2-q}{q-1}} \, e^{-\frac{\alpha}{\beta(q-1)}} \, e^{-\alpha z} \qquad\qquad (\text{A}.43)$$

$$(\alpha > 0; \, \beta > 0; \, 1 < q < 2)$$

The following relations are useful for the Fourier transform of q-Gaussians (with $\beta > 0$):

$$F_q(p) \equiv \int_{-\infty}^\infty dx \, \frac{e^{ixp}}{[1 + (q-1)\beta x^2]^{1/(q-1)}}$$

$$= 2 \int_0^\infty dx \, \frac{\cos(xp)}{[1 + (q-1)\beta x^2]^{1/(q-1)}}$$

$$= \begin{cases} \sqrt{\frac{\pi}{(1-q)\beta}} \, \Gamma\left(\frac{2-q}{1-q}\right) \left(\frac{2\sqrt{\beta(1-q)}}{p}\right)^{\frac{3-q}{2(1-q)}} J_{\frac{3-q}{1-q}}\left(\frac{p}{\sqrt{\beta(1-q)}}\right) & \text{if } q < 1, \\[3ex] \sqrt{\frac{\pi}{\beta}} \, e^{-\frac{p^2}{4\beta}} & \text{if } q = 1, \qquad (\text{A}.44) \\[3ex] \frac{2}{\Gamma\left(\frac{1}{q-1}\right)} \sqrt{\frac{\pi}{\beta(q-1)}} \left(\frac{|p|}{2\sqrt{\beta(q-1)}}\right)^{\frac{3-q}{2(q-1)}} K_{\frac{3-q}{2(q-1)}}\left(\frac{|p|}{\sqrt{\beta(q-1)}}\right) & \text{if } 1 < q < 3, \end{cases}$$

where $J_\nu(z)$ and $K_\nu(z)$ are, respectively, the Bessel and the modified Bessel functions. For the three successive regions of q, we have, respectively, used formulae 3.387-2 (page 346), 3.323-2 (page 333), and 8.432-5 (page 905) of [1252] (see also [1253]). For the $q < 1$ result, we have taken into account the fact that the q-Gaussian identically vanishes for $|x| > \frac{1}{\sqrt{\beta(1-q)}}$.

$$F_q[f](\xi) \equiv \int_{-\infty}^\infty dx \, e_q^{i\xi x} \otimes_q f(x) = \int_{-\infty}^\infty dx \, e_q^{i\xi x [f(x)]^{q-1}} f(x) \qquad (q \geq 1) \quad (\text{A}.45)$$

$$F_q[f](0) = \int_{-\infty}^{\infty} dx \, f(x) \qquad (q \geq 1) \qquad \text{(A.46)}$$

$$\left. \frac{dF_q[f](\xi)}{d\xi} \right|_{\xi=0} = i \int_{-\infty}^{\infty} dx \, x \, [f(x)]^q \qquad (q \geq 1) \qquad \text{(A.47)}$$

$$\left. \frac{d^2 F_q[f](\xi)}{d\xi^2} \right|_{\xi=0} = -q \int_{-\infty}^{\infty} dx \, x^2 \, [f(x)]^{2q-1} \qquad (q \geq 1) \qquad \text{(A.48)}$$

$$\left. \frac{d^3 F_q[f](\xi)}{d\xi^3} \right|_{\xi=0} = -i \, q \, (2q-1) \int_{-\infty}^{\infty} dx \, x^3 \, [f(x)]^{3q-2} \quad (q \geq 1) \qquad \text{(A.49)}$$

$$\left. \frac{d^{(n)} F_q[f](\xi)}{d\xi^n} \right|_{\xi=0} = (i)^n \left\{ \prod_{m=0}^{n-1} [1 + m(q-1)] \right\} \int_{-\infty}^{\infty} dx \, x^n \, [f(x)]^{1+n(q-1)}$$

$$(q \geq 1; \, n = 1, 2, 3...) \text{ (A.50)}$$

$$F_q[af(ax)](\xi) = F_q[f](\xi/a^{2-q}) \quad (a > 0; \, 1 \leq q < 2). \qquad \text{(A.51)}$$

$$\langle x^\eta \rangle_q \equiv \frac{\int_0^{x_{max}} dx \, x^\eta \, [e_q^{-(x/\sigma)^\eta}]^q}{\int_0^{x_{max}} dx \, [e_q^{-(x/\sigma)^\eta}]^q} = \sigma^\eta \frac{1}{1 + \eta - q}, \qquad \text{(A.52)}$$

with $\sigma > 0$, $\eta > 0$, $q \in (-\infty, 1 + \eta)$, and $x_{max} = \frac{\sigma}{(1-q)^{1/\eta}}$ if $q < 1$, and ∞ if $1 \leq q < 1 + \eta$.

This relation can be generalized into

$$\langle x^\eta \rangle_Q \equiv \frac{\int_0^{x_{max}} dx \, x^\eta \, [e_q^{-(x/\sigma)^\eta}]^Q}{\int_0^{x_{max}} dx \, [e_q^{-(x/\sigma)^\eta}]^Q} = \sigma^\eta \frac{1}{(1 + \eta)(Q - q + 1) - Q} > 0, \quad \text{(A.53)}$$

with $\sigma > 0$, $\eta > 0$, $q < 1 + \frac{\eta}{1+\eta} Q$, and $x_{max} = \frac{\sigma}{(1-q)^{1/\eta}}$ if $q < 1$, and ∞ otherwise.

Also

$$\langle x^{n\eta} \rangle_q \equiv \frac{\int_0^{x_{max}} dx \, x^{n\eta} (e_q^{-(x/\sigma)^\eta})^q}{\int_0^{x_{max}} dx \, (e_q^{-(x/\sigma)^\eta})^q}$$

$$= \sigma^{n\eta} \frac{(-1)^n \prod_{k=0}^{n-1} (k\eta + 1)}{\prod_{k=1}^{n} [-k\eta + (k-1)\eta q - 1 + q]}, \qquad \text{(A.54)}$$

with $\eta > 0, n = 1, 2, 3...$, and $x_{max} = \frac{\sigma}{(1-q)^{1/\eta}}$ if $q < 1$, and ∞ otherwise.

This expression can be further generalized as follows:

$$\langle x^{n\eta}\rangle_Q \equiv \frac{\int_0^{x_{max}} dx\, x^{n\eta}(e_q^{-(x/\sigma)^\eta})^Q}{\int_0^{x_{max}} dx\, (e_q^{-(x/\sigma)^\eta})^Q}$$

$$= \sigma^{n\eta} \frac{(-1)^n \prod_{k=0}^{n-1}(k\eta+1)}{\prod_{k=1}^{n}[-(1+Q-q)k\eta+(k-1)\eta Q-1+q]}, \qquad (A.55)$$

with $\eta > 0, n = 1, 2, 3...$, and $x_{max} = \frac{\sigma}{(1-q)^{1/\eta}}$ if $q < 1$, and ∞ otherwise.

The differential equation

$$\frac{dy}{dx} = \frac{y^{q-\alpha}}{[1+(1-q)x]^{\frac{\alpha}{q-1}}} \quad (y(0)=1), \qquad (A.56)$$

yields the solution $y = e_q^x$ ($\forall\alpha$); $\alpha = 0$ corresponds to $\frac{dy}{dx} = y^q$ and $\alpha = (q-1)$ corresponds to $\frac{dy}{dx} = \frac{y}{[1+(1-q)x]}$.[2]

We can define

$$\langle g(x)\rangle_{f(q)} \equiv \frac{\int dx\, g(x)[P(x)]^{f(q)}}{\int dx\, [P(x)]^{f(q)}}, \qquad (A.57)$$

$P(x)$ being a normalized distribution of probabilities, and $f(q)$ a function satisfying $f(1) = 1$. Therefore $q = 1$ yields $\langle g(x)\rangle_1 = \langle g(x)\rangle \equiv \int dx f(x)P(x)$.

A possible choice is given by $g(x) = |x|^m$ and $f(q) = 1 + m(q-1)$ with $m = 0, 1, 2, \ldots$.

If $P(x) \propto e_q^{-\beta x^2}$ ($\beta > 0$; $x \in \mathbb{R}$) we have that the q-kurtosis κ_q is given by

$$\kappa_q \equiv \frac{1}{3}\frac{\langle x^4\rangle_{2q-1}}{\langle x^2\rangle_q^2} = \frac{3-q}{1+q} \quad (q<3). \qquad (A.58)$$

Consequently, if we have another distribution, noted $P'(x)$, the discrepancy of its q-kurtosis κ'_q from $(3-q)/(1+q)$ is a measure of the non-q-Gaussianity of $P'(x)$, in the same sense that the discrepancy of its standard kurtosis $\kappa' \equiv \langle x^4\rangle/3\langle x^2\rangle^2$ from unity is a measure of its non-Gaussianity.

---

[2] This differential equation and its solution are used in [1254, 1255].

If $P(x) \propto e_q^{-\beta x}$ ($\beta > 0$; $x \geq 0$) we have that the *q-ratio* ρ_q is given by

$$\rho_q \equiv \frac{1}{2} \frac{\langle x^2 \rangle_{2q-1}}{\langle x \rangle_q^2} = 2 - q \quad (q < 2). \tag{A.59}$$

Consequently, if we have another distribution, noted $P'(x)$, the discrepancy of its *q*-ratio ρ_q' from $(2 - q)$ is a measure of the non-*q*-exponentiality of $P'(x)$, in the same sense that the discrepancy of its standard ratio $\rho' \equiv \langle x^2 \rangle / 2\langle x \rangle^2$ from unity is a measure of its non-exponentiality.

Appendix B
Escort Distributions and q-Expectation Values

B.1 First Example

In order to illustrate the practical utility and peculiar properties of escort distributions and their associated q-expectation values, we introduce and analyze here a pedagogical example [1256].[3]

Let us assume that we have a set of empirical distributions $\{f_n(x)\}$ ($n = 1, 2, 3, ...$) defined as follows:

$$f_n(x) = \frac{A_n}{(1 + \lambda x)^\alpha} \quad (\lambda > 0; \ \alpha \geq 0), \tag{B.1}$$

if $0 \leq x \leq n$, and zero otherwise. Normalization of $f_n(x)$ immediately yields

$$A_n = \frac{\lambda(\alpha - 1)}{1 - (1 + \lambda n)^{1-\alpha}} . \tag{B.2}$$

In order to have finite values for A_n, $\forall n$, including $n \to \infty$ (i.e., $0 < A_\infty < \infty$), $\alpha > 1$ is needed. Consequently

$$A_\infty = \lambda(\alpha - 1). \tag{B.3}$$

By identifying

$$\alpha \equiv \frac{1}{q - 1}, \tag{B.4}$$

$$\lambda \equiv \beta(q - 1), \tag{B.5}$$

---

[3] The present illustration has greatly benefited from lengthy discussions with S. Abe, who launched [1257] interesting questions regarding q-expectation values, and with E.M.F. Curado.

© Springer Nature Switzerland AG 2023
C. Tsallis, *Introduction to Nonextensive Statistical Mechanics*,
https://doi.org/10.1007/978-3-030-79569-6

Fig. B.1 The distributions $f_n(x)$ for $n = 1, 2, 3, \infty$ (from top to bottom) for $(\lambda, \alpha) = (2, 3/2)$. From [1256]

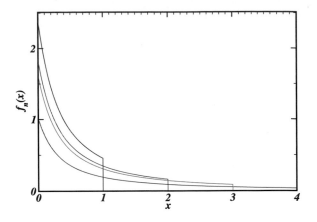

Eq. (B.1) can be rewritten as

$$f_n(x) = A_n \, e_q^{-\beta x} \quad (\beta > 0; \; q \geq 1). \tag{B.6}$$

The variable $x \geq 0$ could be a physical quantity, say earthquake intensity, measured along small intervals, say 10^{-6}, so small that sums can be replaced by integrals within an excellent approximation. The empiric distribution $f_n(x)$ could correspond to different seismic regions, say region 1 (for $n = 1$), region 2 (for $n = 2$), and so on. See Fig. B.1. Suppose we want to characterize the distribution $f_n(x)$ through its mean value. A straightforward calculation yields

$$\langle x \rangle^{(n)} \equiv \int_0^n dx \, x f_n(x) = \frac{1 - (1 + \lambda n)^\alpha + \lambda n [\alpha + (\alpha - 1)\lambda n]}{(\alpha - 2)\lambda(1 + \lambda n)[1 - (1 + \lambda n)^{\alpha-1}]}. \tag{B.7}$$

This quantity is *finite* for all n (including $n \to \infty$) for $\alpha > 2$, *but* $\langle x \rangle^{(\infty)}$ diverges for $1 < \alpha \leq 2$. In other words, we can use it to characterize $f_n(x)$, $\forall n$, for $\alpha > 2$, but we can *not* for $1 < \alpha \leq 2$. The problem is illustrated in Fig. B.2 for $\alpha = 3/2$. This difficulty disappears if we use instead the q-expectation value, defined as follows:

$$\langle x \rangle_q^{(n)} \equiv \frac{\int_0^n dx \, x \, [f_n(x)]^q}{\int_0^n dx \, [f_n(x)]^q} = \frac{(1 + \lambda n)^\alpha - 1 - \lambda \alpha n}{\lambda(\alpha - 1)[(1 + \lambda n)^\alpha - 1]}, \tag{B.8}$$

which equals of course the standard mean value but calculated with the *escort distribution* (introduced in chaos theory [163])

$$F_n(x) \equiv \frac{[f_n(x)]^q}{\int_0^n dx \, [f_n(x)]^q}. \tag{B.9}$$

instead of with the original distribution $f_n(x)$. It follows immediately that

$$\langle x \rangle_q^{(\infty)} = \frac{1}{\lambda(\alpha - 1)},$$ (B.10)

which is *finite* for *all* values $\alpha > 1$, i.e., as long as the norm itself is finite. The problem that we exhibited with the standard mean value reappears, and even worse, if we are interested in the second moment of $f_n(x)$. We have that

$$[\sigma^{(n)}]^2 \equiv \langle x^2 \rangle^{(n)} - [\langle x \rangle^{(n)}]^2$$ (B.11)

is *finite* for all values of n (including for $n \to \infty$) only if $\alpha > 3$, but $\sigma^{(\infty)}$ diverges for $1 < \alpha \leq 3$: see Fig. B.2. This divergence can be regularized by considering [505][4]

$$[\sigma_{2q-1}^{(n)}]^2 \equiv \langle x^2 \rangle_{2q-1}^{(n)} - [\langle x \rangle_{2q-1}^{(n)}]^2 = \frac{\int_0^n dx\, x^2\, [f_n(x)]^{2q-1}}{\int_0^n dx\, [f_n(x)]^{2q-1}} - \left[\frac{\int_0^n dx\, x\, [f_n(x)]^{2q-1}}{\int_0^n dx\, [f_n(x)]^{2q-1}} \right]^2,$$ (B.12)

whose $n \to \infty$ limit is given by

$$[\sigma_{2q-1}^{(\infty)}]^2 = \frac{1+\alpha}{(\alpha - 1)\alpha^2 \lambda^2}.$$ (B.13)

This quantity, like the norm and $\langle x \rangle_q^{(\infty)}$, is finite for all $\alpha > 1$: see Fig. B.2. As a matter of fact, the moments of *all* orders are finite for $\alpha > 1$ if, instead of the original distribution $f_n(x)$, we use the appropriate escort distributions [505]. Indeed, if we consider the *m-th* order moment $\langle x^m \rangle_{q_m}^{(n)}$ with $q_m = mq - (m-1)$ and $m = 0, 1, 2, 3, \ldots$, *all* these moments are *finite* for any $\alpha > 1$ and any n, and they *all diverge* for $\alpha \leq 1$ and $n \to \infty$ (see also Sect. 4.9).

Summarizing:

(i) If we want to characterize, for *all* values of n (including $n \to \infty$), the functional density form (B.1) for all $\alpha > 1$, we can perfectly well do so by using the appropriate escort distributions, whereas the standard mean value is admissible only for $\alpha > 2$, and the standard variance is admissible only for $\alpha > 3$;

(ii) If we only want to characterize, for all $\alpha > 1$ and finite n, which seismic region (in our example with earthquakes) is more dangerous, we can do so either with the standard mean value or with the q-mean value; obviously, the larger n is, the more seismically dangerous the region is;

(iii) If we only want to characterize, for all $\alpha > 1$ and finite n, the size of the fluctuations, we can do so either with the standard variance or with the q-variance; obviously, the larger n is, the larger the fluctuations are.

As we have illustrated, the problem of the empirical verification of a specific *analytic form* for a distribution of probabilities theoretically argued is quite different

---

[4] For the present purpose, we can also use $\langle (x - \langle x \rangle_q)^2 \rangle_{2q-1}^{(n)} = \langle x^2 \rangle_{2q-1}^{(n)} - 2\langle x \rangle_q^{(n)} \langle x \rangle_{2q-1}^{(n)} + (\langle x \rangle_q^{(n)})^2$. In contrast, we cannot use $\langle x^2 \rangle_{2q-1}^{(n)} - (\langle x \rangle_q^{(n)})^2$; indeed, it becomes negative for n large enough.

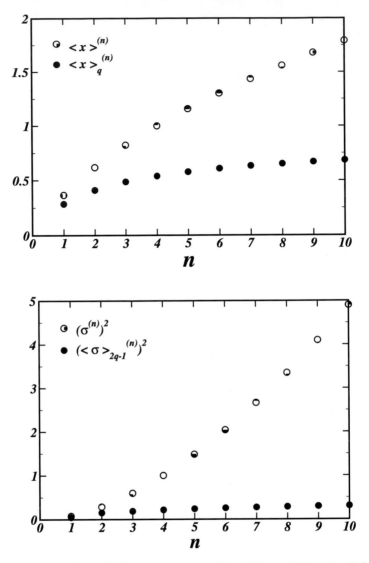

Fig. B.2 The n-dependences of relevant average quantitites of the model $(\lambda, \alpha) = (2, 3/2)$; $q = 1 + \frac{1}{\alpha}$. *Top:* Expectation value $\langle x \rangle^{(n)} \equiv \int_0^n dx\, x f_n(x)$ ($\lim_{n \to \infty} \langle x \rangle^{(n)} = \infty$), and q-expectation value $\langle x \rangle_q^{(n)} \equiv \frac{\int_0^n dx\, x\, [f_n(x)]^q}{\int_0^n dx\, [f_n(x)]^q}$ ($\lim_{n \to \infty} \langle x \rangle_q^{(n)} = \frac{1}{\lambda(\alpha-1)}$). *Bottom:* Variance $[\sigma^{(n)}]^2 \equiv \langle x^2 \rangle^{(n)} - [\langle x \rangle^{(n)}]^2$ ($\lim_{n \to \infty} [\sigma^{(n)}]^2 = \infty$), and $(2q - 1)$-variance $[\sigma_{2q-1}^{(n)}]^2 \equiv \langle x^2 \rangle_{2q-1}^{(n)} - [\langle x \rangle_{2q-1}^{(n)}]^2$ ($\lim_{n \to \infty} [\sigma_{2q-1}^{(n)}]^2 = \frac{1+\alpha}{\lambda^2 \alpha^2 (\alpha-1)}$). From [1256]

from the problem on how *successive experimental data keep filling this functional form*. In particular, the problem of its largest empirical values constitutes an entire branch of mathematical statistics, usually referred to as *extreme value statistics* (or *extreme value theory*) (see, for instance, [1258]), and remains out of the scope of the present book.

B.2 Second Example

In the previous example, we have used academically-constructed "empiric" distributions. However, exactly the same scenario is encountered if we use random models like the one introduced in [961]. The variance of q-Gaussian distributions is finite for $q < 5/3$, and diverges for $5/3 \leq q < 3$; their norm is finite for $q < 3$. Two typical cases are shown in Fig. B.3, one of them for $q < 5/3$, and the other one for $q > 5/3$. In both cases, the fluctuations of the variance $V[X] \equiv \sigma^2$ are considerably larger than those of the q-variance $V_q[X] \equiv \sigma_q^2$. For $q < 5/3$, the variance converges very slowly to its exact asymptotic value; for $q > 5/3$ does not converge at all. In all situations, the q-variance quickly converges to its asymptotic value, which is always finite, thus constituting a very satisfactory characterization. The reasons for precisely considering in this example the q-variance $V_q[X]$, and not any other, are the same that have been indicated in the previous example (see [505] and Sect. 4.9).

B.3 Remarks

Let us end by some general remarks. Abe has shown [1257] that the q-expectation value $\langle Q \rangle_q \equiv \frac{\sum_{i=1}^{W} Q_i p_i^q}{\sum_{i=1}^{W} p_i^q}$, where $\{Q_i\}$ corresponds to any physical quantity, is unstable (in a uniform continuity sense, i.e., similar to the criterion introduced by Lesche for any entropic functional [96], *not* in the thermodynamic sense) for $q \neq 1$, whereas it is stable $q = 1$. If we consider the particular case $Q_i = (1 - q)_{i,j}$, where we use Kroenecker's delta function, we obtain as a corollary that the escort distribution itself is unstable for $q \neq 1$.[5] This fact illustrates a simple property, namely that two quantities can be Lesche-stable, and nevertheless their ratio can be Lesche-unstable. In the present example, both p_i^q and $\sum_{i=1}^{W} p_i^q$ are stable, $\forall q > 0$, but $\frac{p_i^q}{\sum_{j=1}^{W} p_j^q}$ is unstable for $q \neq 1$. The possible epistemological implications of such subtle properties for the 1998 formulation [69] of nonextensive statistical mechanics might benefit from further analysis. The fact stands however that the characterization of the (asymptotic) *power laws* which naturally emerge within this theory undoubtedly is very conveniently done through q-expectation values, whereas it is not so through standard expectation values (which *necessarily* diverge for *all* moments whose order exceeds

---

[5] This special property was also directly established by Curado [1259].

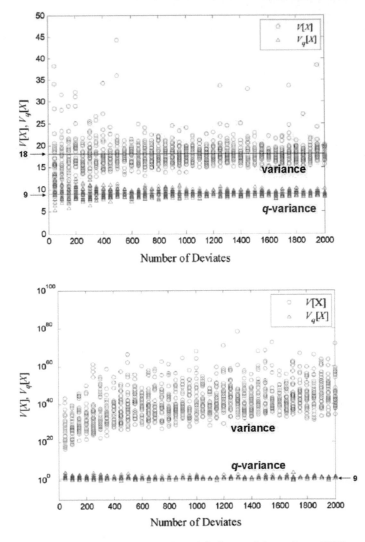

Fig. B.3 Behavior, as functions of the number of deviates, of the variance $V[X] \equiv \sigma^2$ and the q-variance $V_q[X] \equiv \sigma_q^2$ of typical, stochastically generated, q-Gaussians. *Top:* For $q = 1.4$ ($< 5/3$); $\lim_{\# \, of \, deviates \to \infty} V[X] = 18$, and $\lim_{\# \, of \, deviates \to \infty} V_{1.4}[X] = 9$. Notice that the level of fluctuations of $V[X]$ for 2000 deviates is similar to that of $V_{1.4}[X]$ for only 200 deviates. *Bottom:* For $q = 2.75$ ($> 5/3$); $\lim_{\# \, of \, deviates \to \infty} V[X] = \infty$, and $\lim_{\# \, of \, deviates \to \infty} V_{2.75}[X] = 9$. Notice how huge is the ordinate scale. From [961]

some specific one, which depends on the exponent of the power-law). The situation is well illustrated for the constraints to be used for the canonical ensemble (the system being in contact with some thermostat). If, together with the norm constraint $\sum_{i=1}^{W} p_i = 1$, we impose the energy constraint as $\langle \mathcal{H} \rangle_q \equiv \frac{\sum_{i=1}^{W} E_i p_i^q}{\sum_{i=1}^{W} p_i^q} = U_q$, where $\{E_i\}$ are the energy eigenvalues and U_q a fixed *finite* real number, we are dealing (unless we provide some additional qualification) with an unstable quantity. But, interestingly enough, if we impose it in the slightly different form $\sum_{i=1}^{W} E_i \, p_i^q - U_q \sum_{i=1}^{W} p_i^q = 0$, we are simply dealing with two quantities which are both stable. Moreover, the generic discontinuity property focused in [1257] might disappear, in all cases or in the physically relevant ones (see [1260]), if we address *relative* (to some physically sound reference value) quantities, as done for Lesche-stability.

Not a single indication exists that such considerations modify whatsoever at the operational or practical level for the experimentally falsifiable predictions or validations of nonextensive statistical mechanics. However, at the mathematical level, they are kind of intriguing, and would deserve further clarification.

References

1. J.W. Gibbs, *Elementary Principles in Statistical Mechanics-Developed with Especial Reference to the Rational Foundation of Thermodynamics* (C. Scribner's Sons, New York, 1902)
2. J.W. Gibbs, *The Collected Works*, vol. 1 (Yale University Press, New Haven, Thermodynamics, 1948)
3. J.W. Gibbs, *Elementary Principles in Statistical Mechanics* (OX Bow Press, Woodbridge, Connecticut, 1981)
4. G. Nicolis, D. Daems, Probabilistic and thermodynamics aspects of dynamical systems. Chaos **8**, 311–320 (1998)
5. M. Tribus, E.C. McIrvine, Energy and information. Sci. Am. **224**, 178–184 (1971). (September)
6. M. Martin, J. England, *Mathematical Theory of Entropy* (Addison-Wesley, 1981)
7. L. Boltzmann, Weitere Studien über das Wärmegleichgewicht unter Gas molekúlen [Further Studies on Thermal Equilibrium Between Gas Molecules]. Wien. Ber. **66**, 275 (1872)
8. L. Boltzmann, *Uber die Beziehung eines allgemeine mechanischen Satzes zum zweiten Haupsatze der Warmetheorie*, Sitzungsberichte, K. Akademie der Wissenschaften in Wien. Math. Naturwissenschaften **75**, 67–73 (1877)
9. L. Boltzmann, On the relation of a general mechanical theorem to the second law of thermodynamics [English translation], in *Kinetic Theory*, vol. 2, ed. by S. Brus Irreversible Processes (Pergamon Press, Oxford, 1966), pp. 188–193
10. K. Sharp, F. Matschinsky, *Translation of Ludwig Boltzmann's Paper "On the Relationship between the Second Fundamental Theorem of the Mechanical Theory of Heat and Probability Calculations Regarding the Conditions for Thermal Equilibrium" Sitzungberichte der Kaiserlichen Akademie der Wissenschaften. Mathematisch-Naturwissen Classe. Abt. II, LXXVI 1877, pp 373–435 (Wien. Ber. 1877, 76:373-435). Reprinted in Wiss. Abhandlungen, Vol. II, reprint 42, p. 164–223, Barth, Leipzig, 1909.* Entropy **17**, 1971–2009 (2015)
11. J.C. Maxwell, Philos. Mag. (Ser. 4) **19**, 19 (1860)
12. G. Casati, T. Prosen, Triangle map: a model of quantum chaos. Phys. Rev. Lett. **85**, 4261–4264 (2000)
13. Aristotle, *Poetics* (335 BCE)
14. E. Poisson, C.M. Will, *Gravity—Newtonian, Post-Newtonian* (Cambridge University Press, Relativistic, 2014)
15. P.W. Kasteleyn and C.M. Fortuin, *Phase transitions in lattice systems with random local properties*. J. Phys. Soc. Jpn. **26**(Suppl.), 11 (1969)
16. P.G. de Gennes, Exponents for the excluded volume problem as derived by the Wilson method. Phys. Lett. A **38**, 339 (1972)

© Springer Nature Switzerland AG 2023
C. Tsallis, *Introduction to Nonextensive Statistical Mechanics*,
https://doi.org/10.1007/978-3-030-79569-6

17. O. Penrose, *Foundations of Statistical Mechanics: A Deductive Treatment* (Pergamon, Oxford, 1970), p. 167
18. M. Gell-Mann, C. Tsallis (eds.), *Nonextensive Entropy—Interdisciplinary Applications* (Oxford University Press, New York, 2004)
19. M. Gell-Mann, *The Quark and the Jaguar* (W.H. Freeman, New York, 1994)
20. L.O. Chua, Local activity is the origin of complexity. Int. J. Bifurc. Chaos **15**(11), 3435–3456 (2005)
21. M. Baranger, *Chaos, complex and entropy—A physics talk for non-physicists* (2000). https://issuu.com/gfbertini/docs/chaos__complexity_and_entropy_-_a_physics_talk_for
22. M. Baranger, Why Tsallis statistics? Physica A **305**, 27–31 (2002)
23. P. Bak, *How Nature Works: the Science of Self-Organized Criticality* (Springer, New York, 1996)
24. H.J. Jensen, *Self-Organized Criticality: emergent Complex Behavior in Physical and Biological Systems Lecture Notes in Physics*. (Cambridge University Press, Cambridge, 1998)
25. D. Sornette, *Critical Phenomena in Natural Sciences: Chaos, Fractals, Selforganization and Disorder, Series in Synergetics* (Springer, New York, 2001)
26. F. Mallamace, H.E. Stanley (eds.), *The Physics of Complex Systems*, Vol. 155 International School of Physics Enrico Fermi (Italian Physical Society, 2004), 640 pp
27. L. Boltzmann, *Vorlesungen uber Gastheorie* (Leipzig, 1896) [*Lectures on Gas Theory*, transl. S. Brush (Univ. California Press, Berkeley, 1964], Part II, Chapter I, Paragraph 1, page 217
28. A. Einstein, Theorie der Opaleszenz von homogenen Flüssigkeiten und Flüssigkeitsgemischen in der Nähe des kritischen Zustandes. Annalen der Physik **33**, 1275 (1910)
29. E.G.D. Cohen, Boltzmann and Einstein: statistics and dynamics—An unsolved problem, Boltzmann Award Lecture at Statphys-Bangalore-2004. Pramana **64**, 635–643 (2005)
30. E. Fermi, *Thermodynamics*, (Dover, New York, 1936), p.53
31. E. Majorana, *The value of statistical laws in physics and social sciences*, in G. Gentile Jr. in *Scientia* **36**, 58 (1942), in Italian. Translated into English by R.N. Mantegna, Quantitative Finance **5**, 133–140 (2005)
32. C.E. Shannon, Bell Syst. Tech. J. **27**, 379 and 623 (1948); A mathematical theory of communication. Bell Syst. Tech. J. **27**, 379–423 and 623–656 (1948); and *The Mathematical Theory of Communication* (University of Illinois Press, Urbana, 1949)
33. L. Tisza, *Generalized Thermodynamics*, (MIT Press, Cambridge, Massachusetts, 1961), p. 123
34. P.T. Landsberg, *Thermodynamics and Statistical Mechanics (Oxford University Press, New York, 1978) and* (Dover, New York, 1990)
35. P.T. Landsberg, Is equilibrium always an entropy maximum? J. Stat. Phys. **35**, 159 (1984)
36. N.G. van Kampen, *Stochastic Processes in Physics and Chemistry* (North-Holland, Amsterdam, 1981)
37. L.G. Taff, *Celestial Mechanics* (Wiley, New York, 1985)
38. W.C. Saslaw, *Gravitation Physics of Stellar and Galactic Systems* (Cambridge University Press, Cambridge, 1985)
39. R. Balescu, *Equilibrium and Nonequilibrium Statistical Mechanics* (Wiley, 1975, New York)
40. D. Ruelle, *Thermodynamic Formalism—The Mathematical Structures of Classical Equilibrium Statistical Mechanics, Encyclopedia of Mathematics and its Applications*, vol. 5 (Addison-Wesley Publishing Company, Reading, Massachusetts, 1978)
41. D. Ruelle, *Thermodynamic Formalism—The Mathematical Structures of Equilibrium Statistical Mechanics*, 2nd edn. (Cambridge University Press, Cambridge, 2004)
42. F. Takens, in *Structures in Dynamics—Finite Dimensional Deterministic Studies*, eds. by H.W. Broer, F. Dumortier, S.J. van Strien, F. Takens (North-Holland, Amsterdam, 1991), p. 253
43. R. Balian, *From Microphysics to Macrophysics* (Springer-Verlag, Berlin, 1991), pages 205 and 206. Original French edition: *Du microscopique au macroscopique*, Cours de l' Ecole Polytechnique (Ellipses, Paris, 1982)

44. J. Maddox, When entropy does not seem extensive. Nature **365**, 103 (1993)
45. M. Srednicki, Entropy and area. Phys. Lett. **71**, 666–669 (1993)
46. A.C.D. van Enter, R. Fernandez, A.D. Sokal, regularity properties and pathologies of position-space renormalization-group transformations: scope and limitations of Gibbsian theory. J. Stat. Phys. **72**, 879–1167 (1993)
47. C. Tsallis, Possible generalization of Boltzmann-Gibbs statistics. J. Stat. Phys. **52**, 479–487 (1988) [First appeared as preprint in 1987: CBPF-NF-062/87, ISSN 0029-3865, Centro Brasileiro de Pesquisas Fisicas, Rio de Janeiro]
48. E.G.D. Cohen, Statistics and dynamics. Phys. A **305**, 19 26 (2002)
49. C. Tsallis, Nonextensive statistical mechanics and nonlinear dynamics, in *Interdisciplinary aspects of Turbulence, Lecture Notes in Physics*, vol. 756, eds. by W. Hillebrandt and F. Kupka, (Springer, Berlin, 2009), pp. 21–48
50. W. Braun, K. Hepp, The Vlasov dynamics and Its fluctuations in the $1/N$ limit of interacting classical particles. Commun. Math. Phys. **56**, 125–146 (1977)
51. C. Jarzynski, Nonequilibrium equality for free energy differences. Phys. Rev. Lett. **78**, 2690–2693 (1997)
52. C. Beck, E.G.D. Cohen, Superstatistical generalization of the work fluctuation theorem. Phys. A **344**, 393 (2004)
53. S. Abe, C. Beck, E.G.D. Cohen, Superstatistics, thermodynamics, and fluctuations. Phys. Rev. E **76**, 031102 (2007)
54. M. Ponmurugan, Tsallis statistics generalization of nonequilibrium work relations. Phys. Rev. E **93**, 032107 (2016)
55. H. Miyahara, K. Aihara, Work relations with measurement and feedback control on nonuniform temperature systems. Phys. Rev. E **98**, 042138 (2018)
56. E. Korkmazhan, Fluctuation relations for non-Markovian and heterogeneous temperature systems. Phys. A **537**, 122615 (2020)
57. J. Korbel, D.H. Wolpert, Stochastic thermodynamics and fluctuation theorems for non-linear systems. New J. Phys. **23**, 033049 (2021)
58. A. Pluchino, V. Latora, A. Rapisarda, Dynamics and thermodynamics of a model with long-range interactions. Continuum Mech. Thermodyn. **16**, 245–255 (2004)
59. A. Rapisarda, A. Pluchino, *Nonextensive thermodynamics and glassy behavior*, in *Nonextensive Statistical Mechanics: new Trends, New perspectives*, vol. 36. eds. by J.P. Boon, C. Tsallis (Europhysics News, 2005), p. 202
60. F.A. Tamarit, C. Anteneodo, Relaxation and ageing in long-range interacting systems, in *Nonextensive Statistical Mechanics: new Trends, New perspectives*, vol. 36, eds. by J.P. Boon, C. Tsallis (Europhysics News 2005), p. 194
61. A. Pluchino, A. Rapisarda, C. Tsallis, Nonergodicity and central limit behavior in long-range Hamiltonians. Europhys. Lett. **80**, 26002 (2007)
62. S. Thurner, Nonextensive statistical mechanics and complex networks, in *Nonextensive Statistical Mechanics: new Trends, New perspectives*, vol. 36, eds. J.P. Boon, C. Tsallis (Europhysics News, 2005), p. 218
63. A. Carati, Thermodynamics and time averages. Phys. A **348**, 110 (2005)
64. A. Carati, Time-averages and the heat theorem, in *Complexity, Metastability and Nonextensivity*, eds. by C. Beck, G. Benedek, A. Rapisarda, C. Tsallis (World Scientific, Singapore, 2005), p. 55
65. A. Carati, On the fractal dimension of orbits compatible with Tsallis statistics. Phys. A **387**, 1491–1503 (2008)
66. S. Abe, Y. Nakada, Temporal extensivity of Tsallis' entropy and the bound on entropy production rate. Phys. Rev. E **74**, 021120 (2006)
67. E.M.F. Curado, C. Tsallis, Generalized statistical mechanics: connection with thermodynamics. J. Phys. A **24**, L69 (1991)
68. E.M.F. Curado, C. Tsallis, Generalized statistical mechanics: connection with thermodynamics. Corrigenda: J. Phys. A **24**, 3187 (1991); J. Phys. A **25**, 1019 (1992)

69. C. Tsallis, R.S. Mendes, A.R. Plastino, The role of constraints within generalized nonextensive statistics. Phys. A **261**, 534 (1998)
70. S.R.A. Salinas, C. Tsallis (eds.), Nonextensive statistical mechanics and thermodynamics. Braz. J. Phys. **29**(1) (1999)
71. C. Tsallis, Nonextensive statistics: theoretical, experimental and computational evidences and connections, in *Nonextensive Statistical Mechanics and Thermodynamics*, eds. by S.R.A. Salinas, C. Tsallis, Braz. J. Phys. **29**(1), 1–35 (1999)
72. C. Tsallis, Talk at the IMS winter school on nonextensive generalization of Boltzmann-Gibbs statistical mechanics and its applications
73. S. Abe, Y. Okamoto (eds.), *Nonextensive Statistical Mechanics and Its Applications, Series Lecture Notes in Physics* (Springer, Berlin, 2001)
74. P. Grigolini, C. Tsallis, B.J. West (eds.), Classical and quantum complexity and nonextensive. Thermodyn. Chaos Solit. Fract. **13**(3) (2002)
75. G. Kaniadakis, M. Lissia, A. Rapisarda (eds.), Non extensive thermodynamics and physical applications. Phys. A **305**, 1/2 (2002)
76. M. Sugiyama (ed.), *Nonadditive Entropy and Nonextensive Statistical Mechanics, Continuum Mechanics and Thermodynamics* (Springer, Heidelberg, 2004)
77. H.L. Swinney, C. Tsallis (eds.), Anomalous distributions, nonlinear dynamics and nonextensivity. Phys. D **193**(1–4) (2004)
78. G. Kaniadakis, M. Lissia (eds.), News and expectations in thermostatistics. Phys. A **340**(1–3) (2004)
79. H.J. Herrmann, M. Barbosa, E.M.F. Curado (eds.), Trends and perspectives in extensive and non-extensive statistical mechanics. Phys. A **344**(3/4) (2004)
80. C. Beck, G. Benedek, A. Rapisarda, C. Tsallis (eds.), *Complexity, Metastability and Nonextensivity* (World Scientific, Singapore, 2005)
81. J.P. Boon, C. Tsallis (eds.), Nonextensive statistical mechanics: new trends, new perspectives. Europhys. News **36**(6), 183–231 (EDP Sciences, Paris, 2005)
82. G. Kaniadakis, A. Carbone, M. Lissia (eds.), News, expectations and trends in statistical physics. Phys. A **365**(1) (2006)
83. S. Abe, M. Sakagami, N. Suzuki (eds.), Complexity and nonextensivity—New trends in statistical mechanics, in *Progress of Theoretical Physics Supplement*, vol. 162 (Physical Society of Japan, 2006)
84. S. Abe, H. Herrmann, P. Quarati, A. Rapisarda, C. Tsallis (eds.), Complexity, metastability and nonextensivity. Am. Inst. Phys. Conf. Proc. **965** (New York, 2007)
85. C. Tsallis, The nonadditive entropy S_q and its applications in physics and elsewhere: some remarks, in Special Issue *Tsallis Entropy*, vol. 13, ed. by A. Anastasiadis (2011), pp. 1765–1804
86. C. Tsallis, The nonadditive entropy SQ: a door open to the nonuniversality of the mathematical expression of the Clausius thermodynamic entropy in terms of the probabilities of the microscopic configurations, in *Concepts and Recent Advances in Generalized Information Measures and Statistics*, eds. by A.M. Kowalski, R. Rossignoli, E.M.F. Curado (Bentham Science Publishers, 2013), pp. 56–79
87. C. Tsallis, An introduction to nonadditive entropies and a thermostatistical approach of inanimate and living matter. Contemp. Phys. **55**(3), 179–197 (2014)
88. C. Tsallis, Thermodynamics and statistical mechanics for complex systems—Foundations and applications, Lecture presented at the XXVIIth Marian Smoluchowski Symposium on Statistical Physics (Zakopane, Poland, 22–26 September 2014). Acta Phys. Polonica B **46**(6), 1089–1101 (2015)
89. C. Tsallis, Approach of complexity in nature: entropic nonuniqueness. Axioms **5**, 20 (2016)
90. C. Tsallis, On the foundations of statistical mechanics. Eur. Phys. J. Spec. Top. **226**, 1433–1443 (2017)
91. A. Rapisarda, S. Thurner, C. Tsallis, Nonadditive entropies and complex systems. Entropy **21**, 538 (2019)

92. C. Beck, G. Benedek, G. Livadiotis, A. Rapisarda, U. Tirnakli, C. Tsallis, Nonextensive statistical mechanics, superstatistics and beyond: theory and applications in astrophysical and other complex systems. Eur. Phys. J. Spec. Top. **229**, 707–709 (2020)

93. J. von Neumann, Thermodynamik quantenmechanischer Gesamtheiten, Nachrichten von der Gesellschaft der Wissenschaften zu Gottingen S. 273–291 (1927)

94. J. von Neumann, Mathematische Grundlagen der Quantenmechanik (Springer, Berlin, 1932)

95. J. von Neumann, Mathematical Foundations of Quantum Mechanics (Princeton University Press, 1955)

96. D. Lesche, Instabilities of Renyi entropies. J. Stat. Phys. **27**, 419–422 (1982)

97. S. Abe, Stability of Tsallis entropy and instabilities of Renyi and normalized Tsallis entropies. Phys. Rev. E **66**, 046134 (2002)

98. C. Tsallis, E. Brigatti, Nonextensive statistical mechanics: a brief introduction, in *Continuum Mechanics and Thermodynamics*, vol. 16, ed. by M. Sugiyama (Springer-Verlag, Heidelberg, 2004), p. 223

99. S. Abe, B. Lesche, J. Mund, How should the distance of probability assignments be judged? J. Stat. Phys. **128**, 1189–1196 (2007)

100. A.I. Khinchin, Uspekhi Matem. Nauk **8**, 3 (1953); (R.A. Silverman, M.D. Friedman (Trans.), *Mathematical Foundations of Information Theory* (Dover, New York, 1957)

101. C. Tsallis, Nonextensive statistical mechanics and thermodynamics: historical background and present status, in *Nonextensive Statistical Mechanics and Its Applications, Lecture Notes in Physics*, eds. by S. Abe, Y. Okamoto (Springer, Heidelberg, 2001)

102. M. Hotta, I. Joichi, Composability and generalized entropy. Phys. Lett. A **262**, 302 (1999)

103. C. Tsallis, *Nonextensive Statistical Mechanics—Approaching a Complex World*, 1st edn. (Springer, New York, 2009)

104. C. Tsallis, Entropy. Encyclopedia **2**, 264–300 (MDPI Publishers, 2022). https://doi.org/10. 3390/encyclopedia2010018

105. V.I. Arnold, Russian Math. Surv. **18**, 85 (1964)

106. G.M. Zaslavsky, R.Z. Sagdeev, D.A. Usikov, A.A. Chernikov, *Weak Chaos and Quasi-Regular Patterns* (Cambridge University Press, Cambridge, 2001)

107. G.M. Zaslavsky, Chaos, fractional kinetics, and anomalous transport. Phys. Rep. **371**, 461 (2002)

108. G. Benettin, L. Galgani, A. Giorgilli, J.M. Strelcyn, Meccanica **15**, 21 (1980)

109. S.H. Strogatz, *Nonlinear Dynamics and Chaos: with Applications to Physics, Biology, Chemistry and Engineering* (Perseus Books Publishing, Cambridge, MA, 1994)

110. E. Ott, *Chaos in Dynamical Systems*, 2nd edn. (Cambridge University Press, Cambridge, 2002)

111. A.N. Kolmogorov, A new metric invariant of transitive dynamical systems and of automorphisms of Lebesgue spaces. Dok. Acad. Nauk SSSR **119**, 861 (1958); Y.G. Sinai, On the concept of entropy for a dynamic system. Dok. Acad. Nauk SSSR **124**, 768 (1959)

112. R.C. Hilborn, *Chaos and Nonlinear Dynamics: an Introduction for Scientists and Engineers*, 2nd edn. (Oxford University Press, Oxford, 2000)

113. V. Latora, M. Baranger, Kolmogorov-Sinai entropy rate versus physical entropy. Phys. Rev. Lett. **82**, 520 (1999)

114. Y. Pesin, Russ. Math. Surv. **32**, 55 (1977); Y. Pesin, in *Hamiltonian Dynamical Systems: A Reprint Selection*, eds. by R.S. MacKay, J.D. Meiss (Adam Hilger, Bristol, 1987)

115. S. Kullback, R.A. Leibler, On information and sufficiency. Ann. Math. Stat. **22**, 79 (1951); S.L. Braunstein, Geometry of quantum inference. Phys. Lett. A **219**, 169 (1996)

116. C. Tsallis, Generalized entropy-based criterion for consistent testing. Phys. Rev. E **58**, 1442 (1998)

117. L. Borland, A.R. Plastino, C. Tsallis, Information gain within nonextensive thermostatistics. J. Math. Phys. **39**, 6490 (1998); [Errata: J. Math. Phys. **40**, 2196 (1999)]

118. P.W. Lamberti, A.P. Majtey, Non-logarithmic Jensen-Shannon divergence. Phys. A **329**, 81 (2003)

119. A.P. Majtey, P.W. Lamberti, A. Plastino, A monoparametric family of metrics for statistical mechanics. Phys. A **344**, 547 (2004)

120. S. Amari, H. Nagaoka, Methods of information geometry, in *Translations of Mathematical Monographs*, vol. 191 (Oxford University Press, Oxford, 2000); [Eq. (3.80) in page 72 for the triangle equality]

121. A. Dukkipati, M.N. Murty, S. Bhatnagar, Nonextensive triangle equality and other properties of Tsallis relative-entropy minimization. Phys. A **361**, 124–138 (2005)

122. S. Amari, A. Ohara, H. Matsuzoe, Geometry of deformed exponential families: invariant, dually-flat and conformal geometries. Phys. A **391**(18), 4308–4319 (2012)

123. D. Felice, N. Ay, Canonical divergence for flat α-connections: classical and quantum. Entropy **21**, 831 (2019)

124. N. Ay, J. Jost, H.V. Le, L. Schwachhofer, *Information Geometry* (Springer, 2017)

125. H. Umegaki, Conditional expectation in an operator algebra, IV (entropy and information). Kodai Math. Sem. Rep. **14**, 59–85 (1962)

126. M.E. Fisher, Arch. Rat. Mech. Anal. **17**, 377 (1964); J. Chem. Phys. **42**, 3852 (1965); J. Math. Phys. **6**, 1643 (1965)

127. M.E. Fisher, D. Ruelle, J. Math. Phys. **7**, 260 (1966)

128. M.E. Fisher, J.L. Lebowitz, Commun. Math. Phys. **19**, 251 (1970)

129. F. Baldovin, L.G. Moyano, C. Tsallis, Boltzmann-Gibbs thermal equilibrium distribution for classical systems and Newton law: a computational discussion. Europhys. J. B **52**, 113–117 (2006)

130. F. Jüttner, Das Maxwellsche Gesetz der Geschwindigkeitsverteilung in der Relativtheorie. Annalen der Physik **339**, 856–882 (1911)

131. G. Gentile, Nuovo Cimento **17**, 493–497 (1940); Nuovo Cimento **19**, 109 (1942)

132. M.C.S. Vieira, C. Tsallis, D-dimensional ideal gas in parastatistics: thermodynamic properties. J. Stat. Phys. **48**, 97 (1987)

133. H.S. Robertson, *Statistical Thermophysics* (Prentice-Hall, Englewood Cliffs, New Jersey, 1993)

134. C. Tsallis, What are the numbers that experiments provide? Quimica Nova **17**, 468 (1994)

135. E.P. Borges, On a q-generalization of circular and hyperbolic functions. J. Phys. A **31**, 5281–5288 (1998)

136. S. Watanabe, *Knowing and Guessing* (Wiley, New York, 1969)

137. H. Barlow, Conditions for versatile learning, Helmholtz's unconscious inference, and the task of perception. Vision. Res. **30**, 1561 (1990)

138. J. Havrda, F. Charvat, Quantification method of classification processes—Concept of structural α-entropy. Kybernetika **3**, 30–35 (1967)

139. I. Vajda, Axioms of α-entropy of generalized probability distribution. Kybernetika **4**, 105–112 (1968)

140. Z. Daroczy, Generalized information measures. Inf. Control **16**, 36–51 (1970)

141. J. Lindhard, V. Nielsen, Studies in statistical mechanics. Det Kongelige Danske Videnskabernes Selskab Matematisk-fysiske Meddelelser (Denmark) **38**(9), 1–42 (1971)

142. J. Aczel, Z. Daroczy, On measures of information and their characterization, in *Mathematics in Science and Engineering*, ed. by R. Bellman (Academic Press, New York, 1975)

143. A. Wehrl, General properties of entropy. Rev. Mod. Phys. **50**, 221 (1978)

144. S.M. Stigler, *Statistics on the Table—The History of Statistical Concepts and Methods* (Harvard University Press, Cambridge, MA, 1999)

145. I. Csiszar, Information measures: a critical survey, in *Transactions of the Seventh Prague Conference on Information Theory, Statistical Decision Functions, Random Processes, and the European Meeting of Statisticians, 1974* (Reidel, Dordrecht, 1978), p. 73

146. P.M. Schutzenberger, Contributions aux applications statistiques de la théorie de l' information. Publ. Inst. Statist. Univ. Paris **3**, 3 (1954)

147. B.D. Sharma, D.P. Mittal, New non-additive measures of entropy for discrete probability distributions. J. Math. Sci. **10**, 28 (1975)

148. B.D. Sharma, I.J. Taneja, Entropy of type (α, β) and other generalized measures in information theory. Metrika **22**(1075), 205
149. D.P. Mittal, On some functional equations concerning entropy, directed divergence and inaccuracy. Metrika **22**, 35 (1975)
150. A. Renyi, On measures of information and entropy, in *Proceedings of the Fourth Berkeley Symposium*, vol. 1 (University of California Press, Berkeley, Los Angeles, 1961), p. 547
151. A. Renyi, *Probability Theory* (North-Holland, Amsterdam, 1970)
152. J. Balatoni, A. Renyi, Remarks on entropy, in *Publications of the Mathematical Institute of the Hungarian Academy of Sciences*, vol. 1, pp. 9–40 (1956)
153. A. Renyi, On the dimension and entropy of probability distributions. Acta Math. Acad. Sci. Hungaricae **10**, 193–215 (1959)
154. N. Cressie, T.R.C. Read, Multinomial goodness-of-fit tests. J. R. Stat. Soc. B **46**(3), 440–464 (1984)
155. A.N. Gorban, P.A. Gorban, G. Judge, Entropy: the Markov ordering approach. Entropy **12**, 1145–1193 (2010)
156. G. Hardy, J.E. Littlewood, G. Polya, *Inequalities* (Cambridge University Press, Cambridge, 1952)
157. G.P. Patil, C. Taillie, A study of diversity profiles and orderings for a bird community in the vicinity of Colstrip, Montana, in *Contemporary Quantitative Ecology and Related Ecometrics*, eds. by G.P. Patil, M. Rosenzweig (International Co-operative Publishing House, Fairland, Maryland, USA, 1979), pp. 23–48
158. G.P. Patil, C. Taillie, Diversity as a concept and its measurement. J. Am. Stat. Assoc. **77**, 548–561 (1982)
159. G.P. Patil, Diversity profiles, in *Encyclopedia of Environmetrics*, eds. by A.H. El-Shaarawi, W.W. Piegorsch (Wiley, Chichester, UK, 2002)
160. F. Jackson, Mess. Math. **38**, 57 (1909); Quart. J. Pure Appl. Math. **41**, 193 (1910)
161. S. Abe, A note on the q-deformation theoretic aspect of the generalized entropies in nonextensive physics. Phys. Lett. A **224**, 326–330 (1997)
162. C. Tsallis, P.W. Lamberti, D. Prato, A nonextensive critical phenomenon scenario for quantum entanglement. Phys. A **295**, 158 (2001)
163. C. Beck, F. Schlogl, *Thermodynamics of Chaotic Systems* (Cambridge University Press, Cambridge, 1993)
164. S. Abe, Axioms and uniqueness theorem for Tsallis entropy. Phys. Lett. A **271**, 74 (2000)
165. R.J.V. Santos, Generalization of Shannon' s theorem for Tsallis entropy. J. Math. Phys. **38**, 4104 (1997)
166. A.R. Plastino, A. Plastino, *Generalized entropies*, in *Condensed Matter Theories*, vol. 11, eds. by E.V. Ludeña, P. Vashista, R.F. Bishop (Nova Science Publishers, New York, 1996), p. 327
167. A.R. Plastino, A. Plastino, Tsallis entropy and Jaynes' information theory formalism. Braz. J. Phys. **29**, 50 (1999)
168. P. Jizba, J. Korbel, When Shannon and Khinchin meet Shore and Johnson: equivalence of information theory and statistical inference axiomatics. Phys. Rev. E **101**, 042126 (2020)
169. J.E. Shore, R.W. Johnson, IEEE Trans. Inf. Theory **IT-26**, 26 (1980); **IT-27**, 472 (1981); **IT-29**, 942 (1983)
170. J. Uffink, Can the maximum entropy principle be explained as a consistency requirement? Hist. Phil. Mod. Phys. **26**, 223 (1995)
171. S. Presse, K. Ghosh, J. Lee, K.A. Dill, Nonadditive entropies yield probability distributions with biases not warranted by the data. Phys. Rev. Lett. **111**, 180604 (2013)
172. S. Presse, K. Ghosh, J. Lee, K.A. Dill, Principles of maximum entropy and maximum caliber in statistical physics. Rev. Mod. Phys. **85**, 1115–1141 (2013)
173. S. Presse, Nonadditive entropy maximization is inconsistent with Bayesian updating. Phys. Rev. E **90**, 052149 (2014)

174. C. Tsallis, Conceptual inadequacy of the Shore and Johnson axioms for wide classes of complex systems. Entropy **17**, 2853–2861 (2015); A more precise title would have been *Conceptual inadequacy of the S. Presse et al interpretation of the Shore and Johnson axioms for wide classes of complex systems*

175. S. Presse, K. Ghosh, J. Lee, K.A. Dill, C. Tsallis, Conceptual inadequacy of the Shore and Johnson axioms for wide classes of complex systems. Entropy **17**, 5043–5046 (2015)

176. P. Jizba, J. Korbel, Maximum entropy principle in statistical inference: case for non-Shannonian entropies. Phys. Rev. Lett. **122**, 120601 (2019)

177. S. Abe, General pseudoadditivity of composable entropy prescribed by the existence of equilibrium. Phys. Rev. E **63**, 061105 (2001)

178. E.M.F. Curado, General aspects of the thermodynamical formalism, in *Nonextensive Statistical Mechanics and Thermodynamics*, eds. by S.R.A. Salinas, C. Tsallis, Braz. J. Phys. **29**, 36 (1999)

179. E.M.F. Curado, F.D. Nobre, On the stability of analytic entropic forms. Phys. A **335**, 94 (2004)

180. C. Anteneodo, A.R. Plastino, Maximum entropy approach to stretched exponential probability distributions. J. Phys. A **32**, 1089 (1999)

181. P. Grassberger, M. Scheunert, Some more universal scaling laws for critical mappings. J. Stat. Phys. **26**, 697 (1981)

182. T. Schneider, A. Politi, D. Wurtz, Resistance and eigenstates in a tight-binding model with quasi-periodic potential. Z. Phys. B **66**, 469 (1987)

183. G. Anania, A. Politi, Dynamical behavior at the onset of chaos. Europhys. Lett. **7**, 119 (1988)

184. H. Hata, T. Horita, H. Mori, Dynamic description of the critical 2^∞ attractor and 2^m-band chaos. Progr. Theor. Phys. **82**, 897 (1989)

185. H. Mori, H. Hata, T. Horita, T. Kobayashi, Statistical mechanics of dynamical systems. Progr. Theor. Phys. Suppl. **99**, 1 (1989)

186. C. Tsallis, A.R. Plastino, W.-M. Zheng, Chaos, power-law sensitivity to initial conditions—New entropic representation. Chaos Solitons Fractals **8**, 885–891 (1997)

187. U.M.S. Costa, M.L. Lyra, A.R. Plastino, C. Tsallis, Power-law sensitivity to initial conditions within a logistic-like family of maps: fractality and nonextensivity. Phys. Rev. E **56**, 245 (1997)

188. M.L. Lyra, C. Tsallis, Nonextensivity and multifractality in low-dimensional dissipative systems. Phys. Rev. Lett. **80**, 53 (1998)

189. C.R. da Silva, H.R. da Cruz, M.L. Lyra, Low-dimensional non-linear dynamical systems and generalized entropy. Braz. J. Phys. **29**, 144–152 (1999)

190. U. Tirnakli, C. Tsallis, M.L. Lyra, Circular-like maps: sensitivity to the initial conditions, multifractality and nonextensivity. Eur. Phys. J. B **11**, 309 (1999)

191. V. Latora, M. Baranger, A. Rapisarda, C. Tsallis, The rate of entropy increase at the edge of chaos. Phys. Lett. A **273**, 97 (2000)

192. U. Tirnakli, G.F.J. Ananos, C. Tsallis, Generalization of the Kolmogorov-Sinai entropy: logistic -like and generalized cosine maps at the chaos threshold. Phys. Lett. A **289**, 51 (2001)

193. E.P. Borges, C. Tsallis, G.F.J. Ananos, P.M.C. de Oliveira, Nonequilibrium probabilistic dynamics at the logistic map edge of chaos. Phys. Rev. Lett. **89**, 254103 (2002)

194. F. Baldovin, A. Robledo, Sensitivity to initial conditions at bifurcations in one-dimensional nonlinear maps: rigorous nonextensive solutions. Europhys. Lett. **60**, 518 (2002)

195. F. Baldovin, A. Robledo, Universal renormalization-group dynamics at the onset of chaos in logistic maps and nonextensive statistical mechanics. Phys. Rev. E **66**, R045104 (2002)

196. F. Baldovin, C. Tsallis, B. Schulze, Nonstandard entropy production in the standard map. Phys. A **320**, 184 (2003)

197. G.F.J. Ananos, C. Tsallis, Ensemble averages and nonextensivity at the edge of chaos of one-dimensional maps. Phys. Rev. Lett. **93**, 020601 (2004)

198. F. Baldovin, A. Robledo, Nonextensive Pesin identity—Exact renormalization group analytical results for the dynamics at the edge of chaos of the logistic map. Phys. Rev. E **69**, 045202(R) (2004)

199. U. Tirnakli, C. Tsallis, Chaos thresholds of the z-logistic map: connection between the relaxation and average sensitivity entropic indices. Phys. Rev. E **73**, 037201 (2006)
200. A. Robledo, Incidence of nonextensive thermodynamics in temporal scaling at Feigenbaum points. Phys. A **370**, 449–460 (2006)
201. G.F.J. Ananos, F. Baldovin, C. Tsallis, Anomalous sensitivity to initial conditions and entropy production in standard maps: nonextensive approach. Eur. Phys. J. B **46**, 409–417 (2005)
202. M.A. Fuentes, Y. Sato, C. Tsallis, Sensitivity to initial conditions, entropy production, and escape rate at the onset of chaos. Phys. Lett. A **375**, 2988–2991 (2011)
203. K. Herzfeld, Ann. Phys. **51**, 251 (1926)
204. H.C. Urey, Astrophys. J. **49**, 1 (1924)
205. E. Fermi, Z. Phys. **26**, 54 (1924)
206. M. Planck, Ann. Phys. **75**, 673 (1924)
207. R.H. Fowler, Philos. Mag. **1**, 845 (1926)
208. R.H. Fowler, *Statistical Mechanics* (Cambridge University, Cambridge, 1936), p. 572
209. E.P. Wigner, Phys. Rev. **94**, 77 (1954)
210. R.E. Treves, Phys. Rev. **102**, 1533 (1956)
211. A.I. Larkin, Zh. Exper. Teoret. Fiz. **38**, 1896 (1960); English translation: Sov. Phys.-JETP **11**, 1363 (1960)
212. B.F. Gray, J. Chem. Phys. **36**, 1801 (1962)
213. M. Grabowsky, Rep. Math. Phys. **23**, 19 (1986)
214. S.M. Blinder, Canonical partition function for the hydrogen atom via the Coulomb propagator. J. Math. Phys. **36**, 1208–1216 (1995)
215. L.S. Lucena, L.R. da Silva, C. Tsallis, Departure from Boltzmann-Gibbs statistics makes the hydrogen-atom specific heat a computable quantity. Phys. Rev. E **51**, 6247 (1995)
216. N.M. Oliveira-Neto, E.M.F. Curado, F.D. Nobre, M.A. Rego-Monteiro, Approach to equilibrium of the hydrogen atom at low temperature. Phys. A **374**, 251–262 (2007)
217. J.P. Bouchaud, A. Georges, Anomalous diffusion in disordered media: statistical mechanisms, models and physical applications. Phys. Rep. **195**, 127 (1990)
218. M. Antoni, S. Ruffo, Clustering and relaxation in Hamiltonian long-range dynamics. Phys. Rev. E **52**, 2361–2374 (1995)
219. C. Anteneodo, C. Tsallis, Breakdown of the exponential sensitivity to the initial conditions: role of the range of the interaction. Phys. Rev. Lett. **80**, 5313 (1998)
220. C. Tsallis, Nonextensive thermostatistics and fractals. Fractals **3**, 541 (1995)
221. R.F.S. Andrade, S.T.R. Pinho, Tsallis scaling and the long-range Ising chain: a transfer matrix approach. Phys. Rev. E **71**, 026126 (2005)
222. C. Tsallis, L.J.L. Cirto, Black hole thermodynamical entropy. Eur. Phys. J. C **73**, 2487 (2013)
223. C. Tsallis, L.J.L. Cirto, Thermodynamics is more powerful than the role to it reserved by Boltzmann-Gibbs statistical mechanics. Euro. Phys. J. Spec. Top. **223**, 2161 (2014)
224. C. Tsallis, Beyond Boltzmann-Gibbs-Shannon in physics and elsewhere. Entropy **21**, 696 (2019)
225. S. Umarov, C. Tsallis, *Mathematical Foundations of Nonextensive Statistical Mechanics* (World Scientific, 2022)
226. P. Jund, S.G. Kim, C. Tsallis, Crossover from extensive to nonextensive behavior driven by long-range interactions. Phys. Rev. B **52**, 50 (1995)
227. J.R. Grigera, *Extensive and non-extensive thermodynamics. A molecular dynamic test.* Phys. Lett. A **217**, 47 (1996)
228. S.A. Cannas, F.A. Tamarit, Long-range interactions and non-extensivity in ferromagnetic spin systems. Phys. Rev. B **54**, R12661 (1996)
229. L.C. Sampaio, M.P. de Albuquerque, F.S. de Menezes, Nonextensivity and Tsallis statistics in magnetic systems. Phys. Rev. B **55**, 5611 (1997)
230. L.A. del Pino, P. Troncoso, S. Curilef, Thermodynamics from a scaling Hamiltonian. Phys. Rev. B **76**, 172402 (2007)
231. C.A. Condat, J. Rangel, P.W. Lamberti, Anomalous diffusion in the nonasymptotic regime. Phys. Rev. E **65**, 026138 (2002)

232. H.H.A. Rego, L.S. Lucena, L.R. da Silva, C. Tsallis, Crossover from extensive to nonextensive behavior driven by long-range $d = 1$ bond percolation. Phys. A **266**, 42 (1999)

233. U.L. Fulco, L.R. da Silva, F.D. Nobre, H.H.A. Rego, L.S. Lucena, Effects of site dilution on the one-dimensional long-range bond-percolation problem. Phys. Lett. A **312**, 331 (2003)

234. S. Abe, A.K. Rajagopal, Scaling relations in equilibrium nonextensive thermostatistics. Phys. Lett. A **337**, 292–295 (2005)

235. C. Tsallis, H.J. Haubold, Boltzmann-Gibbs entropy is sufficient but not necessary for the likelihood factorization required by Einstein. EPL **110**, 30005 (2015)

236. A. Enciso, P. Tempesta, Uniqueness and characterization theorems for generalized entropies. J. Stat. Mech. 123101 (2017)

237. J. Korbel, Calibration invariance of the MaxEnt distribution in the maximum entropy principle. Entropy **23**, 96 (2021)

238. L. Nivanen, A. Le Mehaute, Q.A. Wang, Generalized algebra within a nonextensive statistics. Rep. Math. Phys. **52**, 437 (2003)

239. E.P. Borges, A possible deformed algebra and calculus inspired in nonextensive thermostatistics. Phys. A **340**, 95–101 (2004)

240. C. Tsallis, S.M.D. Queiros, *Nonextensive statistical mechanics and central limit theorems I - Convolution of independent random variables and q-product*, in *Complexity, Metastability and Nonextensivity*, eds. by S. Abe, H.J. Herrmann, P. Quarati, A. Rapisarda, C. Tsallis, Am. Inst. Phys. Conf. Proc. **965**, 8–20 (New York, 2007)

241. H. Suyari, Mathematical structures derived from the q-multinomial coefficient in Tsallis statistics. Phys. A **386**, 63 (2006)

242. V. Schwammle, C. Tsallis, Two-parameter generalization of the logarithm and exponential functions and Boltzmann-Gibbs-Shannon entropy. J. Math. Phys. **48**, 113301 (2007)

243. C.B. Corcino, R.B. Corcino, Three-parameter logarithm and entropy. J. Funct. Spac. 9791789 (2020)

244. C.B. Corcino, R.B. Corcino, Logarithmic generalization of the Lambert W function and its applications to adiabatic thermostatistics of the three-parameter entropy. Adv. Math. Phys. 6695559 (2021)

245. E.P. Borges, B.G. da Costa, Deformed mathematical objects stemming from the q-logarithm function. Axioms **11**, 138 (2022)

246. E.P. Borges, T. Kodama, C. Tsallis, Along the lines of nonadditive entropies: q-prime numbers and q-zeta functions. Entropy **24**, 60 (2022)

247. J. Korbel, R. Hanel, S. Thurner, Classication of complex systems by their sample-space scaling exponents. New Jo. Phys. **20**, 093007 (2018)

248. J. Korbel, R. Hanel, S. Thurner, Information geometry of scaling expansions of nonexponentially growing configuration spaces. Euro. Phys. J. Spec. Top. **229**, 787–807 (2020)

249. J. Korbel, S.D. Lindner, R. Hanel, S. Thurner, Thermodynamics of small systems with emergent structures. arxiv:2004.06491 [cond-mat.stat-mech]

250. H.J. Jensen, R.H. Pazuki, G. Pruessner, P. Tempesta, Statistical mechanics of exploding phase spaces: ontic open systems. J. Phys. A: Math. Theor. **51**, 375002 (2018)

251. P. Tempesta, H.J. Jensen, Universality classes and information-theoretic measures of complexity via group entropies. Sci. Rep. **10**, 5952 (2020)

252. J.M. Amigo, R. Dale, P. Tempesta, A generalized permutation entropy for noisy dynamics and random processes. Chaos **31**, 013115 (2021)

253. S. Nors Nielsen, R.E. Ulanowicz, Ontic openness: an absolute necessity for all developmental processes. Ecol. Modell. **222**, 2908 (2011)

254. S.A. Kauffman, Children of Newton and modernity. Edge (2013). https://edge.org/response-detail/23801

255. C. Tsallis, Nonextensive statistical mechanics: construction and physical interpretation, in *Nonextensive Entropy—Interdisciplinary Applications*. ed. by M. Gell-Mann, C. Tsallis (Oxford University Press, New York, 2004), pp.1–54

256. H.S. Lima, C. Tsallis, Exploring the neighborhood of q-exponentials. Entropy **22**, 1402 (2020)

257. M.R. Ubriaco, Entropies based on fractional calculus. Phys. Lett. A **373**, 2516 (2009)
258. N. Komatsu, S. Kimura, Entropic cosmology for a generalized black-hole entropy. Phys. Rev. D **88**, 083534 (2013)
259. N. Komatsu, S. Kimura, Non-adiabatic-like accelerated expansion of the late universe in entropic cosmology. Phys. Rev. D **87**, 043531 (2013)
260. N. Komatsu, S. Kimura, Evolution of the universe in entropic cosmologies via different formulations. Phys. Rev. D **89**, 123501 (2014)
261. N. Komatsu, S. Kimura, Entropic cosmology in a dissipative universe. Phys. Rev. D **90**, 123516 (2014)
262. N. Komatsu, S. Kimura, General form of entropy on the horizon of the universe in entropic cosmology. Phys. Rev. D **93**, 043530 (2016)
263. H. Moradpour, A. Bonilla, E.M.C. Abreu, J. Ananias Neto, Accelerated cosmos in a nonextensive setup. Phys. Rev. D **96**, 123504 (2017)
264. M. Tavayef, A. Sheykhi, K. Bamba, H. Moradpour, Tsallis holographic dark energy. Phys. Lett. B **781**, 195–200 (2018)
265. M. Sharif, S. Saba, Tsallis holographic dark energy in $f(G, T)$ gravity. Symmetry **11**, 92 (2019)
266. Q. Huang, H. Huang, J. Chen, L. Zhang, F. Tu, Stability analysis of the Tsallis holographic dark energy model. Class. Q. Grav. **36**, 175001 (2019)
267. S. Waheeda, Reconstruction paradigm in a class of extended teleparallel theories using Tsallis holographic dark energy. Eur. Phys. J. Plus **135**, 11 (2020)
268. D.J. Zamora, C. Tsallis, Thermodynamically consistent entropic-force cosmology. Phys. Lett. B **827**, 136967 (2022). https://doi.org/10.1016/j.physletb.2022.136967
269. D.J. Zamora, C. Tsallis, Thermodynamically consistent entropic late-time cosmological acceleration. Eur. Phys. J. C **82**, 689 (2022)
270. E.M.F. Curado, Private communication (2020)
271. C. Tsallis, Nonextensive statistical mechanics, anomalous diffusion and central limit theorems. Milan J. Math. **73**, 145 (2005)
272. Y. Sato, C. Tsallis, On the extensivity of the entropy S_q for $N \leq 3$ specially correlated binary subsystems, in *Proceedings of the Summer School and Conference on Complexity in Science and Society* (Patras and Ancient Olympia, 2004); Complexity: an unifying direction in science, eds. by T. Bountis, G. Casati, I. Procaccia, Int. J. Bifurc. Chaos **16**, 1727–1738 (2006)
273. C. Tsallis, Is the entropy S_q extensive or nonextensive?, in *Complexity, Metastability and Nonextensivity*, eds. by C. Beck, G. Benedek, A. Rapisarda, C. Tsallis (World Scientific, Singapore, 2005), p. 13; [cond-mat/0409631]
274. C. Tsallis, M. Gell-Mann, Y. Sato, Asymptotically scale-invariant occupancy of phase space makes the entropy S_q extensive. Proc. Natl. Acad. Sc. USA **102**, 15377 (2005)
275. C. Tsallis, M. Gell-Mann, Y. Sato, Extensivity and entropy production, in *Nonextensive Statistical Mechanics: new Trends, New perspectives*, eds. by J.P. Boon, C. Tsallis, Europhys. News **36**, 186 (2005)
276. G. Polya, *Mathematical Discovery*, vol. 1 (Wiley, New York, 1962), p. 88
277. A.N. Kolmogorov, *Foundations of the Theory of Probabilities* (1933); [In English: Chelsea Pub. Co., 2nd edn, 1960]
278. J. Lamperti, *Probability* (Benjamin, New York, 1966)
279. E.C.G. Stueckelberg, A. Petermann, La normalisation des constantes dans la théorie des quanta. Helv. Phys. Acta **26**, 499 (1953)
280. M. Gell-Mann, F.E. Low, Quantum electrodynamics at small distances. Phys. Rev. **95**, 1300 (1954)
281. K.G. Wilson, Renormalization group and critical phenomena. I. Renormalization group and the Kadanoff scaling picture. Phys. Rev. B **4**, 3174 (1971); Renormalization group and critical phenomena. II. Phase-space cell analysis of critical behavior. Phys. Rev. B **4**, 3184 (1971)

282. C. Tsallis, A.C.N. de Magalhaes, Pure and random Potts-like models: real-space renormalization-group approach. Phys. Rep. **268**, 305 (1996)
283. J.A. Marsh, M.A. Fuentes, L.G. Moyano, C. Tsallis, Influence of global correlations on central limit theorems and entropic extensivity. Phys. A **372**, 183–202 (2006)
284. F. Caruso, C. Tsallis, Extensive nonadditive entropy in quantum spin chains, in *Complexity, Metastability and Nonextensivity*, eds. by S. Abe, H.J. Herrmann, P. Quarati, A. Rapisarda, C. Tsallis, Ame. Inst. Phys. Conf. Proc. **965**, 51 (New York, 2007)
285. F. Caruso, C. Tsallis, Nonadditive entropy reconciles the area law in quantum systems with classical thermodynamics. Phys. Rev. E **78**, 021102 (2008)
286. A. Saguia, M.S. Sarandy, Nonadditive entropy for random quantum spin-S chains. Phys. Lett. A **374**, 3384–3388 (2010)
287. J.A. Carrasco, F. Finkel, A. Gonzalez-Lopez, M.A. Rodriguez, P. Tempesta, Generalized isotropic Lipkin-Meshkov-Glick models: ground state entanglement and quantum entropies. J. Stat. Mech. 033114 (2016)
288. A.M.C. Souza, P. Rapcan, C. Tsallis, Area-law-like systems with entangled states can preserve ergodicity. Euro. Phys. J. Spec. Top. **229**(5), 759–772 (2020)
289. F.C. Alcaraz, The critical behaviour of self-dual $Z(N)$ spin systems: finite-size scaling and conformal invariance. J. Phys. A **20**, 2511 (1987)
290. F.C. Alcaraz, M.J. Martins, The operator content of the exactly integrable $SU(N)$ magnets. J. Phys. A **23**, L1079 (1990)
291. S. Sachdev, *Quantum Phase Transitions* (Cambridge University Press, Cambridge, 2000)
292. P. Calabrese, J. Cardy, Entanglement entropy and quantum field theory. J. Stat. Mech.: Theor. Exp. P06002 (2004)
293. P. Ginsparg, G. Moore, *Lectures on 2D String Theory* (Cambridge University Press, Cambridge, 1993)
294. T. Barthel, M.-C. Chung, U. Schollwock, Entanglement scaling in critical two-dimensional fermionic and bosonic systems. Phys. Rev. A **74**, 022329 (2006)
295. J. Eisert, M. Cramer, M.B. Plenio, Colloquium: area laws for the entanglement entropy. Rev. Mod. Phys. **82**, 277 (2010)
296. C. Tsallis, Black hole entropy: a closer look, in Special Issue *Entropy and Gravitation*, ed. by A. Plastino, Entropy **22**, 17 (2020)
297. A. Rodriguez, C. Tsallis, Connection between Dirichlet distributions and a scale-invariant probabilistic model based on Leibniz-like pyramids. J. Stat. Mech. P12027 (2014)
298. W.K. Wootters, Phys. Rev. D **23**, 357 (1981)
299. Rev Robinson, Econ. Stud. **58**, 437 (1991)
300. W.A. Broock, J.A. Scheinkman, W.D. Dechert, B. LeBaron, A test for independence based on the correlation dimension. Econom. Rev. **15**(3), 197–235 (2010)
301. J.H. Binsack, *Plasma Studies with the Imp-2 Satellite, Ph.D.-Thesis* (Cambridge, MA, USA, MIT, 1966), p.200
302. S. Olbert, Summary of experimental results from M.I.T. Detector on IMP-1, in *Physics of the Magnetosphere*, eds. by R.L. Carovillano, J.F. McClay, H.R. Radoski, *Proceedings of Conference at Boston College* (1967); *Astrophysics and Space Science Library*, vol. 40 (Springer, New York, NY, USA, 1968), pp. 641–659
303. V.M. Vasyliunas, A survey of low-energy electrons in the evening sector of the magnetosphere with OGO 1 and OGO 3. J. Geophys. Res. **73**, 2839–2884 (1968)
304. A.M.C. de Souza, C. Tsallis, Student's t- and r-distributions: unified derivation from an entropic variational principle. Phys. A **236**, 52 (1997)
305. C. Tsallis, A.R. Plastino, R.F. Alvarez-Estrada, Escort mean values and the characterization of power-law-decaying probability densities. J. Math. Phys. **50**, 043303 (2009)
306. E. Mayoral, A. Robledo, Tsallis' q index and Mori's q phase transitions at edge of chaos. Phys. Rev. E **72**, 026209 (2005)
307. E. Mayoral, A. Robledo, A recent appreciation of the singular dynamics at the edge of chaos, in *Verhulst 200 on Chaos—Understanding Complex Systems*. ed. by M. Ausloss, M. Dirickx (Springer, Berlin, 2006), pp.339–354

308. S. Martinez, F. Nicolas, F. Pennini, A. Plastino, Tsallis' entropy maximization procedure revisited. Phys. A **286**, 489 (2000)
309. S. Abe, S. Martinez, F. Pennini, A. Plastino, Nonextensive thermodynamic relations. Phys. Lett. A **281**, 126 (2001)
310. G.L. Ferri, S. Martinez, A. Plastino, Equivalence of the four versions of Tsallis' statistics. JSTAT—J. Stat. Mech.: Theory Exp. 1742-5468/05/P04009 (2005)
311. G.L. Ferri, S. Martinez, A. Plastino, The role of constraints in Tsallis' nonextensive treatment revisited. Phys. A **345**, 493 (2005)
312. T. Wada, A.M. Scarfone, Connections between Tsallis' formalisms employing the standard linear average energy and ones employing the normalized q-average energy. Phys. Lett. A **335**, 351 (2005)
313. T. Wada, A.M. Scarfone, A non self-referential expression of Tsallis' probability distribution function. Eur. Phys. J. B **47**, 557–561 (2005)
314. J.S. Andrade Jr., G.F.T. da Silva, A.A. Moreira, F.D. Nobre, E.M.F. Curado, Thermostatistics of overdamped motion of interacting particles. Phys. Rev. Lett. **105**, 260601 (2010)
315. M. Jauregui, *Grandes desvios e independencia assintotica em sistemas fortemente correlacionados*, Doctor Thesis (Centro Brasileiro de Pesquisas Fisicas, Rio de Janeiro, 2015) (see Eqs. (4.69) and (4.71) of the Thesis, in Portuguese)
316. A.M. Mariz, On the irreversible nature of the Tsallis and Renyi entropies. Phys. Lett. A **165**, 409 (1992)
317. J.D. Ramshaw, H-theorems for the Tsallis and Renyi entropies. Phys. Lett. A **175**, 169 (1993)
318. J.A.S. Lima, R. Silva, A.R. Plastino, Nonextensive thermostatistics and the H-theorem. Phys. Rev. Lett. **86**, 2938–2941 (2001)
319. A. Lavagno, Relativistic nonextensive thermodynamics. Phys. Lett. A **301**, 13 (2002)
320. V. Schwammle, E.M.F. Curado, F.D. Nobre, A general nonlinear Fokker-Planck equation and its associated entropy. Eur. Phys. J. B **58**, 159 (2007)
321. V. Schwammle, F.D. Nobre, E.M.F. Curado, Consequences of the H-theorem from nonlinear Fokker-Planck equations. Phys. Rev. E **76**, 041123 (2007)
322. G.A. Casas, F.D. Nobre, E.M.F. Curado, Entropy production and nonlinear Fokker-Planck equations. Phys. Rev. E **86**, 061136 (2012)
323. M.S. Ribeiro, C. Tsallis, F.D. Nobre, Probability distributions extremizing the nonadditive entropy S_δ and stationary states of the corresponding nonlinear Fokker-Planck equation. Phys. Rev. E **88**, 052107 (2013)
324. M.S. Ribeiro, F.D. Nobre, C. Tsallis, Probability distributions and associated nonlinear Fokker-Planck equation for the two-index entropic form $S_{q,\delta}$. Phys. Rev. E **89**, 052135 (2014)
325. S. Abe, A.K. Rajagopal, Validity of the second law in nonextensive quantum thermodynamics. Phys. Rev. Lett. **91**, 120601 (2003)
326. A.R. Plastino, A. Plastino, Tsallis' entropy, Ehrenfest theorem and information theory. Phys. Lett. A **177**, 177 (1993)
327. M.O. Caceres, Irreversible thermodynamics in the framework of Tsallis entropy. Phys. A **218**, 471 (1995)
328. A.K. Rajagopal, Dynamic linear response theory for a nonextensive system based on the Tsallis prescription. Phys. Rev. Lett. **76**, 3469 (1996)
329. A. Chame, E.V.L. de Mello, The Onsager reciprocity relations within Tsallis statistics. Phys. Lett. A **228**, 159 (1997)
330. M.O. Caceres, C. Tsallis, Unpublished private discussion (1993)
331. A. Chame, E.V.L. de Mello, The fluctuation-dissipation theorem in the framework of the Tsallis statistics. J. Phys. A **27**, 3663 (1994)
332. C. Tsallis, Some comments on Boltzmann-Gibbs statistical mechanics. Chaos Solitons Fractals **6**, 539 (1995)
333. G. Casati, C. Tsallis, F. Baldovin, Linear instability and statistical laws of physics. Europhys. Lett. **72**, 355–361 (2005)

334. G. Wilk, Z. Wlodarczyk, Interpretation of the nonextensivity parameter q in some applications of Tsallis statistics and Lévy distributions. Phys. Rev. Lett. **84**, 2770 (2000)
335. C. Beck, Dynamical foundations of nonextensive statistical mechanics. Phys. Rev. Lett. **87**, 180601 (2001)
336. T. Kodama, H.-T. Elze, C.E. Aguiar, T. Koide, Dynamical correlations as origin of nonextensive entropy. Europhys. Lett. **70**, 439 (2005)
337. C. Beck, E.G.D. Cohen, Superstatistics. Phys. A **322**, 267 (2003)
338. A. Plastino, C. Tsallis, Variational method in generalized statistical mechanics. J. Phys. A **26**, L893 (1993)
339. E.K. Lenzi, L.C. Malacarne, R.S. Mendes, Perturbation and variational methods in nonextensive Tsallis statistics. Phys. Rev. Lett. **80**, 218 (1998)
340. R.S. Mendes, C.A. Lopes, E.K. Lenzi, L.C. Malacarne, Variational methods in nonextensive Tsallis statistics: a comparative study. Phys. A **344**, 562 (2004)
341. E.K. Lenzi, R.S. Mendes, A.K. Rajagopal, Green functions based on Tsallis nonextensive statistical mechanics: normalized q-expectation value formulation. Phys. A **286**, 503 (2000)
342. A. Taruya, M. Sakagami, Gravothermal catastrophe and Tsallis' generalized entropy of self-gravitating systems III. Quasi-equilibrium structure using normalized q-values. Phys. A **322**, 285 (2003)
343. A. Taruya, M. Sakagami, Gravothermal catastrophe and Tsallis' generalized entropy of self-gravitating systems II. Thermodynamic properties of stellar polytrope. Phys. A **318**, 387 (2003)
344. A.K. Rajagopal, R.S. Mendes, E.K. Lenzi, Quantum statistical mechanics for nonextensive—Prediction for possible experimental tests. Phys. Rev. Lett. **80**, 3907 (1998)
345. S. Abe, The thermal Green functions in nonextensive quantum statistical mechanics. Eur. Phys. J. B **9**, 679 (1999)
346. A. Cavallo, F. Cosenza, L. De Cesare, Two-time Green's functions and the spectral density method in nonextensive classical statistical mechanics. Phys. Rev. Lett. **87**, 240602 (2001)
347. A. Cavallo, F. Cosenza, L. De Cesare, Two-time Green's functions and spectral density method in nonextensive quantum statistical mechanics. Phys. Rev. E **77**, 051110 (2008)
348. C.G. Darwin, R.H. Fowler, Phil. Mag. J. Sci. **44**, 450 (1922); R.H. Fowler, Phil. Mag. J. Sci. **45**, 497 (1923)
349. A.I. Khinchin, *Mathematical Foundations of Statistical Mechanics* (Dover, New York, 1949)
350. R. Balian, N.L. Balazs, Equiprobability, inference, and entropy in quantum-theory. Ann. Phys. (NY) **179**, 97 (1987); R. Kubo, H. Ichimura, T. Usui, N. Hashitsume, *Statistical Mechanics* (North-Holland, Amsterdam, 1988)
351. S. Abe, A.K. Rajagopal, Microcanonical foundation for systems with power-law distributions. J. Phys. A **33**, 8733 (2000)
352. S. Abe, A.K. Rajagopal, Justification of power-law canonical distributions based on generalized central limit theorem. Europhys. Lett. **52**, 610 (2000)
353. S. Abe, A.K. Rajagopal, Nonuniqueness of canonical ensemble theory arising from microcanonical basis. Phys. Lett. A **272**, 341 (2000)
354. S. Abe, A.K. Rajagopal, Macroscopic thermodynamics of equilibrium characterized by power law canonical distributions. Europhys. Lett. **55**, 6 (2001)
355. A. Rodriguez, C. Tsallis, A generalization of the cumulant expansion. Application to a scale-invariant probabilistic model. J. Math. Phys. **51**, 073301 (2010)
356. S. Abe, G.B. Bagci, Necessity of q-expectation value in nonextensive statistical mechanics. Phys. Rev. E **71**, 016139 (2005)
357. A.R. Lima, T.J.P. Penna, Tsallis statistics with normalized q-expectation values is thermodynamically stable: illustrations. Phys. Lett. A **256**, 221 (1999)
358. L.R. da Silva, E.K. Lenzi, J.S. Andrade, J. Mendes Filho, Tsallis nonextensive statistics with normalized q-expectation values: thermodynamical stability and simple illustrations. Phys. A **275**, 396 (2000)
359. C. Tsallis, A.M.C. Souza, Constructing a statistical mechanics for Beck-Cohen superstatistics. Phys. Rev. E **67**, 026106 (2003)

360. A.M.C. Souza, C. Tsallis, Stability of the entropy for superstatistics. Phys. Lett. A **319**, 273 (2003)
361. A.M.C. Souza, C. Tsallis, Stability analysis of the entropy for superstatistics. Phys. A **342**, 132 (2004)
362. R. Hanel, S. Thurner, M. Gell-Mann, Generalized entropies and the transformation group of superstatistics. PNAS **110**, 3539108 (2011)
363. R. Hanel, S. Thurner, M. Gell-Mann, Generalized entropies and logarithms and their duality relations. PNAS **109**(47), 19151–19154 (2012)
364. G. Cohen-Tannoudji, *Les Constantes Universelles, Collection Pluriel* (Hachette, Paris, 1998)
365. C. Tsallis, S. Abe, Advancing Faddeev: math can deepen Physics understanding. Phys. Today **51**, 114 (1998)
366. S. Abe, Temperature of nonextensive systems: Tsallis entropy as Clausius entropy. Phys. A **368**, 430–434 (2006)
367. L.B. Okun, The fundamental constants of physics. Sov. Phys. Usp. **34**, 818 (1991)
368. L. Okun, Cube or hypercube of natural units?, hep-th/0112339 (2001)
369. M.J. Duff, L.B. Okun, G. Veneziano, Trialogue on the number of fundamental constants. JHEP03 (2002) 023; [physics/0110060]
370. M.J. Duff, Comment on time-variation of fundamental constants (2004); [hep-th/0208093]
371. M. Planck, Uber irreversible Strahlugsforgange. Ann d. Phys. (4) (1900) 1, S. 69–122 (see [370])
372. M. Planck, Verhandlungen der Deutschen Physikalischen Gessellschaft **2**, 202 and 237 (1900) (English translation: *On an Improvement of Wien's Equation for the Spectrum*, in *The Old Quantum Theory*, D. ter Haar, Ed. (Pergamon Press, 1967), p. 79
373. D. ter Haar, S.G. Brush, *Planck's Original Papers in Quantum Physics* (Taylor and Francis, London, 1972)]
374. M.J. Duff, *How fundamental are fundamental constants?*, Contemporary Physics (2014)
375. L.D. Faddeev, *40 years in Mathematical Physics* (World Scientific, Singapore, 1995), p. 463
376. S. Arimoto, Information-theoretical considerations on estimation problems. Inf. Control **19**(3), 181–194 (1971)
377. T.D. Frank, A. Daffertshofer, Exact time-dependent solutions of the Renyi Fokker-Planck equation and the Fokker-Planck equations related to the entropies proposed by Sharma and Mittal. Phys. A **285**, 351–366 (2000)
378. E.P. Borges, I. Roditi, A family of non-extensive entropies. Phys. Lett. A **246**, 399–402 (1998)
379. G. Kaniadakis, Non linear kinetics underlying generalized statistics. Phys. A **296**, 405 (2001); Statistical mechanics in the context of special relativity. Phys. Rev. E **66**, 056125 (2002); Statistical mechanics in the context of special relativity. II. Phys. Rev. E **72**, 036108 (2005)
380. G. Kaniadakis, M. Lissia, A.M. Scarfone, Deformed logarithms and entropies. Phys. A **340**, 41 (2004)
381. F. Shafee, Lambert function and a new non-extensive form of entropy. IMA J. Appl. Math. **72**, 785–800 (2007)
382. M.D. Esteban, D. Morales, A summary on entropy statistics. Kybernetika **31**(4), 337–346 (1995)
383. G.A. Tsekouras, C. Tsallis, Generalized entropy arising from a distribution of q-indices. Phys. Rev. E **71**, 046144 (2005)
384. P.T. Landsberg, V. Vedral, Distributions and channel capacities in generalized statistical mechanics. Phys. Lett. A **247**, 211 (1998); P.T. Landsberg, Entropies galore!, in *Nonextensive Statistical Mechanics and Thermodynamics*, eds. by S.R.A. Salinas, C. Tsallis, Braz. J. Phys. **29**, 46 (1999)
385. A.K. Rajagopal, S. Abe, Implications of form invariance to the structure of nonextensive entropies. Phys. Rev. Lett. **83**, 1711 (1999)
386. R. Hanel, S. Thurner, A comprehensive classification of complex statistical systems and an axiomatic derivation of their entropy and distribution functions. Europhys. Lett. **93**, 20006 (2011)

387. R. Hanel, S. Thurner, When do generalised entropies apply? How phase space volume determines entropy. Europhys. Lett. **96**, 50003 (2011)

388. E.M.F. Curado, P. Tempesta, C. Tsallis, A new entropy based on a group-theoretical structure. Annals Phys. **366**, 22–31 (2016)

389. M. Masi, A step beyond Tsallis and Renyi entropies. Phys. Lett. A **338**, 217 (2005)

390. P. Tempesta, Beyond the Shannon-Khinchin formulation: the composability axiom and the universal-group entropy. Annals Phys. **365**, 180 (2016)

391. P. Tempesta, Group entropies, correlation laws, and zeta functions. Phys. Rev. E **84**, 021121 (2011)

392. J. Naudts, Generalized thermostatistics and mean-field theory. Phys. A **332**, 279 (2003)

393. J. Naudts, Generalized thermostatistics based on deformed exponential and logarithmic functions. Phys. A **340**, 32 (2004)

394. J. Naudts, Estimators, escort probabilities, and ϕ-exponential families in statistical physics. J. Ineq. Pure Appl. Math. **5**, 102 (2004)

395. J. Naudts, *Generalised Thermostatistics* (Universiteit Antwerpen, 2007)

396. P. Tempesta, Formal groups and Z-entropies. Proc. R. Soc. A **472**, 20160143 (2016)

397. P. Tempesta, Multivariate group entropies, super-exponentially growing complex systems, and functional equations. Chaos **30**, 123119 (2020)

398. P. Tempesta, A theorem on the existence of trace-form generalized entropies. Proc. R. Soc. A **471**, 20150165 (2015)

399. S. Zozor, J.F. Bercher, Φ-Informational measures: some results and interrelations. Entropy **23**, 911 (2021)

400. H.G. Miller, A. Plastino, A.R. Plastino, Some remarks on Fisher information, the Cramer-Rao inequality, and their applications to physics, in *Handbook of Statistics*, vol. 45, Chapter 11, (2021) pp. 217–228

401. B.D. Hughes, *Random Walks and Random Environments*, vols. I and II (Clarendon Press, Oxford, 1995)

402. A.M. Mariz, F. van Wijland, H.J. Hilhorst, S.R. Gomes Junior, C. Tsallis, Statistics of the one-dimensional Riemann walk. J. Stat. Phys. **102**, 259 (2001)

403. A.R. Plastino, A. Plastino, Non-extensive statistical mechanics and generalized Fokker-Planck equation. Phys. A **222**, 347 (1995)

404. C. Tsallis, D.J. Bukman, Anomalous diffusion in the presence of external forces: exact time-dependent solutions and their thermostatistical basis. Phys. Rev. E **54**, R2197 (1996)

405. M. Muskat, *The Flow of Homogeneous Fluids Through Porous Media* (McGraw-Hill, New York, 1937)

406. T.D. Frank, *Nonlinear Fokker-Planck Equations: fundamentals and Applications, Series Synergetics* (Springer, Berlin, 2005)

407. D. Prato, C. Tsallis, Nonextensive foundation of Levy distributions. Phys. Rev. E **60**, 2398 (1999)

408. E.M.F. Curado, F.D. Nobre, Derivation of nonlinear Fokker-Planck equations by means of approximations to the master equation. Phys. Rev. E **67**, 021107 (2003)

409. F.D. Nobre, E.M.F. Curado, G. Rowlands, A procedure for obtaining general nonlinear Fokker-Planck equations. Phys. A **334**, 109 (2004)

410. V. Schwammle, F.D. Nobre, C. Tsallis, q-Gaussians in the porous-medium equation: stability and time evolution, 0804.3362 [cond-mat.stat-mech] (2008)

411. I.T. Pedron, R.S. Mendes, T.J. Buratta, L.C. Malacarne, E.K. Lenzi, Logarithmic diffusion and porous media equations: an unified description. Phys. Rev. E **72**, 031106 (2005)

412. L.C. Malacarne, R.S. Mendes, I.T. Pedron, E.K. Lenzi, Nonlinear equation for anomalous diffusion: unified power-law and stretched exponential exact solution. Phys. Rev. E **63**, R030101 (2001)

413. A.C. McBride, *Fractional Calculus* (Halsted Press, New York, 1986)

414. K. Nishimoto, *Fractional Calculus* (University of New Haven Press, New Haven, CT, 1989)

415. S.G. Samko, A.A. Kilbas, O.I. Marichev, *Fractional Integrals and Derivatives* (Gordon and Breach, Yverdon, Switzerland, 1993)

416. R. Hilfer (ed.), *Applications of Fractional Calculus in Physics* (World Scientific, Singapore, 2000)
417. A.S. Chaves, A fractional diffusion equation to describe Lévy flights. Phys. Lett. A **239**, 13 (1998)
418. P.A. Alemany, D.H. Zanette, Fractal random walks from a variational formalism for Tsallis entropies. Phys. Rev. E **49**, R956 (1994)
419. D.H. Zanette, P.A. Alemany, Thermodynamics of anomalous diffusion. Phys. Rev. Lett. **75**, 366 (1995); M.O. Caceres, C.E. Budde, Phys. Rev. Lett. **77**, 2589 (1996); D.H. Zanette, P.A. Alemany, Phys. Rev. Lett. **77**, 2590 (1996)
420. C. Tsallis, A.M.C. de Souza, R. Maynard, in *Lévy Flights and Related Topics in Physics*, eds. by M.F. Shlesinger, G.M. Zaslavsky, U. Frisch (Springer, Berlin, 1995), p. 269
421. C. Tsallis, S.V.F Levy, A.M.C. de Souza, R. Maynard, *Statistical-mechanical foundation of the ubiquity of Lévy distributions in nature*. Phys. Rev. Lett. **75**, 3589 (1995); [Erratum: Phys. Rev. Lett. **77**, 5442 (1996)]
422. C. Tsallis, Z.G. Arenas, Nonextensive statistical mechanics and high energy physics. Euro. Phys. J **71**, 00132 (2014)
423. C. Tsallis, C. Anteneodo, L. Borland, R. Osorio, Nonextensive statistical mechanics and economics. Phys. A **324**, 89 (2003)
424. M. Bologna, C. Tsallis, P. Grigolini, Anomalous diffusion associated with nonlinear fractional derivative Fokker-Planck-like equation: exact time-dependent solutions. Phys. Rev. E **62**, 2213 (2000)
425. F. Gharari, K. Arias-Calluari, F. Alonso-Marroquin, M.N. Najafi, Space-time fractional porous media equation: application on modeling of S&P500 price return. Phys. Rev. E **104**, 054140 (2021)
426. B.C.C. dos Santos, C. Tsallis, Time evolution towards q-Gaussian stationary states through unified Itô-Stratonovich stochastic equation. Phys. Rev. E **82**, 061119 (2010)
427. Z. Gonzalez Arenas, D.G. Barci, C. Tsallis, Nonlinear inhomogeneous Fokker-Planck equation within a generalized Stratonovich prescription. Phys. Rev. E **90**, 032118 (2014)
428. M. Jauregui, A.L.F. Lucchi, J.H.Y. Passos, R.S. Mendes, Stationary solution and H theorem for a generalized Fokker-Planck equation. Phys. Rev. E **104**, 034130 (2021)
429. S. Curilef, D. Gonzalez, C. Calderon, Analyzing the 2019 Chilean social outbreak: modelling Latin American economies. PLoS ONE **16**(8) (2021)
430. U. Tirnakli, H.J. Jensen, C. Tsallis, Restricted random walk model as a new testing ground for the applicability of q-statistics. Europhys. Lett. **96**, 40008 (2011)
431. P.-H. Chavanis, Nonlinear mean field Fokker-Planck equations. Application to the chemotaxis of biological population. Eur. Phys. J. B **62**, 179–208 (2008)
432. R.S. Zola, M.K. Lenzi, L.R. Evangelista, E.K. Lenzi, L.S. Lucena, L.R. Silva, Exact solutions for a diffusion equation with a nonlinear external force. Phys. Lett. A **372**, 2359–2363 (2008)
433. S.M.D. Queiros, On superstatistical multiplicative-noise processes. Braz. J. Phys. **38**, 203–209 (2008)
434. A.R. Plastino, R.S. Wedemann, C. Tsallis, Nonlinear Fokker-Planck equation for an overdamped system with drag depending on direction. Symmetry **13**, 1621 (2021)
435. M.S. Ribeiro, F.D. Nobre, E.M.F. Curado, Classes of N-dimensional nonlinear Fokker-Planck equations associated to Tsallis entropy. in *Tsallis Entropy*, ed. by A. Anastasiadis, Entropy **13**, 1928–1944 (2011)
436. M.S. Ribeiro, F.D. Nobre, E.M.F. Curado, Time evolution of interacting vortices under overdamped motion. Phys. Rev. E **85**, 021146 (2012)
437. E.M.F. Curado, A.M.C. Souza, F.D. Nobre, R.F.S. Andrade, Carnot cycle for interacting particles in the absence of thermal noise. Phys. Rev. E **89**, 022117 (2014)
438. R.F.S. Andrade, A.M.C. Souza, E.M.F. Curado, F.D. Nobre, A thermodynamical formalism describing mechanical interactions. EPL **108**, 20001 (2014)
439. F.D. Nobre, E.M.F. Curado, A.M.C. Souza, R.F.S. Andrade, Consistent thermodynamic framework for interacting particles by neglecting thermal noise. Phys. Rev. E **91**, 022135 (2015)

440. M.S. Ribeiro, G.A. Casas, F.D. Nobre, Second law and entropy production in a nonextensive system. Phys. Rev. E **91**, 012140 (2015)

441. C.M. Vieira, H.A. Carmona, J.S. Andrade Jr., A.A. Moreira, General continuum approach for dissipative systems of repulsive particles. Phys. Rev. E **93**, 060103(R) (2016)

442. M.S. Ribeiro, F.D. Nobre, Repulsive particles under a general external potential: thermodynamics by neglecting thermal noise. Phys. Rev. E **94**, 022120 (2016)

443. A.A. Moreira, C.M. Vieira, H.A. Carmona, J.S. Andrade Jr., C. Tsallis, Overdamped dynamics of particles with repulsive power-law interactions. Phys. Rev. E **98**, 032138 (2018)

444. G.A. Casas, F.D. Nobre, E.M.F. Curado, New type of equilibrium distribution for a system of charges In a spherically-symmetric electric field. EPL **126**, 10005 (2019)

445. A.R. Plastino, R.S. Wedemann, E.M.F. Curado, F.D. Nobre, C. Tsallis, Nonlinear drag forces and the thermostatistics of overdamped motion. Phys. Rev. E **98**, 012129 (2018)

446. A.R. Plastino, E.M.F. Curado, F.D. Nobre, C. Tsallis, From the nonlinear Fokker-Planck to the Vlasov description and back: Confined interacting particles with drag. Phys. Rev. E **97**, 022120 (2018)

447. G. Sicuro, D. Bagchi, C. Tsallis, On the connection between linear combination of entropies and linear combination of extremizing distributions. Phys. Lett. A **380**, 2025–2030 (2016)

448. N.G. van Kampen, Phys. Rep. **24**, 171 (1976); *Stochastic Processes in Physics and Chemistry* (North-Holland, Amsterdam, 1981)

449. H. Risken, *The Fokker-Planck Equation* (Methods of Solution and Applications) (Springer-Verlag, New York, 1984)

450. C. Anteneodo, C. Tsallis, Multiplicative noise: a mechanism leading to nonextensive statistical mechanics. J. Mat. Phys. **44**, 5194–5203 (2003)

451. M.A. Fuentes, M.O. Caceres, Computing the nonlinear anomalous diffusion equation from first principles. Phys. Lett. A **372**, 1236–1239 (2008)

452. M.O. Caceres, Computing a non-Maxwellian velocity distribution from first principles. Phys. Rev. E **67**, 016102 (2003)

453. N.G. van Kampen, J. Stat. Phys. **24**, 175 (1981)

454. A. Celikoglu, U. Tirnakli, S.M.D. Queiros, Analysis of return distributions in the coherent noise model. Phys. Rev. E **82**, 021124 (2010)

455. A. Celikoglu, U. Tirnakli, Earthquakes, model systems and connections to q-statistics. Acta Geophys. **60**, 535–546 (2012)

456. J.J. Wei, X.F. Wu, Z.G. Dai, F.Y. Wang, P. Wang, D. Li, B. Zhang, Similar scale-invariant behaviors between soft gamma-ray repeaters and an extreme epoch from FRB 121102. Astrophys. J. **920**, 153 (2021)

457. S.-R.G. Christopoulos, N.V. Sarlis, q-exponential relaxation of the expected avalanche size in the coherent noise model. Phys. A **407**, 216–225 (2014)

458. S.-R.G. Christopoulos, N.V. Sarlis, An Application of the coherent noise model for the prediction of aftershock magnitude time series. Complexity, 6853892 (2017)

459. C. Tsallis, G. Bemski, R.S. Mendes, Is re-association in folded proteins a case of nonextensivity? Phys. Lett. A **257**, 93–98 (1999)

460. F. Caruso, A. Pluchino, V. Latora, S. Vinciguerra, A. Rapisarda, Analysis of self-organized criticality in the Olami-Feder-Christensen model and in real earthquakes. Phys. Rev. E **75**, 055101(R) (2007)

461. B. Bakar, U. Tirnakli, Analysis of self-organized criticality in Ehrenfest's dog-flea model. Phys. Rev. E **79**, 040103(R) (2009)

462. B. Bakar, U. Tirnakli, Return distributions in dog-flea model revisited. Phys. A **389**, 3382–3386 (2010)

463. M.A. Stapleton, K. Christensen, J. Phys. A **39**, 9107 (2006)

464. U. Tirnakli, Unpublished (2020)

465. Y. Kuramoto, *Chemical Oscillations, Waves, and Turbulence*, Springer Series in Synergetics (Berlin, 1984)

466. A. Pluchino, A. Rapisarda, Metastability in the Hamiltonian Mean Field model and Kuramoto model. Phys. A **365**, 184–189 (2006)

467. G. Miritello, A. Pluchino, A. Rapisarda, Central limit behavior in the Kuramoto model at the "edge of chaos." Phys. A **388**, 4818–4826 (2009)
468. L.G. Moyano, C. Tsallis, M. Gell-Mann, Numerical indications of a q-generalised central limit theorem. Europhys. Lett. **73**, 813–819 (2006)
469. W.J. Thistleton, J.A. Marsh, K.P. Nelson, C. Tsallis, q-Gaussian approximants mimic non-extensive statistical-mechanical expectation for many-body probabilistic model with long-range correlations. Cent. Eur. J. Phys. **7**, 387–394 (2009)
470. A. Rodriguez, V. Schwammle, C. Tsallis, Strictly and asymptotically scale-invariant probabilistic models of N correlated binary random variables having q-Gaussians as $N \to \infty$ limiting distributions. J. Stat. Mech: Theory Exp. P09006 (2008)
471. H.J. Hilhorst, G. Schehr, A note on q Gaussians and non-Gaussians in statistical mechanics. J. Stat. Mech. P06003 (2007)
472. D.B. Ion, M.L.D. Ion, Entropic lower bound for the quantum scattering of spinless particles. Phys. Rev. Lett. **81**, 5714 (1998)
473. M.L.D. Ion, D.B. Ion, Optimal bounds for Tsallis-like entropies in quantum scattering. Phys. Rev. Lett. **83**, 463 (1999)
474. D.B. Ion, M.L.D. Ion, Angle—Angular-momentum entropic bounds and optimal entropies for quantum scattering of spinless particles. Phys. Rev. E **60**, 5261 (1999)
475. D.B. Ion, M.L.D. Ion, Evidences for nonextensivity conjugation in hadronic scattering systems. Phys. Lett. B **503**, 263 (2001)
476. A. Robledo, Criticality in nonlinear one-dimensional maps: RG universal map and nonextensive entropy. Phys. D **193**, 153–160 (2004)
477. S. Umarov, C. Tsallis, S. Steinberg, On a q-central limit theorem consistent with nonextensive statistical mechanics. Milan J. Math. **76**, 307–328 (2008)
478. C. Budde, D. Prato, M. Re, Superdiffusion in decoupled continuous time random walks. Phys. Lett. A **283**, 309 (2001)
479. J.-P. Gazeau, C. Tsallis, Moebius transforms, cycles and q-triplets in statistical mechanics. Entropy **21**, 1155 (2019)
480. L.F. Burlaga, A.F. Vinas, Triangle for the entropic index q of non-extensive statistical mechanic observed by Voyager 1 in the distant heliosphere. Phys. A **356**, 375 (2005)
481. H. Suyari, M. Tsukada, Law of error in Tsallis statistics. IEEE Trans. Inf. Theory **51**, 753 (2005)
482. H. Christodoulidi, C. Tsallis, T. Bountis, Fermi-Pasta-Ulam model with long-range interactions: dynamics and thermostatistics. EPL **108**, 40006 (2014)
483. R. Hanel, S. Thurner, C. Tsallis, Limit distributions of scale-invariant probabilistic models of correlated random variables with the q-Gaussian as an explicit example. Eur. Phys. J. B **72**, 263–268 (2009)
484. B. Mandelbrot, J. Van Ness, Fractional Brownian motions, fractional noises and applications. SIAM Rev. **10**, 422–437 (1968)
485. C. Tsallis, Nonadditive entropy: the concept and its use, in *Statistical Power Law Tails in High Energy Phenomena*, ed. by T.S. Biro, Eur. Phys. J. A **40**, 257–266 (2009)
486. J. Ruseckas, Probabilistic model of N correlated binary random variables and non-extensive statistical mechanics. Phys. Lett. A **379**, 654–659 (2015)
487. H. Bergeron, E.M.F. Curado, J.P. Gazeau, L.M.C.S. Rodrigues, Symmetric generalized binomial distributions. J. Math. Phys. **54**, 123301 (2013)
488. G. Ruiz, C. Tsallis, Emergence of q-statistical functions in a generalized binomial distribution with strong correlations. J. Math. Phys. **56**, 053301 (2015)
489. J. Naudts, H. Suyari, Large deviation estimates involving deformed exponential functions. Phys. A **436**, 716–728 (2015)
490. H. Bergeron, E.M.F. Curado, J.P. Gazeau, L.M.C.S. Rodrigues, Symmetric deformed binomial distributions: an analytical example where the Boltzmann-Gibbs entropy is not extensive. J. Math. Phys. **57**, 023301 (2016)

491. S. Umarov, C. Tsallis, On multivariate generalizations of the q-central limit theorem consistent with nonextensive statistical mechanics, in *Complexity, Metastability and Nonextensivity*, eds. by S. Abe, H.J. Herrmann, P. Quarati, A. Rapisarda, C. Tsallis, Am. Inst. Phys. Conf. Proc. **965**, 34–42 (New York, 2007)

492. S.M.D. Queiros, C. Tsallis, Nonextensive statistical mechanics and central limit theorems II—Convolution of q-independent random variables, in *Complexity, Metastability and Nonextensivity*, eds. by S. Abe, H.J. Herrmann, P. Quarati, A. Rapisarda, C. Tsallis, Am. Inst. Phys. Conf. Proc. **965**, 21–33 (New York, 2007)

493. C. Vignat, A. Plastino, Central limit theorem and deformed exponentials. J. Phys. A **40**, F969–F978 (2007)

494. J.-F. Bercher, C. Vignat, A new look at q-exponential distributions via excess statistics. Phys. A **387**, 5422–5432 (2008)

495. S. Umarov, C. Tsallis, M. Gell-Mann, S. Steinberg, Generalization of symmetric α-stable Lévy distributions for $q>1$. J. Math. Phys. **51**, 033502 (2010)

496. M. Jauregui, C. Tsallis, q-Generalization of the inverse Fourier transform. Phys. Lett. A **375**, 2085–2088 (2011)

497. M. Jauregui, C. Tsallis, E.M.F. Curado, q-Moments remove the degeneracy associated with the inversion of the q-Fourier transform. J. Stat. Mech.: Theory Exp. 10016 (2011)

498. A. Plastino, M.C. Rocca, A direct proof of Jauregui-Tsallis' conjecture. J. Math. Phys. **52**(10), 103503 (2011)

499. A. Plastino, M.C. Rocca, Inversion of Umarov-Tsallis-Steinberg's q-Fourier transform and the complex-plane generalization. Phys. A **391**, 4740–4747 (2012)

500. G. Sicuro, P. Tempesta, A. Rodriguez, C. Tsallis, On the robustness of the q-Gaussian family. Annals Phys. **363**, 316–336 (2015)

501. S. Umarov, C. Tsallis, The limit distribution in the q-CLT for $q \geq 1$ is unique and can not have a compact support. J. Phys. A: Math. Theor. **49**, 415204 (2016)

502. S. Umarov, C. Tsallis, On a representation of the inverse F_q-transform. Phys. Lett. A **372**, 4874–4876 (2008)

503. H.J. Hilhorst, Central limit theorems for correlated variables: some critical remarks. Brazilian J. Phys. **39**(2A), 371–379 (2009)

504. H.J. Hilhorst, Note on a q-modified central limit theorem. J. Stat. Mech. Theory Exp. 10023 (2010)

505. C. Tsallis, A.R. Plastino, R.J. Alvarez-Estrada, Escort mean values and the characterization of power-law-decaying probability densities. J. Math. Phys. **50**, 043303 (2009)

506. S. Umarov, S.M.D. Queiros, Functional-differential equations for the q-Fourier transform of q-Gaussians. J. Phys. A **43**, 095202 (2010)

507. K.P. Nelson, S. Umarov, Nonlinear statistical coupling. Phys. A **389**, 2157–2163 (2010)

508. H. Touchette, The large deviation approach to statistical mechanics. Phys. Rep. **478**, 1–69 (2009)

509. G. Ruiz, C. Tsallis, Towards a large deviation theory for strongly correlated systems. Phys. Lett. A **376**, 2451–2454 (2012)

510. H. Touchette, Comment on "Towards a large deviation theory for strongly correlated systems." Phys. Lett. A **377**(5), 436–438 (2013)

511. G. Ruiz, C. Tsallis, Reply to Comment on "Towards a large deviation theory for strongly correlated systems." Phys. Lett. A **377**, 491–495 (2013)

512. U. Tirnakli, C. Tsallis, N. Ay, Approaching a large deviation theory for complex systems. Nonlinear Dyn. **106**, 2537–2546 (2021)

513. U. Tirnakli, E.P. Borges, The standard map: from Boltzmann-Gibbs statistics to Tsallis statistics. Nat. Sci. Rep. **6**, 23644 (2016)

514. B.V. Chirikov, A universal instability of many-dimensional oscillator systems. Phys. Rep. **52**, 263 (1979)

515. U. Tirnakli, C. Tsallis, Extensive numerical results for integrable case of standard map. Nonlinear Phenom. Complex Syst. **23**(2), 149–152 (2020)

516. A. Bountis, J.J.P. Veerman, F. Vivaldi, Cauchy distributions for the integrable standard map. Phys. Lett. A **384**, 126659 (2020)
517. G. Ruiz, U. Tirnakli, E.P. Borges, C. Tsallis, Statistical characterization of the standard map. J. Stat. Mech. 063403 (2017)
518. G.F. Mazenko, Vortex velocities in the $O(n)$ symmetric time-dependent Ginzburg-Landau model. Phys. Rev. Lett. **78**, 401 (1997)
519. H. Qian, G.F. Mazenko, Vortex dynamics in a coarsening two-dimensional XY model. Phys. Rev. E **68**, 021109 (2003)
520. O. Afsar, U. Tirnakli, Probability densities for the sums of iterates of the sinc-circle map in the vicinity of the quasiperiodic edge of chaos. Phys. Rev. E **82**, 046210 (2010)
521. H. Hernandez-Saldana, A. Robledo, Dynamics at the quasiperiodic onset of chaos, Tsallis q-statistics and Mori's q-phase transitions. Phys. A **370**, 286–300 (2006)
522. D. Broadhurst (1999). http://pi.lacim.uqam.ca/piDATA/feigenbaum.txt
523. R. Tonelli, G. Mezzorani, F. Meloni, M. Lissia, M. Coraddu, Entropy production and Pesin-like identity at the onset of chaos. Progr. Theor. Phys. **115**, 23–29 (2006)
524. A. Celikoglu, U. Tirnakli, Sensitivity function and entropy increase rates for z-logistic map family at the edge of chaos. Phys. A **372**, 238–242 (2006)
525. K. Cetin, O. Afsar, U. Tirnakli, Generalized Pesin-like identity and scaling relations at chaos threshold of the Rossler system. Entropy **20**, 216 (2018)
526. F.A.B.F. de Moura, U. Tirnakli, M.L. Lyra, Convergence to the critical attractor of dissipative maps: log-periodic oscillations, fractality and nonextensivity. Phys. Rev. E **62**, 6361 (2000)
527. P. Grassberger, Temporal scaling at Feigenbaum points and nonextensive thermodynamics. Phys. Rev. Lett. **95**, 140601 (2005)
528. A. Robledo, L.G. Moyano, q-deformed statistical-mechanical property in the dynamics of trajectories EN route to the Feigenbaum attractor. Phys. Rev. E **77**, 032613 (2008)
529. M.C. Mackey, M. Tyran-Kaminska, Phys. Rep. **422**, 167–222 (2006)
530. U. Tirnakli, C. Beck, C. Tsallis, Central limit behavior of deterministic dynamical systems. Phys. Rev. E **75**, 040106(R) (2007)
531. U. Tirnakli, C. Tsallis, C. Beck, A closer look at time averages of the logistic map at the edge of chaos. Phys. Rev. E **79**, 056209 (2009)
532. C. Tsallis, U. Tirnakli, Nonadditive entropy and nonextensive statistical mechanics—Some central concepts and recent applications. J. Phys. C Ser. **201**, 012001 (2010)
533. B. Luque, L. Lacasa, A. Robledo, Feigenbaum graphs at the onset of chaos. Phys. Lett. A **376**, 3625–3629 (2012)
534. U. Tirnakli, Two-dimensional maps at the edge of chaos: numerical results for the Henon map. Phys. Rev. E **66**, 066212 (2002)
535. E.P. Borges, U. Tirnakli, Mixing and relaxation dynamics of the Henon map at the edge of chaos, in *Anomalous Distributions, Nonlinear Dynamics and Nonextensivity*, eds. by H.L. Swinney, C. Tsallis, Phys. D **193**, 148 (2004)
536. E.P. Borges, U. Tirnakli, Two-dimensional dissipative maps at chaos threshold: sensitivity to initial conditions and relaxation dynamics. Phys. A **340**, 227–233 (2004)
537. P.R. Hauser, E.M.F. Curado, C. Tsallis, On the universality classes of the Henon map. Phys. Lett. A **108**, 308–310 (1985)
538. A. Pluchino, A. Rapisarda, C. Tsallis, Noise, synchrony, and correlations at the edge of chaos. Phys. Rev. E **87**, 022910 (2013)
539. F. Baldovin, E. Brigatti, C. Tsallis, Quasi-stationary states in low dimensional Hamiltonian systems. Phys. Lett. A **320**, 254 (2004)
540. F. Baldovin, L.G. Moyano, A.P. Majtey, A. Robledo, C. Tsallis, Ubiquity of metastable-to-stable crossover in weakly chaotic dynamical systems. Phys. A **340**, 205 (2004)
541. G. Ruiz, C. Tsallis, Roundoff-induced attractors and reversibility in conservative two-dimensional maps. Phys. A **386**, 720–728 (2007)
542. G. Casati, T. Prosen, Mixing property of triangular billiards. Phys. Rev. Lett **83**, 4729–4732 (1999)

543. M. Horvat, M.D. Esposti, S. Isola, T. Prosen, L. Bunimovich, On ergodic and mixing properties of the triangle map. Phys. D **238**, 395–415 (2009)
544. G. Ruiz, T. Bountis, C. Tsallis, Time-evolving statistics of chaotic orbits of conservative maps in the context of the central limit theorem. Int. J. Bifurc. Chaos **22**(9), 1250208 (2012)
545. G. Ruiz, U. Tirnakli, E.P. Borges, C. Tsallis, Statistical characterization of discrete conservative systems: the web map. Phys. Rev. E **96**, 042158 (2017)
546. U. Tirnakli, C. Tsallis, K. Cetin, Dynamical robustness of discrete conservative systems: Harper and generalized standard maps. J. Stat. Mech. 063206 (2020)
547. C. Moore, Unpredictability and undecidability in dynamical systems. Phys. Rev. Lett. **64**, 2354–2357 (1990)
548. C. Moore, Generalized shifts: unpredictability and undecidability in dynamical systems. Nonlinearity **4**, 199–230 (1991)
549. A. Politi, R. Badii, J. Phys. A: Math. Gen. **30**, L627 (1997)
550. C. Moore, Private Communication (2006)
551. L.G. Moyano, A. Majtey, C. Tsallis, Weak chaos and metastability in a symplectic system of many long-range-coupled standard maps. Euro. Phys. J. B **52**(4), 493–500 (2006)
552. M. Kac, G. Uhlenbeck, P.C. Hemmer, On the van der Waals theory of vapour-liquid equilibrium. I. Discussion of a one-dimensional model. J. Math. Phys. **4**, 216 (1963)
553. S. Inagaki, Thermodynamic stability of modified Konishi-Kaneko system. Progr. Theor. Phys. **90**, 577–584 (1993)
554. S. Inagaki, T. Konishi, Dynamical stability of a simple model similar to self-gravitating systems. Publ. Astron. Soc. Jpn. **45**, 733–735 (1993)
555. Y.Y. Yamaguchi, Slow relaxation at critical point of second order phase transition in a highly chaotic Hamiltonian system. Progr. Theor. Phys. **95**, 717–731 (1996)
556. F.A. Tamarit, C. Anteneodo, Long-range interacting rotators: connection with the mean-field approximation. Phys. Rev. Lett. **84**, 208 (2000)
557. V. Latora, A. Rapisarda, C. Tsallis, Non-Gaussian equilibrium in a long-range Hamiltonian system. Phys. Rev. E **64**, 056134 (2001)
558. B.J.C. Cabral, C. Tsallis, Metastability and weak mixing in classical long-range many-rotator system. Phys. Rev. E **66**, 065101(R) (2002)
559. F.D. Nobre, C. Tsallis, Classical infinite-range-interaction Heisenberg ferromagnetic model: metastability and sensitivity to initial conditions. Phys. Rev. E **68**, 036115 (2003)
560. V. Latora, A. Rapisarda, S. Ruffo, Lyapunov instability and finite size effects in a system with long-range forces. Phys. Rev. Lett. **80**, 692–695 (1998)
561. V. Latora, A. Rapisarda, S. Ruffo, Chaos and statistical mechanics in the Hamiltonian mean field model. Phys. D **131**, 38–54 (1999)
562. V. Latora, A. Rapisarda, S. Ruffo, Superdiffusion and out-of-equilibrium chaotic dynamics with many degrees of freedoms. Phys. Rev. Lett. **83**, 2104–2107 (1999)
563. V. Latora, A. Rapisarda, S. Ruffo, Chaotic dynamics and superdiffusion in a Hamiltonian system with many degrees of freedom. Phys. A **280**, 81–86 (2000)
564. A. Campa, A. Giansanti, D. Moroni, Metastable states in a class of long-range Hamiltonian systems. Phys. A **305**, 137–143 (2002)
565. A. Pluchino, A. Rapisarda, C. Tsallis, A closer look at the indications of q-generalized Central Limit Theorem behavior in quasi-stationary states of the HMF model. Phys. A **387**, 3121–3128 (2008)
566. L.J.L. Cirto, V.R.V. Assis, C. Tsallis, Influence of the interaction range on the thermostatistics of a classical many-body system. Phys. A **393**, 286–296 (2014)
567. L.J.L. Cirto, A. Rodriguez, F.D. Nobre, C. Tsallis, Validity and failure of the Boltzmann weight. EPL **123**, 30003 (2018)
568. A. Campa, A. Giansanti, D. Moroni, C. Tsallis, Classical spin systems with long-range interactions: universal reduction of mixing. Phys. Lett. A **286**, 251–256 (2001)
569. M.C. Firpo, Analytic estimation of the Lyapunov exponent in a mean-field model undergoing a phase transition. Phys. Rev. E **57**, 6599–6603 (1998)

570. V. Latora, A. Rapisarda, C. Tsallis, Fingerprints of nonextensive thermodynamics in a long-range hamiltonian system. Phys. A **305**, 129–136 (2002)
571. M.-C. Firpo, S. Ruffo, Chaos suppression in the large size limit for long-range systems. J. Phys. A **34**, L511–L518 (2001)
572. C. Anteneodo, R.O. Vallejos, On the scaling laws for the largest Lyapunov exponent in long-range systems: a random matrix approach. Phys. Rev. E **65**, 016210 (2002)
573. M.A. Montemurro, F. Tamarit, C. Anteneodo, Ageing in an infinite-range Hamiltonian system of coupled rotators. Phys. Rev. E **67**, 031106 (2003)
574. U. Tirnakli, S. Abe, Ageing in coherent noise models and natural time. Phys. Rev. E **70**, 056120 (2004)
575. Y. Li, N. Li, B. Li, Temperature dependence of thermal conductivities of coupled rotator lattice and the momentum diffusion in standard map. Eur. Phys. J. B **88**, 182 (2015)
576. Y. Li, N. Li, U. Tirnakli, B. Li, C. Tsallis, Thermal conductance of the coupled-rotator chain: influence of temperature and size. EPL **117**, 60004 (2017)
577. C. Tsallis, H.S. Lima, U. Tirnakli, D. Eroglu, First-principle validation of Fourier's law in $d = 1, 2, 3$, preprint (2022), 2203.00102 [cond-mat.stat-mech]
578. H.S. Lima, C. Tsallis, U. Tirnakli, D. Eroglu, *Study of Fourier's Law for the Planar-rotator Chain with Generic-Range Coupling* (2022). In progress
579. L.G. Moyano, C. Anteneodo, Diffusive anomalies in a long-range Hamiltonian system. Phys. Rev. E **74**, 021118 (2006)
580. A. Robledo, Universal glassy dynamics at noise-perturbed onset of chaos. A route to ergodicity breakdown. Phys. Lett. A **328**, 467–472 (2004)
581. F. Baldovin, A. Robledo, Parallels between the dynamics at the noise-perturbed onset of chaos in logistic maps and the dynamics of glass formation. Phys. Rev. E **72**, 066213 (2005)
582. F.D. Nobre, C. Tsallis, Metastable states of the classical inertial infinite-range-interaction Heisenberg ferromagnet: role of initial conditions. Phys. A **344**, 587 (2004)
583. L.J.L. Cirto, L.S. Lima, F.D. Nobre, Controlling the range of interactions in the classical inertial ferromagnetic Heisenberg model: analysis of metastable states. J. Stat. Mech. P04012 (2015)
584. A. Rodriguez, F.D. Nobre, C. Tsallis, d-Dimensional classical Heisenberg model with arbitrarily-ranged interactions: Lyapunov exponents and distributions of momenta and energies. Entropy **21**, 31 (2019)
585. A. Rodriguez, F.D. Nobre, C. Tsallis, Quasi-stationary-state duration in d-dimensional long-range model. Phys. Rev. Res. **2**, 023153 (2020)
586. E. Fermi, J. Pasta, S. Ulam, *Studies of Nonlinear Problems* (Los Alamos, Report No. LA-1940, 1955)
587. T. Dauxois, Fermi, Pasta, Ulam, and a mysterious lady. Phys. Today **6**(1), 55–57 (2008)
588. D. Bagchi, C. Tsallis, Fermi-Pasta-Ulam-Tsingou problems: passage from Boltzmann to q-statistics. Phys. A **491**, 869–873 (2018)
589. M. Leo, R.A. Leo, P. Tempesta, Thermostatistics in the neighborhood of the π-mode solution for the Fermi-Pasta-Ulam β system: from weak to strong chaos. J. Stat. Mech. P04021 (2010)
590. C.G. Antonopoulos, H. Christodoulidi, Weak chaos detection in the Fermi-Pasta-Ulam-α system using q-Gaussian statistics, in Special Issue edited by G. Nicolis, M. Robnik, V. Rothos, H. Skokos, Int. J. Bifurc. Chaos **21**(8), 2285–2296 (2011)
591. M. Leo, R.A. Leo, P. Tempesta, C. Tsallis, Non Maxwellian behaviour and quasi-stationary regimes near the modal solutions of the Fermi-Pasta-Ulam β-system. Phys. Rev. E **85**, 031149 (2012)
592. M. Leo, R.A. Leo, P. Tempesta, A non-Boltzmannian behavior of the energy distribution for quasi-stationary regimes of the Fermi-Pasta-Ulam system. Ann. Phys. **333**, 12–18 (2013)
593. H. Christodoulidi, T. Bountis, C. Tsallis, L. Drossos, Dynamics and Statistics of the Fermi–Pasta–Ulam β–model with different ranges of particle interactions. JSTAT, 123206 (2016)
594. D. Bagchi, C. Tsallis, Sensitivity to initial conditions of d-dimensional long-range-interacting quartic Fermi-Pasta-Ulam model: universal scaling. Phys. Rev. E **93**, 062213 (2016)

595. D. Bagchi, C. Tsallis, Long-ranged Fermi-Pasta-Ulam systems in thermal contact: crossover from q-statistics to Boltzmann-Gibbs statistics. Phys. Lett. A **381**, 1123–1128 (2017)

596. D. Bagchi, Thermal transport in the Fermi-Pasta-Ulam model with long-range interactions. Phys. Rev. E **95**, 032102 (2017)

597. J.E. Macias-Diaz, A. Bountis, Supratransmission in β-Fermi-Pasta-Ulam chains with different ranges of interactions. Commun. Nonlinear Sci. Numer. Simul. **63**, 307–321 (2018)

598. A. Carati, L. Galgani, F. Gangemi, R. Gangemi, Relaxation times and ergodicity properties in a realistic ionic-crystal model, and the modern form of the FPU problem. Phys. A **532**, 121911 (2019)

599. A. Carati, L. Galgani, F. Gangemi, R. Gangemi, Approach to equilibrium via Tsallis distributions in a realistic ionic-crystal model and in the FPU model. Eur. Phys. J. Spec. Top. **229**, 743–749 (2020)

600. F. Gangemi, R. Gangemi, A. Carati, L. Galgani, Thermal fluctuations in a realistic ionic-crystal model. Phys. A **586**, 126463 (2022)

601. U. Tirnakli, C. Tsallis, Epidemiological model with anomalous kinetics—Early stages of the Covid-19 pandemic. Front. Phys. **8**, 613168 (2020)

602. C. Tsallis, Dynamical scenario for nonextensive statistical mechanics. Phys. A **340**, 1–10 (2004)

603. L.F. Burlaga, N.F. Ness, M.H. Acuna, Multiscale structure of magnetic fields in the heliosheath. J. Geophys. Res.-Space Phys. **111**, A09112 (2006)

604. L.F. Burlaga, A.F. Vinas, N.F. Ness, M.H. Acuna, Tsallis statistics of the magnetic field in the heliosheath. Astrophys. J. **644**, L83–L86 (2006)

605. L.F. Burlaga, N.F. Ness, M.H. Acuna, Magnetic fields in the heliosheath and distant heliosphere: voyager 1 and 2 observations during 2005 and 2006. Astrophys. J. **668**, 1246–1258 (2007)

606. L.F. Burlaga, D.B. Berdichevsky, L.K. Jian, A. Koval, N.F. Ness, J. Park, J.D. Richardson, A. Szabo, Magnetic fields observed by Voyager 2 in the heliosheath. Astrophys. J. **906**, 119 (2021)

607. M.P. Leubner, Z. Voros, A nonextensive entropy approach to solar wind intermittency. Astrophys. J. **618**, 547 (2005)

608. L.F. Burlaga, A. F-Vinas, C. Wang, Tsallis distributions of magnetic field strength variations in the heliosphere: 5–90 AU. J. Geophys. Res.-Space Phys. **112**, A07206 (2007)

609. L.F. Burlaga, A.F.-Vinas, Tsallis distribution functions in the solar wind: magnetic field and velocity observations, in *Complexity, Metastability and Nonextensivity*, eds. by S. Abe, H.J. Herrmann, P. Quarati, A. Rapisarda, C. Tsallis, Am. Inst. Phys. Conf. Proc. vol. 965 (New York, 2007), pp. 259–266

610. T. Nieves-Chinchilla, A.F. Vinas, Solar wind electron distribution functions inside magnetic clouds. J. Geophys. Res. **113**, A02105 (2008)

611. L.P. Chimento, F. Pennini, A. Plastino, Naudts-like duality and the extreme Fisher information principle. Phys. Rev. E **62**, 7462 (2000)

612. J. Naudts, Dual description of nonextensive ensembles, in *Classical and Quantum Complexity and Nonextensive Thermodynamics*, eds. by P. Grigolini, C. Tsallis, B.J. West, Chaos Solitons Fractals **13** (3), 445–450 (2002)

613. J. Naudts, *Generalised Thermostatistics* (Springer, London, 2011)

614. C. Tsallis, Generalization of the possible algebraic basis of q-triplets. Eur. Phys. J. Spec. Top. **226**, 455–466 (2017)

615. C. Tsallis, Statistical mechanics for complex systems: on the structure of q-triplets, in *Physical and Mathematical Aspects of Symmetries, Proceedings of the 31st International Colloquium in Group Theoretical Methods in Physics*, eds. by S. Duarte, J.-P. Gazeau, S. Faci, T. Micklitz, R. Scherer, F. Toppan (Springer, Berlin, 2017), pp. 51–60

616. N.O. Baella Pajuelo, Private Communication (2008)

617. T. Stosic, B. Stosic, V.P. Singh, Q-triplet for Brazos river discharge: the edge of chaos? Phys. A **495**, 137–142 (2018)

618. D. Stosic, D. Stosic, T.B. Ludermir, T. Stosic, Nonextensive triplets in cryptocurrency exchanges. Phys. A **505**, 1069–1074 (2018)
619. G.L. Ferri, M.F. Reynoso Savio, A. Plastino, *Tsallis' q*-triplet and the ozone layer. Phys. A **389**, 1829–1833 (2010)
620. D.B. de Freitas, J.R. de Medeiros, Nonextensivity in the solar magnetic activity during the increasing phase of solar Cycle 23. Europhys. Lett. **88**, 19001 (2009)
621. C. Tsallis, Nonadditive entropy and nonextensive statistical mechanics—An overview after 20 years. Braz. J. Phys. **39**, 337 (2009)
622. A.A. Tateishi, R. Hanel, S. Thurner, The transformation groupoid structure of the *q*-Gaussian family. Phys. Lett. A **377**, 1804–1809 (2013)
623. A. Robledo, The renormalization group, optimization of entropy and non-extensivity at criticality. Phys. Rev. Lett. **83**, 2289 (1999)
624. A. Robledo, Unorthodox properties of critical clusters. Mol. Phys. **103**, 3025–3030 (2005)
625. A. Robledo, q-statistical properties of large critical clusters. Int. J. Mod. Phys. B **21**, 3947–3953 (2007)
626. P.R.S. Carvalho, Critical exponents and amplitude ratios of scalar nonextensive q-field theories. Phys. Rev. D **98**, 085019 (2018)
627. V.N. Borodikhin, Dynamic critical behavior of the two-dimensional Ising model with nonextensive statistics. Phys. Rev. E **102**, 012116 (2020)
628. P.R.S. Carvalho, Nonextensive statistical field theory. Phys. Lett. B (2022). In press
629. G. Combe, V. Richefeu, M. Stasiak, A.P.F. Atman, Experimental validation of nonextensive scaling law in confined granular media. Phys. Rev. Lett. **115**, 238301 (2015)
630. S. Gheorghiu, J.R. van Ommen, M.-O. Coppens, Power-law distribution of pressure fluctuations in multiphase flow. Phys. Rev. E **67**, 041305 (2003)
631. P. Douglas, S. Bergamini, F. Renzoni, Tunable Tsallis distributions in dissipative optical lattices. Phys. Rev. Lett. **96**, 110601 (2006)
632. E. Lutz, F. Renzoni, Beyond Boltzmann-Gibbs statistical mechanics in optical lattices. Nat. Phys. **9**, 615–619 (2013)
633. E.A. Bender, E.R. Canfield, J. Comb. Theory A **24**, 296–307 (1978)
634. S. Thurner, C. Tsallis, Nonextensive aspects of self-organized scale-free gas-like networks. Europhys. Lett. **72**, 197–203 (2005)
635. S.G.A. Brito, L.R. da Silva, C. Tsallis, Role of dimensionality in complex networks. Nat. Sci. Rep. **6**, 27992 (2016)
636. T.C. Nunes, S. Brito, L.R. da Silva, C. Tsallis, Role of dimensionality in preferential attachment growth in the Bianconi-Barabasi model. J. Stat. Mech. 093402 (2017)
637. N. Cinardi, A. Rapisarda, C. Tsallis, A generalised model for asymptotically-scale-free geographical networks. J. Stat. Mech. 043404 (2020)
638. C.Y. Wong, G. Wilk, Tsallis fits to p_T spectra and relativistic hard scattering in pp collisions at LHC. Phys. Rev. D **87**, 114007 (2013)
639. F.D. Nobre, M.A. Rego-Monteiro, C. Tsallis, Nonlinear generalizations of relativistic and quantum equations with a common type of solution. Phys. Rev. Lett. **106**, 140601 (2011)
640. A.R. Plastino, C. Tsallis, Dissipative effects in nonlinear Klein-Gordon dynamics. EPL **113**, 50005 (2016)
641. A differential equation of the Bernoulli type is written as $dy/dx + p(x)\,y = q(x)\,y^n$. If we consider $p(x)$ and $q(x)$ to be constants, we precisely recover the form of Eq. (6.1)
642. C. Tsallis, J.C. Anjos, E.P. Borges, Fluxes of cosmic rays: a delicately balanced stationary state. Phys. Lett. A **310**, 372–376 (2003)
643. C. Beck, Superstatistics: theory and applications, in *Nonadditive Entropy and Nonextensive Statistical Mechanics*, ed. by M. Sugiyama, Contin. Mech. Thermodyn. **16**, 293 (Springer, Heidelberg, 2004)
644. E.G.D. Cohen, Superstatistics. Phys. D **193**, 35 (2004)
645. C. Beck, Superstatistics, escort distributions, and applications. Phys. A **342**, 139 (2004)
646. C. Beck, Superstatistics in hydrodynamic turbulence. Phys. D **193**, 195 (2004)

647. C. Beck, E.G.D. Cohen, S. Rizzo, Atmospheric turbulence and superstatistics, in *Nonextensive Statistical Mechanics: new Trends, New perspectives*, eds. by J.P. Boon, C. Tsallis, Europhys. News **36**, 189 (European Physical Society, 2005)

648. C. Beck, Superstatistics: recent developments and applications, in *Complexity, Metastability and Nonextensivity*, eds. by C. Beck, G. Benedek, A. Rapisarda, C. Tsallis (World Scientific, Singapore, 2005), p. 33

649. H. Touchette, C. Beck, Asymptotics of superstatistics. Phys. Rev. E **71**, 016131 (2005)

650. P.H. Chavanis, Coarse-grained distributions and superstatistics. Phys. A **359**, 177–212 (2006)

651. C. Beck, Stretched exponentials from superstatistics. Phys. A **365**, 96–101 (2006)

652. C. Beck, Correlations in superstatistical systems, in *Complexity, Metastability and Nonextensivity*, vol. 965, eds. by S. Abe, H.J. Herrmann, P. Quarati, A. Rapisarda, C. Tsallis, Am. Inst. Phys. Conf. Proc. (New York, 2007), pp. 60–67

653. C. Beck, Generalized information and entropy measures in physics. Contemp. Phys. **50**, 495–510 (2009)

654. C. Beck, Superstatistics: theory and applications, in *Nonadditive Entropy and Nonextensive Statistical Mechanics*, ed. by M. Sugiyama, Continuum Mech. Thermodyn. **16**, 293 (Springer, Heidelberg, 2004)

655. E. Lutz, Anomalous diffusion and Tsallis statistics in an optical lattice. Phys. Rev. A **67**, 051402(R) (2003)

656. R.G. DeVoe, Power-law distributions for a trapped ion interacting with a classical buffer gas. Phys. Rev. Lett. **102**, 063001 (2009)

657. M. Notzold, S.Z. Hassan, J. Tauch, E. Endres, R. Wester, M. Weidemuller, Thermometry in a multipole ion trap. Appl. Sci. **10**, 5264 (2020)

658. F. Sattin, Derivation of Tsallis' statistics from dynamical equations for a granular gas. J. Phys. A **36**, 1583 (2003)

659. R. Arevalo, A. Garcimartin, D. Maza, Anomalous diffusion in silo drainage. Eur. Phys. J. E **23**, 191–198 (2007)

660. R. Arevalo, A. Garcimartin, D. Maza, A non-standard statistical approach to the silo discharge, in *Complex Systems—New Trends and Expectations*, eds. by H.S. Wio, M.A. Rodriguez, L. Pesquera, Eur. Phys. J.-Spec. Top. **143** (2007)

661. A. Baldassari, U.M. Bettolo Marconi, A. Puglisi, Influence of correlations on the velocity statistics of scalar granular gases. Europhys. Lett. **58**, 14–20 (2002)

662. E. Fermi, Angular distribution of the pions produced in high energy nuclear collisions. Phys. Rev. **81**, 683–687 (1951)

663. R. Hagedorn, Statistical thermodynamics of strong interactions at high energy. Suppl. Nuovo Cimento **3**, 147 (1965)

664. R.D. Field, R.P. Feynman, A parametrization of the properties of quark jets. N. Phys. B **136**, 1–76 (1978)

665. I. Bediaga, E.M.F. Curado, J. Miranda, A thermodynamical equilibrium approach in $e^+e^- \rightarrow hadrons$. Phys. A **286**, 156 (2000)

666. C. Beck, Non-extensive statistical mechanics and particle spectra in elementary interactions. Physica A **286**, 164 (2000)

667. W.M. Alberico, A. Lavagno, P. Quarati, Non-extensive statistics, fluctuations and correlations in high energy nuclear collisions. Eur. Phys. J C **12**, 499 (2000)

668. T.S. Biro, A. Jakovac, Power-law tails from multiplicative noise. Phys. Rev. Lett. **94**, 132302 (2005)

669. T.S. Biro, G. Purcsel, Nonextensive Boltzmann equation and hadronization. Phys. Rev. Lett. **95**, 162302 (2005)

670. M. Biyajima, M. Kaneyama, T. Mizoguchi, G. Wilk, Analyses of κ_t distributions at RHIC by means of some selected statistical and stochastic models. Eur. Phys. J. C **40**, 243 (2005)

671. O.V. Utyuzh, G. Wilk, Z. Wlodarczyk, Multiparticle production processes from the information theory point of view. Acta Phys. Hung. A **25**, 65 (2006)

672. C.Y. Wong, G. Wilk, L.J.L. Cirto, C. Tsallis, From QCD-based hard-scattering to nonextensive statistical mechanical descriptions of transverse momentum spectra in high-energy pp and $p\bar{p}$ collisions. Phys. Rev. D **91**, 114027 (2015)

673. G. Wilk, Z. Wlodarczyk, Tsallis distribution with complex nonextensivity parameter q. Phys. A **413**, 53–58 (2014)

674. G. Wilk, Z. Wlodarczyk, Tsallis distribution decorated with log-periodic oscillation. Entropy **17**(1), 384–400 (2015)

675. G. Wilk, Z. Wlodarczyk, Oscillations in multiparticle production processes. Entropy **19**(12), 670 (2017)

676. D.B. Walton, J. Rafelski, Equilibrium distribution of heavy quarks in Fokker-Planck dynamics. Phys. Rev. Lett. **84**, 31 (2000)

677. PHENIX Collaboration, Measurement of neutral mesons in $p + p$ collisions at $\sqrt{s} = 200\ GeV$ and scaling properties of hadron production. Phys. Rev. D **83**, 052004 (2011)

678. M. Biyajima, T. Mizoguchi, N. Nakajima, N. Suzuki, G. Wilk, Modified Hagedorn formula including temperature fluctuation—Estimation of temperature at RHIC experiments. Eur. Phys. J. C **48**, 597–603 (2007)

679. T. Osada, G. Wilk, Nonextensive hydrodynamics for relativistic heavy-ion collision. Phys. Rev. C **77**, 044903 (2008)

680. L. Marques, E. Andrade-II, A. Deppman, Nonextensivity of hadronic systems. Phys. Rev. D **87**, 114022 (2013)

681. L. Marques, J. Cleymans, A. Deppman, Description of high-energy pp collisions using Tsallis thermodynamics: Transverse momentum and rapidity distributions. Phys. Rev. D **91**, 054025 (2015)

682. J. Chen, J. Deng, Z. Tang, Z. Xu, L. Yi, Nonequilibrium kinetic freeze-out properties in relativistic heavy ion collisions from energies employed at the RHIC beam energy scan to those available at the LHC. Phys. Rev. C **104**, 034901 (2021)

683. A. Deppman, Self-consistency in non-extensive thermodynamics of highly excited hadronic states. Phys. A **391**, 6380–6385 (2012); A. Deppman, Corrigendum to "Self-consistency in non-extensive thermodynamics of highly excited hadronic states" [Physica A 391 (2012) 6380–6385], Phys. A **400**, 207–208 (2014)

684. A. Deppman, Thermodynamics with fractal structure, Tsallis statistics, and hadrons. Phys. Rev. D **93**, 054001 (2016)

685. A. Deppman, E. Megias, D.P. Menezes, Fractals, non-extensive statistics, and QCD. Phys. Rev. D **101**, 034019 (2020)

686. R. Rath, A. Khuntia, R. Sahoo, J. Cleymans, Event multiplicity, transverse momentum and energy dependence of charged particle production, and system thermodynamics in pp collisions at the Large Hadron Collider. J. Phys. G: Nucl. Part. Phys. **47**, 055111 (2020)

687. F.S. Navarra, O.V. Utyuzh, G. Wilk, Z. Wlodarczyk, Estimating inelasticity with the information theory approach. Phys. Rev. D **67**, 114002 (2003)

688. M. Rybczynski, Z. Wlodarczyk, Inelasticity resulting from rapidity spectra analysis. New J. Phys. **22**, 113002 (2020)

689. P.A. Zyla et al.(Particle Data Group), Review of particle physics. Prog. Theor. Exp. Phys. (8), 083C01 (2020)

690. E. Megias, E. Andrade II, A. Deppman, A. Gammal, D.P. Menezes, T.N. da Silva, V.S. Timoteo, Tsallis statistics and thermofractals: applications to high energy and hadron physics. arxiv:2201.08771 [hep-ph]

691. R. Hagedorn, Hadronic matter near the boiling point. Nuovo Cimento LVI A (4), 1027 (1968)

692. C. Beck, Generalized statistical mechanics of cosmic rays. Phys. A **331**, 173 (2004)

693. G.C. Yalcin, C. Beck, Generalized statistical mechanics of cosmic rays: application to positron-electron spectral indices. Sci. Rep. **8**, 1764 (2018)

694. M. Smolla, B. Schafer, H. Lesch, C. Beck, Universal properties of primary and secondary cosmic ray energy spectra. New J. Phys. **22**, 093002 (2020)

695. G. Kaniadakis, A. Lavagno, P. Quarati, Generalized statistics and solar neutrinos. Phys. Lett. B **369**, 308 (1996)

696. P. Quarati, A. Carbone, G. Gervino, G. Kaniadakis, A. Lavagno, E. Miraldi, Constraints for solar neutrinos fluxes. Nucl. Phys. A **621**, 345c (1997)

697. G. Kaniadakis, A. Lavagno, P. Quarati, Non-extensive statistics and solar neutrinos. Astrophys. Space Sci. **258**, 145 (1998)
698. B.M. Boghosian, P.J. Love, P.V. Coveney, I.V. Karlin, S. Succi, J. Yepez, Galilean-invariant lattice-Boltzmann models with H-theorem. Phys. Rev. E **68**, 025103(R) (2003)
699. B.M. Boghosian, P. Love, J. Yepez, P.V. Coveney, Galilean-invariant multi-speed entropic lattice Boltzmann models. Phys. D **193**, 169 (2004)
700. K.E. Daniels, C. Beck, E. Bodenschatz, Defect turbulence and generalized statistical mechanics. Phys. D **193**, 208 (2004)
701. A.M. Reynolds, M. Veneziani, Rotational dynamics of turbulence and Tsallis statistics. Phys. Lett. A **327**, 9 (2004)
702. S. Rizzo, A. Rapisarda, Environmental atmospheric turbulence at Florence airport. Am. Inst. Phys. Conf. Proc. **742**, 176–181 (2004)
703. S. Rizzo, A. Rapisarda, Application of superstatistics to atmospheric turbulence, in *Complexity, Metastability and Nonextensivity*, eds. by C. Beck, G. Benedek, A. Rapisarda and C. Tsallis (World Scientific, Singapore, 2005), p. 246
704. M.J.A. Bolzan, F.M. Ramos, L.D.A. Sa, C. Rodrigues Neto, R.R. Rosa, Analysis of fine-scale canopy turbulence within and above an Amazon forest using Tsallis' generalized thermostatistics. J. Geophys. Res.-Atmos. **107**, 8063 (2002)
705. F.M. Ramos, M.J.A. Bolzan, L.D.A. Sa, R.R. Rosa, Atmospheric turbulence within and above an Amazon forest. Phys. D **193**, 278 (2004)
706. C. Beck, G.S. Lewis, H.L. Swinney, Measuring non-extensivity parameters in a turbulent Couette-Taylor flow. Phys. Rev. E **63**, 035303 (2001)
707. C. Tsallis, E.P. Borges, F. Baldovin, Mixing and equilibration: protagonists in the scene of nonextensive statistical mechanics. Phys. A **305**, 1–18 (2002)
708. C.N. Baroud, H.L. Swinney, Nonextensivity in turbulence in rotating two-dimensional and three-dimensional flows. Phys. D **184**, 21 (2003)
709. S. Jung, B.D. Storey, J. Aubert, H.L. Swinney, Nonextensive statistical mechanics for rotating quasi-two-dimensional turbulence. Phys. D **193**, 252 (2004)
710. S. Jung, H.L. Swinney, Velocity difference statistics in turbulence. Phys. Rev. E **72**, 026304 (2005)
711. T. Arimitsu, N. Arimitsu, Analysis of fully developed turbulence in terms of Tsallis statistics. Phys. Rev. E **61**, 3237 (2000)
712. N. Arimitsu, T. Arimitsu, Analysis of velocity derivatives in turbulence based on generalized statistics. Europhys. Lett. **60**, 60 (2002)
713. A.M. Reynolds, On the application of nonextensive statistics to Lagrangian turbulence. Phys. Fluids **15**, L1 (2003)
714. B.M. Boghosian, Thermodynamic description of the relaxation of two-dimensional turbulence using Tsallis statistics. Phys. Rev. E **53**, 4754 (1996)
715. C. Anteneodo, C. Tsallis, Two-dimensional turbulence in pure-electron plasma: a nonextensive thermostatistical description. J. Mol. Liq. **71**, 255 (1997)
716. M. Peyrard, I. Daumont, Statistical properties of one dimensional "turbulence." Europhys. Lett. **59**, 834 (2002)
717. M. Peyrard, The statistical distributions of one-dimensional "turbulence", in *Anomalous Distributions, Nonlinear Dynamics and Nonextensivity*, eds. by H.L. Swinney, C. Tsallis, Phys. D **193**, 265 (2004)
718. T. Gotoh, R.H. Kraichnan, Turbulence and tsallis statistics, in *Anomalous Distributions, Nonlinear Dynamics and Nonextensivity*, eds. by H.L. Swinney, C. Tsallis, Phys. D **193**, 231 (2004)
719. A. La Porta, G.A. Voth, A.M. Crawford, J. Alexander, E. Bodenschatz, Fluid particle accelerations in fully developed turbulence. Nature (London) **409**, 1017 (2001)
720. B.M. Boghosian, J.P. Boon, Lattice Boltzmann equation and nonextensive diffusion, in *Nonextensive Statistical Mechanics: new Trends, New Perspectives*, eds. by J.P. Boon, C. Tsallis, Europhys. News **36**, 192 (2005)

721. P. Grosfils, J.P. Boon, Nonextensive statistics in viscous fingering. Phys. A **362**, 168–173 (2006)
722. P. Grosfils, J.P. Boon, Statistics of precursors to fingering processes. Europhys. Lett. **74**, 609–615 (2006)
723. M.S. Reis, J.C.C. Freitas, M.T.D. Orlando, E.K. Lenzi, I.S. Oliveira, Evidences for Tsallis non-extensivity on CMR manganites. Europhys. Lett. **58**, 42 (2002)
724. M.S. Reis, J.P. Araújo, V.S. Amaral, E.K. Lenzi, I.S. Oliveira, Magnetic behavior of a non-extensive S-spin system: possible connections to manganites. Phys. Rev. B **66**, 134417 (2002)
725. M.S. Reis, V.S. Amaral, J.P. Araujo, I.S. Oliveira, Magnetic phase diagram for a non-extensive system: experimental connection with manganites. Phys. Rev. B **68**, 014404 (2003)
726. F.A.R. Navarro, M.S. Reis, E.K. Lenzi, I.S. Oliveira, A study on composed nonextensive magnetic systems. Phys. A **343**, 499 (2004)
727. M.S. Reis, V.S. Amaral, J.P. Araujo, I.S. Oliveira, Non-extensivity of inhomogeneous magnetic systems, in *Complexity, Metastability and Nonextensivity*, eds. by C. Beck, G. Benedek, A. Rapisarda, C. Tsallis (World Scientific, Singapore, 2005), p. 230
728. M.S. Reis, V.S. Amaral, R.S. Sarthour, I.S. Oliveira, Experimental determination of the nonextensive entropic parameter q. Phys. Rev. B **73**, 092401 (2006)
729. R.M. Pickup, R. Cywinski, C. Pappas, B. Farago, P. Fouquet, Generalized spin glass relaxation. Phys. Rev. Lett. **102**, 097202 (2009)
730. B. Liu, J. Goree, Superdiffusion and non-Gaussian statistics in a driven-dissipative 2D dusty plasma. Phys. Rev. Lett. **100**, 055003 (2008)
731. J.A.S. Lima, R. Silva Jr., J. Santos, Plasma oscillations and nonextensive statistics. Phys. Rev. E **61**, 3260 (2000)
732. M.P. Leubner, Fundamental issues on kappa-distributions in space plasmas and interplanetary proton distributions. Phys. Plasmas **11**, 1308 (2004)
733. M.P. Leubner, Core-halo distribution functions: a natural equilibrium state in generalized thermostatistics. Astrophys. J. **604**, 469 (2004)
734. J. Du, Nonextensivity in nonequilibrium plasma systems with Coulombian long-range interactions. Phys. Lett. A **329**, 262 (2004)
735. S.I. Kononenko, V.M. Balebanov, V.P. Zhurenko, O.V. Kalantar'yan, V.I. Karas, V.T. Kolesnik, V.I. Muratov, V.E. Novikov, I.F. Potapenko, R.Z. Sagdeev, Nonequilibrium electron distribution functions in a smiconductor plasma irradiated with fast ions. Plasma Phys. Rep. **30**, 671 (2004)
736. P. Brault, A. Caillard, A.L. Thomann, J. Mathias, C. Charles, R.W. Boswell, S. Escribano, J. Durand, T. Sauvage, Plasma sputtering deposition of platinum into porous fuel cell electrodes. J. Phys. D **37**, 3419 (2004)
737. F. Valentini, Nonlinear Landau damping in nonextensive statistics. Phys. Plasmas **12**, 072106 (2005)
738. P.H. Yoon, T. Rhee, C.-M. Ryu, Self-consistent generation of superthermal electrons by beam-plasma interaction. Phys. Rev. Lett. **95**, 215003 (2005)
739. S. Bouzat, R. Farengo, Effects of varying the step particle distribution on a probabilistic transport model. Phys. Plasmas **12**, 122303 (2005)
740. S. Bouzat, R. Farengo, Probabilistic transport model with two critical mechanisms for magnetically confined plasmas. Phys. Rev. Lett. **97**, 205008 (2006)
741. M.P. Leubner, Z. Voros, W. Baumjohann, Nonextensive entropy approach to space plasma fluctuations and turbulence. Adv. Geosci. **2**, Chapter 4 (2006)
742. V. Munoz, A nonextensive statistics approach for Langmuir waves in relativistic plasmas. Nonlinear Process. Geophys. **13**, 237–241 (2006)
743. I.D. Dubinova, A.E. Dubinov, The theory of ion-sound solitons in plasma with electrons featuring the Tsallis distribution. Tech. Phys. Lett. **32**, 575–578 (2006)
744. P.H. Yoon, T. Rhee, C.M. Ryu, Self-consistent formation of electron kappa distribution:1. Theory. J. Geophys. Res.-Space Phys. **111**, A09106 (2006)

745. F. Valentini, R. D'Agosta, Electrostatic Landau pole for kappa-velocity distributions. Phys. Plasmas **14**, 092111 (2007)

746. C.-M. Ryu, T. Rhee, T. Umeda, P.H. Yoon, Y. Omura, Turbulent acceleration of superthermal electrons. Phys. Plasmas **14**, 100701 (2007)

747. L. Guo, J. Du, The κ parameter and κ-distribution in κ-deformed statistics for the systems in an external field. Phys. Lett. A **362**, 368–370 (2007)

748. T. Cattaert, M.A. Hellberg, L. Mace, Oblique propagation of electromagnetic waves in a kappa-Maxwellian plasma. Phys. Plasmas **14**, 082111 (2007)

749. L. Liu, J. Du, Ion acoustic waves in the plasma with the power-law q-distribution in nonextensive statistics. Phys. A **387**, 4821–4827 (2008)

750. J.Y. Kim, H.C. Lee, G. Go, Y.H. Choi, Y.S. Hwang, K.J. Chung, Exploring the nonextensive thermodynamics of partially ionized gas in magnetic field. Phys. Rev. E **104**, 045202 (2021)

751. H. Qiu, D. Xiao, J. Wu, S. Wu, C. Zhong, X. Li, X. Peng, Y. Yuan, Q. Cai, J. Chang, T. Hu, Z. Hu, Y. Zhu, Initial measurement of ion nonextensive parameter with geodesic acoustic mode theory. Sci. Rep. (2022). In press

752. A.R. Plastino, A. Plastino, Stellar polytropes and Tsallis' entropy. Phys. Lett. A **174**, 384 (1993)

753. A.R. Plastino, A. Plastino, Tsallis entropy and the Vlasov-Poisson equations, in *Nonextensive Statistical Mechanics and Thermodynamics*, eds. by S.R.A. Salinas, C. Tsallis, Braz. J. Phys. **29**, 79 (1999)

754. A.R. Plastino, S_q entropy and self-gravitating systems, in *Nonextensive Statistical Mechanics: New Trends, New perspectives*, eds. by J.P. Boon, C. Tsallis, Europhys. News **36**, 208 (2005) (European Physical Society, 2005)

755. A. Plastino, D. Monteoliva, M.C. Rocca, Tsallis' statistics for long range interactions: gravity. Phys. A (2022). In press. https://doi.org/10.1016/j.physa.2021.126597

756. D.B. de Freitas, Stellar age dependence of the nonextensive magnetic braking index: a test for the open cluster αPer. EPL **135**, 19001 (2021)

757. V.H. Hamity, D.E. Barraco, Generalized nonextensive thermodynamics applied to the cosmic background radiation in a Robertson-Walker universe. Phys. Rev. Lett. **76**, 4664 (1996)

758. D.F. Torres, H. Vucetich, Cosmology in a non-standard statistical background. Phys. A **259**, 397 (1998)

759. D.F. Torres, Precision cosmology as a test for statistics. Phys. A **261**, 512 (1998)

760. H.P. de Oliveira, S.L. Sautu, I.D. Soares, E.V. Tonini, Chaos and universality in the dynamics of inflationary cosmologies. Phys. Rev. D **60**, 121301 (1999)

761. K.S. Fa, E.K. Lenzi, Exact equation of state for 2-dimensional gravitating systems within Tsallis statistical mechanics. J. Math. Phys. **42**, 1148 (2001); [Erratum: **43**, 1127 (2002)]

762. H.P. de Oliveira, I.D. Soares, E.V. Tonini, Universality in the chaotic dynamics associated with saddle-centers critical points. Phys. A **295**, 348 (2001)

763. M.E. Pessah, D.F. Torres, H. Vucetich, Statistical mechanics and the description of the early universe I: Foundations for a slightly non-extensive cosmology. Phys. A **297**, 164 (2001)

764. M.E. Pessah, D.F. Torres, Statistical mechanics and the description of the early universe II: Principle of detailed balance and primordial ^4He formation. Phys. A **297**, 201 (2001)

765. C. Hanyu, A. Habe, The differential energy distribution of the universal density profile of dark halos. Astrophys. J. **554**, 1268 (2001)

766. J.A.S. Lima, R. Silva, J. Santos, Jeans' gravitational instability and nonextensive kinetic theory. Astron. Astrophys. **396**, 309 (2002)

767. R. Silva, J.S. Alcaniz, Non-extensive statistics and the stellar polytrope index. Phys. A **341**, 208 (2004)

768. A. Taruya, M. Sakagami, Gravothermal catastrophe and Tsallis' generalized entropy of self-gravitating systems. Phys. A **307**, 185 (2002)

769. P.-H. Chavanis, Gravitational instability of polytropic spheres and generalized thermodynamics. Astro. Astrophys. **386**, 732 (2002)

770. A. Taruya, M. Sakagami, Long-term evolution of stellar self-gravitating systems away from the thermal equilibrium: connection with nonextensive statistics. Phys. Rev. Lett. **90**, 181101 (2003)

771. W.H. Siekman, The entropic index of the planets of the solar system. Chaos Solitons Fractals **16**, 119 (2003)
772. P.H. Chavanis, Gravitational instability of isothermal and polytropic spheres. Astron. Astrophys. **401**, 15 (2003)
773. A. Taruya, M. Sakagami, Gravothermal catastrophe and Tsallis' generalized entropy of self-gravitating systems III. Quasi-equilibrium structure using normalized q-values. Phys. A **322**, 285 (2003)
774. H.P. de Oliveira, I.D. Soares. E.V. Tonini, Chaos and universality in the dynamics of inflationary cosmologies. II. The role of nonextensive statistics. Phys. Rev. D **67**, 063506 (2003)
775. C. Castro, A note on fractal strings and $\mathcal{E}^{(\infty)}$ spacetime. Chaos Solitons Fractals **15**, 797 (2003)
776. C. Beck, Nonextensive scalar field theories and dark energy models. Phys. A **340**, 459 (2004)
777. T. Matos, D. Nunez, R.A. Sussman, A general relativistic approach to the Navarro-Frenk-White galactic halos. Class. Q. Grav. **21**, 5275 (2004)
778. H.P. de Oliveira, I.D. Soares, E.V. Tonini, Role of the nonextensive statistics in a three-degrees of freedom gravitational system. Phys. Rev. D **70**, 084012 (2004)
779. A. Taruya, M. Sakagami, Fokker-Planck study of stellar self-gravitating system away from the thermal equilibrium: connection with nonextensive statistics. Phys. A **340**, 453 (2004)
780. M. Sakagami, A. Taruya, Description of quasi-equilibrium states of self-gravitating systems based on non-extensive thermostatistics. Phys. A **340**, 444 (2004)
781. J.A.S. Lima, L. Marassi, Mass function of halos: a new analytical approach. Int. J. Mod. Phys. D **13**, 1345 (2004)
782. J.L. Du, The nonextensive parameter and Tsallis distribution for self-gravitating systems. Europhys. Lett. **67**, 893 (2004)
783. A. Ishikawa, T. Suzuki, Relations between a typical scale and averages in the breaking of fractal distribution. Phys. A **343**, 376 (2004)
784. A. Nakamichi, M. Morikawa, Is galaxy distribution non-extensive and non-Gaussian? Phys. A **341**, 215 (2004)
785. C.A. Wuensche, A.L.B. Ribeiro, F.M. Ramos, R.R. Rosa, Nonextensivity and galaxy clustering in the Universe. Phys. A **344**, 743 (2004)
786. R. Silva, J.A.S. Lima, Relativity, nonextensivity and extended power law distributions. Phys. Rev. E **72**, 057101 (2005)
787. P.H. Chavanis, J. Vatteville, F. Bouchet, Dynamics and thermodynamics of a simple model similar to self-gravitating systems: the HMF model. Eur. Phys. J. B **46**, 61–99 (2005)
788. J.A.S. Lima, R.E. de Souza, Power-law stellar distributions. Phys. A **350**, 303 (2005)
789. S.H. Hansen, D. Egli, L. Hollenstein, C. Salzmann, Dark matter distribution function from non-extensive statistical mechanics. New Astron. **10**, 379 (2005)
790. T. Matos, D. Nunez, R.A. Sussman, The spacetime associated with galactic dark matter halos. Gen. Relativ. Gravit. **37**, 769 (2005)
791. S.H. Hansen, Cluster temperatures and non-extensive thermo-statistics. New Astron. **10**, 371 (2005)
792. H.P. Oliveira, I.D. Soares, Dynamics of black hole formation: evidence for nonextensivity. Phys. Rev. D **71**, 124034 (2005)
793. N.W. Evans, J.H. An, Distribution function of dark matter. Phys. Rev. D **73**, 023524 (2006)
794. S.H. Hansen, B. Moore, A universal density slope—Velocity anisotropy relation for relaxed structures. New Astron. **11**, 333–338 (2006)
795. J.L. Du, The Chandrasekhar's condition of the equilibrium and stability for a star in the nonextensive kinetic theory. New Astron. **12**, 60–63 (2006)
796. A. Taruya, M.A. Sakagami, Quasi-equilibrium evolution in self-gravitating N-body systems, in *Complexity and Nonextensivity: New Trends in Statistical Mechanics*, eds. by M. Sakagami, N. Suzuki, S. Abe, Prog. Theor. Phys. Suppl. **162**, 53–61 (2006)
797. P.-H. Chavanis, Phase transitions in self-gravitating systems. Int. J. Mod. Phys. B **20**, 3113–3198 (2006)

798. P.H. Chavanis, Lynden-Bell and Tsallis distributions for the HMF model. Eur. Phys. J. B **53**, 487–501 (2006)

799. T. Kronberger, M.P. Leubner, E. van Kampen, Dark matter density profiles: a comparison of nonextensive statistics with N-body simulations. Astron. Astrophys. **453**, 21–25 (2006)

800. P.H. Chavanis, Dynamical stability of collisionless stellar systems and barotropic stars: the nonlinear Antonov first law. Astron. Astrophys. **451**, 109–123 (2006)

801. D. Nunez, R.A. Sussman, J. Zavala, L.G. Cabral-Rosetti, T. Matos, Empirical testing of Tsallis' thermodynamics as a model for dark matter halos. AIP Conf. Proc. A **857**, 316–320 (2006)

802. J. Zavala, D. Nunez, R.A. Sussman, L.G. Cabral-Rosetti, T. Matos, Stellar polytropes and Navarro-Frenk-White halo models: comparison with observations. J. Cosmol. Astrop. Phys. **6**(8) (2006)

803. L. Marassi, J.A.S. Lima, Press-Schechter mass function and the normalization problem. Internat. J. Mod. Phys. D **16**, 445–452 (2007)

804. M. Razeira, B.E.J. Bodmann, C.A. Zen Vasconcellos, Strange matter and strange stars with Tsallis statistics. Internat. J. Mod. Phys. D **16**, 365–372 (2007)

805. B.D. Shizgal, Suprathermal particle distributions in space physics: Kappa distributions and entropy. Astrophys. Space Sci. **312**, 227–237 (2007)

806. E.I. Barnes, L.L.R. Williams, A. Babul, J.J. Dalcanton, Velocity distributions from nonextensive thermodynamics. Astrophys. J. **655**, 847–850 (2007)

807. A. Nakamichi, T. Tatekawa, M. Morikawa, Statistical mechanics of $SDSS$ galaxy distribution and cosmological N-body simulations, in *Complexity, Metastability and Nonextensivity*, eds. by S. Abe, H.J. Herrmann, P. Quarati, A. Rapisarda and C. Tsallis, Am. Inst. Phys. Conf. Proc. **965**, 267–272 (New York, 2007)

808. C. Feron, J. Hjorth, Simulated dark-matter halos as a test of nonextensive statistical mechanics. Phys. Rev. E **77**, 022106 (2008)

809. J.D. Vergados, S.H. Hansen, O. Host, Impact of going beyond the Maxwell distribution in direct dark matter detection rates. Phys. Rev. D **77**, 023509 (2008)

810. H.P. de Oliveira, I.D. Soares, E.V. Tonini, Black-hole bremsstrahlung and the efficiency of mass-energy radiative transfer. Phys. Rev. D **78**, 044016 (2008)

811. A. Bernui, C. Tsallis, T. Villela, Deviation from Gaussianity in the cosmic microwave background temperature fluctuations. Europhys. Lett. **78**, 19001 (2007)

812. C. Frigerio Martins, J.A.S. Lima, P. Chimenti, Galaxy rotation curves and nonextensive statistics. MNRAS **449**, 3645–3650 (2015)

813. A.S. Betzler, E.P. Borges, Nonextensive distributions of asteroid rotation periods and diameters. Astron. Astrophys. **539**, A158 (2012)

814. A.S. Betzler, E.P. Borges, Nonextensive statistical analysis of meteor showers and lunar flashes. Mon. Not. R. Astron. Soc. **447**, 765–771 (2015)

815. A.S. Betzler, E.P. Borges, Mass distributions of meteorites. Mon. Not. R. Astron. Soc. **493**, 4058–4064 (2020)

816. G.P. Pavlos, O.E. Malandraki, E.G. Pavlos, A.C. Iliopoulos, L.P. Karakatsanis, Non-extensive statistical analysis of magnetic field during the March 2012 ICME event using a multi-spacecraft approach. Phys. A **464**, 149–181 (2016)

817. M. Baiesi, M. Paczuski, A.L. Stella, Intensity thresholds and the statistics of the temporal occurrence of solar flares. Phys. Rev. Lett. **96**, 051103 (2006)

818. U. Tirnakli, Ageing in earthquake models, in *Complexity, Metastability and Nonextensivity, Proceedings of the 31st Workshop of the International School of Solid State Physics (20–26 July 2004, Erice-Italy)*, eds. by C. Beck, G. Benedek, A. Rapisarda and C. Tsallis (World Scientific, Singapore, 2005), p. 350

819. S. Abe and N. Suzuki, Law for the distance between successive earthquakes. J. Geophys. Res. (Solid Earth) **108**, B2, 2113 (2003)

820. S. Abe, N. Suzuki, Ageing and scaling of earthquake aftershocks. Phys. A **332**, 533–538 (2004)

821. O. Sotolongo-Costa, A. Posadas, Fragment-asperity interaction model for earthquakes. Phys. Rev. Lett. **92**, 048501 (2004)
822. S. Abe, U. Tirnakli, P.A. Varotsos, Complexity of seismicity and nonextensive statistics, in *Nonextensive Statistical Mechanics: new Trends, New perspectives*, eds. by J.P. Boon, C. Tsallis, Europhysics News **36**, 206 (2005)
823. S. Abe, N. Suzuki, Scale-free statistics of time interval between successive earthquakes. Phys. A **350**, 588–596 (2005)
824. G.S. Franca, C.S. Vilar, R. Silva, J.S. Alcaniz, Nonextensivity in geological faults? Phys. A **377**, 285–290 (2007)
825. M. Kalimeri, C. Papadimitriou, G. Balasis, K. Eftaxias, Dynamical complexity detection in pre seismic emissions using nonadditive Tsallis entropy. Phys. A **387**, 1161–1172 (2008)
826. C. Papadimitriou, M. Kalimeri, K. Eftaxias, Nonextensivity and universality in the earthquake preparation process. Phys. Rev. E **77**, 036101 (2008)
827. G. Balasis, I.A. Daglis, C. Papadimitriou, M. Kalimeri, A. Anastasiadis, K. Eftaxias, Dynamical complexity in D_{st} time series using non-extensive Tsallis entropy. Geophys. Res. Lett. **35**, L14102 (2008)
828. A.H. Darooneh, C. Dadashinia, Analysis of the spatial and temporal distributions between successive earthquakes: nonextensive statistical mechanics viewpoint. Phys. A **387**, 3647–3654 (2008)
829. F. Vallianatos, On the non-extensivity in Mars geological faults. EPL **102**, 28006 (2013)
830. S.L.E.F. da Silva, G. Corso, Nonextensive Gutenberg-Richter law and the connection between earthquakes and marsquakes. Eur. Phys. J. B **94**, 25 (2021)
831. G.S. Franca, C.S. Vilar, R. Silva, J.S. Alcaniz, Nonextensivity in geological faults? Phys. A **377**, 285–290 (2007)
832. P. Bak, K. Christensen, L. Danon, T. Scanlon, Unified scaling law for earthquakes. Phys. Rev. Lett. **88**, 178501 (2002)
833. F. Caruso, A. Pluchino, V. Latora, A. Rapisarda, S. Vinciguerra, Self-organized criticality and earthquakes, in *Complexity, Metastability and Nonextensivity*, eds. by S. Abe, H.J. Herrmann, P. Quarati, A. Rapisarda, C. Tsallis, Am. Inst. Phys. Conf. Proc. **965**, 281-284 (New York, 2007)
834. J. Jersblad, H. Ellman, K. Stochkel, A. Kasteberg, L. Sanchez-Palencia, R. Kaiser, Non-Gaussian velocity distributions in optical lattices. Phys. Rev. A **69**, 013410 (2004)
835. M. Ausloos, F. Petroni, Tsallis nonextensive statistical mechanics of El Nino Southern Oscillation Index. Phys. A **373**, 721–736 (2007)
836. F. Petroni, M. Ausloos, High frequency (daily) data analysis of the Southern Oscillation index. Tsallis nonextensive statistical mechanics approach, in *Complex Systems—New Trends and Expectations*, eds. by H.S. Wio, M.A. Rodriguez, L. Pesquera, Eur. Phys. J. Spec. Top. **143**, 201–208 (2007)
837. K. Ivanova, H.N. Shirer, T.P. Ackerman, E.E. Clothiaux, Dynamical model and nonextensive statistical mechanics of liquid water path fluctuations in stratus clouds. J. Geophys. Res.-Atmosp. **112**, D10211 (2007)
838. G. Gervino, C. Cigolini, A. Lavagno, C. Marino, P. Prati, L. Pruiti, G. Zangari, Modelling temperature distributions and radon emission at Stromboli Volcano using a non-extensive statistical approach. Phys. A **340**, 402 (2004)
839. C.S. Barbosa, D.S.R. Ferreira, M.A. do Espirito Santo, A.R.R. Papa, Statistical analysis of geomagnetic field reversals and their consequences. Phys. A **392** (24), 6554–6560 (2013)
840. F. Vallianatos, G. Michas, Complexity of fracturing in terms of non-extensive statistical physics: from earthquake faults to Arctic sea ice fracturing. Entropy **22**, 1194 (2020)
841. Y.S. Weinstein, S. Lloyd, C. Tsallis, Border between between regular and chaotic quantum dynamics. Phys. Rev. Lett. **89**, 214101 (2002)
842. Y.S. Weinstein, C. Tsallis, S. Lloyd, On the emergence of nonextensivity at the edge of quantum chaos, in *Decoherence and Entropy in Complex Systems*, ed. by H.-T. Elze, *Lecture Notes in Physics*, vol. 633 (Springer-Verlag, Berlin, 2004), p. 385

843. S.M.D. Queiros, C. Tsallis, Edge of chaos of the classical kicked top map: sensitivity to initial conditions, in *Complexity, Metastability and Nonextensivity, Proceedings of the 31st Workshop of the International School of Solid State Physics (20–26 July 2004, Erice-Italy)*, eds. by C. Beck, G. Benedek, A. Rapisarda, C. Tsallis (World Scientific, Singapore, 2005), p. 135

844. S. Abe, A.K. Rajagopal, Nonadditive conditional entropy and its significance for local realism. Phys. A **289**, 157–164 (2001)

845. C. Tsallis, S. Lloyd, M. Baranger, Peres criterion for separability through nonextensive entropy. Phys. Rev. A **63**, 042104 (2001)

846. R. Rossignoli, N. Canosa, Non-additive entropies and quantum statistics. Phys. Lett. A **264**, 148 (1999)

847. F.C. Alcaraz, C. Tsallis, Frontier between separability and quantum entanglement in a many spin system. Phys. Lett. A **301**, 105 (2002)

848. C. Tsallis, D. Prato, C. Anteneodo, Separable-entangled frontier in a bipartite harmonic system. Eur. Phys. J. B **29**, 605 (2002)

849. J. Batle, A.R. Plastino, M. Casas, A. Plastino, Conditional q-entropies and quantum separability: a numerical exploration. J. Phys. A **35**, 10311 (2002)

850. J. Batle, M. Casas, A.R. Plastino, A. Plastino, Entanglement, mixedness, and q-entropies. Phys. Lett. A **296**, 251 (2002)

851. K.G.H. Vollbrecht, M.M. Wolf, Conditional entropies and their relation to entanglement criteria. J. Math. Phys. **43**, 4299 (2002)

852. N. Canosa, R. Rossignoli, Generalized nonadditive entropies and quantum entanglement. Phys. Rev. Lett. **88**, 170401 (2002)

853. N. Canosa, R. Rossignoli, Generalized nonadditive entropies and quantum entanglement. Phys. Rev. Lett. **88**, 170401 (2002)

854. R. Rossignoli, N. Canosa, Generalized entropic criterion for separability. Phys. Rev. A **66**, 042306 (2002)

855. R. Rossignoli, N. Canosa, Violation of majorization relations in entangled states and its detection by means of generalized entropic forms. Phys. Rev. A **67**, 042302 (2003)

856. N. Canosa, R. Rossignoli, Generalized entropies and quantum entanglement. Phys. A **329**, 371 (2003)

857. R. Rossignoli, N. Canosa, Limit temperature for entanglement in generalized statistics. Phys. Lett. A **323**, 22 (2004)

858. R. Rossignoli, N. Canosa, Generalized disorder measure and the detection of quantum entanglement. Phys. A **344**, 637 (2004)

859. O. Guhne, M. Lewenstein, Entropic uncertainty relations and entanglement. Phys. Rev. A **70**, 022316 (2004)

860. A.R. Plastino, A. Daffertshofer, Liouville dynamics and the conservation of classical information. Phys. Rev. Lett. **93**, 138701 (2004)

861. N. Canosa, R. Rossignoli, General non-additive entropic forms and the inference of quantum density operstors. Phys. A **348**, 121 (2005)

862. N. Canosa, R. Rossignoli, M. Portesi, Majorization properties of generalized thermal distributions. Phys. A **368**, 435–441 (2006)

863. N. Canosa, R. Rossignoli, M. Portesi, Majorization relations and disorder in generalized statistics. Phys. A **371**, 126–129 (2006)

864. X. Hu, Z. Ye, Generalized quantum entropy. J. Math. Phys. **46**, 023502 (2006)

865. O. Giraud, Distribution of bipartite entanglement for random pure states. J. Phys. A **40**, 2793–2801 (2007)

866. F. Buscemi, P. Bordone, A. Bertoni, Linear entropy as an entanglement measure in two-fermion systems Phys. Rev. A **75**, 032301 (2007)

867. R. Prabhu, A.R. Usha Devi, G. Padmanabha, Separability of a family of one parameter W and Greenberger-Horne-Zeilinger multiqubit states using Abe-Rajagopal q-conditional entropy approach. Phys. Rev. A **76**, 042337 (2007)

868. R. Augusiak, J. Stasinska, P. Horodecki, Beyond the standard entropic inequalities: stronger scalar separability criteria and their applications. Phys. Rev. A **77**, 012333 (2008)

869. H. Yang, Z.Y. Ding, D. Wang, H. Yuan, X.K. Song, J. Yang, C.J. Zhang, L. Ye, Experimental certification of the steering criterion based on a general entropic uncertainty relation. Phys. Rev. A **101**, 022324 (2020)

870. S. Wollmann, R. Uola, A.C.S. Costa, Experimental demonstration of robust quantum steering. Phys. Rev. Lett. **125**, 020404 (2020); Supplemental Material

871. F. Toscano, R.O. Vallejos, C. Tsallis, Random matrix ensembles from nonextensive entropy. Phys. Rev. E **69**, 066131 (2004)

872. M.S. Hussein, M.P. Pato, Fractal structure of random matrices. Phys. A **285**, 383–391 (2000)

873. A.Y. Abul-Magd, Nonextensive random matrix theory approach to mixed regular chaotic dynamics. Phys. Rev. E **71**, 066207 (2005)

874. A.Y. Abul-Magd, Superstatistics in random matrix theory. Phys. A **361**, 41–54 (2005)

875. A.Y. Abul-Magd, Random matrix theory within superstatistics. Phys. Rev. E **72**, 066114 (2005)

876. A.Y. Abul-Magd, Superstatistical random-matrix-theory approach to transition intensities in mixed systems. Phys. Rev. E **73**, 056119 (2006)

877. R. Xie, S. Deng, W. Deng, M.P. Pato, Generalized poisson ensemble. Phys. A **585**, 126427 (2022)

878. R.N. Costa Filho, M.P. Almeida, G.A. Farias, J.S. Andrade, Displacement operator for quantum systems with position-dependent mass. Phys. Rev. A **84**, 050102(R) (2011)

879. S.H. Mazharimousavi, Revisiting the displacement operator for quantum systems with position-dependent mass. Phys. Rev. A **85**, 034102 (2012)

880. F.D. Nobre, M.A. Rego-Monteiro, C. Tsallis, A Generalized nonlinear Schroedinger equation: classical field-theoretic approach. EPL **97**, 41001 (2012)

881. I.V. Toranzo, A.R. Plastino, J.S. Dehesa, A. Plastino, Quasi-stationary states of the NRT nonlinear Schroedinger equation. Phys. A **392**, 3945–3951 (2013)

882. M.A. Rego-Monteiro, F.D. Nobre, Classical field theory for a non-Hermitian Schroedinger equation with position-dependent masses. Phys. Rev. A **88**, 032105 (2013)

883. M.A. Rego-Monteiro, F.D. Nobre, Nonlinear quantum equations: classical field theory. J. Math. Phys. **54**, 103302 (2013)

884. A.R. Plastino, C. Tsallis, Nonlinear Schroedinger equation in the presence of uniform acceleration. J. Math. Phys. **54**(4), 041505 (2013)

885. R.N. Costa Filho, G. Alencar, B.-S. Skagerstam, J.S. Andrade Jr., Morse potential derived from first principles. EPL **101**, 10009 (2013)

886. A.R. Plastino, A.M.C. Souza, F.D. Nobre, C. Tsallis, Stationary and uniformly accelerated states in nonlinear quantum mechanics. Phys. Rev. A **90**, 062134 (2014)

887. F. Pennini, A.R. Plastino, A. Plastino, Pilot wave approach to the NRT nonlinear Schroedinger equation. Phys. A **403**, 195–205 (2014)

888. B.G. da Costa, E.P. Borges, Generalized space and linear momentum operators in quantum mechanics. J. Math. Phys. **55**, 062105 (2014)

889. L.G.A. Alves, H.V. Ribeiro, M.A.F. Santos, R.S. Mendes, E.K. Lenzi, Solutions for a q-generalized Schroedinger equation of entangled interacting particles. Phys. A **429**, 35–44 (2015)

890. A. Plastino, M.C. Rocca, Classical field-theoretical approach to the non-linear q-Klein-Gordon equation. EPL **116**, 41001 (2016)

891. T. Bountis, F.D. Nobre, Travelling-wave and separated variable solutions of a nonlinear Schroedinger equation. J. Math. Phys. **57**, 082106 (2016)

892. F.D. Nobre, M.A. Rego-Monteiro, C. Tsallis, Nonlinear q-generalizations of quantum equations: homogeneous and nonhomogeneous cases - An Overview. Entropy **19**, 39 (2017)

893. M.A. Rego-Monteiro, Lorentzian solitary wave in a generalised nonlinear Schroedinger equation. Phys. Lett. A **384**, 126132 (2020)

894. R.L. Liboff, *Introductory Quantum Mechanics*, 4th edn. (Addison Wesley, San Francisco, 2003)

895. W.I. Fushchich, W.M. Shtelen, J. Phys. A **16**, 271 (1983)
896. E.K. Lenzi, C. Anteneodo, L. Borland, Escape time in anomalous diffusive media. Phys. Rev. E **63**, 051109 (2001)
897. V. Aquilanti, E.P. Borges, N.D. Coutinho, K.C. Mundim, V.H. Carvalho-Silva, From statistical thermodynamics to molecular kinetics: the change, the chance and the choice, in *The Quantum World of Molecules* (Accademia Nazionale dei Lincei, Rome, 27–28 April 2017); Rendiconti Lincei **29**, 787–802, Scienze Fisiche e Naturali (2018)
898. K.C. Mundim, S. Baraldi, H.G. Machado, F.M.C. Vieira, Temperature coefficient (Q10) and its applications in biological systems: beyond the Arrhenius theory. Ecol. Modell. **431**, 109127 (2020)
899. G.A. Tsekouras, A. Provata, C. Tsallis, Nonextensivity of the cyclic lattice Lotka Volterra model. Phys. Rev. E **69**, 016120 (2004)
900. C. Anteneodo, Entropy production in the cyclic lattice Lotka-Volterra. Eur. Phys. J. B **42**, 271 (2004)
901. R.H. Austin, K. Beeson, L. Eisenstein, H. Frauenfelder, I.C. Gunsalus, V.P. Marshall, Phys. Rev. Lett. **32**, 403 (1974); R.H. Austin, K. Beeson, L. Eisenstein, H. Frauenfelder, Biochemistry **14**, 5355 (1975)
902. C.H.S. Amador, L.S. Zambrano, Evidence for energy regularity in the Mendeleev periodic table. Phys. A **389**, 3866–3869 (2010)
903. M. Sekania, W.H. Appelt, D. Benea, H. Ebert, D. Vollhardt, L. Chioncel, Scaling behavior of the Compton profile of alkali metals. Phys. A **489**, 18–27 (2018)
904. J.A.E. Bonart, W.H. Appelt, D. Vollhardt, L. Chioncel, Scaling behavior of the momentum distribution of a quantum Coulomb system in a confining potential. Phys. Rev. B **102**, 024306 (2020)
905. C. Vignat, A. Plastino, A.R. Plastino, J.S. Dehesa, Quantum potentials with q-Gaussian ground states. Phys. A **391**, 1068–1073 (2012)
906. C. Anteneodo, C. Tsallis, A.S. Martinez, Risk aversion in economic transactions. Europhys. Lett. **59**, 635 (2002)
907. L. Borland, Option pricing formulas based on a non-Gaussian stock price model. Phys. Rev. Lett. **89**, 098701 (2002)
908. L. Borland, A theory of non-gaussian option pricing. Quant. Finance **2**, 415 (2002)
909. T. Kaizoji, Phys. A **326**, 256 (2003)
910. G. Ruiz, A.F. de Marcos, Evidence for criticality in financial data. Eur. Phys. J. B **91**, 1 (2018)
911. C. Tsallis, E.P. Borges, Comment on "Pricing of financial derivatives based on the Tsallis statistical theory" by Zhao. Pan Yue Zhang Chaos Solitons Fractals **148**, 111026 (2021)
912. P. Zhao, J. Pan, Q. Yue, J. Zhang, Pricing of financial derivatives based on the Tsallis statistical theory. Chaos Solitons Fractals **142**, 110463 (2021)
913. M. Ausloos, K. Ivanova, Dynamical model and nonextensive statistical mechanics of a market index on large time windows. Phys. Rev. E **68**, 046122 (2003)
914. N. Kozuki, N. Fuchikami, Dynamical model of financial markets: fluctuating "temperature" causes intermittent behavior of price changes. Phys. A **329**, 222 (2003)
915. N. Kozuki, N. Fuchikami, Dynamical model of foreign exchange markets leading to Tsallis distribution, in *Noise in Complex Systems and Stochastic Dynamics*, eds. by L. Schimansky-Geier, D. Abbott, A. Neiman, C. Van den Broeck, Proc. SPIE **5114**, 439 (2003)
916. I. Matsuba, H. Takahashi, Generalized entropy approach to stable Levy distributions with financial application. Phys. A **319**, 458 (2003)
917. F. Michael, M.D. Johnson, Derivative pricing with non-linear Fokker-Planck dynamics. Phys. A **324**, 359 (2003)
918. R. Osorio, L. Borland, C. Tsallis, Distributions of high-frequency stock-market observables, in *Nonextensive Entropy—Interdisciplinary Applications*, eds. by M. Gell-Mann, C. Tsallis (Oxford University Press, New York, 2004)
919. T. Yamano, Distribution of the Japanese posted land price and the generalized entropy. Eur. Phys. J. B **38**, 665 (2004)

920. L. Borland, J.-P. Bouchaud, A non-Gaussian option pricing model with skew. Quant. Finance **4**, 499–514 (2004)
921. L. Borland, The pricing of stock options, in *Nonextensive Entropy—Interdisciplinary Applications*, eds. by M. Gell-Mann and C. Tsallis (Oxford University Press, New York, 2004)
922. E.P. Borges, Empirical nonextensive laws for the county distribution of total personal income and gross domestic product. Phys. A **334**, 255–266 (2004)
923. T. Kaizoji, Inflation and deflation in financial markets. Phys. A **343**, 662 (2004)
924. P. Richmond, L. Sabatelli, Langevin processes, agent models and socio-economic systems. Phys. A **336**, 27 (2004)
925. A.P. Mattedi, F.M. Ramos, R.R. Rosa, R.N. Mantegna, Value-at-risk and Tsallis statistics: risk analysis of the aerospace sector. Phys. A **344**, 554 (2004)
926. L. Borland, Long-range memory and nonextensivity in financial markets, in *Nonextensive Statistical Mechanics: new Trends, New perspectives*, eds. by J.P. Boon and C. Tsallis, Europhysics News **36**, 228 (2005)
927. L. Borland, A non-Gaussian model of stock returns: option smiles, credit skews, and a multi-time scale memory, in *Noise and Fluctuations in Econophysics and Finance*, eds. by D. Abbott, J.-P. Bouchaud, X. Gabaix, J.L. McCauley, Proc. SPIE **5848**, 55 (SPIE, Bellingham, WA, 2005)
928. L. Borland, J. Evnine, B. Pochart, A Merton-like approach to pricing debt based on a non-Gaussian asset model, in *Complexity, Metastability and Nonextensivity*, eds. by C. Beck, G. Benedek, A. Rapisarda, C. Tsallis (World Scientific, Singapore, 2005), p. 306
929. F.D.R. Bonnet, J. van der Hoeck, A. Allison, D. Abbott, Path integrals in fluctuating markets with a non-Gaussian option pricing model, in *Noise and Fluctuations in Econophysics and Finance*, eds. by D. Abbott, J.-P. Bouchaud, X. Gabaix and J.L. McCauley, Proc. SPIE **5848**, 66 (SPIE, Bellingham, WA, 2005)
930. S.M.D. Queiros, On the emergence of a generalised Gamma distribution. Application to traded volume in financial markets. Europhys. Lett. **71**, 339 (2005)
931. S.M.D. Queiros, On non-Gaussianity and dependence in financial in time series: a nonextensive approach. Quant. Finance **5**, 475–487 (2005)
932. J. de Souza, L.G. Moyano, S.M.D. Queiros, On statistical properties of traded volume in financial markets. Eur. Phys. J. B **50**, 165–168 (2006)
933. L. Borland, A non-Gaussian stock price model: Options, credit and a multi-timescale memory, in *Complexity and Nonextensivity: new Trends in Statistical Mechanics*, eds. by M. Sakagami, N. Suzuki and S. Abe, Prog. Theor. Phys. Suppl. **162**, 155–164 (2006)
934. T. Kaizoji, An interacting-agent model of financial markets from the viewpoint of nonextensive statistical mechanics. Phys. A **370**, 109–113 (2006)
935. B.M. Tabak, D.O. Cajueiro, Assessing inefficiency in euro bilateral exchange rates. Phys. A **367**, 319–327 (2006)
936. D.O. Cajueiro, B.M. Tabak, Is the expression $H = 1/(3 - q)$ valid for real financial data? Phys. A **373**, 593–602 (2007)
937. S.M.D. Queiros, On new conditions for evaluate long-time scales in superstatistical time series. Phys. A **385**, 191–198 (2007)
938. S.M.D. Queiros, L.G. Moyano, J. de Souza, C. Tsallis, A nonextensive approach to the dynamics of financial observables. Eur. Phys. J. B **55**, 161–168 (2007)
939. L. Moriconi, Delta hedged option valuation with underlying non-Gaussian returns. Phys. A **380**, 343–350 (2007)
940. S. Reimann, Price dynamics from a simple multiplicative random process model—Stylized facts and beyond? Eur. Phys. J. B **56**, 381–394 (2007)
941. A.H. Darooneh, Insurance pricing in small size markets. Phys. A **380**, 109–114 (2007)
942. M. Vellekoop, H. Nieuwenhuis, On option pricing models in the presence of heavy tails. Quant. Finance **7**, 563–573 (2007)
943. A.A.G. Cortines, R. Riera Freire, Non-extensive behavior of a stock market index at microscopic time scales. Phys. A **377**, 181–192 (2007)

944. R. Rak, S. Drozdz, J. Kwapien, Nonextensive statistical features of the Polish stock market fluctuations. Phys. A **374**, 315–324 (2007)
945. S. Drozdz, M. Forczek, J. Kwapien, P. Oswiecimka, R. Rak, Stock market return distributions: from past to present. Phys. A **383**, 59–64 (2007)
946. T.S. Biro, R. Rosenfeld, Microscopic origin of non-Gaussian distributions of financial returns. Phys. A **387**, 1603–1612 (2008)
947. A.A.G. Cortines, R. Riera, C. Anteneodo, Measurable inhomogeneities in stock trading volume flow. Europhys. Lett. **83**, 30003 (2008)
948. N. Gradojevic, R. Gencay, Overnight interest rates and aggregate market expectations. Econ. Lett. **100**, 27–30 (2008)
949. J. Ludescher, C. Tsallis, A. Bunde, Universal behaviour of interoccurrence times between losses in financial markets: an analytical description. Europhys. Lett. **95**, 68002 (2011)
950. J. Ludescher, A. Bunde, Universal behavior of the interoccurrence times between losses in financial markets: independence of the time resolution. Phys. Rev. E **90**, 062809 (2014)
951. C. Tsallis, Economics and finance: q-statistical features galore, in the Special Issue *Entropic Applications in Economics and Finance*, eds. by M. Stutzer, S. Bekiros, Entropy **19**, 457 (2017)
952. C.G. Antonopoulos, G. Michas, F. Vallianatos, T. Bountis, Evidence of q-exponential statistics in Greek seismicity. Phys. A **409**, 71–77 (2014)
953. M.I. Bogachev, A.R. Kayumov, A. Bunde, Universal internucleotide statistics in full genomes: a footprint of the DNA structure and packaging? PLoS ONE **9**(12), e112534 (2014)
954. I. Stavrakas, D. Triantis, S.K. Kourkoulis, E.D. Pasiou, I. Dakanali, Latin Am. J. Sol. Struct. **13**, 22 (2016)
955. A. Greco, C. Tsallis, A. Rapisarda, A. Pluchino, G. Fichera, L. Contrafatto, Acoustic emissions in compression of building materials: q-statistics enables the anticipation of the breakdown point. Eur. Phys. J. Spec. Top. **229**(5), 841–849 (2020)
956. P. Manshour, M. Anvari, N. Reinke, M. Sahim, M.R.R. Tabar, Interoccurrence time statistics in fully-developed turbulence. Sci. Rep. **6**, 27452 (2016)
957. S. Kirkpatrick, C.D. Gelatt, M.P. Vecchi, Optimization by simulated annealing. Science **220**, 671–680 (1983)
958. H. Szu, R. Hartley, Fast simulated annealing. Phys. Lett. A **122**, 157–162 (1987)
959. C. Tsallis, D.A. Stariolo, Generalized simulated annealing. Notas de Fisica/CBPF 026 (1994)
960. C. Tsallis, D.A. Stariolo, Generalized simulated annealing. Phys. A **233**, 395 (1996)
961. W. Thistleton, J.A. Marsh, K. Nelson, C. Tsallis, Generalized Box-Muller method for generating q-Gaussian random deviates. IEEE Trans. Inf. Theory **53**, 4805–4810 (2007)
962. P. Serra, A.F. Stanton, S. Kais, Pivot method for global optimization. Phys. Rev. E **55**(1162), x (1997)
963. P. Serra, A.F. Stanton, S. Kais, R.E. Bleil, Comparison study of pivot methods for global optimization. J. Chem. Phys. **106**, 7170 (1997)
964. K.C. Mundim, C. Tsallis, Geometry optimization and conformational analysis through generalized simulated annealing. Int. J. Quantum Chem. **58**, 373 (1996)
965. J. Schulte, Nonpolynomial fitting of multiparameter functions. Phys. Rev. E **53**, R1348 (1996)
966. I. Andricioaei, J.E. Straub, Generalized simulated annealing algorithms using Tsallis statistics: application to conformational optimization of a tetrapeptide. Phys. Rev. E **53**, R3055 (1996)
967. I. Andricioaei, J.E. Straub, On Monte Carlo and molecular dynamics inspired by Tsallis statistics: methodology, optimization and applications to atomic clusters. J. Chem. Phys. **107**, 9117 (1997)
968. I. Andricioaei, J.E. Straub, An efficient Monte Carlo algorithm for overcoming broken ergodicity in the simulation of spin systems. Phys. A **247**, 553 (1997)
969. Y. Xiang, D.Y. Sun, W. Fan, X.G. Gong, Generalized simulated annealing algorithm and its application to the Thomson model. Phys. Lett. A **233**, 216 (1997)

970. U.H.E. Hansmann, Simulated annealing with Tsallis weights: a numerical comparison. Phys. A **242**, 250 (1997)
971. U.H.E. Hansmann, Parallel tempering algorithm for conformational studies of biological molecules. Chem. Phys. Lett. **281**, 140 (1997)
972. U.H.E. Hansmann, Y. Okamoto, Generalized-ensemble Monte Carlo method for systems with rough energy landscape. Phys. Rev. E **56**, 2228 (1997)
973. U.H.E. Hansmann, M. Masuya, Y. Okamoto, Characteristic temperatures of folding of a small peptide. Proc. Natl. Acad. Sci. USA **94**, 10652 (1997)
974. M.R. Lemes, C.R. Zacharias, A. Dal Pino, Jr., Generalized simulated annealing: application to silicon clusters. Phys. Rev. B **56**, 9279 (1997)
975. P. Serra, S. Kais, Symmetry breaking and of binery clusters. Chem. Phys. Lett. **275**, 211 (1997)
976. B.J. Berne, J.E. Straub, Novel methods of sampling phase space in the simulation of biological systems. Curr. Opin. Struct. Biol. **7**, 181 (1997)
977. Y. Okamoto, Protein folding problem as studied by new simulation algorithms. Recent Res. Dev. Pure Appl. Chem. **2**, 1 (1998)
978. H. Nishimori, J. Inoue, Convergence of simulated annealing using the generalized transition probability. J. Phys. A **31**, 5661 (1998)
979. A. Linhares, J.R.A. Torreao, Microcanonical optimization applied to the traveling salesman problem. Int. J. Mod. Phys. C **9**, 133 (1998)
980. K.C. Mundim, T. Lemaire, A. Bassrei, Optimization of nonlinear gravity models through Generalized Simulated Annealing. Phys. A **252**, 405 (1998)
981. M.A. Moret, P.M. Bisch, F.M.C. Vieira, Algorithm for multiple minima search. Phys. Rev. E **57**, R2535 (1998)
982. M.A. Moret, P.G. Pascutti, P.M. Bisch, K.C. Mundim, Stochastic molecular optimization using generalized simulated annealing. J. Comp. Chem. **19**, 647 (1998)
983. U.H.E. Hansmann, F. Eisenmenger, Y. Okamoto, Stochastic dynamics in a new generalized ensemble. Chem. Phys. Lett. **297**, 374 (1998)
984. J.E. Straub, I. Andricioaei, exploiting Tsallis statistics, in *Algorithms for Macromolecular Modeling*, eds. by P. Deuflhard, J. Hermans, B. Leimkuhler, A. Mark, S. Reich, R. D. Skeel, *Lecture Notes in Computational Science and Engineering* **4**, 189 (Springer, Berlin, 1998)
985. J.E. Straub, Protein folding and optimization algorithms, in *The Encyclopedia of Computational Chemistry*, vol. 3, eds. by P.v. R. Schleyer, N.L. Allinger, T. Clark, J. Gasteiger, P.A. Kollman, H.F. Schaefer III, P.R. Schreiner (Wiley, Chichester, 1998), p. 2184
986. T. Kadowaki, H. Nishimori, Quantum annealing in the transverse Ising model. Phys. Rev. E **58**, 5355 (1998)
987. D.E. Ellis, K.C. Mundim, V.P. Dravid, J.W. Rylander, *Hybrid classical and quantum modeling of defects, interfaces and surfaces, in Computer Aided-Design of High-Temperature Materials*, vol. 350 (Oxford University Press, Oxford, 1999)
988. K.C. Mundim, D.E. Ellis, Stochastic classical molecular dynamics coupled to functional density theory: applications to large molecular systems, in *Nonextensive Statistical Mechanics and Thermodynamics*, eds. by S.R.A. Salinas, C. Tsallis, Braz. J. Phys. **29**, 199 (1999)
989. D.E. Ellis, K.C. Mundim, D. Fuks, S. Dorfman, A. Berner, Interstitial carbon in copper: electronic and mechanical properties. Philos. Mag. B **79**, 1615 (1999)
990. L. Guo, D.E. Ellis, K.C. Mundim, Macrocycle-macrocycle interactions within one-dimensional Cu phthalocyanine chains. J. Porphyr. Phthalocyanines **3**, 196 (1999)
991. A. Berner, K.C. Mundim, D.E. Ellis, S. Dorfman, D. Fuks, R. Evenhaim, Microstructure of Cu-C interface in Cu-based metal matrix composite. Sens. Actuat. **74**, 86 (1999)
992. A. Berner, D. Fuks, D.E. Ellis, K.C. Mundim, S. Dorfman, Formation of nano-crystalline structure at the interface in Cu-C composite. Appl. Surf. Sci. **144–145**, 677 (1999)
993. R.F. Gutterres, M.A. de Menezes, C.E. Fellows, O. Dulieu, Generalized simulated annealing method in the analysis of atom-atom interaction. Chem. Phys. Lett. **300**, 131 (1999)
994. U.H.E. Hansmann, Y. Okamoto, Tackling the protein folding problem by a generalized-ensemble approach with Tsallis statistics, in *Nonextensive Statistical Mechanics and Thermodynamics*, eds. by S.R.A. Salinas, C. Tsallis, Braz. J. Phys. **29**, 187 (1999)

995. U.H.E. Hansmann, Y. Okamoto, New Monte Carlo algorithms for protein folding. Curr. Opin. Struc. Biol. **9**, 177 (1999)

996. J.E. Straub, I. Andricioaei, Computational methods inspired by Tsallis statistics: Monte Carlo and molecular dynamics algorithms for the simulation of classical and quantum systems, in *Nonextensive Statistical Mechanics and Thermodynamics*, eds. by S.R.A. Salinas and C. Tsallis, Braz. J. Phys. **29**, 179 (1999)

997. U.H.E. Hansmann, Y. Okamoto, J.N. Onuchic, The folding funnel landscape of the peptide met-enkephalin. Proteins **34**, 472 (1999)

998. Y. Pak, S.M. Wang, Folding of a 16−residue helical peptide using molecular dynamics simulation with Tsallis effective potential. J. Chem. Phys. **111**, 4359 (1999)

999. G. Gielis, C. Maes, A simple approach to time-inhomogeneous dynamics and applications to (fast) simulated annealing. J. Phys. A **32**, 5389 (1999)

1000. M. Iwamatsu, Y. Okabe, Reducing quasi-ergodicity in a double well potential by Tsallis Monte Carlo simulation. Phys. A **278**, 414 (2000)

1001. Y. Xiang, D.Y. Sun, X.G. Gong, Generalized simulated annealing studies on structures and properties of Ni_n ($n = 2 − 55$) clusters. J. Phys. Chem. A **104**, 2746 (2000)

1002. F. Calvo, F. Spiegelmann, Mechanisms of phase transitions in sodium clusters: from molecular to bulk behavior. J. Chem. Phys. **112**, 2888 (2000)

1003. V.R. Ahon, F.W. Tavares, M. Castier, A comparison of simulated annealing algorithms in the scheduling of multiproduct serial batch plants. Brazilian J. Chem. Eng. **17**, 199 (2000)

1004. A. Fachat, K.H. Hoffmann, A. Franz, Simulated annealing with threshold accepting or Tsallis statistics. Comput. Phys. Commun. **132**, 232 (2000)

1005. J. Schulte, J. Ushio, T. Maruizumi, Non-equilibrium molecular orbital calculations of Si/SiO2 interfaces. Thin Solid Films **369**, 285 (2000)

1006. Y. Xiang, X.G. Gong, Efficiency of generalized simulated annealing. Phys. Rev. E **62**, 4473 (2000)

1007. S. Dorfman, V. Liubich, D. Fuks, K.C. Mundim, Simulations of decohesion and slip of the $\Sigma_3 < 111 >$ grain boundary in tungsten with non-empirically derived interatomic potentials: the influence of boron interstitials. J. Phys.: Condens. Matter **13**, 6719 (2001)

1008. D. Fuks, K.C. Mundim, L.A.C. Malbouisson, A. Berner, S. Dorfman, D.E. Ellis, Carbon in copper and silver: diffusion and mechanical properties. J. Mol. Struct. **539**, 199 (2001)

1009. I. Andricioaei, J. Straub, M. Karplus, Simulation of quantum systems using path integrals in a generalized ensemble. Chem. Phys. Lett. **346**, 274 (2001)

1010. A. Franz, K.H. Hoffmann, Best possible strategy for finding ground states. Phys. Rev. Lett. **86**, 5219 (2001)

1011. T. Munakata, Y. Nakamura, Temperature control for simulated annealing. Phys. Rev. E **64**, 046127 (2001)

1012. D. Fuks, S. Dorfman, K.C. Mundim, D.E. Ellis, Stochastic molecular dynamics in simulations of metalloid impurities in metals. Int. J. Quant. Chem. **85**, 354 (2001)

1013. M.A. Moret, P.G. Pascutti, K.C. Mundim, P.M. Bisch, E. Nogueira Jr., Multifractality, Levinthal paradox, and energy hypersurface. Phys. Rev. E **63**, 020901(R) (2001)

1014. I. Andricioaei, J.E. Straub, Computational methods for the simulation of classical and quantum many body systems sprung from the non-extensive thermostatistics, in *Nonextensive Statistical Mechanics and Its Applications*, eds. by S. Abe, Y. Okamoto, Series *Lecture Notes in Physics* (Springer, Heidelberg, 2001) [ISBN 3-540-41208-5]

1015. Y. Okamoto, U.H.E. Hansmann, Protein folding simulations by a generalized-ensemble algorithm based on Tsallis statistics, in *Nonextensive Statistical Mechanics and Its Applications*, eds. by S. Abe, Y. Okamoto, Series *Lecture Notes in Physics* (Springer, Heidelberg, 2001) [ISBN 3-540-41208-5]

1016. U.H.E. Hansmann, Protein energy landscapes as studied by a generalized-ensemble approach with Tsallis statistics, in *Classical and Quantum Complexity and Nonextensive Thermodynamics*, eds. by P. Grigolini, C. Tsallis, B.J. West, Chaos Solitons Fractals **13**,(3), 507 (Pergamon-Elsevier, Amsterdam, 2002)

1017. Y. Pak, S. Jang, S. Shin, Prediction of helical peptide folding in an implicit water by a new molecular dynamics scheme with generalized effective potential. J. Chem. Phys. **116**, 6831 (2002)

1018. A. Franz, K.H. Hoffmann, Optimal annealing schedules for a modified Tsallis statistics. J. Comput. Phys. **176**, 196 (2002)

1019. M. Iwamatsu, Generalized evolutionary programming with Levy-type mutation. Comp. Phys. Comm. **147**, 729 (2002)

1020. T. Takaishi, Generalized ensemble algorithm for $U(1)$ gauge theory. Nucl. Phys. B (Proc. Suppl.) **106**, 1091 (2002)

1021. A.F.A Vilela, J.J. Soares Neto, K.C. Mundim, M.S.P. Mundim, R. Gargano, Fitting potential energy surface for reactive scattering dynamics through generalized simulated annealing. Chem. Phys. Lett. **359**, 420 (2002)

1022. T.W. Whitfield, L. Bu, J.E. Straub, Generalized parallel sampling. Phys. A **305**, 157 (2002)

1023. M.A. Moret, P.M. Bisch, K.C. Mundim, P.G. Pascutti, New stochastic strategy to analyze helix folding. Biophys. J. **82**, 1123 (2002)

1024. L.E. Espinola Lopez, R. Gargano, K.C. Mundim, J.J. Soares Neto, The Na + HF reactive probabilities calculations using two different potential energy surfaces. Chem Phys. Lett. **361**, 271 (2002)

1025. S. Jang, S. Shin, Y. Pak, Replica-exchange method using the generalized effective potential. Phys. Rev. Lett. **91**, 058305 (2003)

1026. Z.X. Yu, D. Mo, Generalized simulated annealing algorithm applied in the ellipsometric inversion problem. Thin Solid Films **425**, 108 (2003)

1027. J.I. Inoue, K. Tabushi, A generalization of the deterministic annealing EM algorithm by means of non-extensive statistical mechanics. Int. J. Mod. Phys. B **17**, 5525 (2003)

1028. I. Fukuda, H. Nakamura, Deterministic generation of the Boltzmann-Gibbs distribution and the free energy calculation from the Tsallis distribution. Chem. Phys. Lett. **382**, 367 (2003)

1029. Z.N. Ding, D.E. Ellis, E. Sigmund, W.P. Halperin, D.F. Shriver, Alkali-ion kryptand interactions and their effects on electrolyte conductivity. Phys. Chem. Chem. Phys. **5**, 2072 (2003)

1030. I. Fukuda, H. Nakamura, Efficiency in the generation of the Boltzmann-Gibbs distribution by the Tsallis dynamics reweighting method. J. Phys. Chem. B **108**, 4162 (2004)

1031. A.D. Anastasiadis, G.D. Magoulas, Nonextensive statistical mechanics for hybrid learning of neural networks. Phys. A **344**, 372 (2004)

1032. J.J. Deng, H.S. Chen, C.L. Chang, Z.X. Yang, A superior random number generator for visiting distributions in GSA. Int. J. Comput. Math. **81**, 103 (2004)

1033. J.G. Kim, Y. Fukunishi, A. Kidera, H. Nakamura, Stochastic formulation of sampling dynamics in generalized ensemble methods. Phys. Rev. E **69**, 021101 (2004)

1034. J.G. Kim, Y. Fukunishi, A. Kidera, H. Nakamura, Generalized simulated tempering realized on expanded ensembles of non-Boltzmann weights. J. Chem. Phys. **121**, 5590 (2004)

1035. A. Dall' Igna Jr., R.S. Silva, K.C. Mundim, L.E. Dardenne, Performance and parameterization of the algorithm simplified generalized simulated annealing. Genet. Mol. Biol. **27**, 616 (2004)

1036. K.C. Mundim, An analytical procedure to evaluate electronic integrals for molecular quantum mechanical calculations. Phys. A **350**, 338 (2005)

1037. E.R. Correia, V.B. Nascimento, C.M.C. Castilho, A.S.C. Esperidiao, E.A. Soares, V.E. Carvalho, The generalized simulated annealing algorithm in the low energy diffraction search problem. J. Phys.: Cond. Matt. **17**, 1–16 (2005)

1038. M. Habeck, W. Rieping, M. Nilges, A replica-exchange Monte Carlo scheme for Bayesian data analysis. Phys. Rev. Lett. **94**, 018105 (2005)

1039. M.A. Moret, P.M. Bisch, E. Nogueira Jr., P.G. Pascutti, Stochastic strategy to analyze protein folding. Phys. A **353**, 353 (2005)

1040. A.F.A. Vilela, R. Gargano, P.R.P. Barreto, Quasi-classical dynamical properties and reaction rate of the Na+HF system on two different potential energy surfaces. Int. J. Quant. Chem. **103**, 695 (2005)

1041. J.J. Deng, C.L. Chang, Z.X. Yang, An exact random number generator for visiting distribution in *GSA*. I. J. Simul. **6**, 54–61 (2005)

1042. I. Fukuda, H. Nakamura, Molecular dynamics sampling scheme realizing multiple distributions. Phys. Rev. E **71**, 046708 (2005)

1043. R. Gangal, P. Sharma, Human pol II promoter prediction: time series descriptors and machine learning. Nucleic Acids Res. **33**, 1332 (2005)

1044. H.B. Liu, K.D. Jordan, On the convergence of parallel tempering Monte Carlo simulations of LJ38. J. Phys. Chem. A **109**, 5203 (2005)

1045. Q. Xu, S.Q. Bao, R. Zhang, R.J. Hu, M. Sbert, Adaptive sampling for Monte Carlo global illumination using Tsallis entropy. in *Computational Intelligence and Security, Part 2, Proceedings Lecture Notes in Artificial Intelligence*, vol. 3802 (Springer, Berlin, 2005), pp. 989–994

1046. I. Fukuda, M. Horie, H. Nakamura, Deterministic design for Tsallis distribution sampling. Chem. Phys. Lett. **405**, 364 (2005)

1047. I. Fukuda, H. Nakamura, Construction of an extended invariant for an arbitrary ordinary differential equation with its development in a numerical integration algorithm. Phys. Rev. E **73**, 026703 (2006)

1048. M.A. Moret, P.G. Pascutti, P.M. Bisch, M.S.P. Mundim, K.C. Mundim, Classical and quantum conformational analysis using generalized genetic algorithm. Phys. A **363**, 260–268 (2006)

1049. P.H. Nguyen, E. Mittag, A.E. Torda, G. Stock, Improved Wang-Landau sampling through the use smoothed potential-energy surfaces. J. Chem. Phys. **124**, 154107 (2006)

1050. J. Hannig, E.K.P. Chong, S.R. Kulkarni, Relative frequencies of generalized simulated annealing. Math. Operat. Res. **31**, 199–216 (2006)

1051. R.K. Niven, q-exponential structure of arbitrary-order reaction kinetics. Chem. Eng. Sci. **61**, 3785–3790 (2006)

1052. A.D. Anastasiadis, G.D. Magoulas, Evolving stochastic learning algorithm based on Tsallis entropic index. Eur. Phys. J. B **50**, 277–283 (2006)

1053. G.D. Magoulas, A. Anastasiadis, Approaches to adaptive stochastic search based on the nonextensive q-distribution. Int. J. Bifurc. Chaos **16**, 2081–2091 (2006)

1054. Y.Q. Gao, L. Yang, On the enhanced sampling over energy barriers in molecular dynamics simulation. J. Chem. Phys. **125**, 114103 (2006)

1055. C.J. Wang, X.F. Wang, Nonextensive thermostatistical investigation of free electronic gas in metal. Acta Phys. Sinica **55**, 2138–2143 (2006)

1056. F.P. Agostini, D.D.O. Soares-Pinto, M.A. Moret, C. Osthoff, P.G. Pascutti, Generalized simulated annealing applied to protein folding studies. J. Comp. Chem. **27**, 1142–1155 (2006)

1057. L.J. Yang, M.P. Grubb, Y.Q. Gao, Application of the accelerated molecular dynamics simulations to the folding of a small protein. J. Chem. Phys. **126**, 125102 (2007)

1058. W.J. Son, S. Jang, Y. Pak, S. Shin, Folding simulations with novel conformational search method. J. Chem. Phys. **126**, 104906 (2007)

1059. T. Morishita, M. Mikami, Enhanced sampling via strong coupling to a heat bath: relationship between Tsallis and multicanonical algorithms. J. Chem. Phys. **127**, 034104 (2007)

1060. S.W. Rick, Replica exchange with dynamical scaling. J. Chem. Phys. **126**, 054102 (2007)

1061. W. Guo, S. Cui, A q-parameterized deterministic annealing EM algorithm based on nonextensive statistical mechanics. IEEE Trans. Signal Process. **56**, 3069–3080 (2008)

1062. L.G. Gamero, A. Plastino, M.E. Torres, Wavelet analysis and nonlinear dynamics in a nonextensive setting. Phys. A **246**, 487 (1997)

1063. A. Capurro, L. Diambra, D. Lorenzo, O. Macadar, M.T. Martin, C. Mostaccio, A. Plastino, E. Rofman, M.E. Torres, J. Velluti, Tsallis entropy and cortical dynamics: the analysis of EEG signals. Phys. A **257**, 149 (1998)

1064. A. Capurro, L. Diambra, D. Lorenzo, O. Macadar, M.T. Martins, C. Mostaccio, A. Plastino, J. Perez, E. Rofman, M.E. Torres, J. Velluti, Human dynamics: the analysis of EEG signals with Tsallis information measure. Phys. A **265**, 235 (1999)

1065. Y. Sun, K.L. Chan, S.M. Krishnan, D.N. Dutt, Tsallis' multiscale entropy for the analysis of nonlinear dynamical behavior of ECG signals, in *Medical Diagnostic Techniques and Procedures*, eds. by M. Singh et al. (Narosa Publishing House, London, 1999), p. 49

1066. M.T. Martin, A.R. Plastino, A. Plastino, Tsallis-like information measures and the analysis of complex signals. Phys. A **275**, 262 (2000)

1067. M.E. Torres, L.G. Gamero, Relative complexity changes in time series using information measures. Phys. A **286**, 457 (2000)

1068. O.A. Rosso, M.T. Martin, A. Plastino, Brain electrical activity analysis using wavelet based informational tools. Phys. A **313**, 587 (2002)

1069. O.A. Rosso, M.T. Martin, A. Plastino, Brain electrical activity analysis using wavelet-based informational tools (II): Tsallis non-extensivity and complexity measures. Phys. A **320**, 497 (2003)

1070. M. Shen, Q. Zhang, P.J. Beadle, Nonextensive entropy analysis of non-stationary ERP signals, in *IEEE International Conference on Neural Networks and Signal Processing* (Nanjing, China, 2003), pp. 806–809

1071. C. Vignat, J.-F. Bercher, Analysis of signals in the Fisher-Shannon information plane. Phys. Lett. **312**, 27 (2003)

1072. M.M. Anino, M.E. Torres, G. Schlotthauer, Slight parameter changes detection in biological models: a multiresolution approach. Phys. A **324**, 645 (2003)

1073. M.E. Torres, M.M. Anino, G. Schlotthauer, Automatic detection of slight parameter changes associated to complex biomedical signals using multiuresolution q-entropy. Med. Eng. Phys. **25**, 859 (2003)

1074. A.L. Tukmakov, Application of the function of the number of states of a dynamic system to investigation of electroencephalographic reaction to photostimulation. Zhurnal Vysshei Nervnoi Deyatelnosti imeni i P Pavlova **53**, 523 (2003)

1075. H.L. Rufiner, M.E. Torres, L. Gamero, D.H. Milone, Introducing complexity measures in nonlinear physiological signals: application to robust speech recognition. Phys. A **332**, 496 (2004)

1076. M. Rajkovic, Entropic nonextensivity as a measure of time series complexity. Phys. A **340**, 327 (2004)

1077. A. Plastino, M.T. Martin, O.A. Rosso, Generalized information measures and the analysis of brain electrical signals, in *Nonextensive Entropy—Interdisciplinary Applications*, eds. by M. Gell-Mann, C. Tsallis (Oxford University Press, New York, 2004)

1078. A. Plastino, O.A. Rosso, Entropy and statistical complexity in brain activity, in *Nonextensive Statistical Mechanics: new Trends, New perspectives*, eds. by J.P. Boon, C. Tsallis, Europhys. News **36**, 224 (2005)

1079. O.A. Rosso, M.T. Martin, A. Figliola, K. Keller, A. Plastino, EEG analysis using wavelet-based information tools. J. Neurosci. Methods **153**, 163–182 (2006)

1080. A. Olemskoi, S. Kokhan, Effective temperature of self-similar time series: analytical and numerical developments. Phys. A **360**, 37–58 (2006)

1081. J.A. Gonzalez, I. Rondon, Cancer and nonextensive statistics. Phys. A **369**, 645–654 (2006)

1082. T. Jiang, W. Xiang, P.C. Richardson, J. Guo, G. Zhu, PAPR reduction of OFDM signals using partial transmit sequences with low computational complexity. IEEE Trans. Broadcast. **53**, 719–724 (2007)

1083. P. Zhao, P. Van Eetvelt, C. Goh, N. Hudson, S. Wimalaratna, E.C. Ifeachor, EEG markers of Alzheimer's disease using Tsallis entropy, in *Communicated at the 3rd International Conference on Computational Intelligence in Medicine and Healthcare (CIMED2007)* (Plymouth, U.K., 2007)

1084. M.E. Torres, H.L. Rufiner, D.H. Milone, A.S. Cherniz, Multiresolution information measures applied to speech recognition. Phys. A **385**, 319–332 (2007)

1085. L. Zunino, D.G. Perez, M.T. Martin, A. Plastino, M. Garavaglia, O.A. Rosso, Characterization of Gaussian self-similar stochastic processes using wavelet-based informational tools. Phys. Rev. E **75**, 021115 (2007)

1086. D.G. Perez, L. Zunino, M.T. Martin, M. Garavaglia, A. Plastino, O.A. Rosso, Model-free stochastic processes studied with q-wavelet-based informational tools. Phys. Lett. A **364**, 259–266 (2007)

1087. J. Poza, R. Hornero, J. Escudero, A. Fernandez, C.I. Sanchez, Regional analysis of spontaneous MEG rhythms in patients with Alzheimer's desease using spectral entropies. Annals Biomed. Eng. **36**, 141–152 (2008)

1088. R. Sneddon, The Tsallis entropy for natural information. Phys. A **386**, 101–128 (2007)

1089. G. Rho, A.L. Callara, G. Petri, M. Nardelli, E.P. Scilingo, A. Greco, V. De Pascalis, Linear and nonlinear quantitative EEG analysis during neutral hypnosis following an opened/closed eye paradigm. Symmetry **13**, 1423 (2021)

1090. S. Saha, K. Jindal, D. Shakti, S. Tewary, V. Sardana, Chirplet transform-based machine-learning approach towards classification of cognitive state change using galvanic skin response and photoplethysmography signals, in *Expert Systems* (Wiley, 2022), https://doi.org/10.1111/exsy.12958

1091. M.P. de Albuquerque, I.A. Esquef, A.R.G. Mello, M.P. de Albuquerque, Image thresholding using Tsallis entropy. Pattern Recogn. Lett. **25**, 1059 (2004)

1092. S. Martin, G. Morison, W. Nailon, T. Durrani, Fast and accurate image registration using Tsallis entropy and simultaneous perturbation stochastic approximation. Electron. Lett. **40**(10), 20040375 (2004)

1093. Y. Li, E.R. Hancock, Face recognition using shading-based curvature attributes, in *Proceedings of the International Conference on Pattern Recognition (ICPR)* (IEEE, Cambridge, 2004), pp. 538–541

1094. Y. Li, E.R. Hancock, Face recognition with generalized entropy measurements, in *Proceedings of International Conference on Image Analysis and Recognition. Lecture Notes in Computer Science*, vol. 3212 (2004), p. 733

1095. W. Tedeschi, H.-P. Muller, D.B. de Araujo, A.C. Santos, U.P.C. Neves, S.N. Erne, O. Baffa, Generalized mutual information $fMRI$ analysis: a study of the Tsallis q parameter. Phys. A **344**, 705 (2004)

1096. W. Tedeschi, H.-P. Muller, D.B. de Araujo, A.C. Santos, U.P.C. Neves, S.N. Erne, O. Baffa, Generalized mutual information tests applied to $fMRI$ analysis. Phys. A **352**, 629 (2005)

1097. W. Shitong, F.L. Chung, Note on the equivalence relationship between Renyi-entropy based and Tsallis-entropy based image thresholding. Pattern Recogn. Lett. **26**, 2309–2312 (2005)

1098. M.P. Wachowiak, R. Smolikova, G.D. Tourassi, A.S. Elmaghraby, Estimation of generalized entropies with sample spacing. Pattern. Anal. Appl. **8**, 95–101 (2005)

1099. S. Liao, W. Fan, A.C.S. Chung, D.Y. Yeung, Facial expression recognition using advanced local binary patterns, Tsallis entropies and global appearance features, in *2006 IEEE International Conference on Image Processing* (2006)pp. 665–668

1100. A. Ben Hamza, Nonextensive information-theoretic measure for image edge detection. J. Electron. Imaging **15**, 013011 (2006)

1101. N. Cvejic, C.N. Canagarajah, D.R. Bull, Image fusion metric based on mutual information and Tsallis entropy. Electron. Lett. **42**(11) (2006)

1102. P.K. Sahoo, G. Arora, Image thresholding using two-dimensional Tsallis-Havrda-entropy. Pattern Recogn. Lett. **27**, 520–528 (2006)

1103. S. Sun, L. Zhang, C. Guo, Medical image registration by minimizing divergence measure based on Tsallis entropy. Int. J. Biomed. Sci. **2**, 75–80 (2007)

1104. Y. Li, X. Fan, G. Li, Image segmentation based on Tsallis-entropy and Renyi-entropy and their comparison, in *2006 IEEE International Conference on Industrial Informatics, INDIN'06* Article 4053516 (2007), pp. 943–948

1105. A. Nakib, H. Oulhadj, P. Siarry, Image histogram thresholding based on multiobjective optimization. Signal Process. **87**, 2516–2534 (2007)

1106. I. Kilic, O. Kayakan, A new nonlinear quantizer for image processing within nonextensive statistics. Phys. A **381**, 420–430 (2007)

1107. S. Wang, F.L. Chung, F. Xiong, A novel image thresholding method based on Parzen window estimate. Pattern Recogn. **41**, 117–129 (2008)

1108. F. Murtagh, J.-L. Starck, Wavelet and curvelet moments for image classification: application to aggregate mixture grading. Pattern Recogn. Lett. **29**, 1557–1564 (2008)

1109. J. Mohanalin, Beenamol, P.K. Kalra, N. Kumar, A novel automatic microcalcification detection technique using Tsallis entropy and a type II fuzzy index, Comput. Math. Appl. **60** (8), 2426–2432 (2010)

1110. P.R.B. Diniz, L.O. Murta, D.G. Brum, D.B de Araujo, A.C. Santos, Brain tissue segmentation using q-entropy in multiple sclerosis magnetic resonance images. Brazilian J. Med. Biol. Res. **43**, 77–84 (2010)

1111. R.J. Al-Azawi, N.M.G. Al-Saidi, H.A. Jalab, H. Kahtan, R.W. Ibrahim, Efficient classification of COVID-19 CT scans by using q-transform model for feature extraction. PeerJ. Comput. Sci. **7**, e553 (2021)

1112. S. Abe, N. Suzuki, Itineration of the Internet over nonequilibrium stationary states in Tsallis statistics. Phys. Rev. E **67**, 016106 (2003)

1113. A. Upadhyaya, J.-P. Rieu, J.A. Glazier, Y. Sawada, Anomalous diffusion and non-Gaussian velocity distribution of Hydra cells in cellular aggregates. Phys. A **293**, 549 (2001)

1114. H. Takagi, M.J. Sato, T. Yanagida, M. Ueda, Functional analysis of spontaneous cell movement under different physiological conditions. PLoS ONE **7**, e2648 (2008)

1115. A.M. Reynolds, Can spontaneous cell movements be modelled as Lévy walks? Phys. A **389**, 273–277 (2010)

1116. S.Z. Lin, P.C. Chen, L.Y. Guan, Y. Shao, Y.K. Hao, Q. Li, B. Li, D.A. Weitz, X.Q. Feng, Universal statistical laws for the velocities of collective migrating cells. Adv. Biosys. **4**, 2000065 (2020)

1117. S.Z. Lin, W.Y. Zhang, D. Bi, B. Li, X.Q. Feng, Energetics of mesoscale cell turbulence in two-dimensional monolayers. Commun. Phys. **4**, 21 (2021)

1118. R.S. Mendes, L.R. Evangelista, S.M. Thomaz, A.A. Agostinho, L.C. Gomes, A unified index to measure ecological diversity and species rarity. Ecography (2008)

1119. A. Bezerianos, S. Tong, N. Thakor, Time-dependent entropy estimation of EEG rhythm changes following brain ischemia. Annals Biomed. Eng. **31**, 221–232 (2003)

1120. S. Tong, Y. Zhu, R.G. Geocadin, D. Hanley, N.V. Thakor, A. Bezerianos, Monitoring brain injury with Tsallis entropy, in *Proceedings of 23rd IEEE/EMBS Conference* (Istambul, 2001)

1121. A. Bezerianos, S. Tong, Y. Zhu, N.V. Thakor, Nonadditive information theory for the analyses of brain rythms, in *Proceedings of 23rd IEEE/EMBS Conference* (Istambul, 2001)

1122. N.V. Thakor, J. Paul, S. Tong, Y. Zhu, A. Bezerianos, Entropy of brain rhythms: normal versus injury EEG, in *Proceedings of 11th IEEE Signal Processing Workshop* (2001), pp. 261–264

1123. A. Bezerianos, S. Tong, J. Paul, Y. Zhu, N.V. Thakor, Information measures of brain dynamics, in *Proceedings of V-th IEEE—EURASIP Biennal International Workshop on Nonlinear Signal and Image Processing (NSP-01)* (Baltimore, MD, 2001)

1124. L. Cimponeriu, S. Tong, A. Bezerianos, N.V. Thakor, Synchronization and information processing across the cerebral cortexfollowing cardiac arrest injury, in *Proceedings of 24th IEEE/EMBS Conference* (San Antonio, Texas, 2002)

1125. S. Tong, Y. Zhu, A. Bezerianos, N.V. Thakor, Information flow across the cerebral cortex of schizofrenics, in *Proceedings of Biosignal Interpretation* (2002)

1126. S. Tong, A. Bezerianos, J. Paul, Y. Zhu, N. Thakor, Nonextensive entropy measure of EEG following brain injury from cardiac arrest. Phys. A **305**, 619–628 (2002)

1127. S. Tong, A. Bezerianos, A. Malhotra, Y. Zhu, N. Thakor, Parameterized entropy analysis of EEG following hypoxic-ischemic brain injury. Phys. Lett. A **314**, 354–361 (2003)

1128. R.G. Geocadin, S. Tong, A. Bezerianos, S. Smith, T. Iwamoto, N.V. Thakor, D.F. Hanley, Approaching brain injury after cardiac arrest: from bench to bedside, in *Proceedings of Neuroengineering Workshop* (Capri, 2003), pp. 277–280

1129. N. Thakor, S. Tong, Advances in quantitative electroencephalogram analysis methods. Annual Rev. Biomed. Eng. **6**, 453 (2004)

1130. J. Gao, W.W. Tung, Y. Cao, J. Hu, Y. Qi, Power-law sensitivity to initial conditions in a time series with applications to epileptic seizure detection. Phys. A **353**, 613 (2005)

1131. S.M. Cai, Z.H. Jiang, T. Zhou, P.L. Zhou, H.J. Yang, B.H. Wang, Scale invariance of human electroencephalogram signals in sleep. Phys. Rev. E **76**, 061903 (2007)

1132. C. Tsallis, U. Tirnakli, Predicting COVID-19 peaks around the world. Front. Phys. **8**, 217 (2020)

1133. G. 't Hooft, On the quantum structure of a black hole. Nucl. Phys. B **256**, 727–745 (1985)

1134. G. 't Hooft, Virtual black holes and space-time structure. Found. Phys. **48**, 1134–1149 (2018)

1135. G. 't Hooft, States, in private communication, that a black-hole basically is a spheric bidimensional surface with radius $\propto 2GM$ (2019)

1136. T. Rohlf, C. Tsallis, Long-range memory elementary 1D cellular automata: dynamics and nonextensivity. Phys. A **379**, 465–470 (2007)

1137. F.A. Tamarit, S.A. Cannas, C. Tsallis, Sensitivity to initial conditions and nonextensivity in biological evolution. Eur. Phys. J. B **1**, 545 (1998)

1138. P.M. Gleiser, F.A. Tamarit, S.A. Cannas, Self-organized criticality in a model of biological evolution with long-range interactions. Phys. A **275**, 272 (2000)

1139. U. Tirnakli, M. Lyra, Damage spreading in the Bak-Sneppen model: sensitivity to the initial conditions and the equilibration dynamics. Int. J. Mod. Phys. C **14**, 805 (2003)

1140. M.L. Lyra, U. Tirnakli, Damage spreading in the Bak-Sneppen and ballistic deposition models: critical dynamics and nonextensivity. Phys. D **193**, 329 (2004)

1141. U. Tirnakli, M.L. Lyra, Critical dynamics of anisotropic Bak-Sneppen model. Phys. A **342**, 151 (2004)

1142. A.R.R. Papa, C. Tsallis, Imitation games: power-law sensitivity to initial conditions and nonextensivity. Phys. Rev. E **57**, 3923 (1998)

1143. M. Rybczynski, Z. Wlodarczyk, G. Wilk, Self-organized criticality in atmospheric cascades. Nucl. Phys. B (Proc. Suppl.) **97**, 81–84 (2001)

1144. S.T.R. Pinho, R.F.S. Andrade, Power law sensitivity to initial conditions for abelian directed self-organized critical models. Phys. A **344**, 601 (2004)

1145. D.J. de Solla Price, Networks of scientific papers. Science **149**, 510–515 (1965)

1146. D.J. Watts, S.H. Strogatz, Collective dynamics of "small-world" networks. Nature **393**, 440 (1998)

1147. R. Albert, A.-L. Barabasi, Statistical mechanics of complex networks. Rev. Mod. Phys. **74**, 47 (2002)

1148. M.E.J. Newman, The structure and function of complex networks. SIAM Rev. **45**, 167–256 (2003)

1149. S. Boccaletti, V. Latora, Y. Moreno, M. Chavez, D.-U. Hwang, Complex networks: structure and dynamics. Phys. Rep. **424**, 175–308 (2006)

1150. C. Tsallis, Connection between scale-free networks and nonextensive statistical mechanics. Eur. Phys. J. Spec. Top. **161**, 175–180 (2008)

1151. A.-L. Barabasi, R. Albert, Emergence of scaling in random networks. Science **286**, 509 (1999)

1152. D.J.B. Soares, C. Tsallis, A.M. Mariz, L.R. da Silva, Preferential attachment growth model and nonextensive statistical mechanics. Europhys. Lett. **70**, 70 (2005)

1153. S. Brito, T.C. Nunes, L.R. da Silva, C. Tsallis, Scaling properties of d-dimensional complex networks. Phys. Rev. E **99**, 012305 (2019)

1154. G. Bianconi, A.L. Barabasi, Competition and multiscaling in evolving networks. Europhys. Lett. **54**, 436–442 (2001)

1155. R. Albert, A.-L. Barabasi, Topology of evolving networks: local events and universality. Phys. Rev. Lett. **85**, 5234 (2000)

1156. T. Emmerich, A. Bunde, S. Havlin, Structural and functional properties of spatially embedded scale-free networks. Phys. Rev. E **89**, 062806 (2014)

1157. B.J. Kim, A. Trusina, P. Minnhagen, K. Sneppen, Self organized scale-free networks from merging and regeneration. Eur. Phys. J. B **43**, 369 (2005)

1158. R.M. de Oliveira, S. Brito, L.R. da Silva, C. Tsallis, Statistical mechanical approach of complex networks with weighted links. JSTAT 063402 (2022)

1159. R.M. de Oliveira, S. Brito, L.R. da Silva, C. Tsallis, Connecting complex networks to non-additive entropies. Sci. Rep. **11**, 1130 (2021)

1160. C. Tsallis, R.M. de Oliveira, Complex network growth model: possible isomorphism between nonextensive statistical mechanics and random geometry. Chaos **32**, 053126 (2022)

1161. J.P.K. Doye, Network topology of a potential energy landscape: a static scale-free network. Phys. Rev. Lett. **88**, 238701 (2002)

1162. J.P.K. Doye, C.P. Massen, Characterization of the network topology of the energy landscapes of atomic clusters. J. Chem. Phys. **122**, 084105 (2005)

1163. G.K. Zipf, *Human Behavior and the Principle of Least Effort* (Addison-Wesley, 1949)

1164. M.A. Montemurro, Beyond the Zipf-Mandelbrot law in quantitative linguistics. Phys. A **300**, 567 (2001); M.A. Montemurro, A generalization of the Zipf-Mandelbrot law in linguistics, in *Nonextensive Entropy—Interdisciplinary Applications*, eds. by M. Gell-Mann and C. Tsallis (Oxford University Press, New York, 2004)

1165. F. Benford, The law of anomalous numbers. Proc. Am. Philos. Soc. **78**, 551–572 (1938)

1166. C. Tsallis, M.P. de Albuquerque, Are citations of scientific papers a case of nonextensivity? Eur. Phys. J. B **13**, 777 (2000)

1167. H.M. Gupta, J.R. Campanha, R.A.G. Pesce, Power-law distributions for the citation index of scientific publications and scientists. Braz. J. Phys. **35**, 981–986 (2005)

1168. S. Picoli, R.S. Mendes, L.C. Malacarne, E.K. Lenzi, Scaling behavior in the dynamics of citations to scientific journals. Europhys. Lett. **75**, 673–679 (2006)

1169. S. Redner, How popular is your paper? An empirical study of the citation distribution. Eur. Phys. J. B **4**, 131–134 (1998)

1170. D. Koutsoyiannis, Z.W. Kundzewicz, Editorial—Quantifying the impact of hydrological studies. Hydrol. Sci. J. 3–17 (2007)

1171. M.P. de Albuquerque, M.P. de Albuquerque, D.B. Mussi, Unpublished (2008)

1172. Z. Neda, L. Varga, T.S. Biro, Science and facebook: the same popularity law! PLoS ONE **12**(7), e0179656 (2017)

1173. K. Briggs, C. Beck, Modelling train delays with q-exponential functions. Phys. A **378**, 498–504 (2007)

1174. E. Mitsokapas, B. Schafer, R.J. Harris, C. Beck, Statistical characterization of airplane delays. Sci. Rep. **11**, 7855 (2021)

1175. J.R.G. Braga, H.F. Campos Velho, E.H. Shiguemori, P. Doherty, Drone autonomous navigation by hardware image processing, in *Mecanica Computacional*, vol. XXXVII, eds. by A. Cardona, L. Garelli, J.M. Gimenez, P.A. Kler, S. Marquez Damian, M.A. Storti (Santa Fe, Argentina, 2019), pp. 2033–2043

1176. G. Williams, B. Schafer, C. Beck, Superstatistical approach to air pollution statistics. Phys. Rev. Res. **2**, 013019 (2020)

1177. L.C. Malacarne, R.S. Mendes, E.K. Lenzi, q-Exponential distribution in urban agglomeration. Phys. Rev. E **65**, 017106 (2002)

1178. S. Picoli Jr., R.S. Mendes, L.C. Malacarne, Statistical properties of the circulation of magazines and newspapers. Europhys. Lett. **72**, 865–871 (2005)

1179. R.S. Mendes, L.C. Malacarne, C. Anteneodo, Statistics of football dynamics. Eur. Phys. J. B **57**, 357–363 (2007)

1180. E.P. Borges, Comment on "The individual success of musicians, like that of physicists, follows a stretched exponential distribution." Eur. Phys. J. B **30**, 593–595 (2002)

1181. A.C. Tsallis, C. Tsallis, A.C.N. Magalhaes, F.A. Tamarit, Human and computer learning: an experimental study. Complexus **1**, 181 (2003)

1182. S.A. Cannas, D. Stariolo, F.A. Tamarit, Learning dynamics of simple perceptrons with non-extensive cost functions. Netw.: Comput. Neural Scie. **7**, 141 (1996)

1183. T. Hadzibeganovic, S.A. Cannas, A Tsallis' statistics based neural network model for novel word learning. Phys. A **388**, 732–746 (2009)

1184. T. Hadzibeganovic, S.A. Cannas, Measuring and modeling the complexity of polysynthetic language learning: a non-extensive neural network approach. Glottotheory **1**, 104–106 (2008)

1185. C. Tsallis, Some thoughts on theoretical physics. Phys. A **344**, 718–736 (2004)

1186. A. Pluchino, V. Latora, A. Rapisarda, Metastable states, anomalous distributions and correlations in the HMF model, in *Anomalous distributions, nonlinear dynamics and nonextensivity*, eds. by H.L. Swinney, C. Tsallis, Phys. D **193**(1–4), 315 (Elsevier, Amsterdam, 2004)

1187. A. Robledo, C. Velarde, How, why and when Tsallis statistical mechanics provides precise descriptions of natural phenomena. Entropy **24**, 1761 (2022)

1188. S. Curilef, C. Tsallis, Critical temperature and nonextensivity in long-range-interacting Lennard-Jones-like fluids. Phys. Lett. A **264**, 270 (1999)

1189. E.P. Borges, C. Tsallis, Negative specific heat in a Lennard-Jones-like gas with long-range interactions, in *Non Extensive Statistical Mechanics and Physical Applications*, eds. by G. Kaniadakis, M. Lissia, A. Rapisarda, Phys. A **305**, 148–151 (2002)

1190. M.N. Kadijani, H. Abbasi, S. Nezamipour, Molecular dynamics simulation of gas models of Lennard-Jones type of interactions: extensivity associated with interaction range and external noise. Phys. A **475**, 35–45 (2017)

1191. S. Curilef, A long-range ferromagnetic spin model with periodic boundary conditions. Phys. Lett. A **299**, 366 (2002)

1192. S. Curilef, Mean field, long-range ferromagnets and periodic boundary conditions. Phys. A **340**, 201 (2004)

1193. S. Curilef, On exact summations in long-range interactions. Phys. A **344**, 456 (2004)

1194. A. Rodriguez, F.D. Nobre, C. Tsallis, Quasi-stationary-state duration in the classical d-dimensional long-range inertial XY ferromagnet. Phys. Rev. E **103**, 042110 (2021)

1195. A. Rodriguez, F.D. Nobre, C. Tsallis, Criticality in the duration of quasistationary state. Phys. Rev. E **104**, 014144 (2021)

1196. A. Rodriguez, F.D. Nobre, C. Tsallis, Finite-size scaling of quasi-stationary-state temperature. Phys. Rev. E **105**, 044111 (2022)

1197. L.F. Santos, F. Borgonovi, G.L. Celardo, Cooperative shielding in many-body systems with long-range interaction. Phys. Rev. Lett. **116**, 250402 (2016)

1198. U. Tirnakli, M. Marques, C. Tsallis, Entropic extensivity and large deviations in the presence of strong correlations. Phys. D **431**, 133132 (2022)

1199. D.J. Zamora, C. Tsallis, Probabilistic models with nonlocal correlations: numerical evidence of q-large deviation theory, 2205.00110 [cond-mat.stat-mech]

1200. C. Tsallis, Enthusiasm and skepticism, pillars of science. Physics (2022), in press

1201. C. Tsallis, What should a statistical mechanics satisfy to reflect nature? Phys. D **193**, 3–34 (2004)

1202. D.R. White, N. Kejzar, C. Tsallis, D. Farmer, S. White, A generative model for feedback networks. Phys. Rev. E **73**, 016119 (2006)

1203. M.D.S. de Meneses, S.D. da Cunha, D.J.B. Soares, L.R. da Silva, Preferential attachment scale-free growth model with random fitness and connection with Tsallis statistics. Prog. Theor. Phys. Suppl. **162**, 131 (2006)

1204. S. Thurner, F. Kyriakopoulos, C. Tsallis, Unified model for network dynamics exhibiting nonextensive statistics. Phys. Rev. E **76**, 036111 (2007)

1205. A. Einstein, in *The Collected Papers of A. Einstein*, vol. 3, no. 26 (Princeton University Press, Princeton, 1993)

1206. A. Carati, L. Galgani, Analog of Planck's formula and effective temperature in classical statistical mechanics far from equilibrium. Phys. Rev. E **61**, 4791 (2000)

1207. N.S. Krylov, Nature **153**, 709 (1944); N.S. Krylov, in *Works on the Foundations of Statistical Physics*, Trans. by A.B. Migdal, Ya. G. Sinai, Yu. L. Zeeman, Princeton Series in Physics (Princeton University Press, Princeton, 1979)

1208. A.M. Mathai, H.J. Haubold, On generalized entropy measures and pathways. Phys. A **385**, 493–500 (2007)

1209. A.M. Mathai, H.J. Haubold, On generalized distributions and pathways. Phys. Lett. A **372**, 2109–2113 (2008)

1210. B.P. Vollmayr-Lee, E. Luijten, A Kac-potential treatment of nonintegrable interactions. Phys. Rev. E **63**, 031108 (2001)

1211. Lj. Milanovic, H.A. Posch, W. Thirring, Statistical mechanics and computer simulation of systems with attractive positive power-law potentials. Phys. Rev. E **57**, 2763 (1998)

1212. M. Antoni, A. Torcini, Anomalous diffusion as a signature of a collapsing phase in two-dimensional self-gravitating systems. Phys. Rev. E **57**, R6233 (1998)

1213. O.N. Mesquita, asked me about the validity, in nonextensive statistical mechanics, of the zeroth principle during a Colloquium I was lecturing in the Physics Department of the Federal University of Minas Gerais (Belo Horizonte, Brazil, 1993)

1214. M. Nauenberg, A critique of q-entropy for thermal statistics. Phys. Rev. E **67**, 036114 (2003)

1215. C. Tsallis, Comment on "Critique of q-entropy for thermal statistics" by M. Nauenberg Phys. Rev. E **69**, 038101 (2004)

1216. M.P. de Albuquerque, et al, Unpublished preliminary results

1217. O.J.E. Maroney, Thermodynamic constraints on fluctuation phenomena. Phys. Rev. E **80**, 061141 (2009)

1218. T.S. Biro, K. Urmossy, Z. Schram, Thermodynamics of composition rules. J. Phys. G **37**, 094027 (2010)

1219. T.S. Biro, P. Van, Zeroth law compatibility of nonadditive thermodynamics. Phys. Rev. E **83**, 061147 (2011)

1220. T.S. Biro, Z. Schram, Lattice gauge theory with fluctuating temperature. EPJ Web Conf. **13**, 05004 (2011)

1221. T. S. Biro, Is there a temperature?—Conceptual challenges at high energy, acceleration and complexity, in *Fundamental Theories in Physics*, vol. 171 (Springer, 2011)

1222. D.H. Zanette, M.A. Montemurro, Dynamics and nonequilibrium states in the Hamiltonian mean-field model: a closer look. Phys. Rev. E **67**, 031105 (2003)

1223. A. Pluchino, V. Latora, A. Rapisarda, Glassy dynamics in the HMF model. Phys. A **340**, 187 (2004)

1224. A. Pluchino, V. Latora, A. Rapisarda, Glassy phase in the Hamiltonian mean field model. Phys. Rev. E **69**, 056113 (2004)

1225. A. Rapisarda, A. Pluchino, Nonextensive thermodynamics and glassy behaviour in Hamiltonian systems. Europhys. News **36**, 202 (2005)

1226. A. Pluchino, A. Rapisarda, Glassy dynamics and nonextensive effects in the HMF model: the importance of initial conditions. Progr. Theor. Phys. Suppl. **162**, 18 (2006)

1227. D.H. Zanette, M.A. Montemurro, Thermal measurements of stationary nonequilibrium systems: a test for generalized thermostatistics. Phys. Lett. A **316**, 184 (2003)

1228. R. Luzzi, A.R Vasconcellos, J. Galvao Ramos, Trying to make sense of disorder. Science **298**, 1171 (2002)

1229. S. Abe, Tsallis entropy: how unique?, in *Nonadditive Entropy and Nonextensive Statistical Mechanics*, ed. by M. Sugiyama, Contin. Mech. Thermodyn. **16**, 237 (Springer, Heidelberg, 2004)

1230. C. Tsallis, D. Prato, A.R. Plastino, Nonextensive statistical mechanics: some links with astronomical phenomena. in *Proceedings of the XIth United Nations / European Space Agency Workshop on Basic Space Science* (Cordoba, Argentina, 2002); eds. by H. Haubold, M. Rabolli, Astrophys. Space Sci. **290**, 259 (Kluwer, 2004); [cond-mat/0301590]

1231. C. Tsallis, E.P. Borges, Nonextensive statistical mechanics—Applications to nuclear and high energy physics, in *Proceedings of the 10th International Workshop on Multiparticle Production—Correlations and Fluctuations in QCD*, eds. by N.G. Antoniou, F.K. Diakonos, C.N. Ktorides (World Scientific, Singapore, 2003), p. 326; [cond-mat/0301521]

1232. S. Abe, A.K. Rajagopal, Revisiting disorder and Tsallis statistics. Science **300**, 249 (2003)

1233. A. Plastino, Revisiting disorder and Tsallis statistics. Science **300**, 250 (2003)

1234. V. Latora, A. Rapisarda, A. Robledo, Revisiting disorder and Tsallis statistics. Science **300**, 250 (2003)

1235. D.H. Zanette, M.A. Montemurro, A note on non-thermodynamical applications of nonextensive statistics. Phys. Lett. A **324**, 383 (2004)

1236. R. Englman, Maximum entropy principles in fragmentation data analysis, in *High-pressure Shock Compression of Solids II—Dynamic Fracture and Fragmentation*, eds. by L. Davison, D.E. Grady, M. Shahinpoor (Springer, Berlin, 1997), pp. 264–281

1237. A.R. Plastino, A. Plastino, B.H. Soffer, Ambiguities in the forms of the entropic functional and constraints in the maximum entropy formalism. Phys. Lett. A **363**, 48–52 (2007)

1238. F. Topsoe, Factorization and escorting in the game-theoretical approach to non-extensive entropy measures. Phys. A **365**, 91–95 (2006)

1239. A.R. Plastino, C. Tsallis, R.S. Wedemann, H.J. Haubold, Entropy optimization, generalized logarithms, and duality relations. Entropy **24**, 1723 (2022)

1240. T.S. Biro, G.G. Barnafoldi, P. Van, New entropy formula with fluctuating reservoir. Phys. A **417**, 215–220 (2015)

1241. A.M. Scarfone, P. Quarati, G. Mezzorani, M. Lissia, Analytical predictions of non-Gaussian distribution parameters for stellar plasmas. Astrophys. Space Sci. **315**, 1–4 (2008)

1242. R.J. Baxter, *Exactly Solved Models in Statistical Mechanics* (Academic Press, London, 1982), p.353

1243. R.S. Mendes, C. Tsallis, Renormalization group approach to nonextensive statistical mechanics. Phys. Lett. A **285**, 273 (2001)

1244. R. Silva Jr., A.R. Plastino, J.A.S. Lima, A Maxwellian path to the q-nonextensive velocity distribution function. Phys. Lett. A **249**, 401 (1998)

1245. C. Vignat, A. Plastino, The p-sphere and the geometric substratum of power-law probability distributions. Phys. Lett. A **343**, 411 (2005)

1246. C. Vignat, J. Naudts, Stability of families of probability distributions under reduction of the number of degrees of freedom. Phys. A **350**, 296–302 (2005)

1247. C. Vignat, A. Plastino, Scale invariance and related properties of q-Gaussian systems. Phys. Lett. A **365**, 370–375 (2007)

1248. C. Vignat, A. Plastino, Poincare's observation and the origin of Tsallis generalized canonical distributions. Phys. A **365**, 167–172 (2006)

1249. T. Dauxois, Non-Gaussian distributions under scrutiny. J. Stat. Mech. (2007) N08001

1250. C. Tsallis, "Non-Gaussian distributions under scrutiny" under scrutiny, eds. by H.J. Haubold, A.M. Mathai, in *Proceedings of the Third UN/ESA/NASA Workshop on the International Heliophysical Year 2007 and Basic Space Science, Astrophysics and Space Science Proceedings*, vol. 1 (Springer, Berlin-Heidelberg, 2010)

1251. E.M.F. Curado, C.E. Cedeño, I.D. Soares, C. Tsallis, Relativistic gas: Lorentz-invariant distribution for the velocities. Chaos **32**, 103110 (2022)

1252. I.S. Gradshteyn, I.M. Ryzhik, in *Table of Integrals, Series, and Products*, Sixth Edition, eds. by A. Jeffrey, D. Zwillinger (Academic Press, San Diego, 2000)

1253. S. Abe, A.K. Rajagopal, Rates of convergence of non-extensive statistical distributions to Lévy distributions in full and half spaces. J. Phys. A **33**, 8723–8732 (2000)

1254. G. Wilk, Z. Wlodarczyk, Nonextensive information entropy for stochastic networks. Acta Phys. Polon. B **35**, 871 (2004)

1255. G. Wilk, Z. Wlodarczyk, Consequences of temperature fluctuations in observables measured in high-energy collisions. Eur. Phys. J. A **48**, 161 (2012)

1256. V. Schwammle, C. Tsallis, Private discussions

1257. S. Abe, Instability of q-averages in nonextensive statistical mechanics. EPL **84**, 60006 (2008)

1258. S. Coles, *An Introduction to Statistical Modeling of Extreme Values, Springer Series in Statistics* (Springer, London, 2001)

1259. E.M.F. Curado, Private Communication (2008)

1260. R. Hanel, S. Thurner, C. Tsallis, On the robustness of q-expectation values and Rényi entropy. Europhys. Lett. **85**, 20005 (2009)

1261. A. Rodriguez, A. Pluchino, U. Tirnakli, A. Rapisarda, C. Tsallis, Non-extensive footprints in dissipative and conservative dynamical systems. Symmetry (2023). In press

Index

© Springer Nature Switzerland AG 2023
C. Tsallis, *Introduction to Nonextensive Statistical Mechanics*,
https://doi.org/10.1007/978-3-030-79569-6

Printed in the United States
by Baker & Taylor Publisher Services